ILLUSTRATIONS

of

ALIEN PLANTS

of the

BRITISH ISLES

ILLUSTRATIONS
of
ALIEN PLANTS
of the
BRITISH ISLES

E. J. CLEMENT

D. P. J. SMITH

I. R. THIRLWELL

incorporating
artwork originally prepared for

D. McCLINTOCK's

A New Illustrated British Flora, vol. 3

with drawings by

M. Godfrey, J. Ironside, W.A. Page, F.J. Rumsey, H.A. Salzen, D.P.J. Smith

and (deceased)

E. Dickson, A.M. Ferguson, K.M. Hollick, B. Keel, D. Long, C. Rowlands, J. Russell

Botanical Society of the British Isles
London
2005

Published by the Botanical Society of the British Isles
Charity number 212560

©BSBI 2005

ISBN 0 901158 32 1

Printed in Great Britain by
J. & P. Davison, 3 James Place, Treforest, Potypridd, Mid Glamorgan, CF37 1SQ

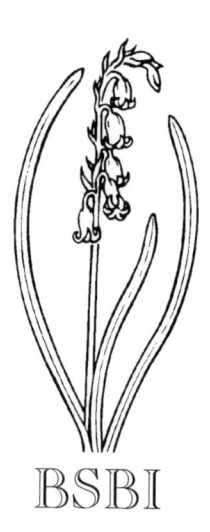

BSBI

The cover illustration shows Apple-of-Peru (*Nicandra physalodes* (L.) Gaertn.)
drawn by D. P. J. Smith

CONTENTS

PREFACE	i
ACKNOWLEDGEMENTS	iii
INTRODUCTION	iv
THE PLATES	iv
THE FUTURE	v
ABBREVIATIONS AND SYMBOLS	vii
PLATES	1
SELECTED BIBLIOGRAPHY	445
INDEX	447

PREFACE

The need for a volume of alien plant illustrations to supplement the comparatively few that are included in Stella Ross-Craig's *Drawings of British Plants* has long been felt.

The late David McClintock began the task of organising a group of seven gifted amateur artists to fill this gap back in the early 1960s. By the end of 1968 the late Mrs E. Dickson (Ipswich), Miss A.M. Ferguson (Quendon), Miss K.M. Hollick (Ashbourne), Miss B. Keel (Bethersden), Miss D. Long (Denton), Mr C. Rowlands (River-in-Dover) and Mrs J. Russell (Dunmow) had, between them, provisionally drawn 260 plants, rising to 350 plants by 1972. In 1981 the work was still continuing, but at a much slower pace as other commitments took over David McClintock's time. For a while the late B.E. Smythies and A.L. Grenfell took over the reins and made a little progress. Finally E.J. Clement picked up and continued the task. Unfortunately the requirement that drawings be of fresh specimens from wild British or Irish localities, that **all** parts, including fruits and seeds, be drawn and that no drawings be inked in until verified by David McClintock and other referees had severely delayed the project. Meanwhile, alas, all the founder artists died.

To push things along Eric Clement began, in 1975, to encourage new artists to submit artwork for publication in *BSBI News*, concentrating on the casual and rarer species. This has continued in one form or another ever since.

Sadly, in November 2001 David McClintock died. His ambition to provide a magnificent descriptive and illustrative alien plant volume in the format of a third volume to R. W. Butcher's *A new illustrated British Flora*, vols. 1-2 was not achieved. He had anticipated this possibility and had requested Eric Clement (pers. comm. 6 March 1994) to complete the work or, failing this, to deposit the drawings at the Natural History Museum (South Kensington, London), "if, if they don't get used…with a note telling folk about them."

Eric Clement could not bear the thought of such an enormous amount of work and botanical skill being confined to a dark, museum cupboard. A rally round for support quickly revealed enthusiasm from Delf Smith and Ian Thirlwell and the idea for the current volume was born. To avoid a delay of perhaps another forty years while 'missing' species and botanical details were hunted for, the decision was made to allow exactly two years in which to produce camera-ready copy of all that we could assemble. Outstanding plants would need to wait for a supplementary volume.

David McClintock's descriptions, few of which were ever completed, have all been omitted. Comparable diagnoses are almost all now available in Stace's *New Flora of the British Isles*, and hence we are able to reproduce the drawings in an attractive large format. Longer descriptions can often be found in Clapham, Tutin & Moore (1987) and/or Sell & Murrell (1996).

| ERIC CLEMENT | DELF SMITH | IAN THIRLWELL |
| Gosport, Hampshire | Portsmouth, Hampshire | Portsmouth, Hampshire |

January 2005

ACKNOWLEDGEMENTS

There would be no book but for the great generosity of so many gifted artists. It has involved thousands of hours of painstaking work. These blessed contributors are, in alphabetical order:
†Evangeline Dickson, †A.M. Ferguson, Margot Godfrey, †Kathleen Hollick, Jane Ironside, †B. Keel, †Dorothy Long, Wendy Page, †Charles Rowlands, †Jocelyn Russell, Fred Rumsey, Heather Salzen and Delf Smith.

The artists could not have drawn without the plants, sent from all corners of the British Isles by very many past and present members of the Botanical Society of the British Isles and the Wildflower Society. The list, dating back over some forty-five years, is very long and has, alas, now largely been obscured by time, but for every contribution we are truly grateful.

Over the last two years help in many ways, from advice to logistics, has been generously supplied by many, most notably by Sally Foster, Gordon Hanson, Rosemary Parslow, Barry Phillips, Colin Pope, Clive Stace and Paul Stanley.

Finally we wish to thank Gwynn Ellis for ably editing this work and for guiding it through publication.

INTRODUCTION

This work makes no pretence of being perfect. It consists of a broad miscellany of line drawings of mostly unusual alien (= non-native) plants that occur in the wild in the British Isles and for which diagnostic, botanical drawings are difficult (or impossible!) to find in the literature available in a typical small botanical library. They have been drawn by a wide spectrum of artists over many (forty-three) years, and all have been donated without cost. A few were drawn by professional artists, but most were the work of very talented amateurs. The composite result is not intended to be an elegant work of art, but to form a serviceable guide to assist accurate identification of problematical plants.

The drawings have been assembled from two main sources. Foremost is the large collection of approximately 400 unpublished drawings that David McClintock oversaw (see Preface). Secondly, Margot Godfrey (Barnehurst), Jane Ironside (Fakenham), Wendy Page (Horsham), Fred Rumsey (Natural History Museum, London), Heather Salzen (Aberdeen) and Delf Smith (Portsmouth) generously volunteered to produce some additional drawings to fill some of the obvious gaps.

None of the very fine drawings from *BSBI News*, the *BSBI Handbook* series or Ryves, Clement and Foster's *Alien Grasses of the British Isles* have been reproduced here. Hence none of the artwork in this book has previously been published; all has been produced anew from live or occasionally pressed specimens. Not one has been copied from previously published works (which is a remarkably common practice!), so any errors will be ours alone.

THE PLATES

Stace's *New Flora of the British Isles* has dominated the scene of British field botany almost immediately from its first appearance in 1991; a second edition was published in 1997. For these reasons our book is arranged exactly in the order of Stace (1997). The family number and name at the top of each plate follow this work without exception. Note that the family numbers in the two editions of Stace differ slightly due to the inclusion of additional families in the second edition.

The plates have been reproduced as large as possible to convey their beauty and to prevent the obscuring of fine details. Each plate shows the habit of the plant or, for larger species, a branch of it together with botanical dissections/enlargements. These show diagnostic features which vary within a genus or family. Fruits and seeds are figured wherever possible, but this is a more exacting requirement. Scale bars showing size are present representing either millimetres or centimetres.

The legend at the foot of the plate gives the full scientific name, a few being updated in Stace (1999) or elsewhere, authority and the common English name if there is one. Almost all conform with Stace (1997). Astute eyes will notice that the abbreviations of author names differ between Stace (1991) and Stace (1997). This is because the former followed the recommendations for consistency provisionally proposed by Meikle (1980), and the latter adopted the current standard of Brummitt & Powell (1992). In order to save space no synonyms are included; the most relevant ones can be found in Stace's *New Flora*. After the name and authority comes a list of the plant parts illustrated in the drawing; the botanical terms are explained, almost without exception, in Stace's glossary (pp 1179-1196 in edition 1, and pp 983-997 in edition 2). Hickey & King's *The Cambridge Illustrated Glossary of Botanical Terms* can often usefully be used for further clarification. Next, in parentheses, is given the origin of the plant that is figured. It consists of the place name followed by the abbreviated vice-county name; a full list of the vice-county numbers and names can be found on the inside back cover of Stace (1991 or 1997). Unfortunately David McClintock and others sometimes failed to document where the plant specimen was collected, in which case the origin is given as 'Source unknown'. Where the origin is a horticultural one growing material of unknown or foreign provenance the source is specified as '*Hort.*'; if the location of the garden is known, this is added, e.g. '*Hort.* Cambridge'. Occasionally one or more parts of the plate have been added from a supplementary source, typically a fruit or seed, and this is indicated.

Initially we had intended to complete the legend with colour notes, as done by Stella Ross-Craig (1948-1973), but we later decided that a reference to a fuller description would be more helpful. Hence we give the page number to both the first and second editions respectively of Stace's *New Flora*, e.g. STACE 130/106 for *Dicentra formosa*, most libraries possessing one or the other. A dash (—) is used to indicate that there is no entry for this species. Very rarely STACE —/— occurs when the species is not mentioned in either edition. Recommended as sources for full descriptions are the two classic works on the wild and cultivated plants of Europe, viz. *Flora Europaea*, vols. 1-5, and *The European garden Flora*, vols. 1-6. Griffiths (1994) can often be used as a comprehensive and inexpensive alternative, but the descriptions are very short; it comprises a surprisingly generous selection of weeds and wild flowers. References to further descriptions, as well as to additional iconography (line drawings, paintings or photographs), may usually be gleaned from Clement & Foster's *Alien plants of the British Isles* and Ryves, Clement & Foster's *Alien grasses of the British Isles*. These last two works also present a fuller account of synonymy.

Pressed, voucher specimens of the plant drawn have only very occasionally been preserved, but material from the same locality can sometimes be found in national herbaria, notably at the Natural History Museum (**BM**), Royal Botanic Gardens, Edinburgh (**E**) and at Reading University (**RNG**). A few also exist in private herbaria, including that of one of the authors (*Herb*. **EJC**). Critical examination of such specimens could in the future call for changes in determinations.

The selection of species in this volume perhaps demands a little explanation. The plants were all drawn on an **opportunity** basis, and fresh material in some families that are well represented was much more easily obtainable than in others. Special interests of both artists and collectors added to this bias. Drawing tiny 'green things' raised little enthusiasm with most artists. Other drawings have, alas, 'gone missing' over the decades: we could not trace them, or even a copy of them.

THE FUTURE

Perusal of Stace's *Flora* soon indicates that more drawings are required. It is hoped that this book will stimulate our artists of the future to remedy the situation by filling the gaps, and a second book of illustrations can be assembled. Let us not wait another forty-five years for such a work!

ABBREVIATIONS AND SYMBOLS

BSBI	Botanical Society of the British Isles
cm	centimetre(s)
Co	County
det.	determined (named by an expert)
Herb.	Herbarium, Herbaria
Hort.	of horticultural (=garden) source (plate legend)
LS	longitudinal section
mm	millimetre(s)
N,S,E,W	North, South, East, West
p	page
pp	pages
pers. comm.	personal communication
RBG	Royal Botanic Garden
sensu lato	in the broad sense
ssp.	subspecies
STACE 130/106	signifies that, for example, a description can be found on p 130 of Stace's *New Flora* (1991) and on p 106 of Stace's *New Flora* (1997). A dash (—) is used to indicate where there is no entry
TS	transverse section
var.	varietas: variety
VC	vice-county (a county or a subdivision of a larger county)
x	hybrid
†	deceased (of person)
♂	male (on plate)
♀	female (on plate)
☿	bisexual (on plate)

2. SELAGINELLACEAE

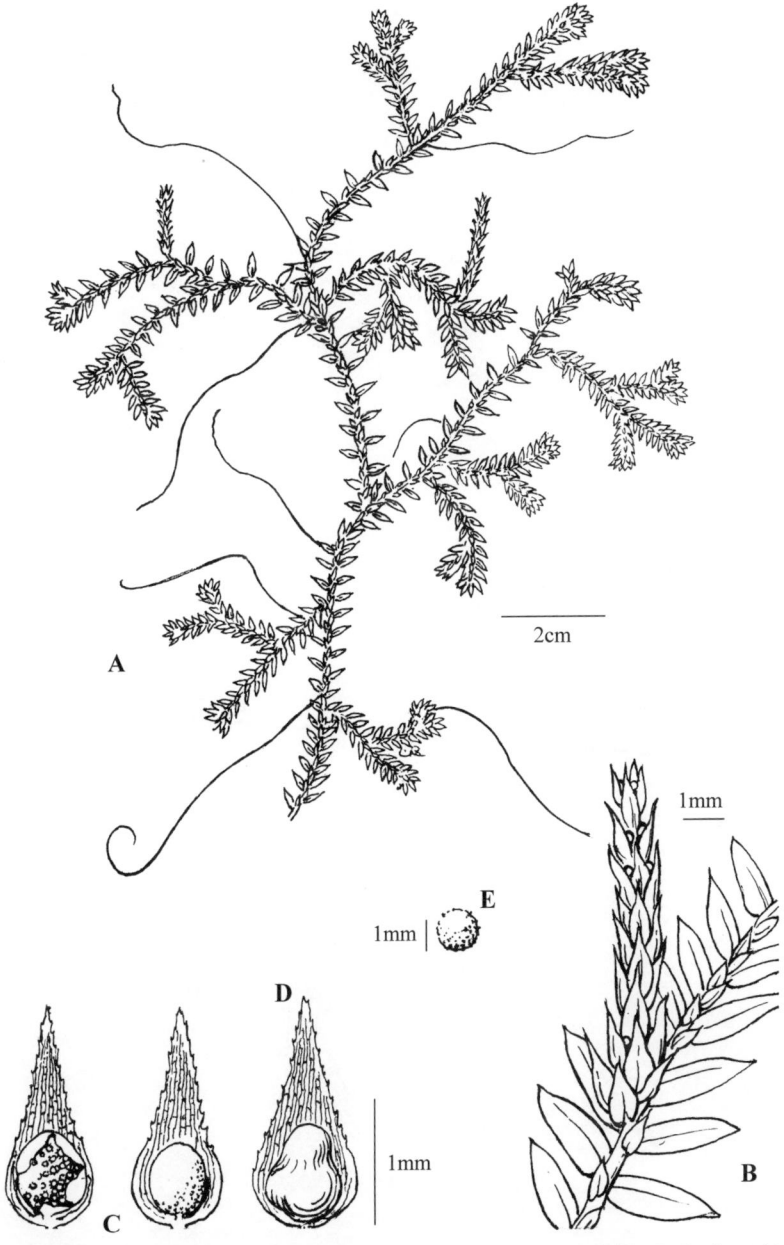

Selaginella kraussiana (Kunze) A. Braun (Krauss's Clubmoss). A habit; B fertile and barren branches; C microsporangia in leaf-axils, one bursting open; D megasporangium in leaf-axil; E megaspore. (Source unknown). STACE 8/7.

8. PTERIDACEAE

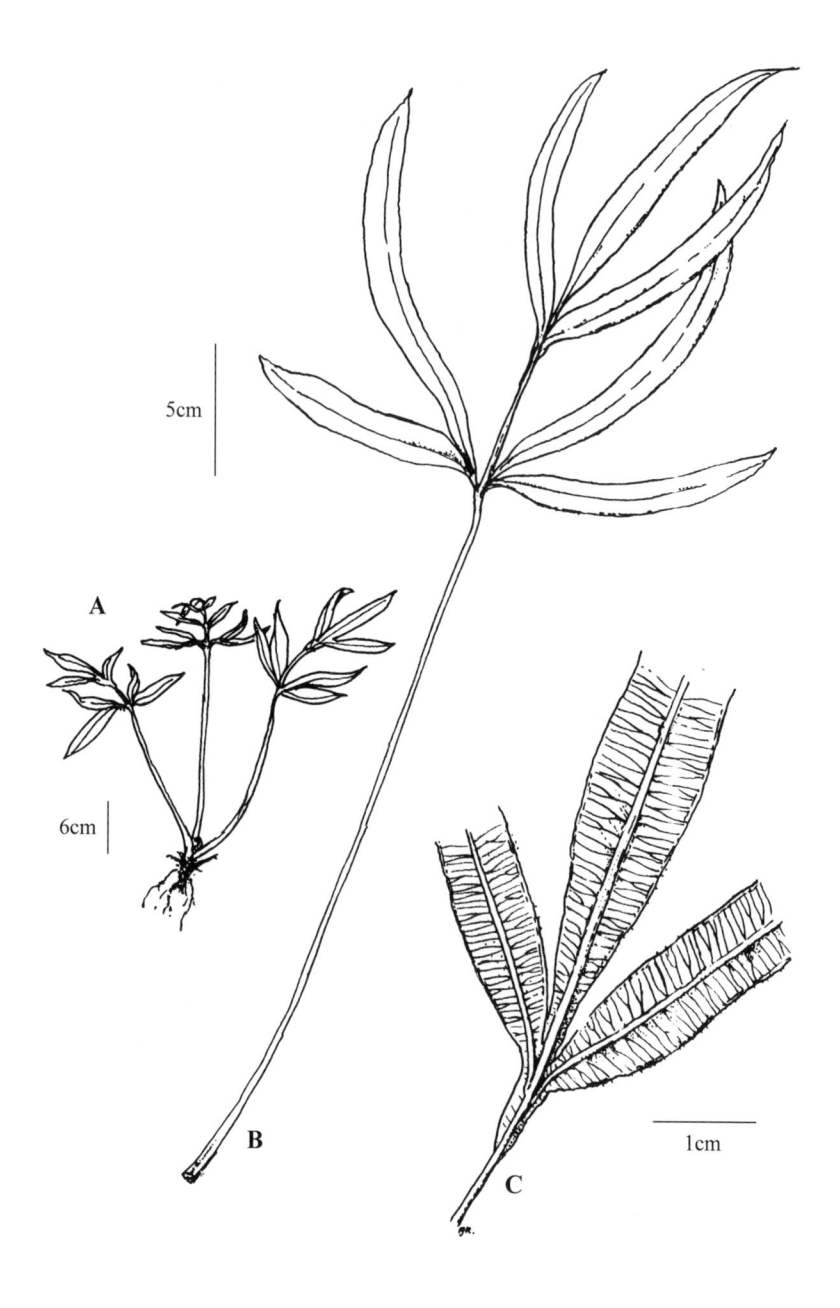

Pteris cretica L. (Ribbon Fern). A habit; B frond; C detail of uppermost pinnules. (Source unknown). STACE 20/16.

11. POLYPODIACEAE

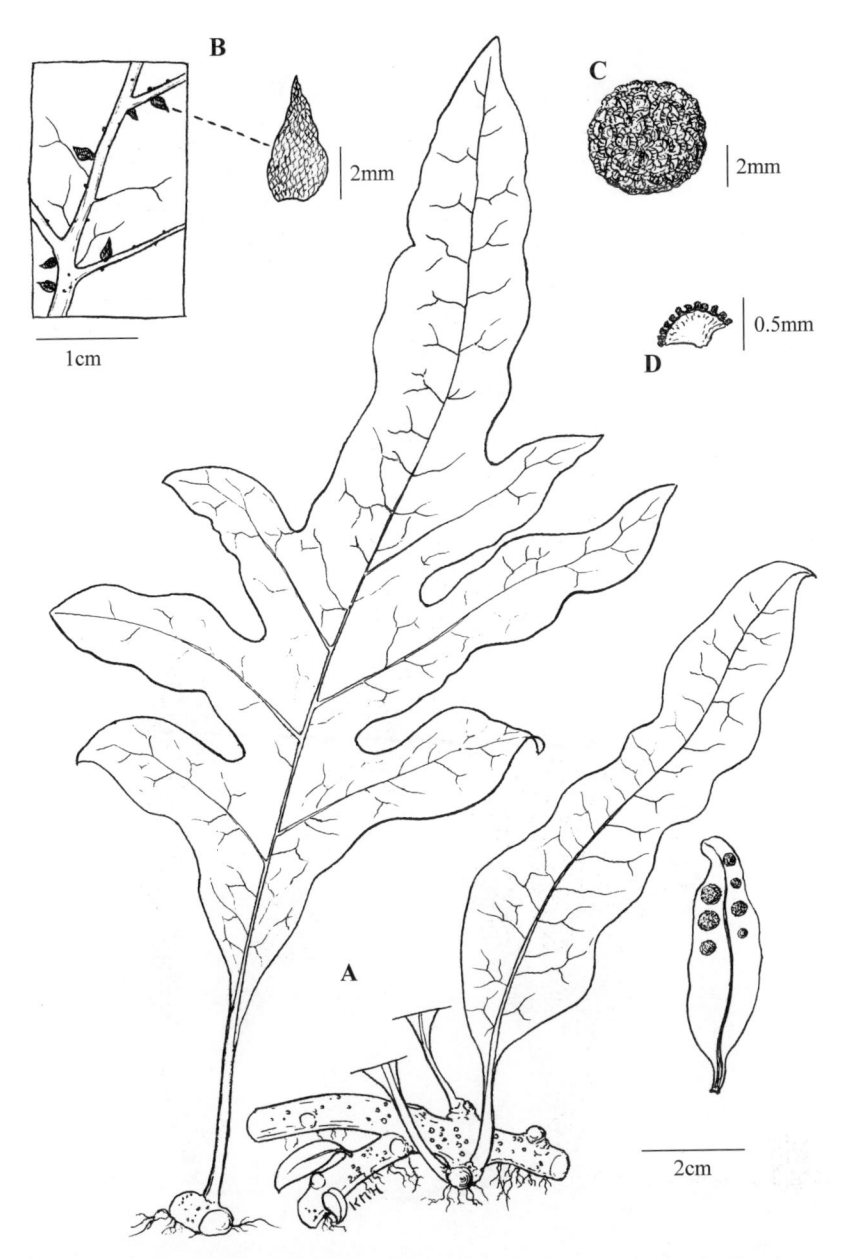

Phymatosorus diversifolius (Willd.) Pic. Serm. (Kangaroo Fern). A habit, showing three variable fronds; B underside of frond, the clathrate scales sparse to absent; C sorus; D sporangium. (Guernsey). STACE 23/20.

16. WOODSIACEAE

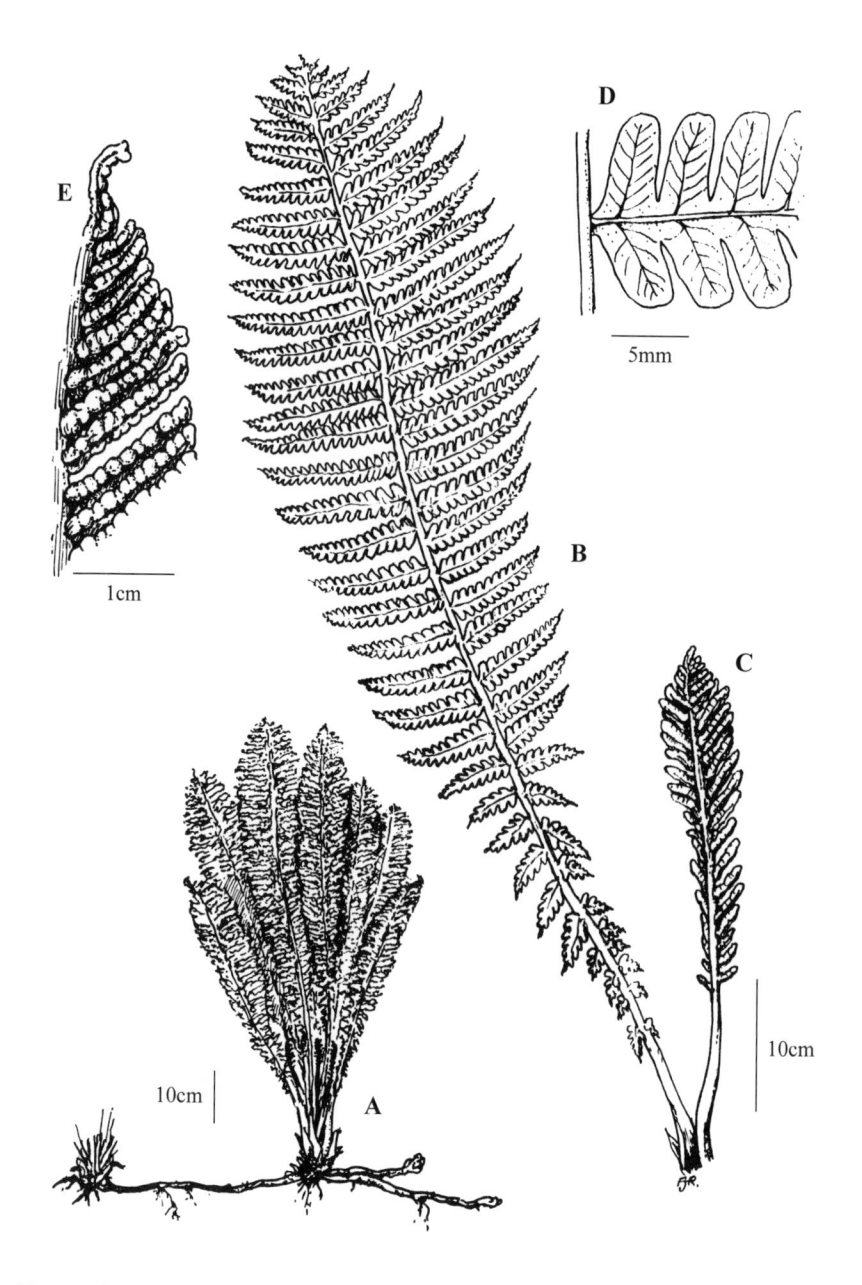

Matteuccia struthiopteris (L.) Tod. (Ostrich Fern). A habit; B sterile frond; C fertile frond; D detail of sterile pinna; E detail of fertile pinna. (Dibden Purlieu, S Hants). STACE 33/26.

17. DRYOPTERIDACEAE

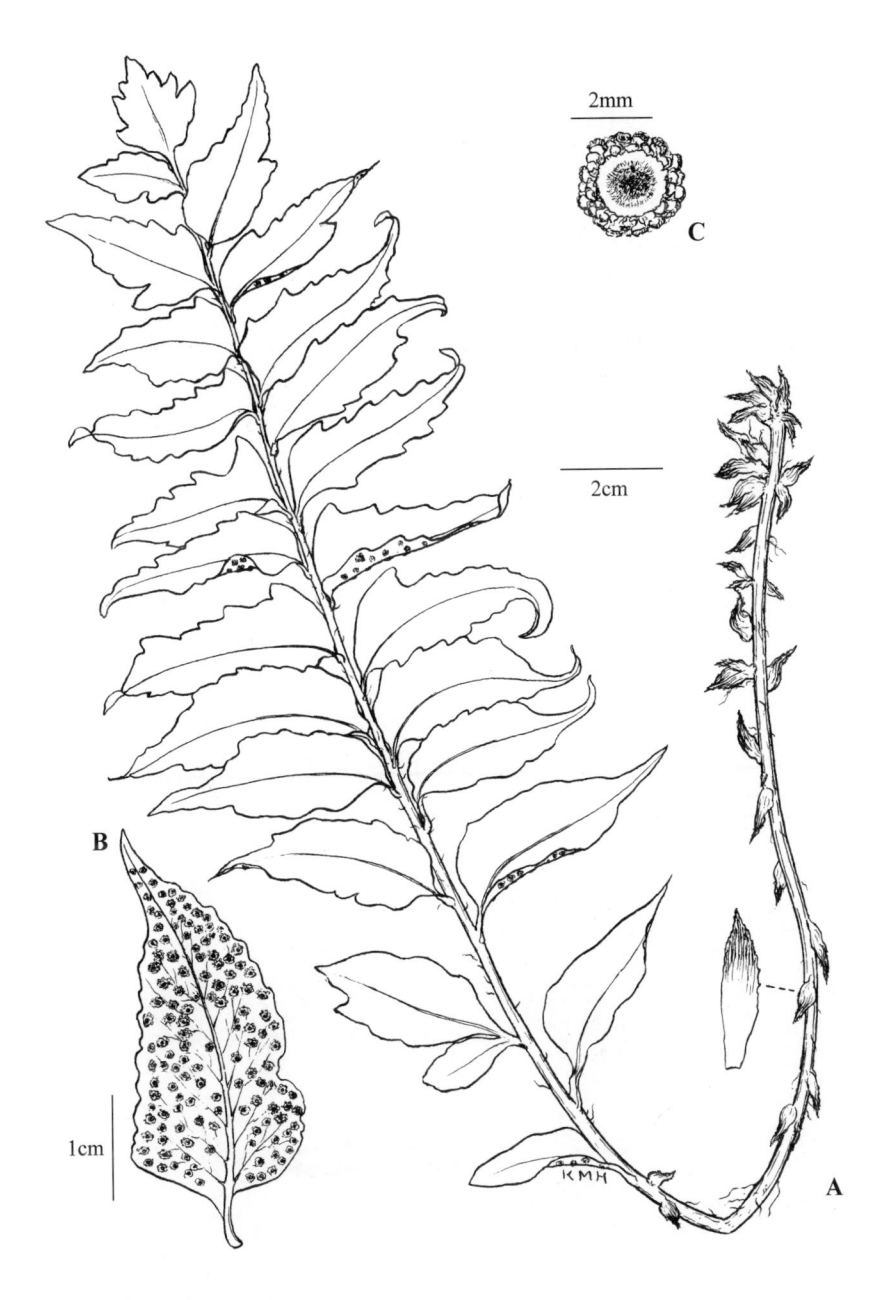

Cyrtomium falcatum (L.f.) C. Presl (House Holly-fern). A frond; B underside of fertile pinna; C indusium partly covering sorus. (Isles of Scilly). STACE 38/31.

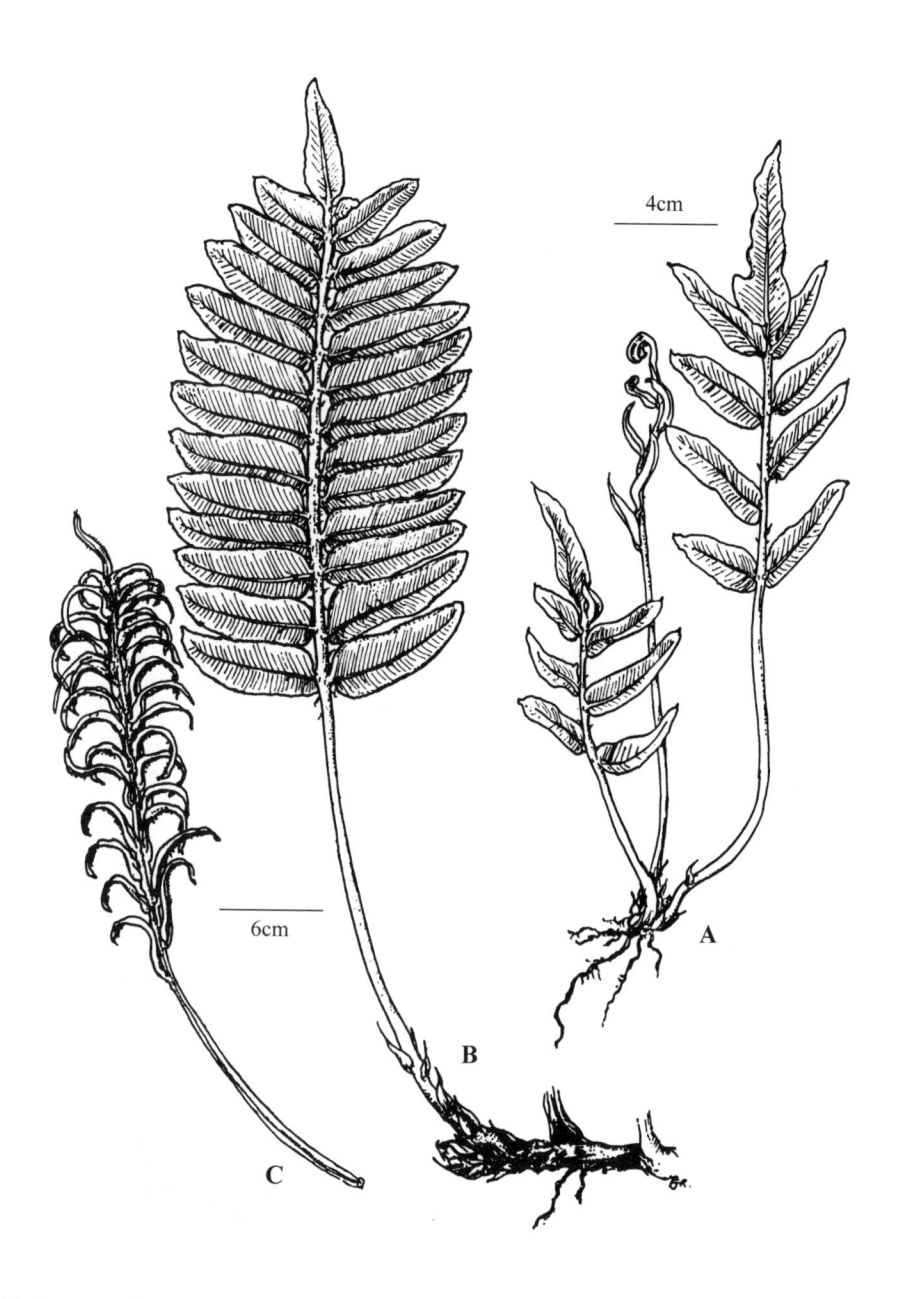

Blechnum cordatum (Desv.) Hieron. (Chilean Hard-fern). A habit of young plant; B sterile frond and rhizome; C fertile frond. (A, Glengariff, W Cork; B, C, Isles of Scilly). STACE 36/45.

26. ARISTOLOCHIACEAE

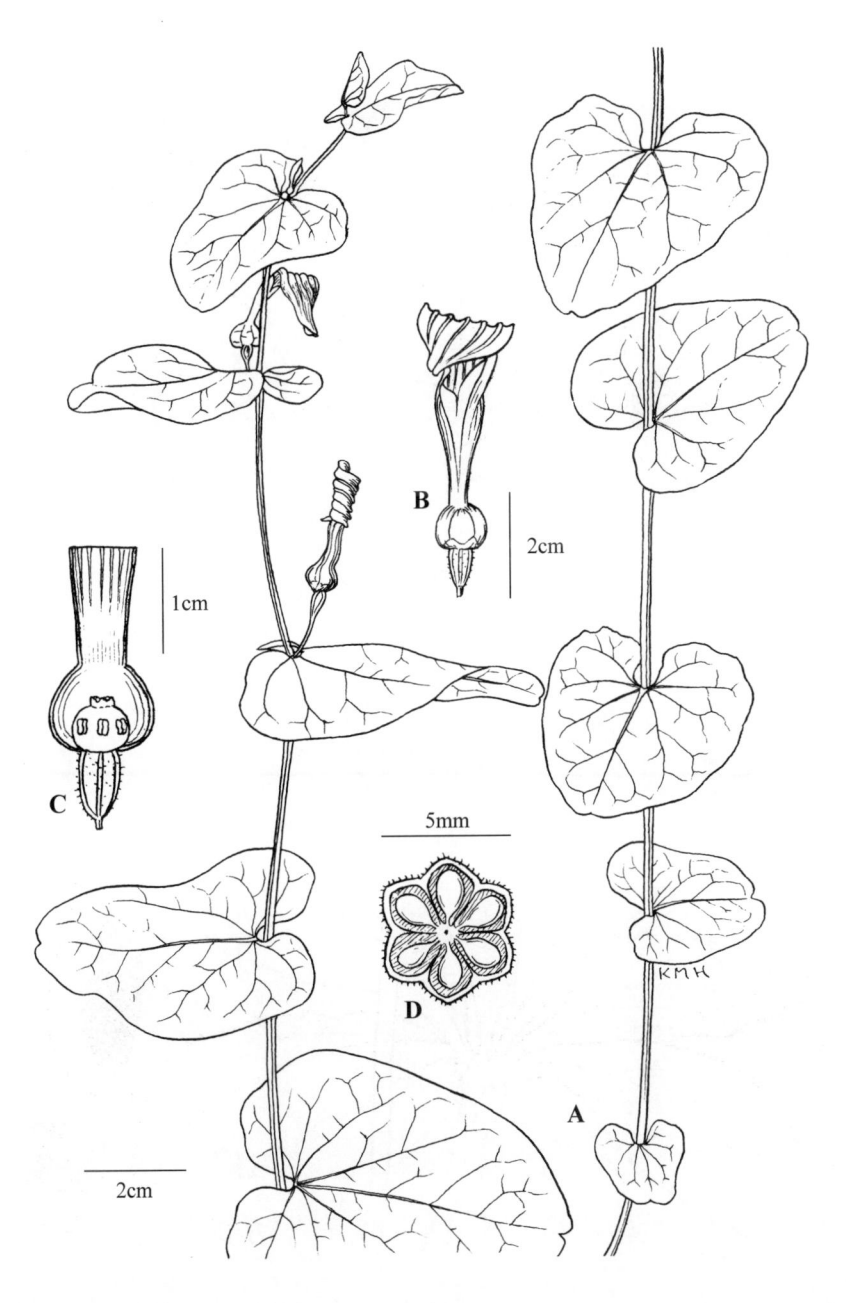

Aristolochia rotunda L. (Smearwort). A habit; B flower; C flower with part of corolla cut away; D TS of ovary. (Woldingham, Surrey). STACE 90/74.

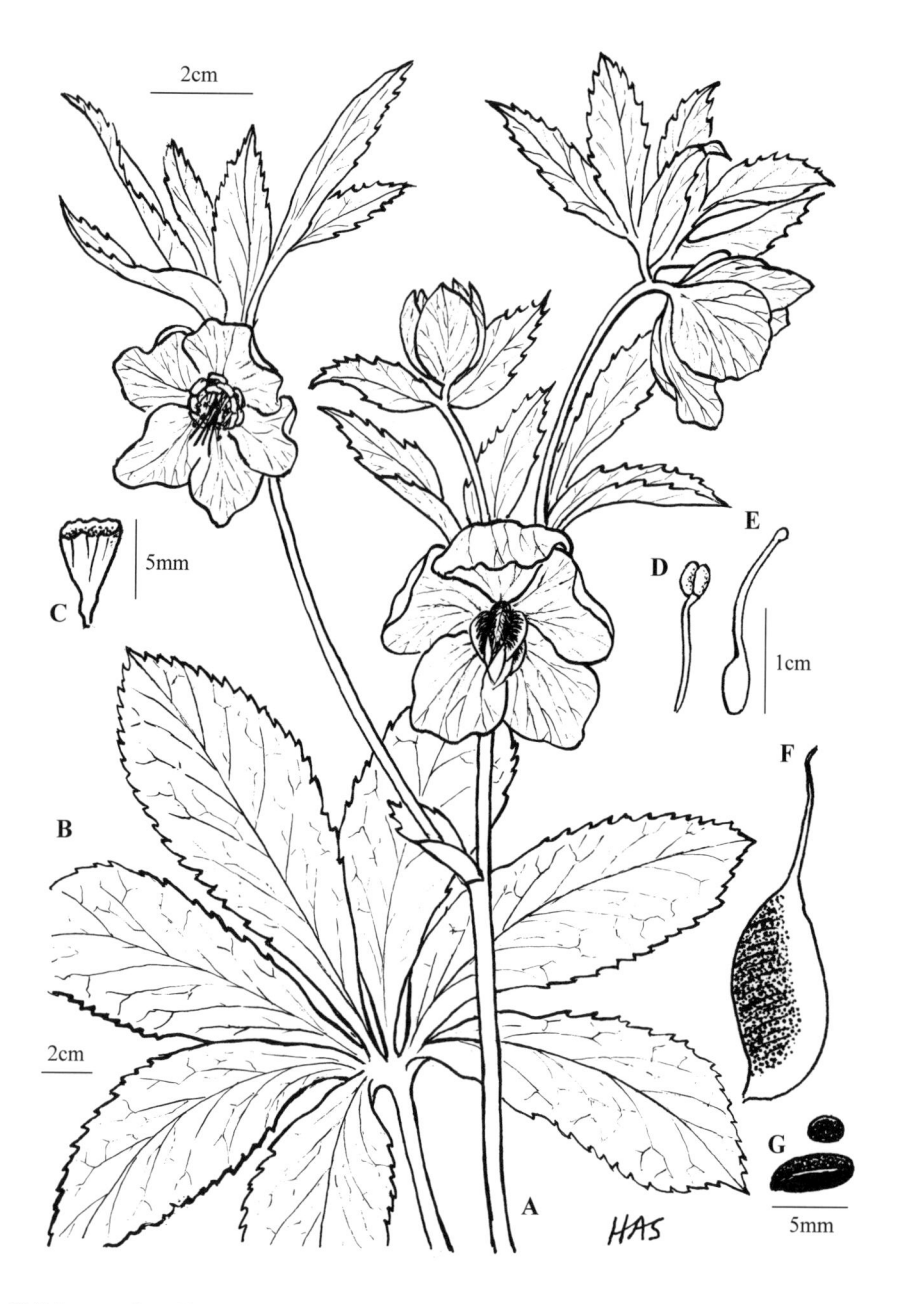

Helleborus orientalis Lam. (Lenten-rose). A flowering stem; B basal leaf; C nectary; D stamen; E carpel; F follicle; G seed and TS of seed. (*Hort.* Aberdeen, S Aberdeen). STACE 98/80.

30. RANUNCULACEAE

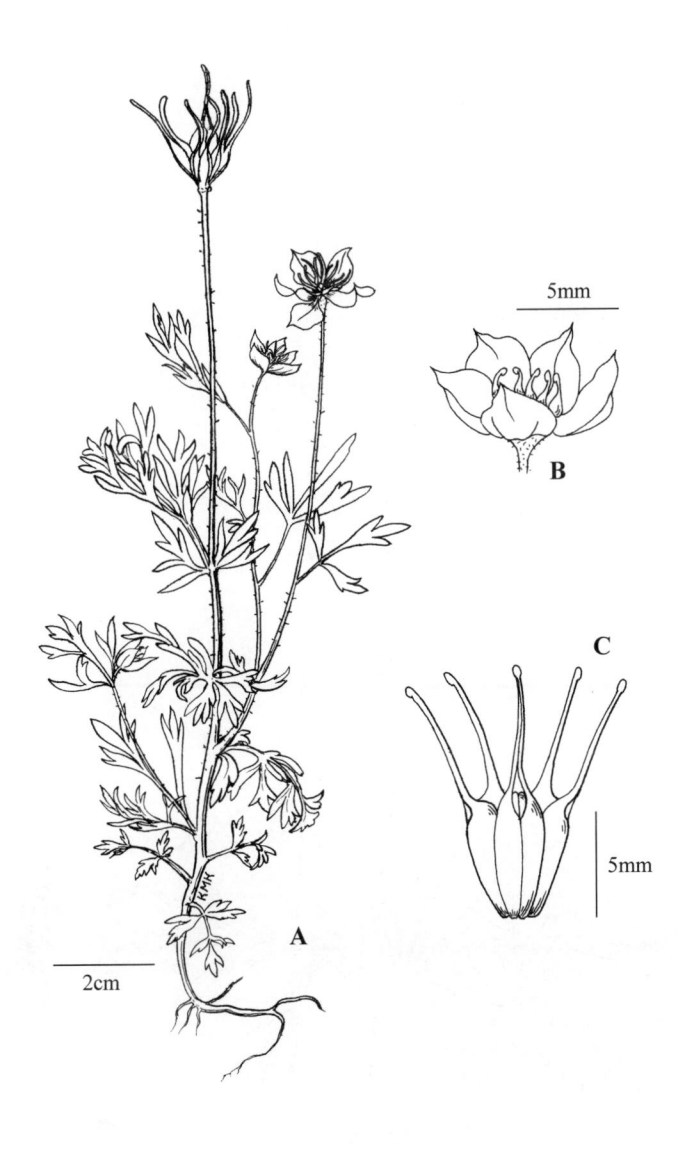

Nigella hispanica L. (Fennel-flower). A habit; B flower; C fruit of five follicles. (Tunbridge Wells, W Kent). STACE —/—.

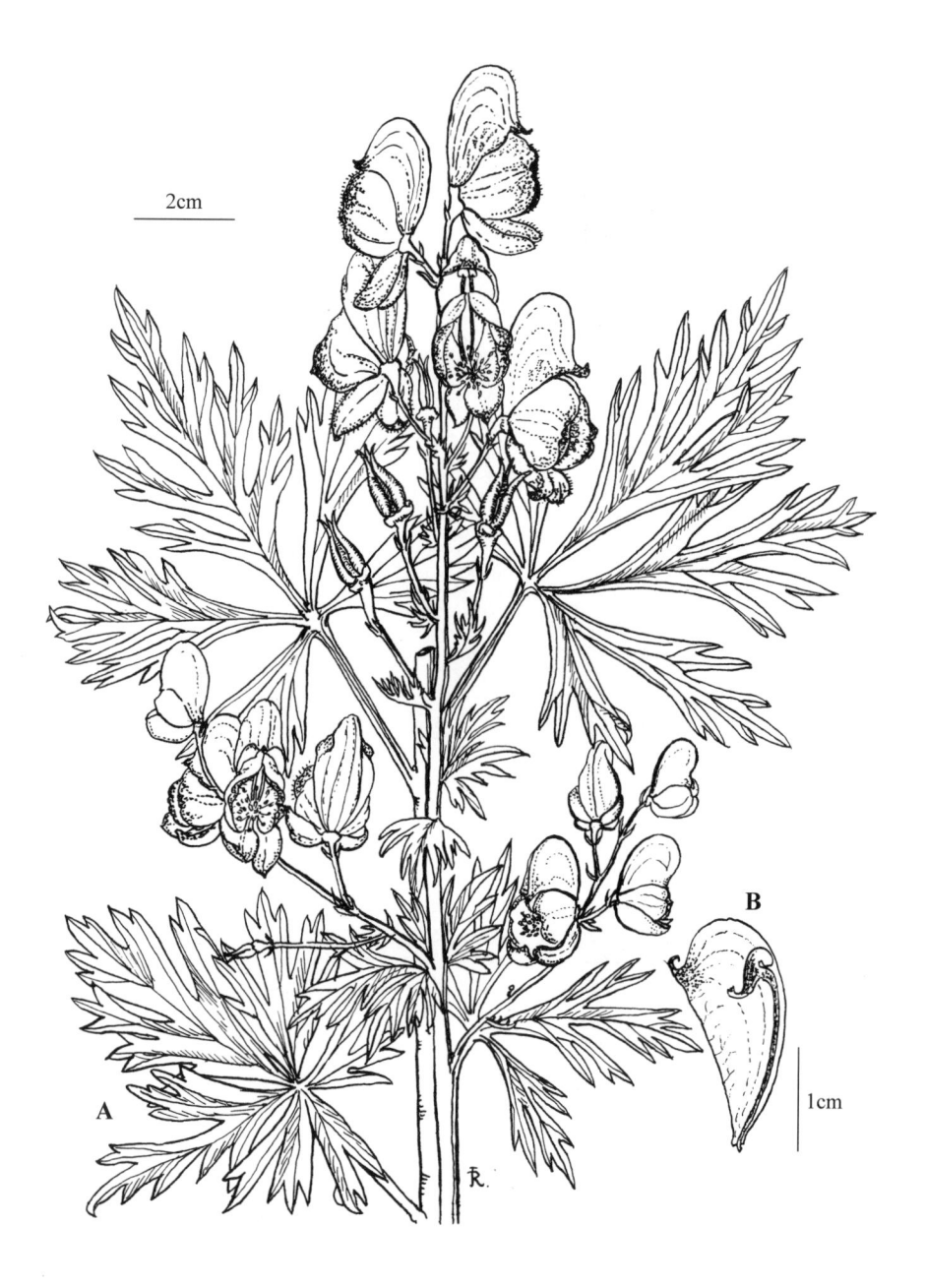

Aconitum x cammarum L. 'Bicolor' (Hybrid Monk's-hood). A habit; B LS of upper, hood-like sepal with tubular petal (nectary) within. (*Hort*. Dunmow, N Essex ex Scotland). STACE 100/81.

30. RANUNCULACEAE 11

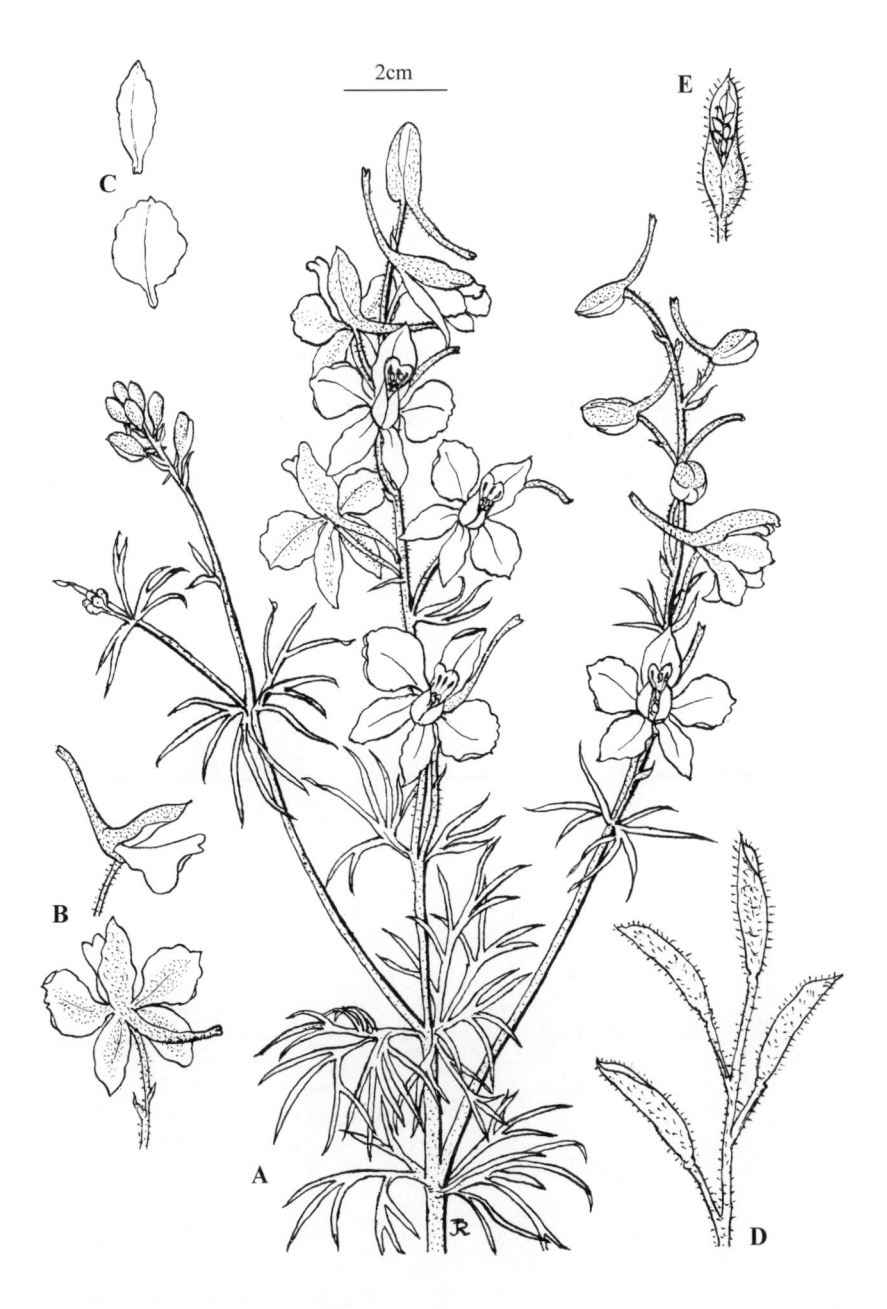

Consolida ajacis (L.) Schur (Larkspur). A habit; B flower; C lateral and lower petaloid sepals; D raceme of fruits; E follicle opening, exposing seeds. (*Hort*. Dunmow, N Essex). STACE 100/82.

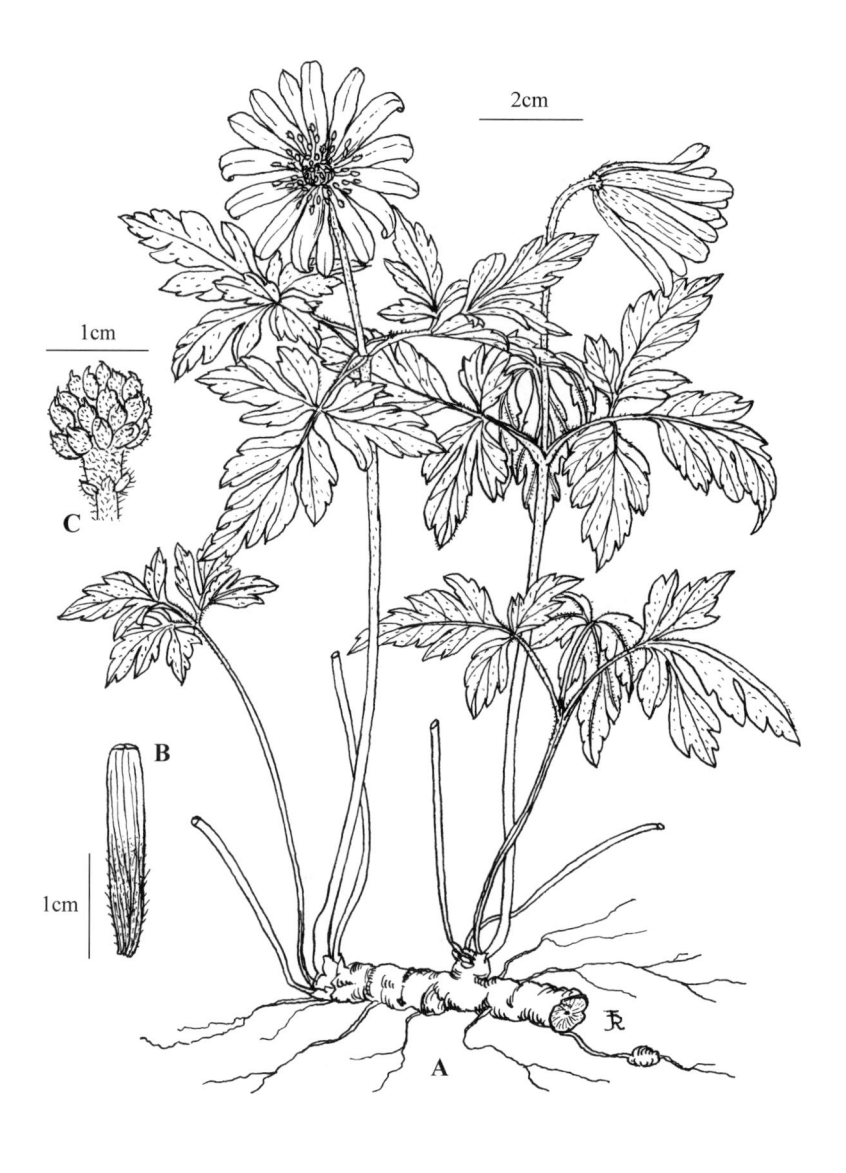

Anemone apennina L. (Blue Anemone). A habit; B underside of petaloid sepal; C head of achenes. (Well, NE Yorks). STACE 102/82.

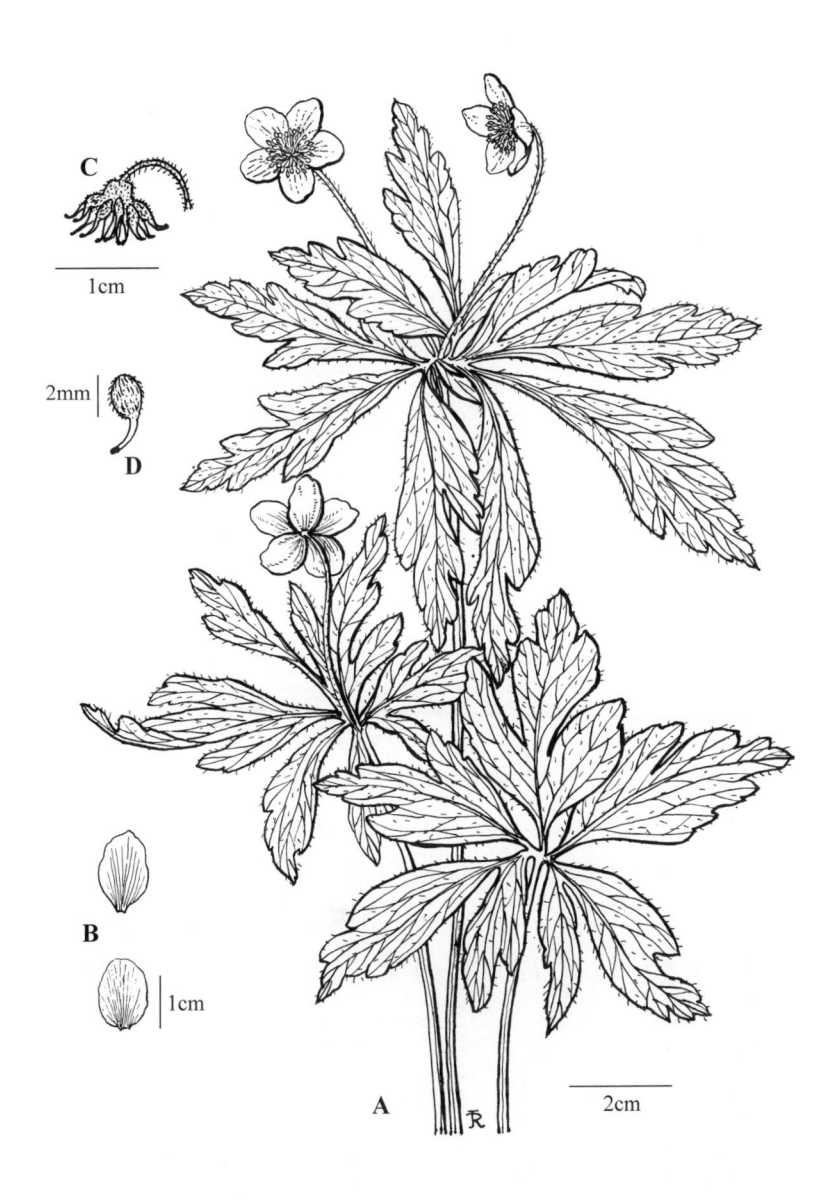

Anemone ranunculoides L. (Yellow Anemone). A habit; B petaloid sepals; C head of achenes; D achene. (Tewkesbury, E Gloucs). STACE 102/83.

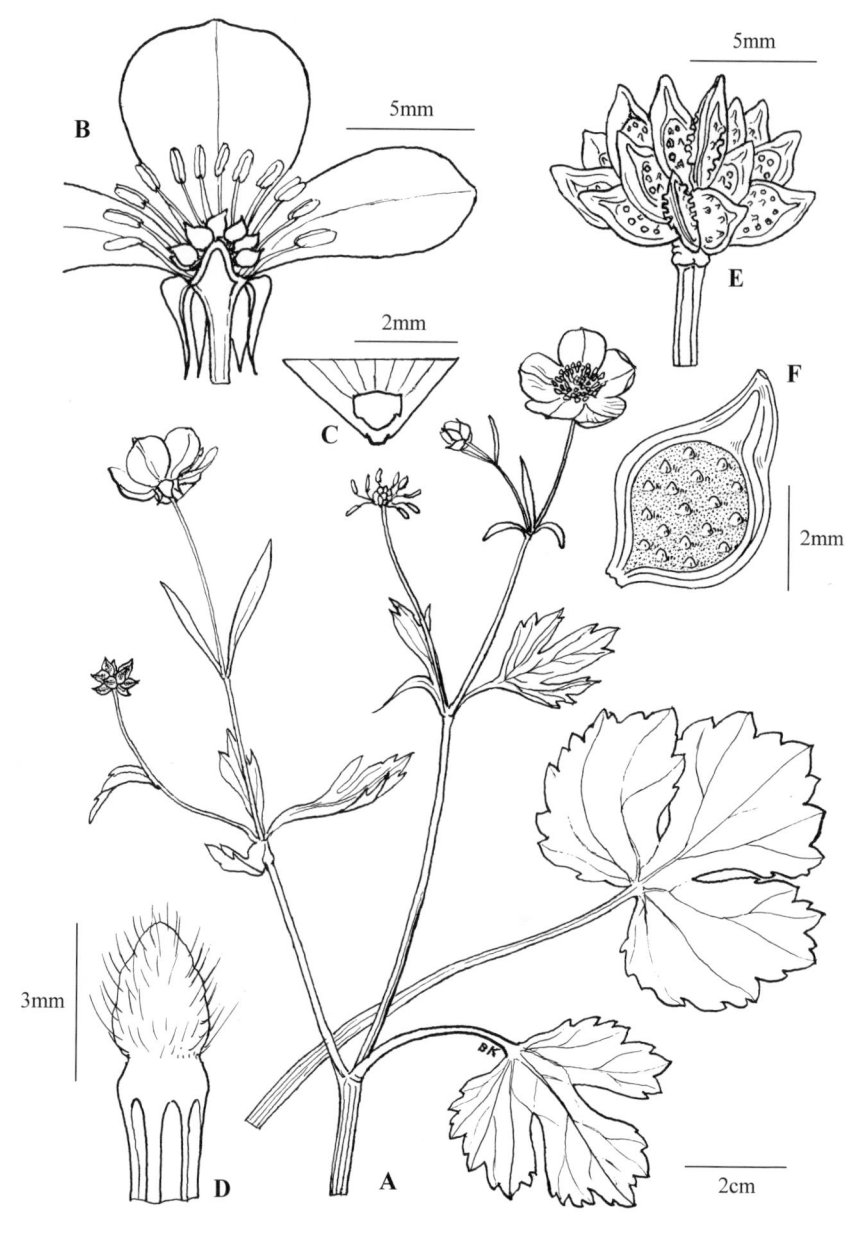

Ranunculus marginatus d'Urv. var. **trachycarpus** (Fisch. & C.A. Mey.) Azn. (St Martin's Buttercup). A habit; B LS of flower; C nectar-pit at base of petal; D receptacle; E head of achenes; F achene. (St Martin's, Isles of Scilly). STACE 109/88.

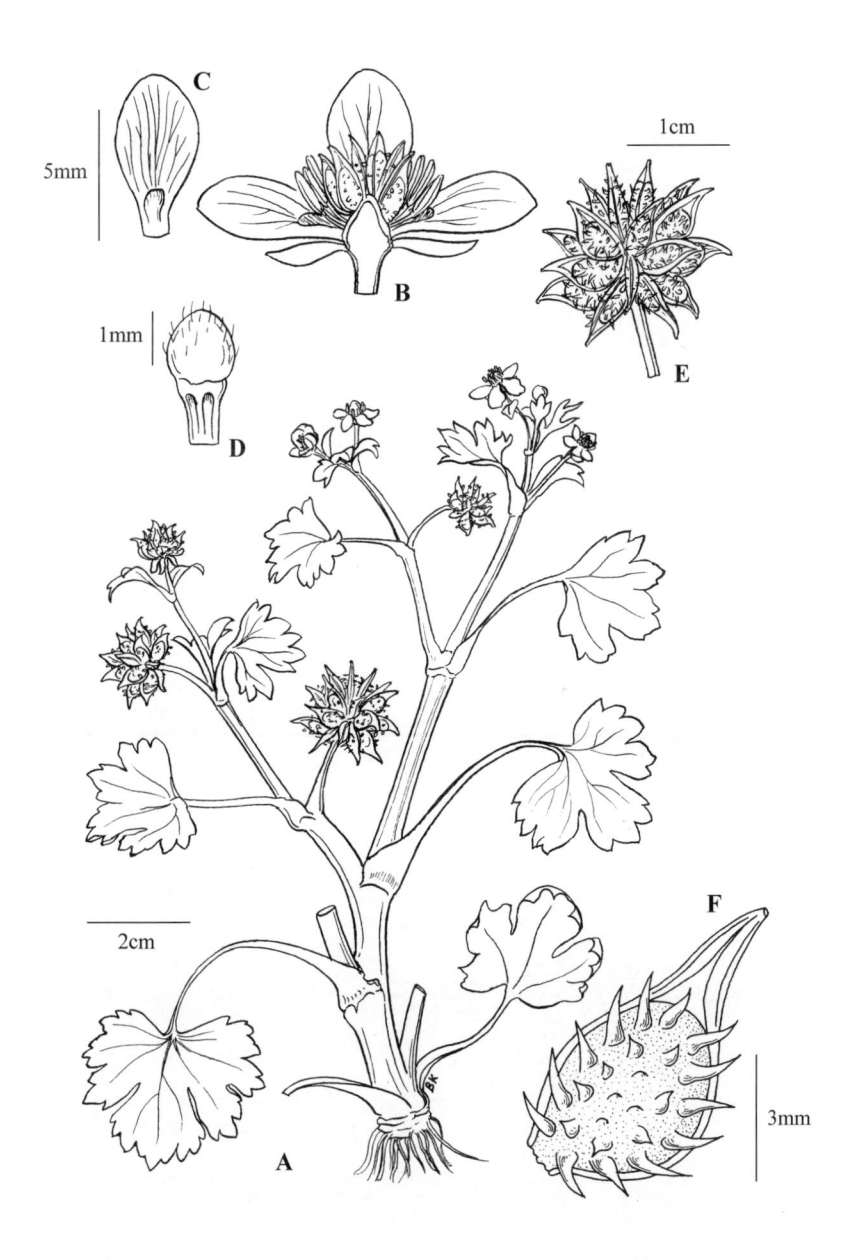

30. RANUNCULACEAE 15

Ranunculus muricatus L. (Rough-fruited Buttercup). A habit; B LS of flower; C petal with nectar-pit; D receptacle; E head of achenes; F achene. (Isles of Scilly). STACE 109/90.

Berberis glaucocarpa Stapf (Great Barberry). A young twig; B barren twig; C flowering and fruiting twigs; D flower with some perianth segments removed; E stamen; F fruit opened out; G seed. (S Somerset). STACE 120/100.

31. BERBERIDACEAE

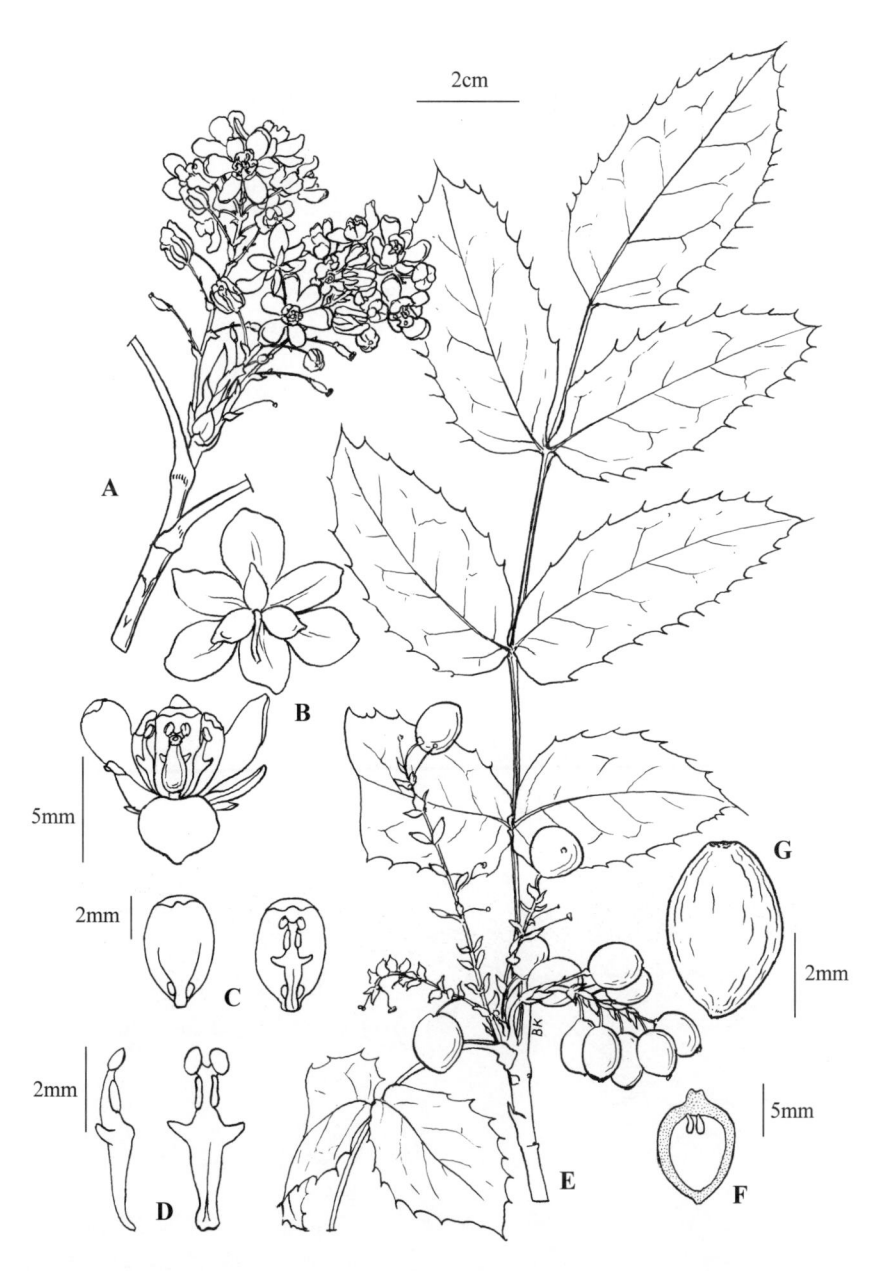

Mahonia aquifolium (Pursh) Nutt. (Oregon-grape). A flowering twig; B flower; C inner perianth segments, one with a stamen; D stamen; E fruiting twig; F LS of young fruit; G seed. (Source unknown). STACE 123/101.

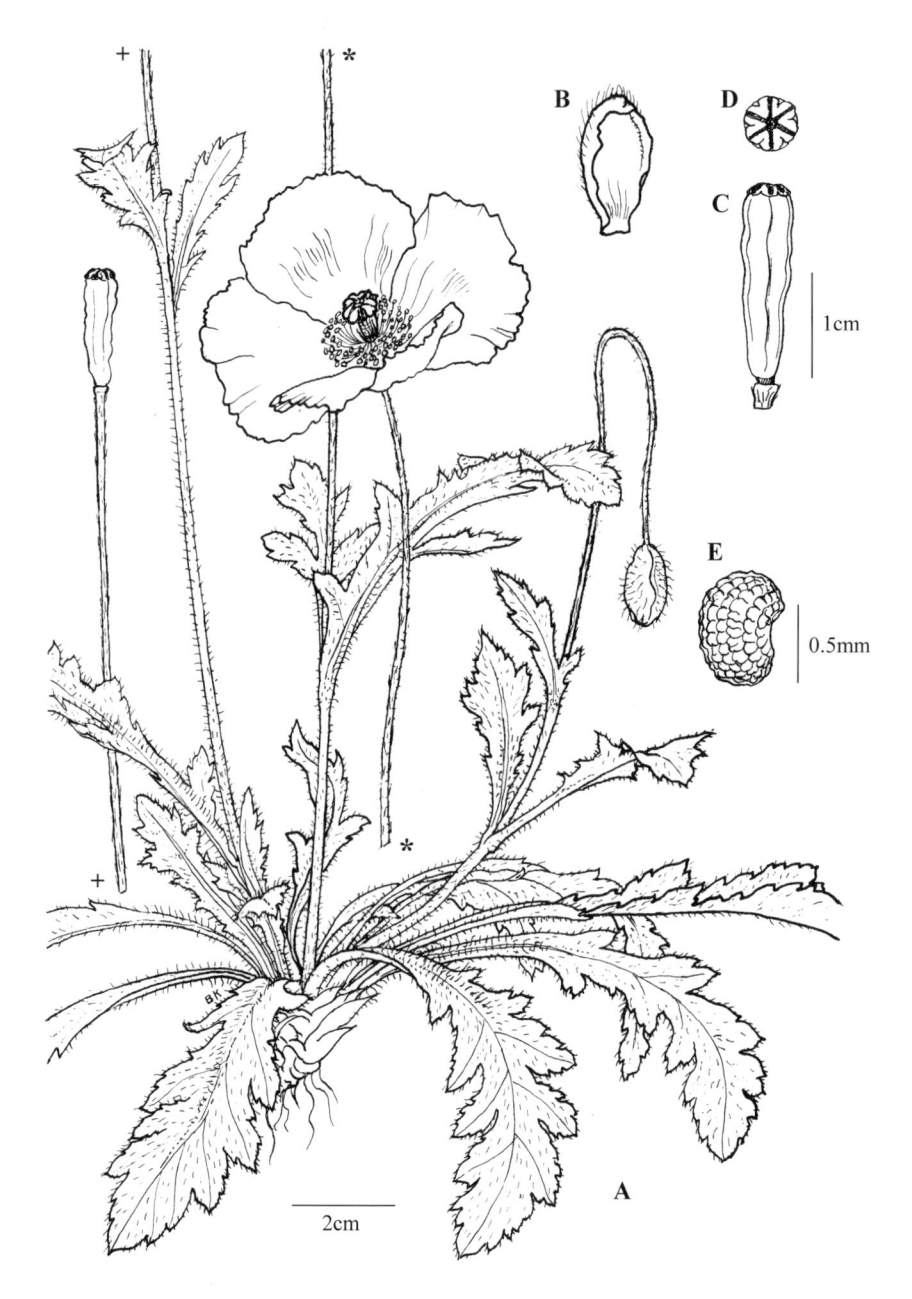

Papaver atlanticum (Ball) Coss. (Atlas Poppy). A habit; B sepal; C capsule; D persistent stigma at apex of fruit; E seed. (Source unknown). STACE 125/102.

32. PAPAVERACEAE

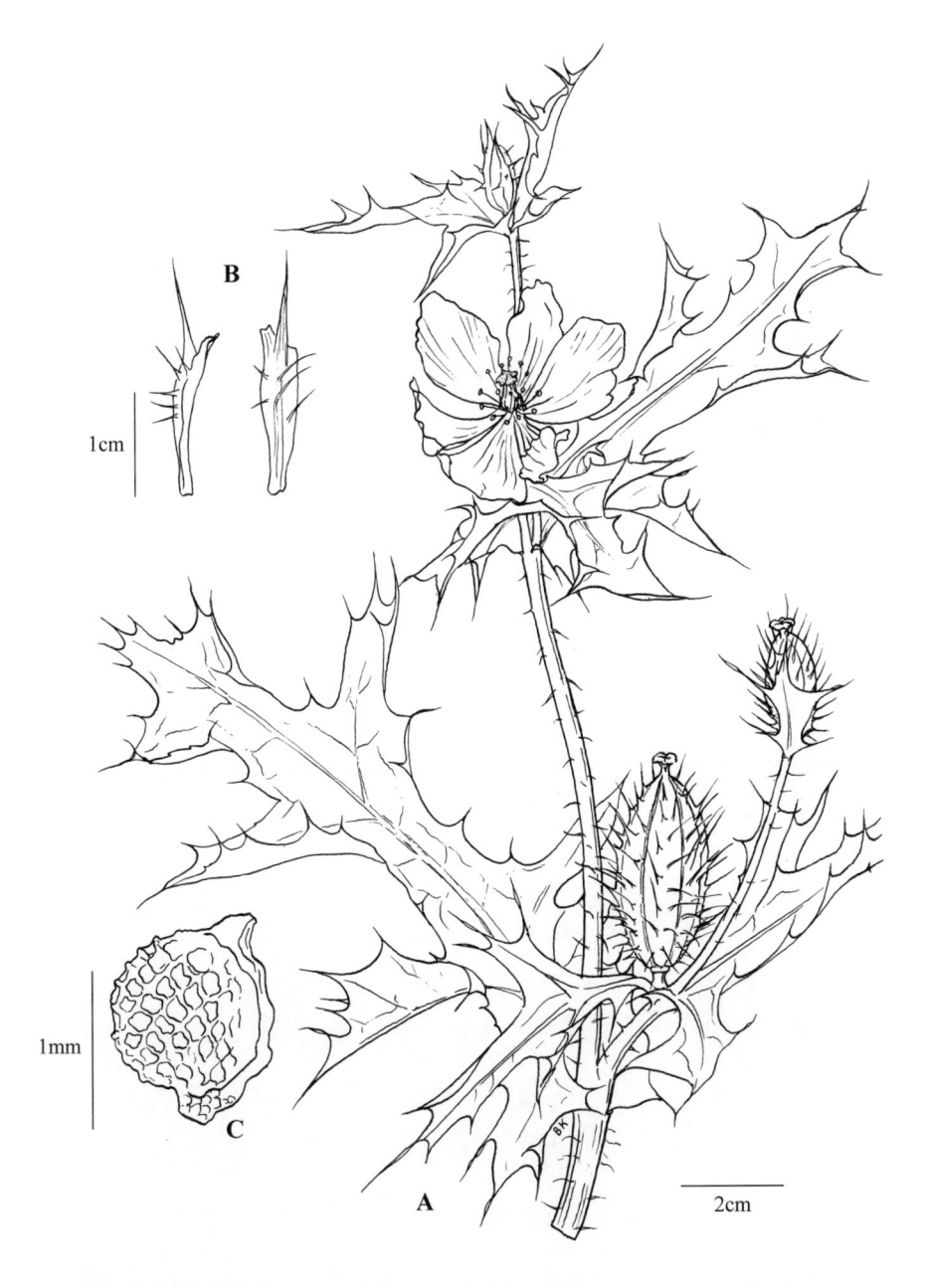

Argemone mexicana L. (Mexican Poppy). A habit; B sepal; C seed. (Source unknown). STACE 127/105.

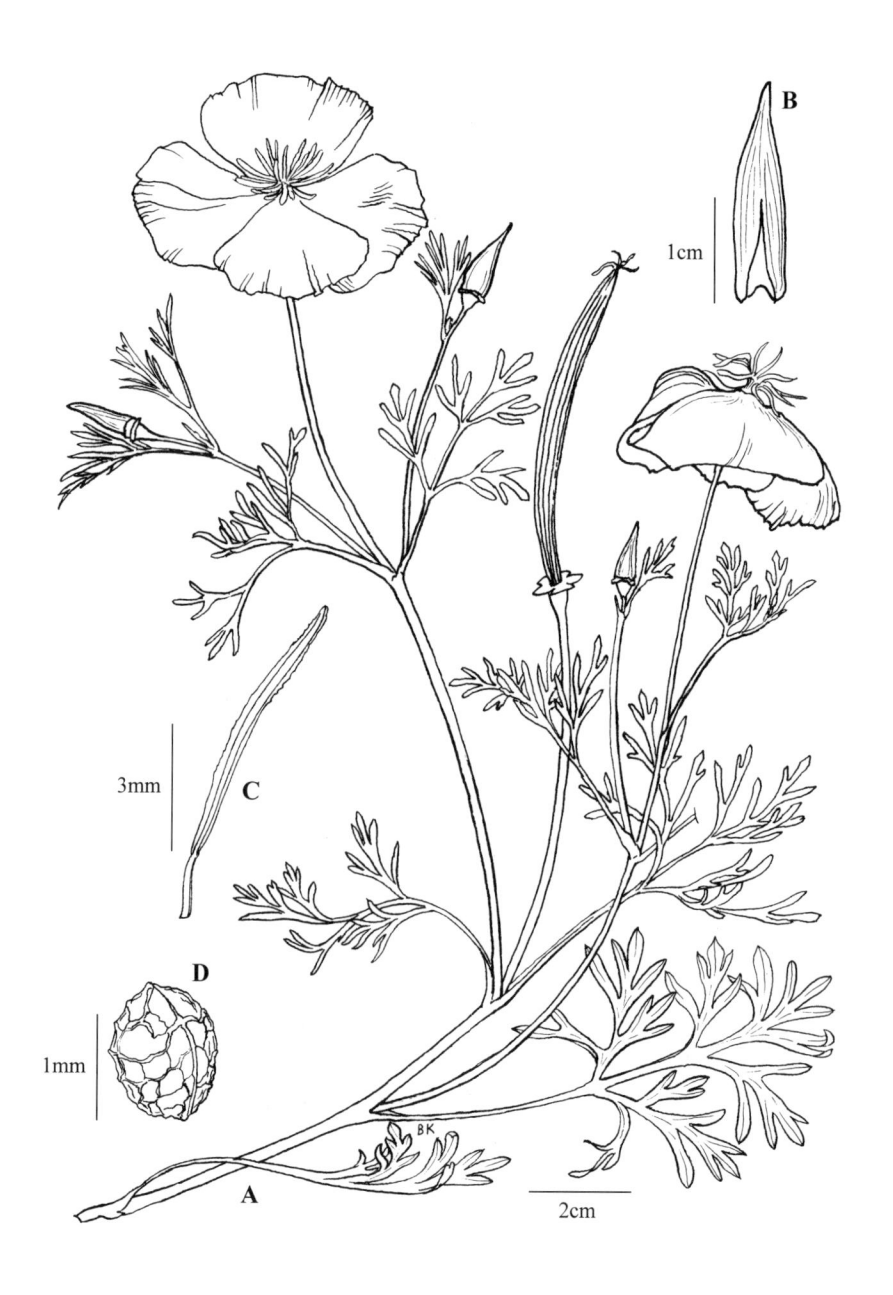

Eschscholzia californica Cham. (Californian Poppy). A habit; B calyx; C stamen; D seed. (Source unknown). STACE 129/105.

32. PAPAVERACEAE

Macleaya × kewensis Turrill (Hybrid Plume-poppy). A habit; B underside of leaf apex; C sepal; D flower minus sepals; E style and stigma; F sterile capsule; G LS of capsule. (Source unknown). STACE 129/105.

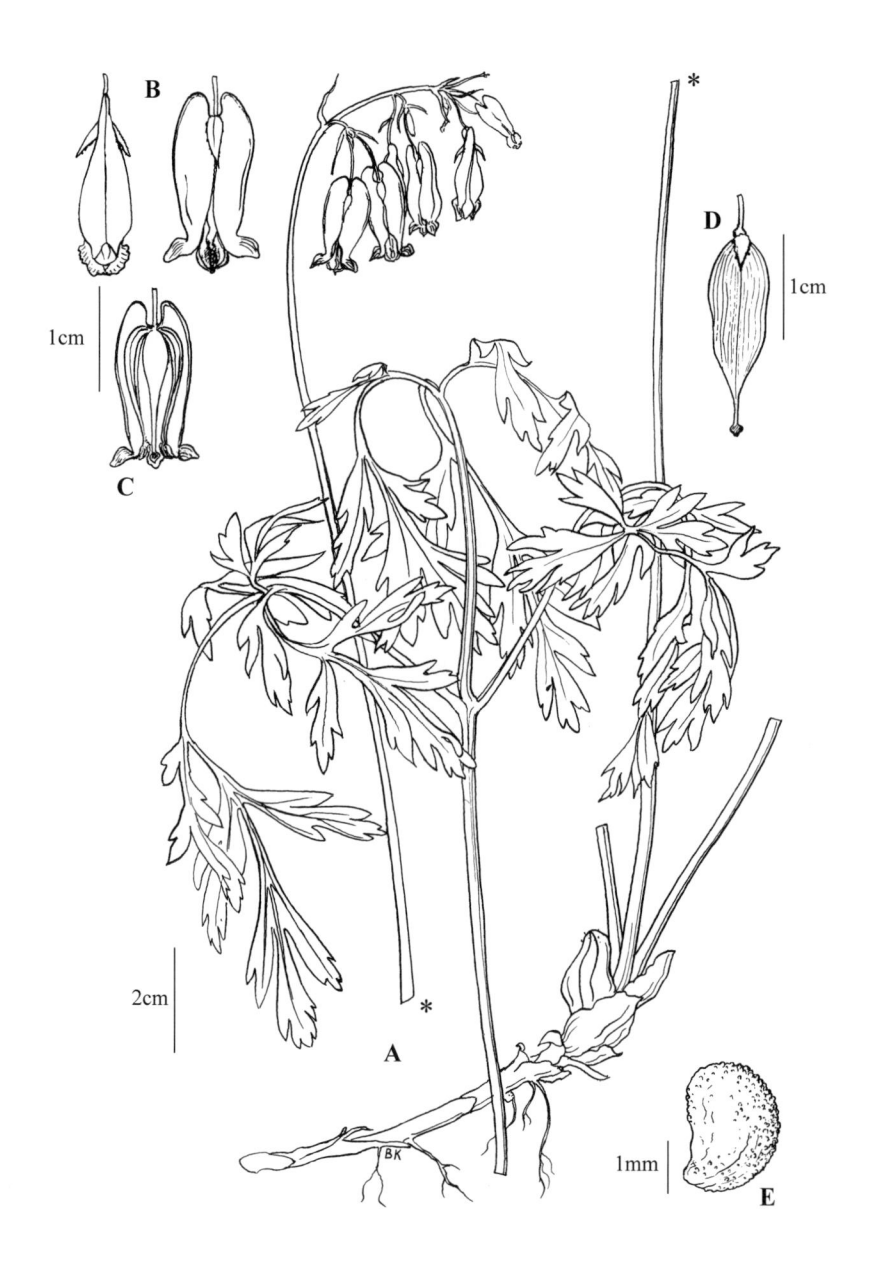

Dicentra formosa (Haw.) Walp. (Bleeding-heart). A habit; B flower; C flower with part of corolla cut away; D capsule; E seed. (Source unknown). STACE 130/106.

33. FUMARIACEAE

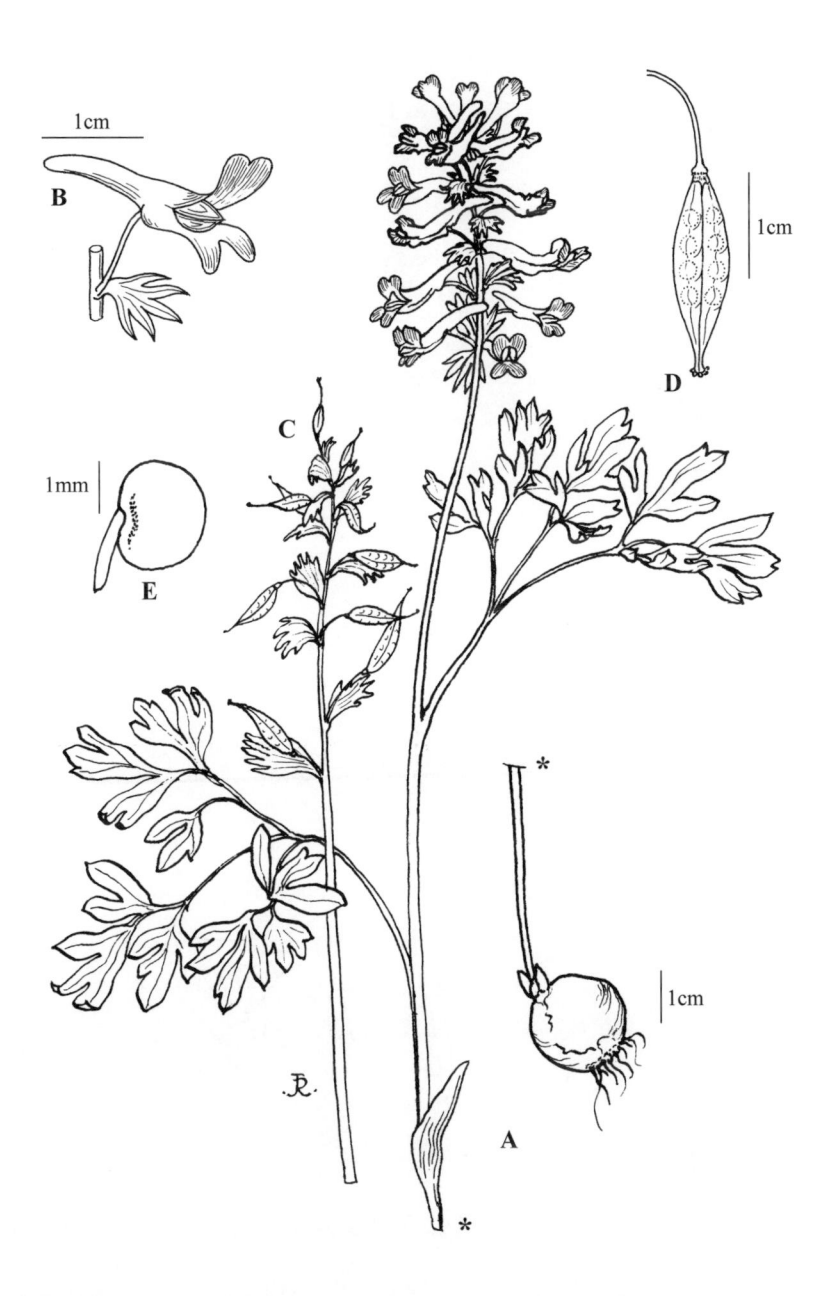

Corydalis solida (L.) Clairv. (Bird-in-a-bush). A habit; B flower showing lobed bract; C fruiting stem; D capsule; E seed. (Source unknown). STACE 131/107.

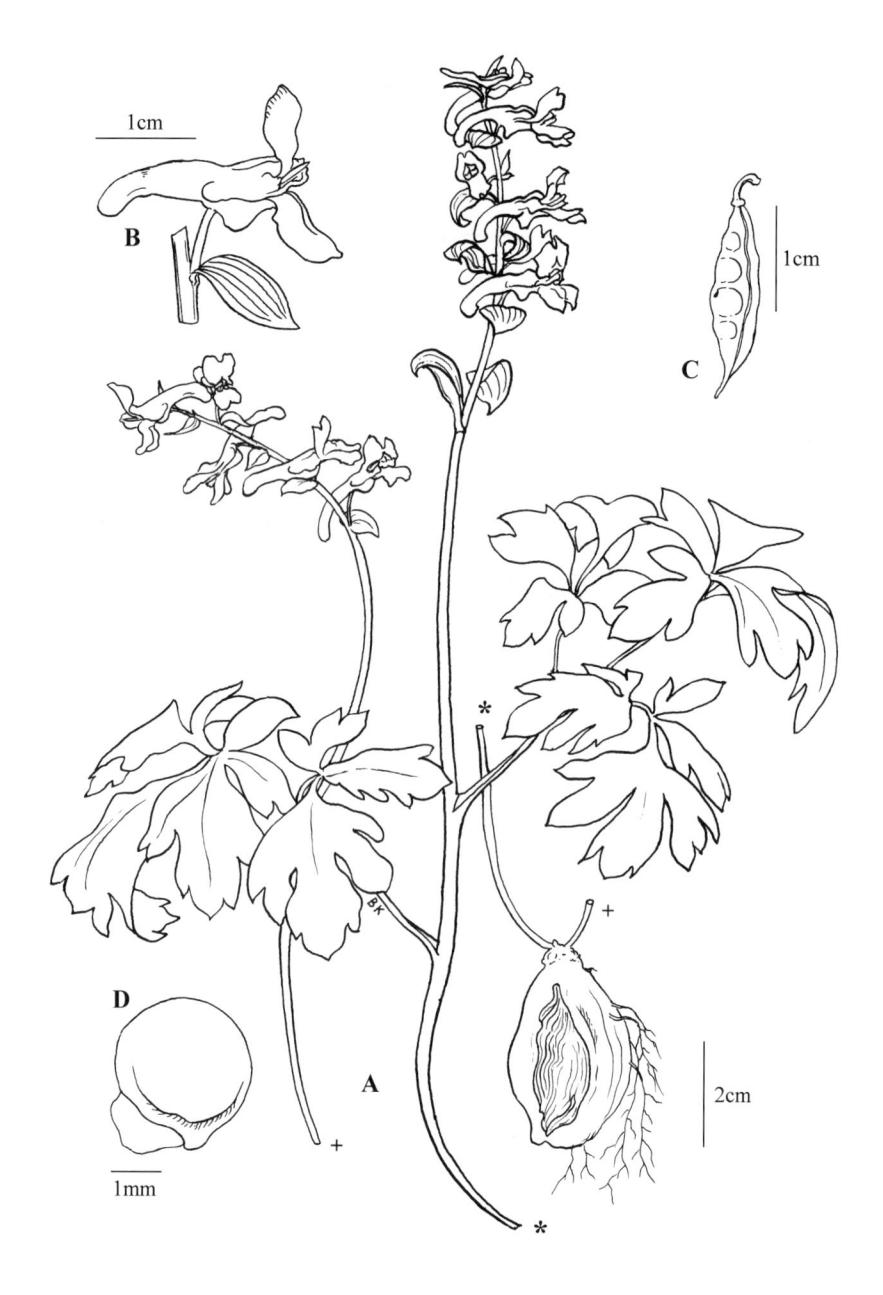

Corydalis cava (L.) Schweigg. & Körte (Hollowroot). A habit; B flower showing entire bract; C capsule; D seed. (Britwell House, Oxon). STACE 131/107.

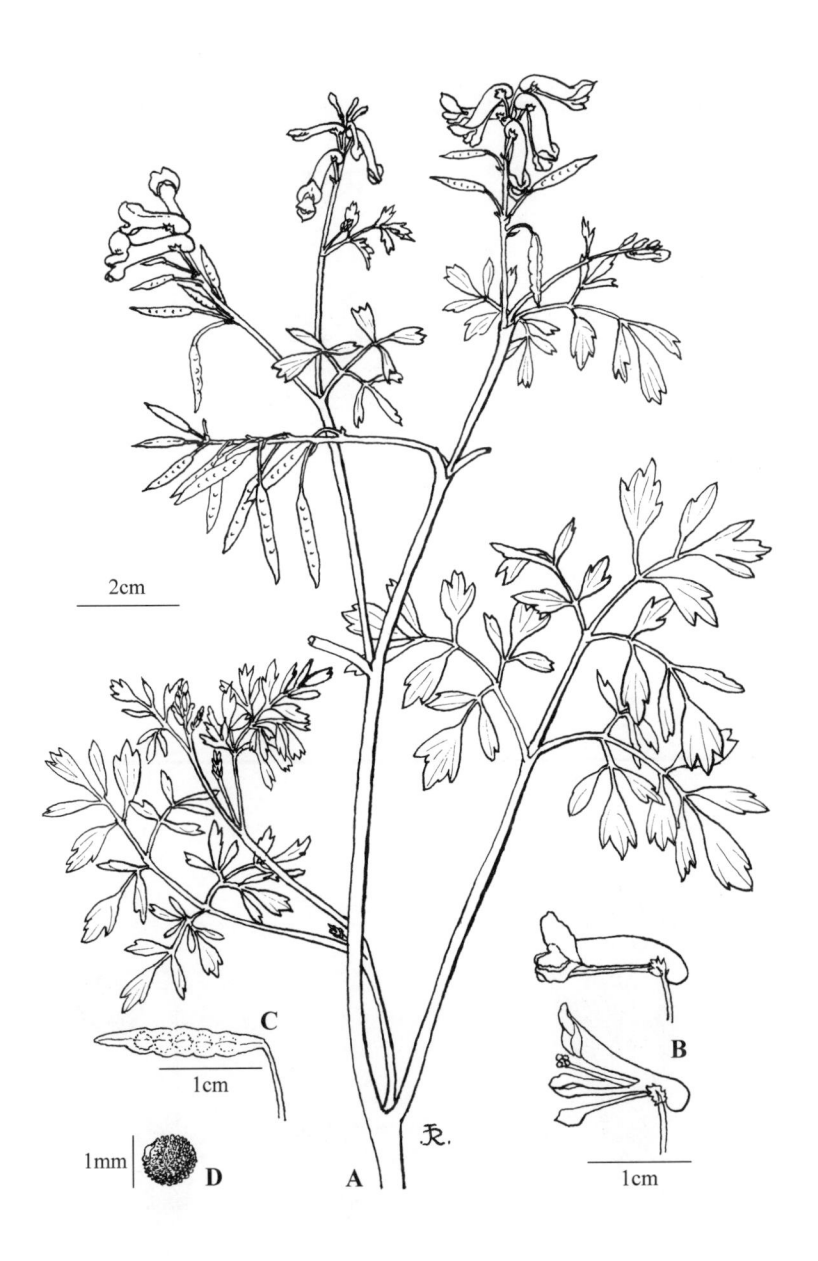

Pseudofumaria alba (Mill.) Lidén (Pale Corydalis). A habit; B flower; C capsule; D seed. (Ipswich, E Suffolk). STACE 131/107.

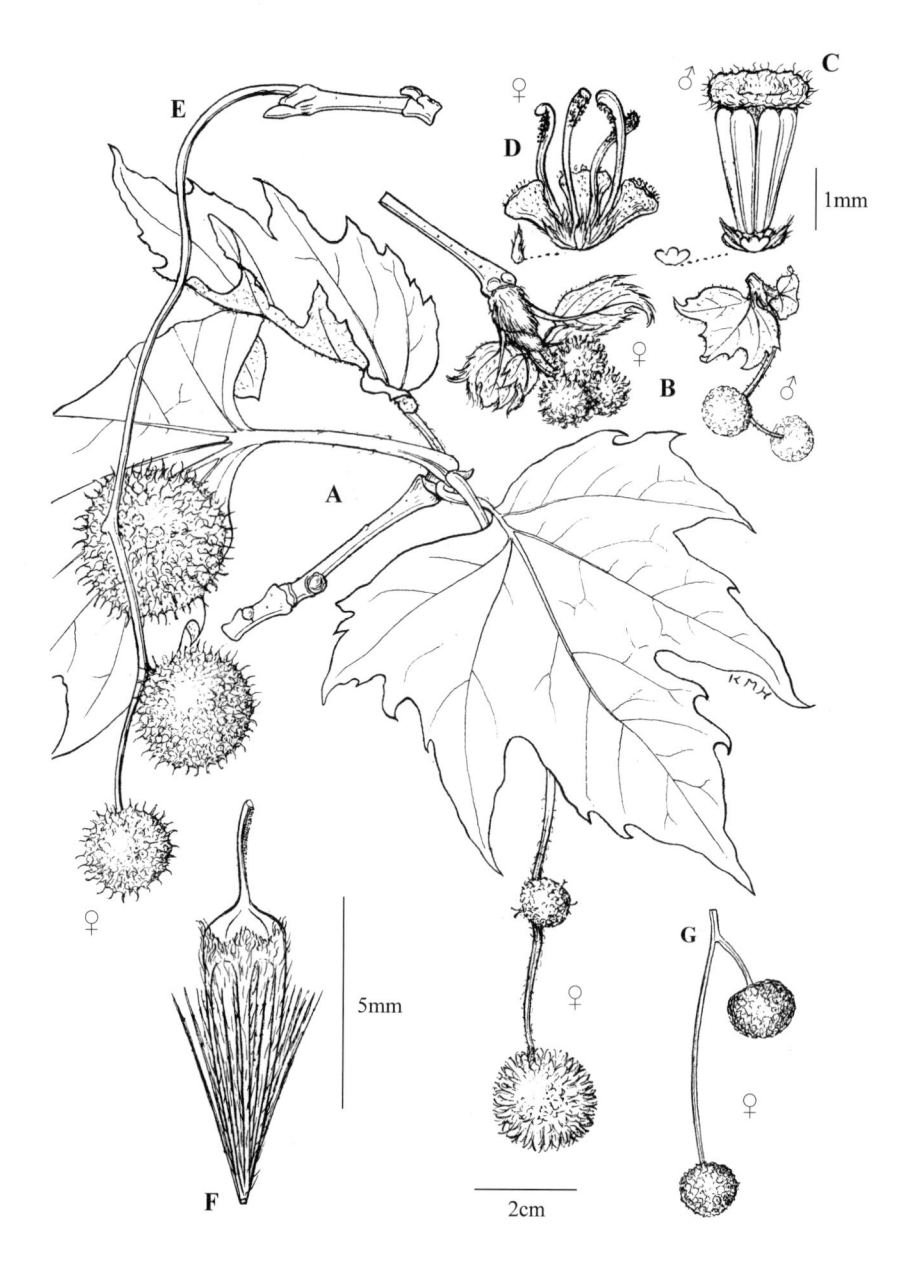

Platanus x hispanica Mill. ex Münchh. (London Plane). A flowering branch; B male and female inflorescences; C male flower; D four female flowers and three staminodes; E mature fruiting branch; F achene; G fruiting receptacles. (Central London, Middlesex). STACE 136/111.

36. CANNABACEAE

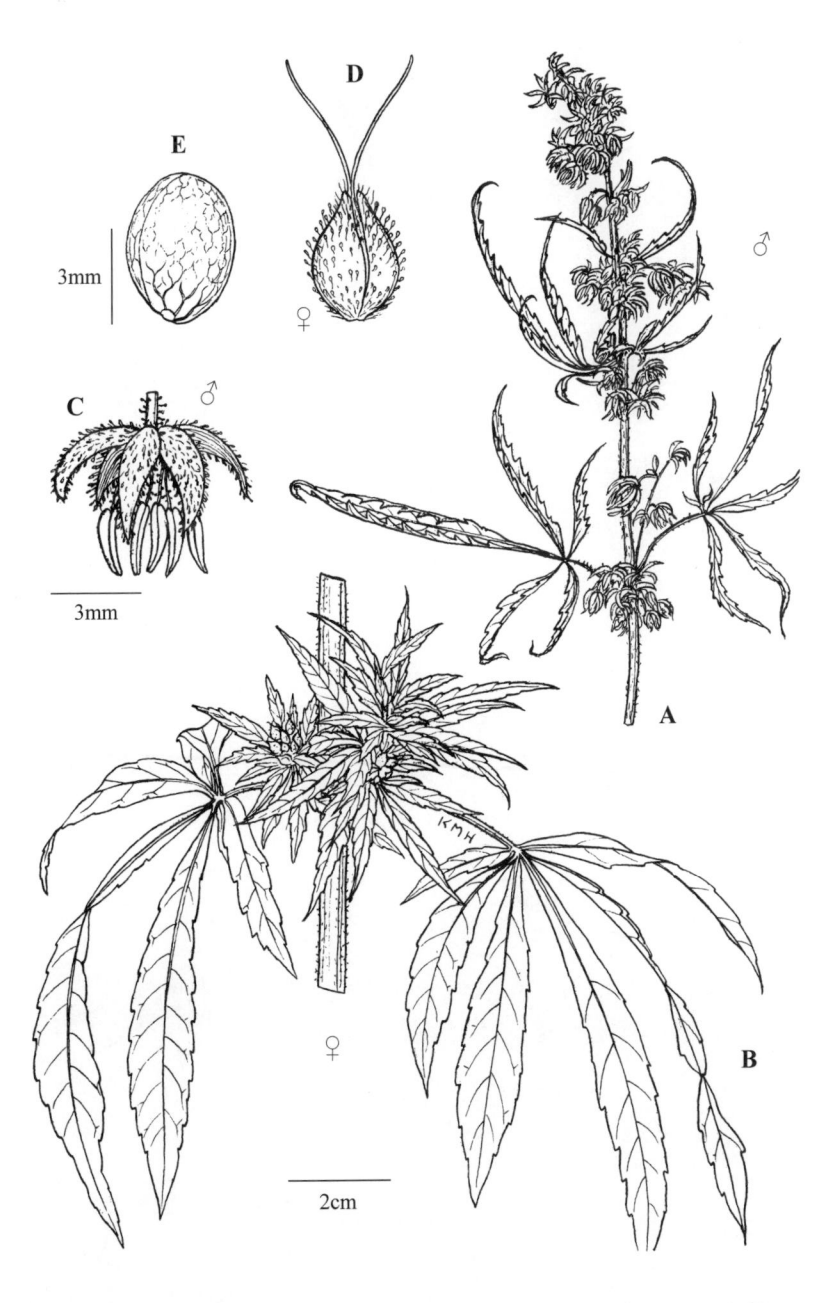

Cannabis sativa L. (Hemp). A part of male plant; B part of female plant; C male flower; D female flower; E seed. (Source unknown). STACE 142/116.

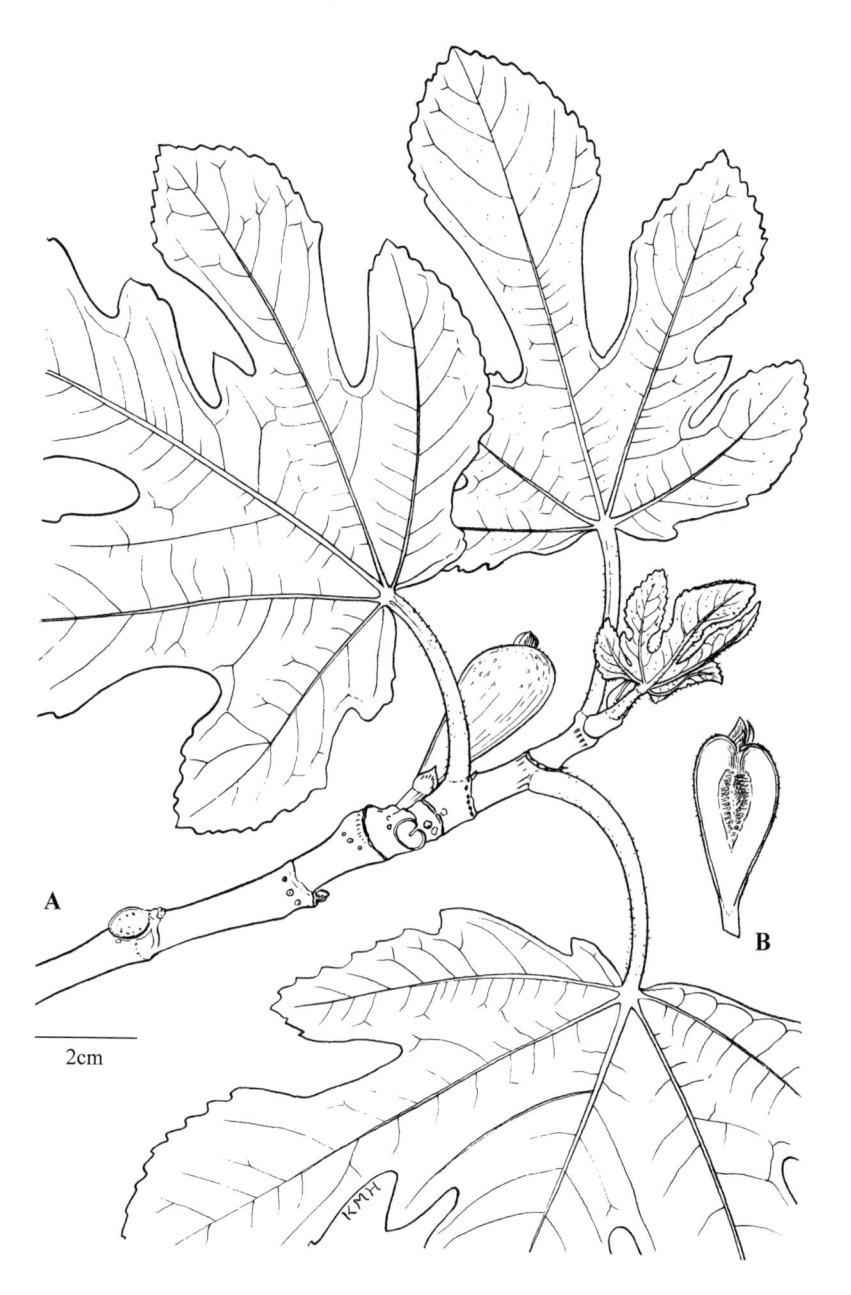

Ficus carica L. (Fig). A young fruiting branch; B LS of unripe fruiting head (a 'fig'). (The Mumbles, Glam). STACE 143/116.

38. URTICACEAE

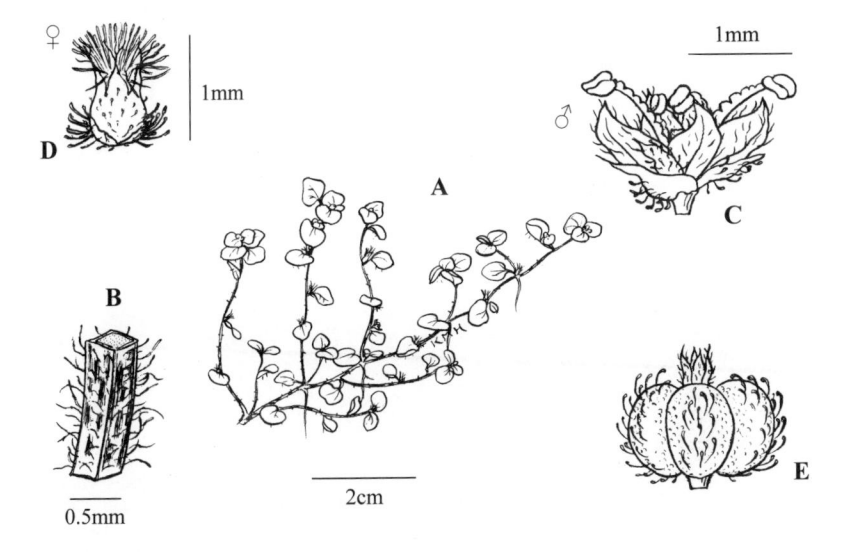

Soleirolia soleirolii (Req.) Dandy (Mind-your-own-business). A habit; B part of stem; C male flower; D female flower; E fruiting perianth. (*Hort*. Ashbourne, Derbys). STACE 144/118.

39. JUGLANDACEAE

Juglans regia L. (Walnut). A flowering branches; B fruiting branch; C male flower; D female flower; E 'nut'. (Source unknown). STACE 145/118.

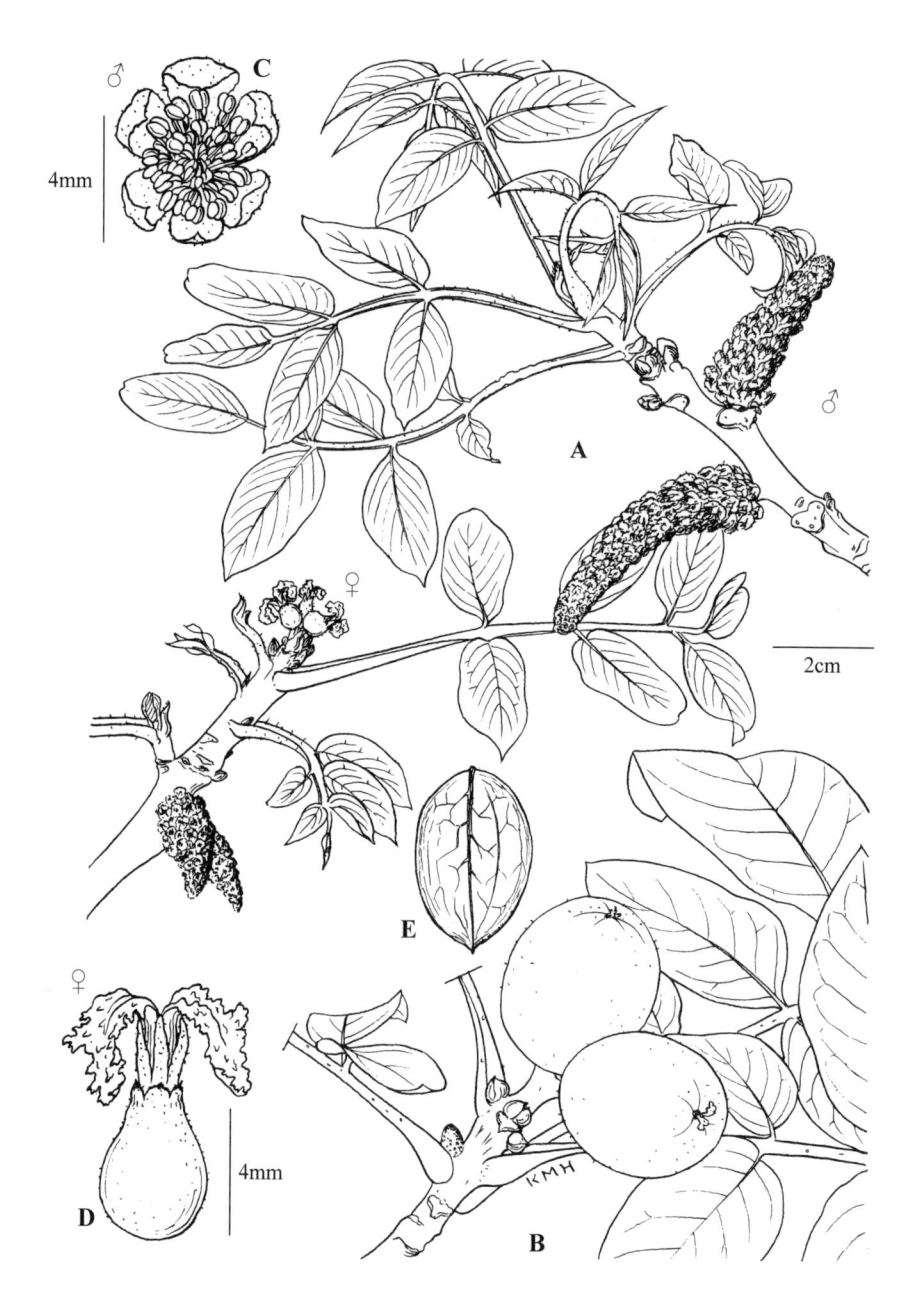

40. MYRICACEAE

Myrica pensylvanica Loisel. ex Duhamel (Bayberry). A flowering branch of female plant; B female flower, one bracteole removed; C fruit with thick coat of wax. (Holmsley, S Hants). STACE 146/119.

32   41. FAGACEAE

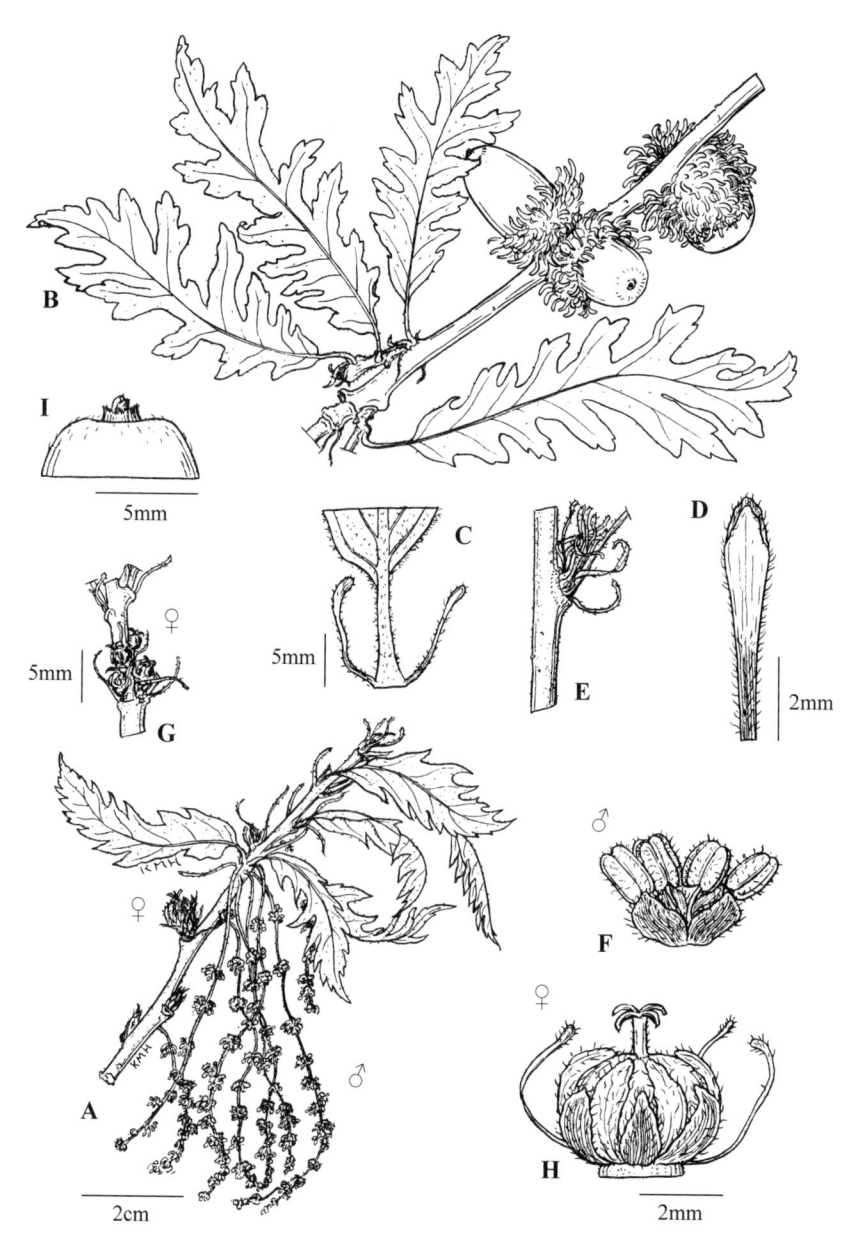

Quercus cerris L. (Turkey Oak). A flowering branch; B fruiting branch; C base of leaf with stipules; D stipule; E bud surrounded by persistent stipules; F male flower; G cluster of female flowers; H female flower; I apex of acorn. (Dover, E Kent). STACE 150/122.

41. FAGACEAE

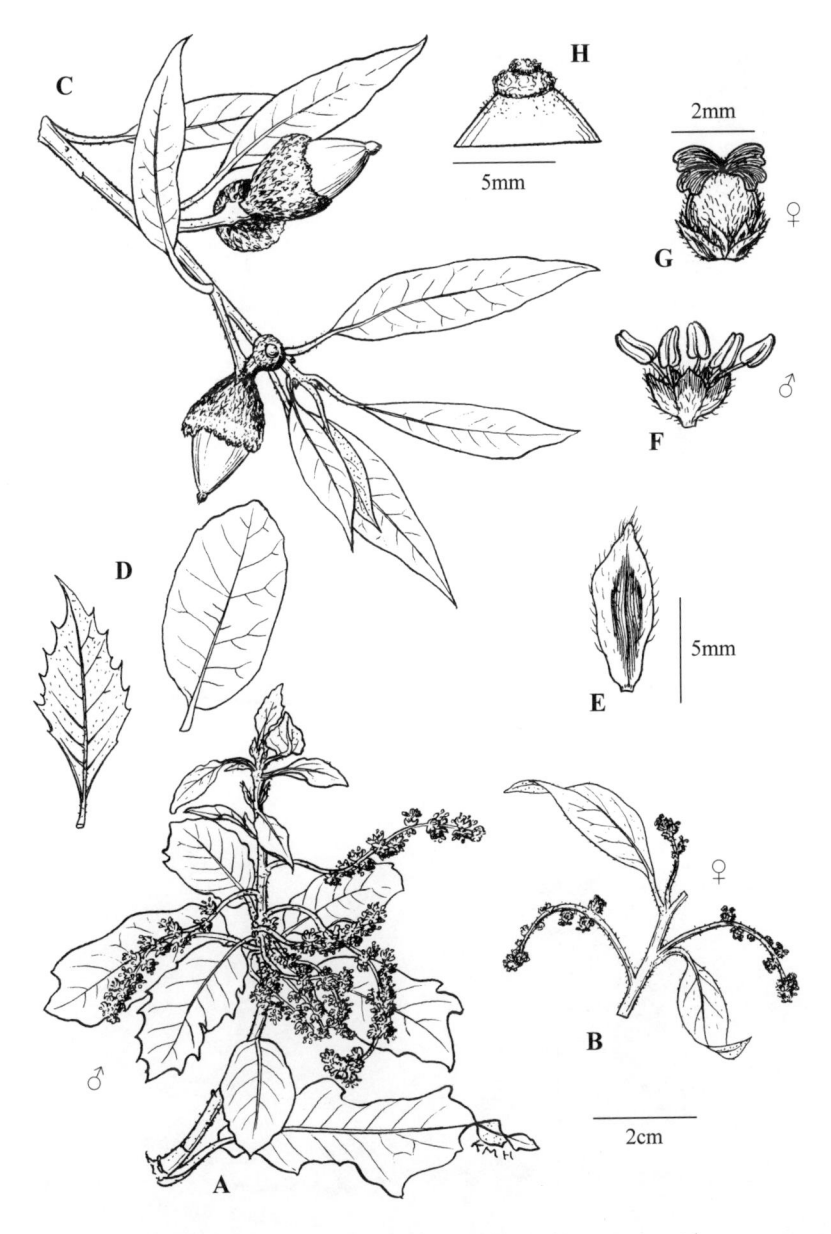

Quercus ilex L. (Evergreen Oak). A male flowering branch; B female flowering branch; C fruiting branch; D variation in leaf shape; E stipule; F male flower; G female flower; H apex of acorn. (Source unknown). STACE 150/122.

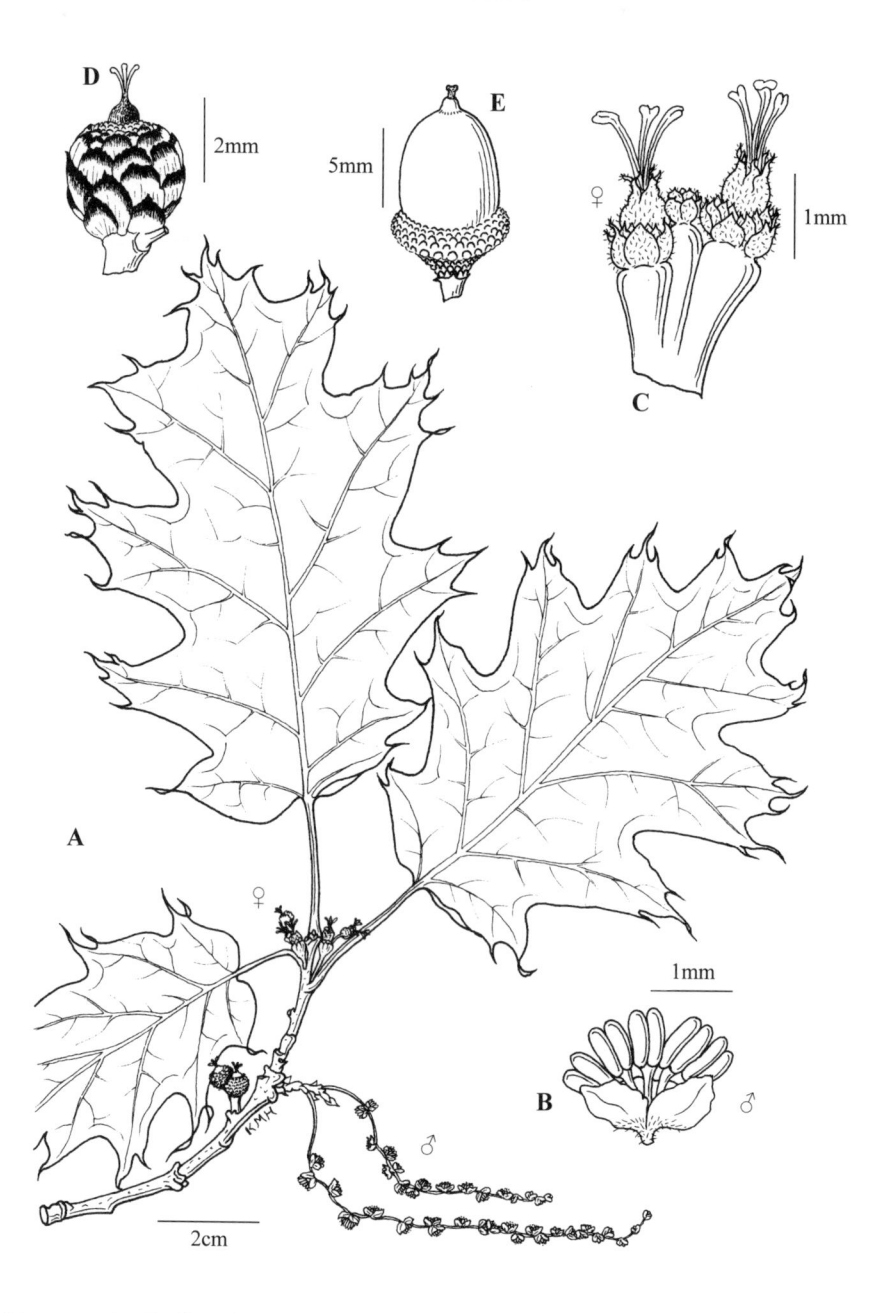

Quercus rubra L. (Red Oak). A flowering branch; B male flower; C cluster of female flowers; D young fruit; E acorn in cupule. (*Hort.* Kew, Surrey). STACE 151/123.

42. BETULACEAE 35

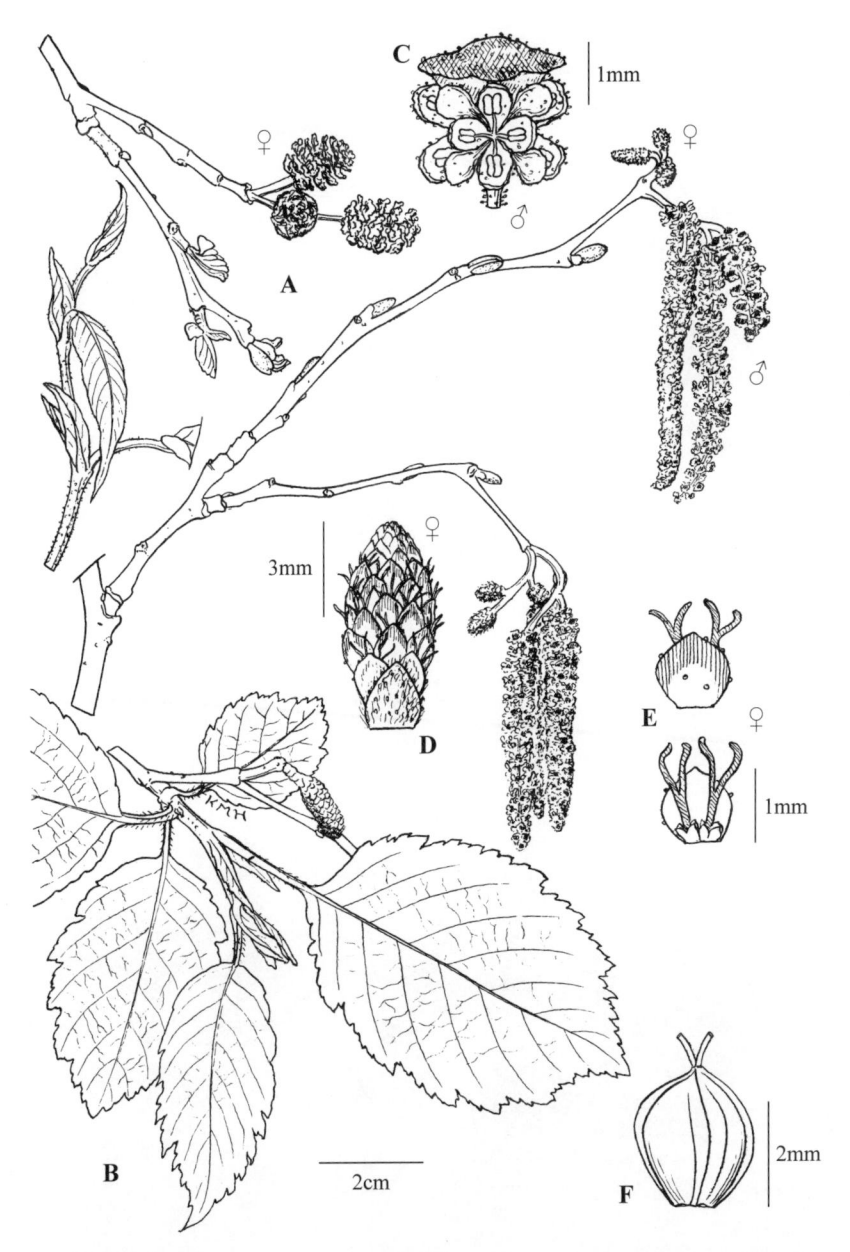

Alnus incana (L.) Moench (Grey Alder). A twigs in spring; B twig in summer; C male flower; D catkin of female flowers; E bract with two female flowers; F nutlet. (Clova, Angus). STACE 155/126.

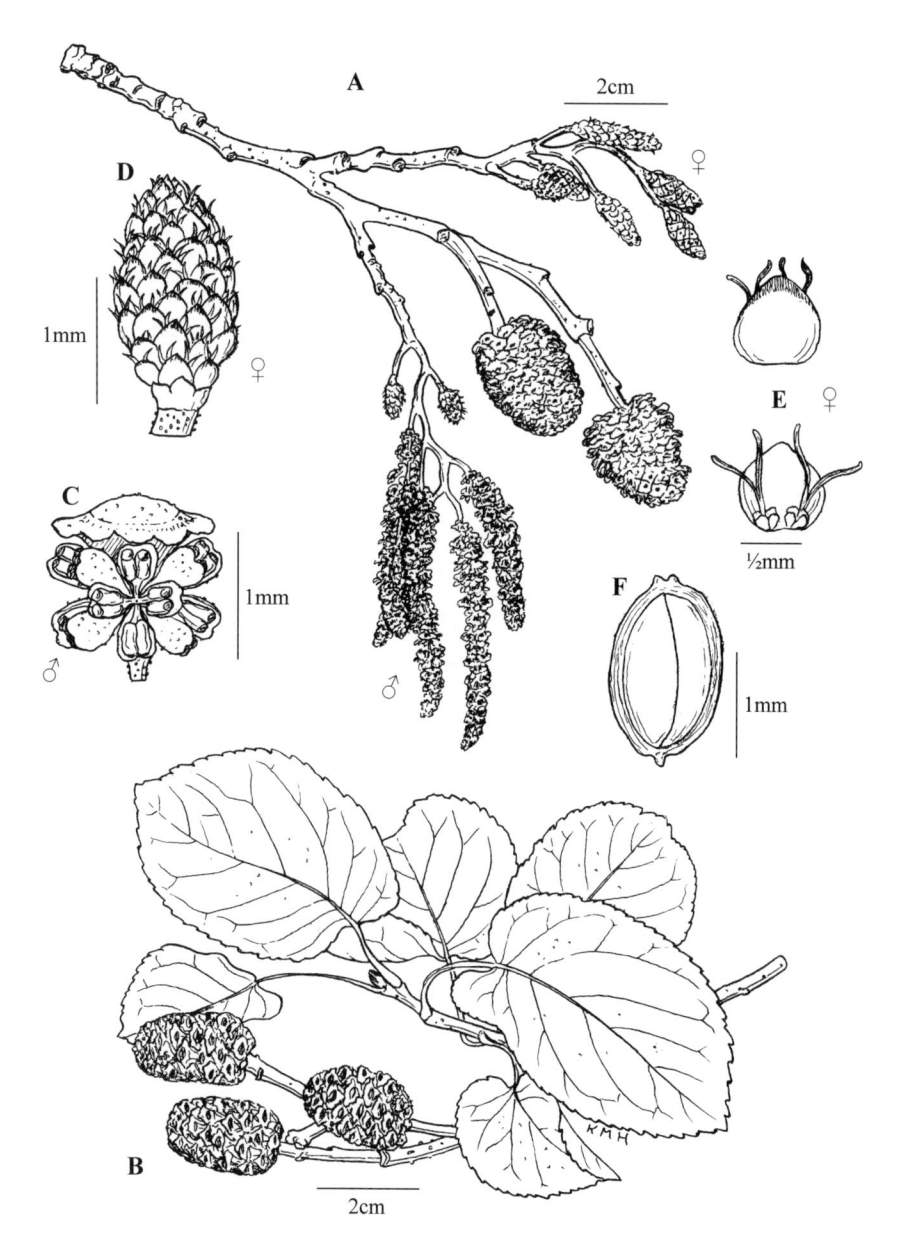

42. BETULACEAE

Alnus cordata (Loisel.) Duby (Italian Alder). A flowering twig in spring; B fruiting twig in summer; C male flower; D catkin of female flowers; E bract with two female flowers; F nutlet. (Rescobie, Angus). STACE 155/126.

43. PHYTOLACCACEAE

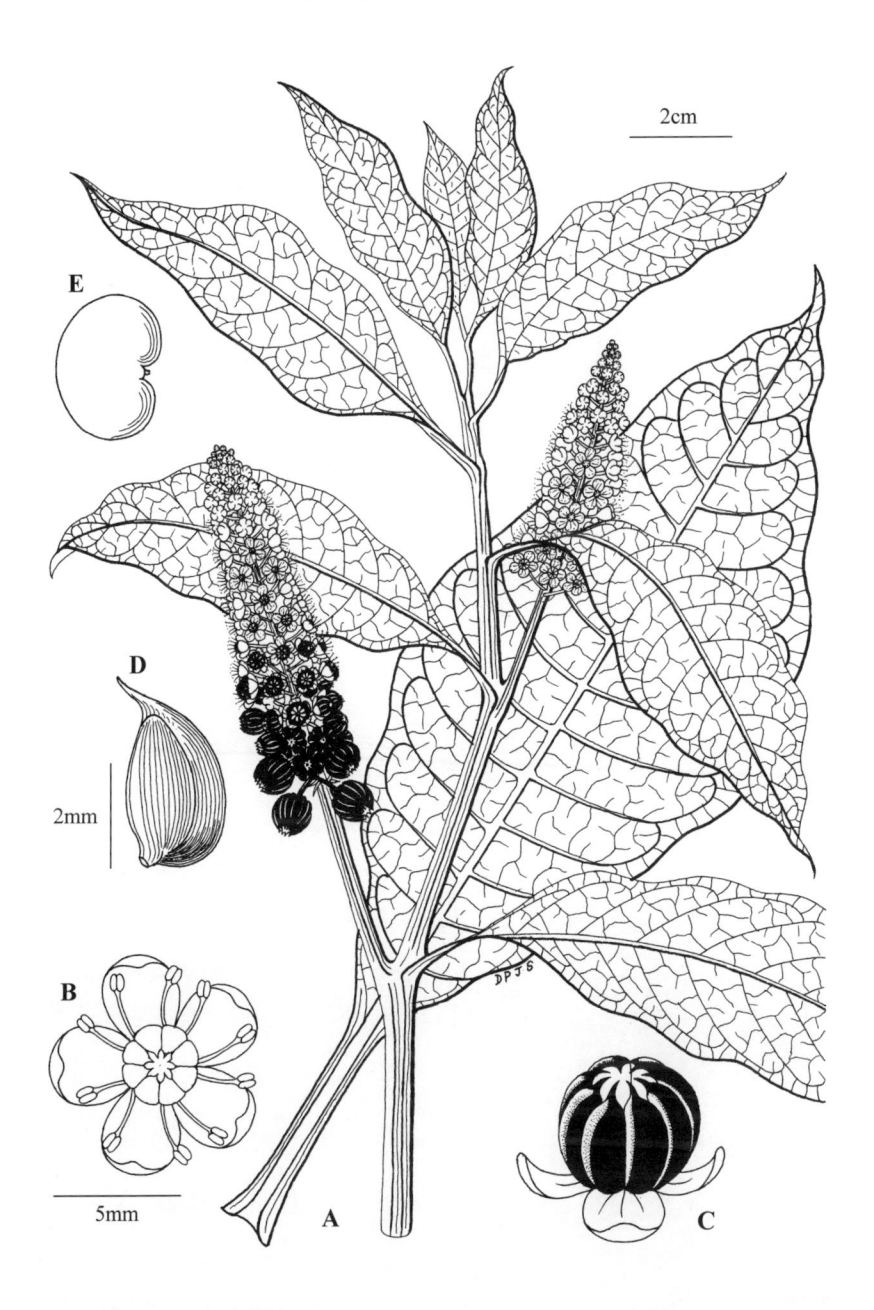

Phytolacca acinosa Roxb. (Indian Pokeweed). A habit; B flower; C fruit; D carpel; E seed. (Source unknown). STACE 157/128.

38   43A. NYCTAGINACEAE

Mirabilis jalapa L. (Marvel-of-Peru). A habit; B LS of calyx; C corolla opened out;
D gynoecium; E fruit with part of calyx removed; F fruit. (Channel Islands). STACE 157/128.

44. AIZOACEAE

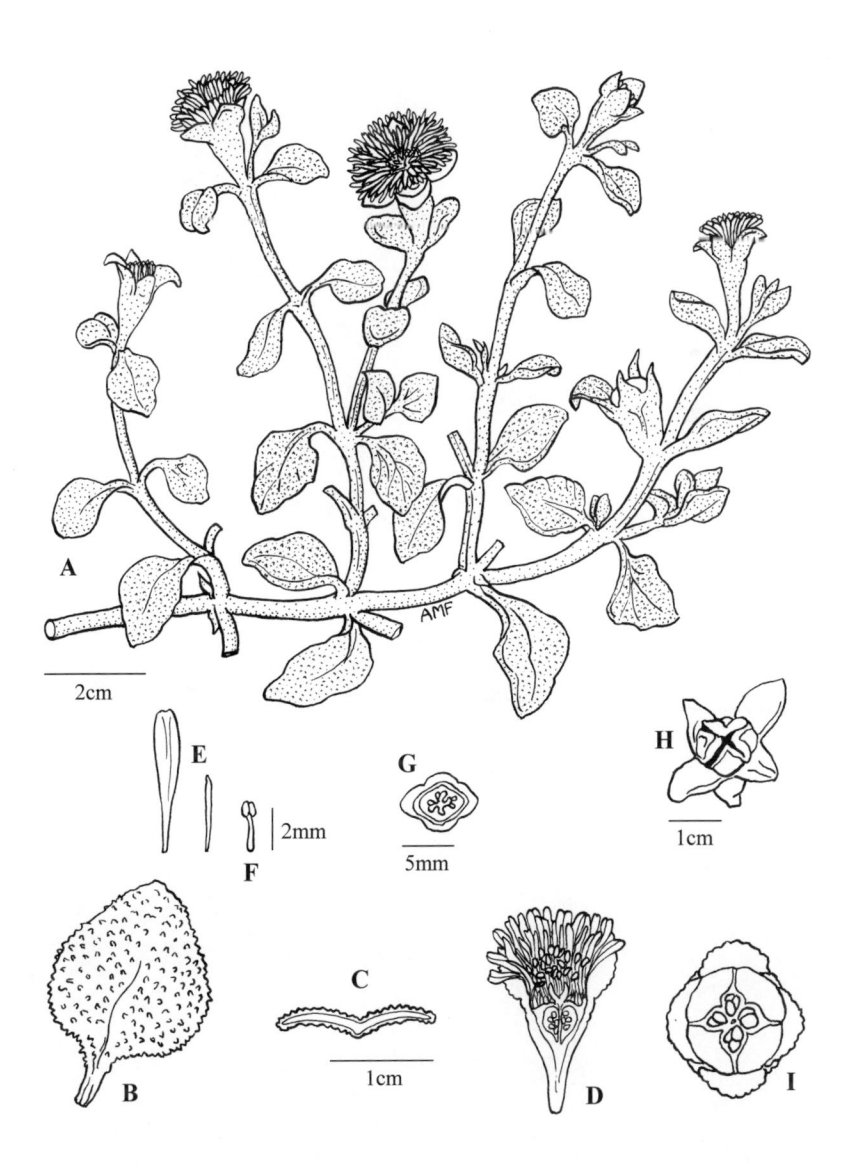

Aptenia cordifolia (L. f.) Schwantes (Heart-leaf Ice-plant). A habit; B leaf; C TS of leaf; D LS of flower; E petaloid staminodes; F stamen; G TS of ovary; H fruit; I TS of fruit. (St Mary's, Scilly). STACE 158/129.

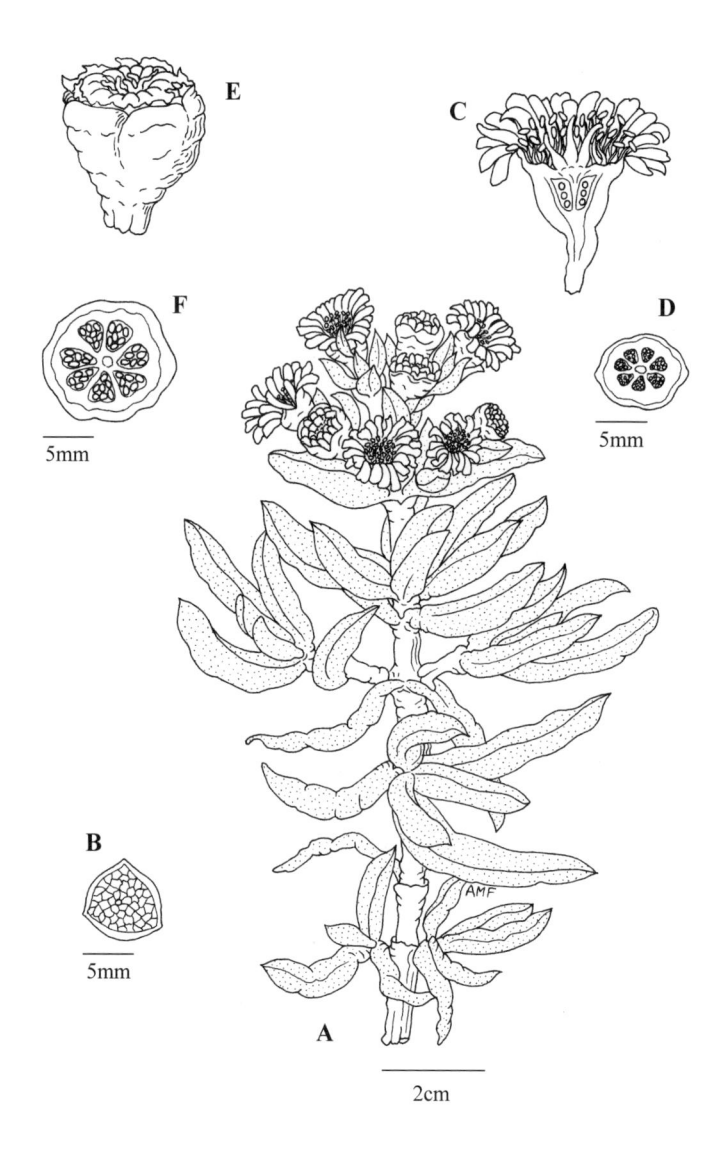

Ruschia caroli (L. Bolus) Schwantes (Shrubby Dewplant). A habit; B TS of leaf; C LS of flower; D TS of ovary; E fruit; F TS of fruit. (Bryher, Scilly). STACE 158/130.

44. AIZOACEAE

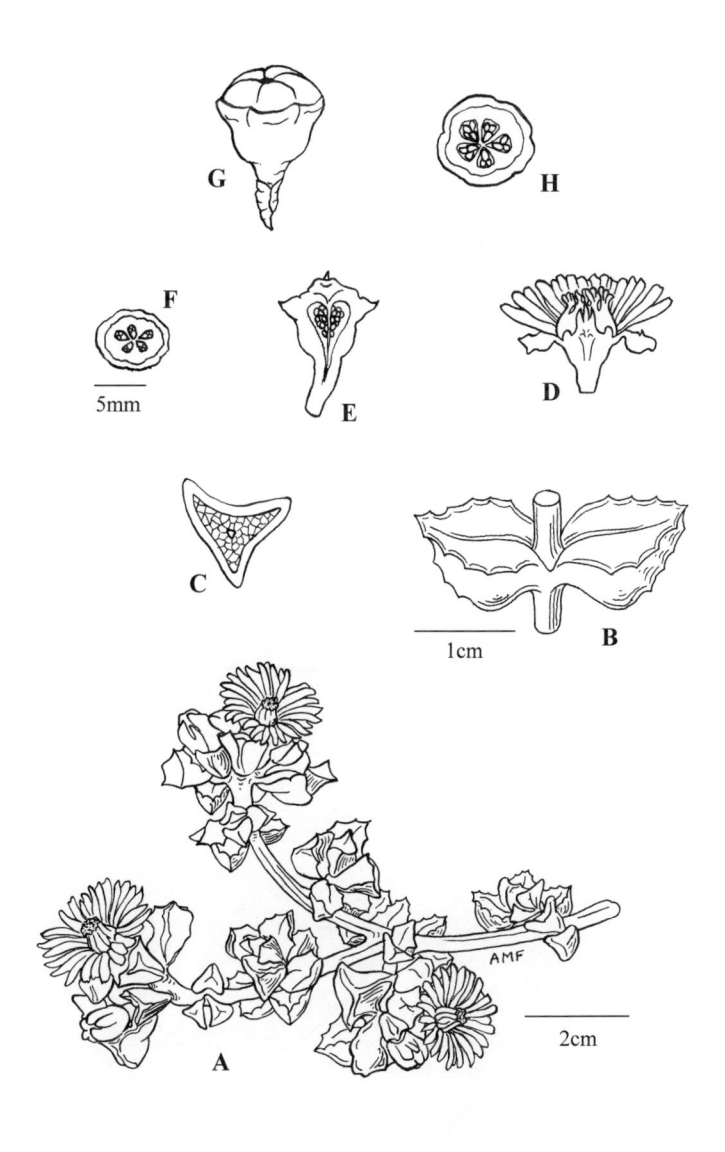

Oscularia deltoides (L.) Schwantes (Deltoid-leaved Dewplant). A habit; B pair of leaves; C TS of leaf; D LS of flower; E LS of ovary; F TS of ovary; G fruit; H TS of fruit. (Source unknown). STACE 159/130.

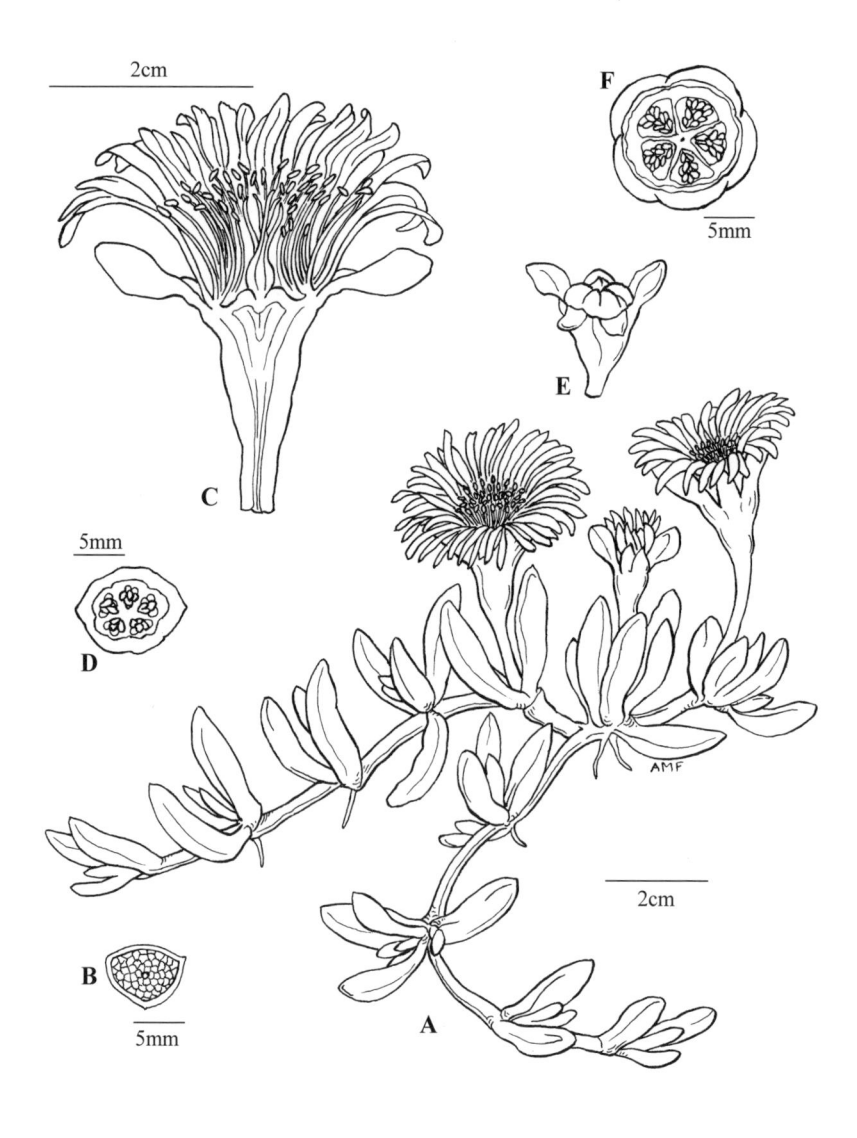

Disphyma crassifolium (L.) L. Bolus (Purple Dewplant). A habit; B TS of leaf; C LS of flower; D TS of ovary; E fruit; F TS of fruit. (Lizard, W Cornwall). STACE 161/130.

44. AIZOACEAE

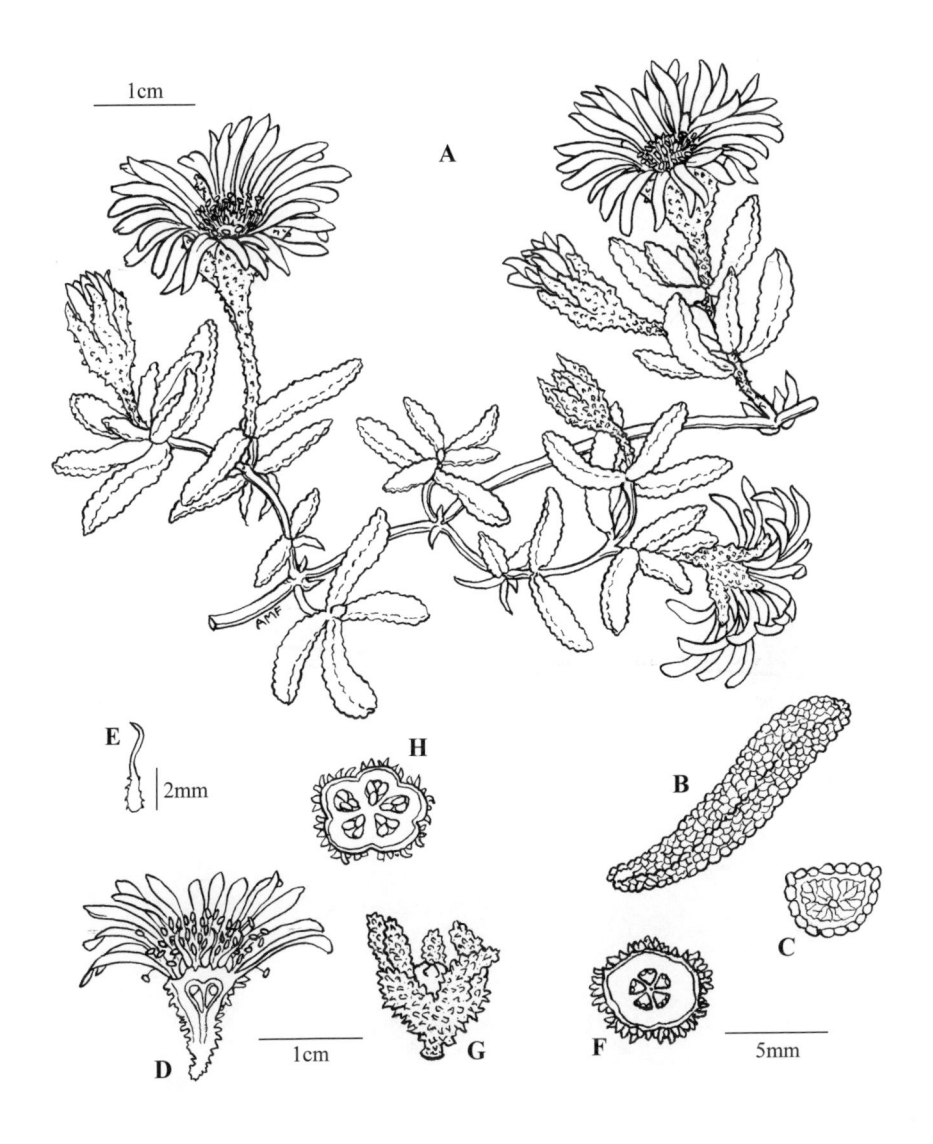

Drosanthemum floribundum (Haw.) Schwantes (Pale Dewplant). A habit; B leaf; C TS of leaf; D LS of flower; E inner petaloid staminode; F TS of ovary; G fruit; H TS of fruit. (Isles of Scilly). STACE 161/132.

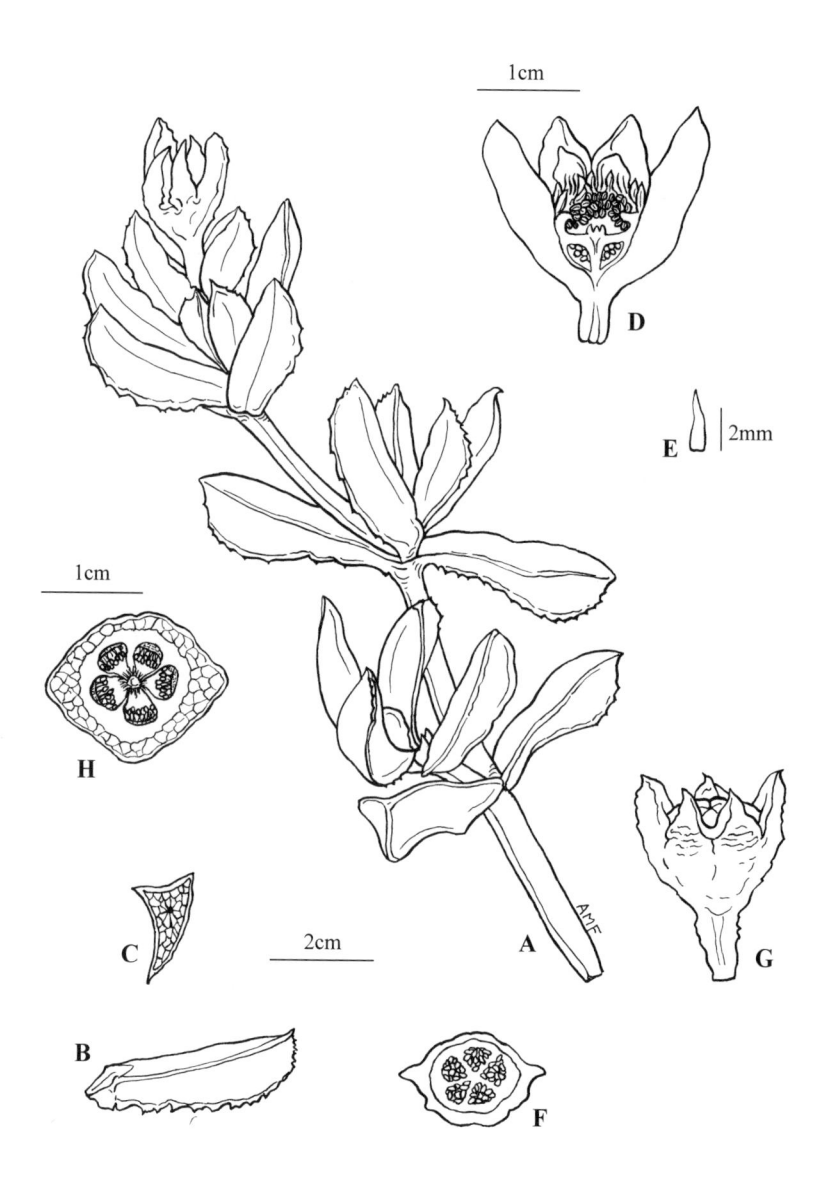

Erepsia heteropetala (Haw.) Schwantes (Lesser Sea-fig). A habit; B leaf; C TS of leaf; D LS of flower; E petaloid staminode; F TS of ovary; G fruit; H TS of fruit. (St Mary's, Scilly). STACE 161/132.

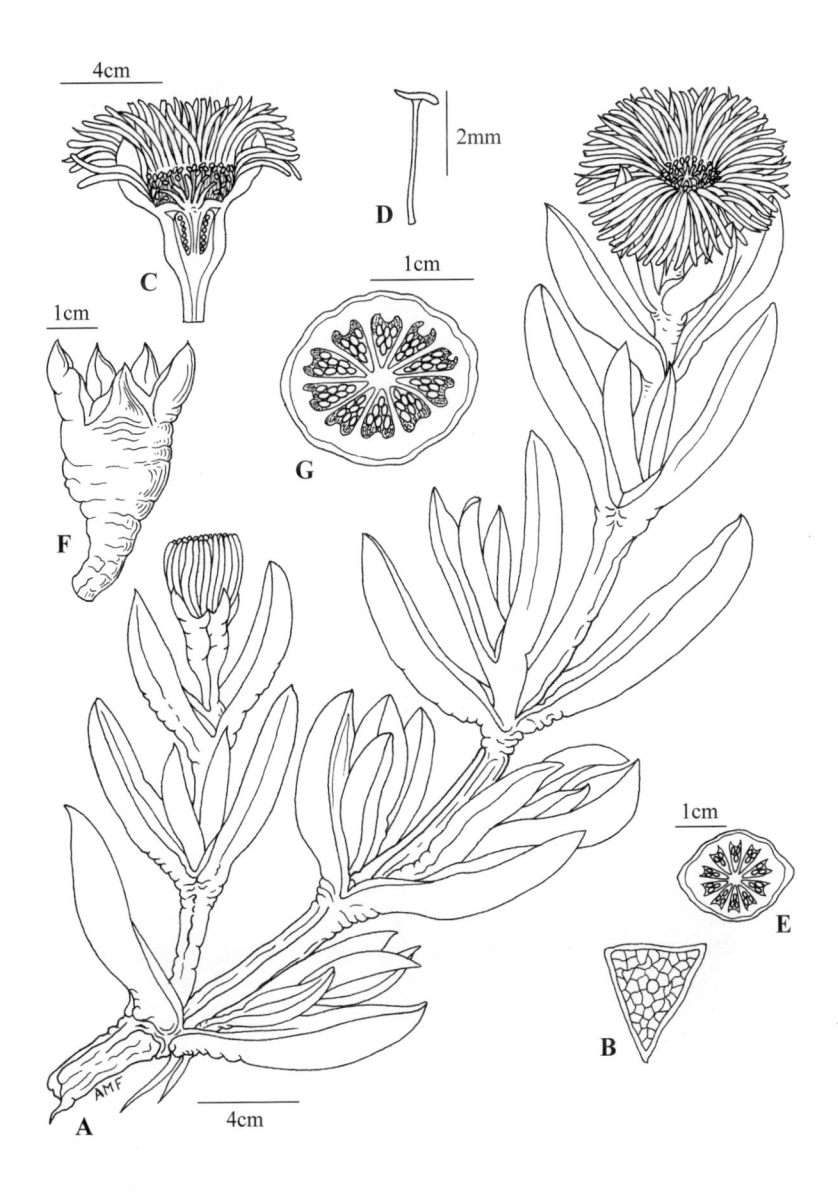

44. AIZOACEAE

Carpobrotus acinaciformis (L.) L. Bolus (Sally-my-handsome). A habit; B TS of leaf; C LS of flower; D stamen; E TS of ovary; F fruit; G TS of fruit. (Source unknown). STACE 162/132.

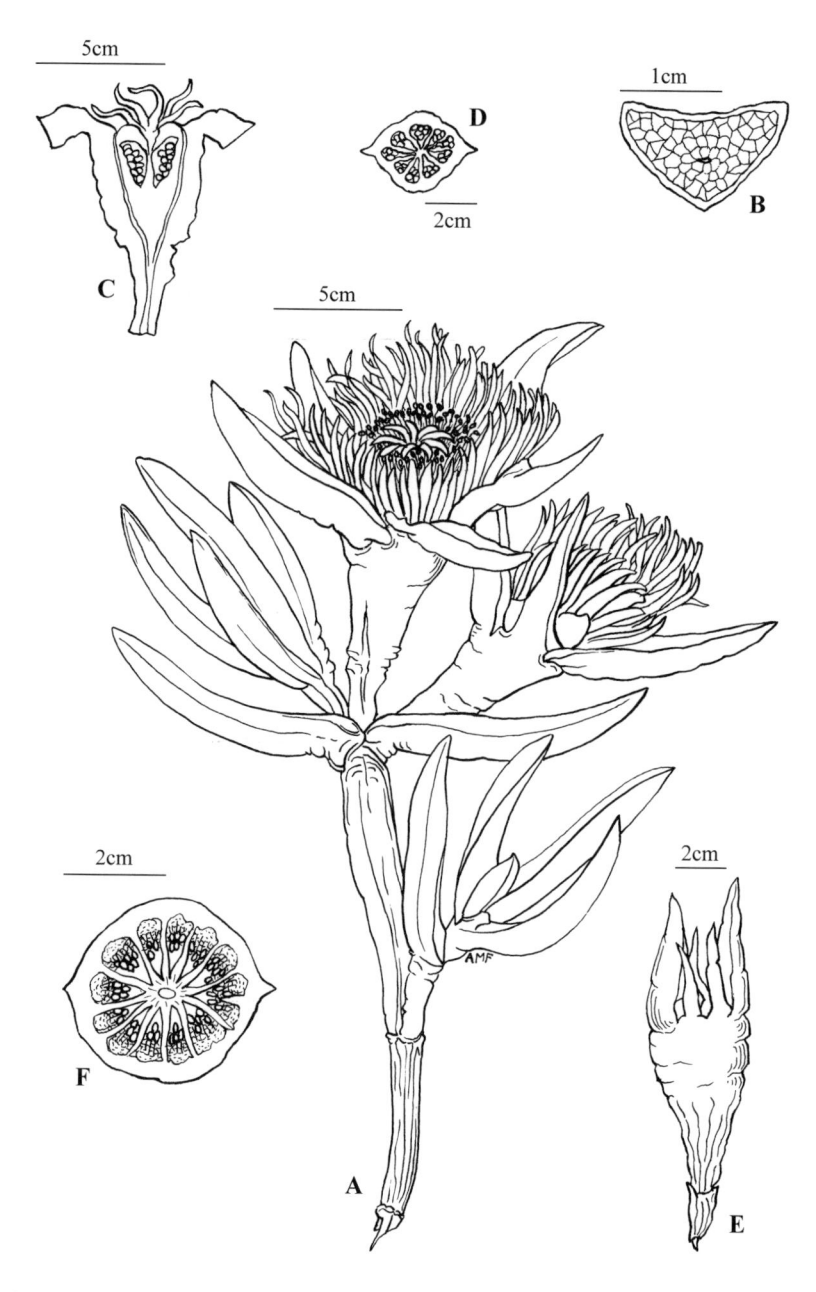

Carpobrotus edulis (L.) N.E. Br. (Hottentot-fig). A habit; B TS of leaf; C LS of flower; D TS of ovary; E fruit; F TS of fruit. (Jersey). STACE 162/133.

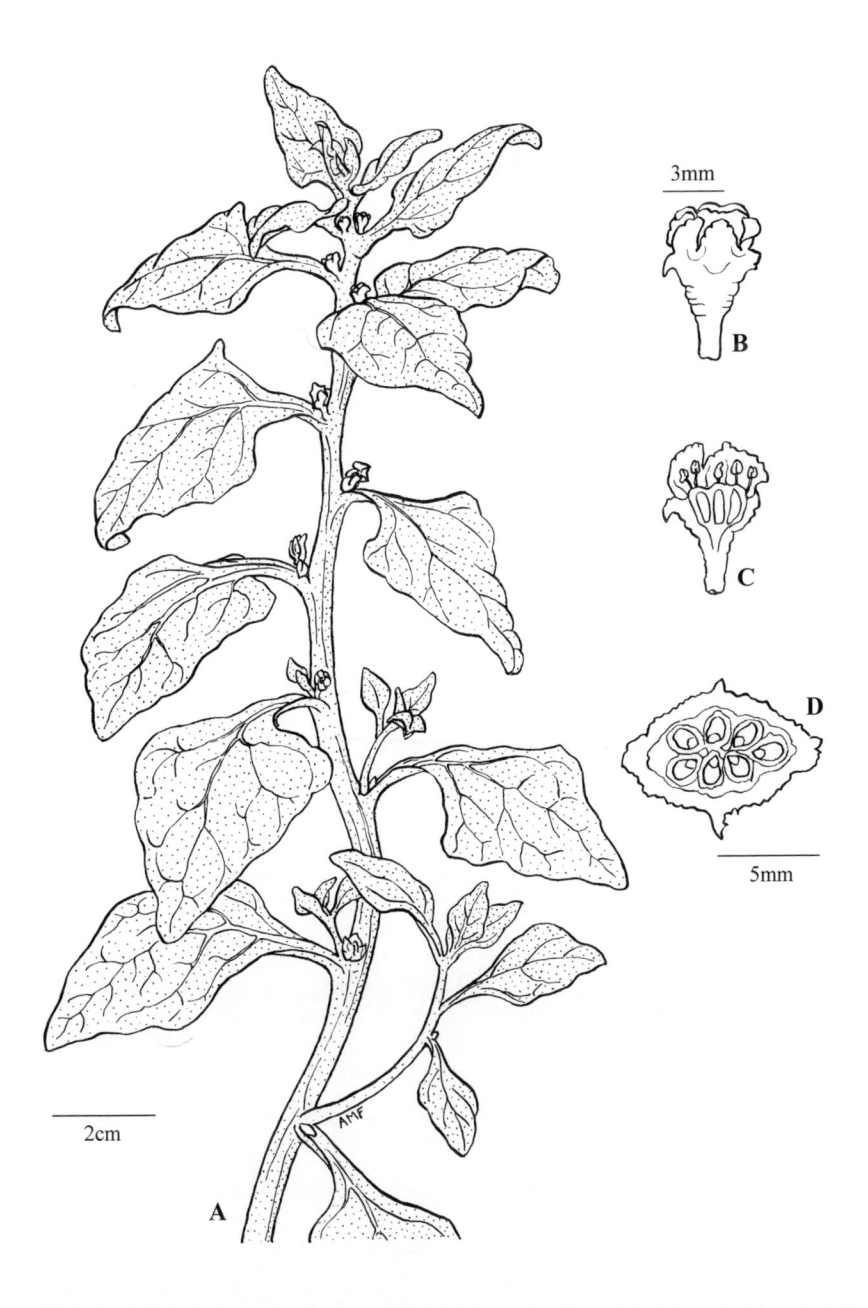

44. AIZOACEAE

Tetragonia tetragonioides (Pall.) Kuntze (New Zealand Spinach). A habit; B flower; C LS of flower; D TS of fruit. (Source unknown). STACE 163/133.

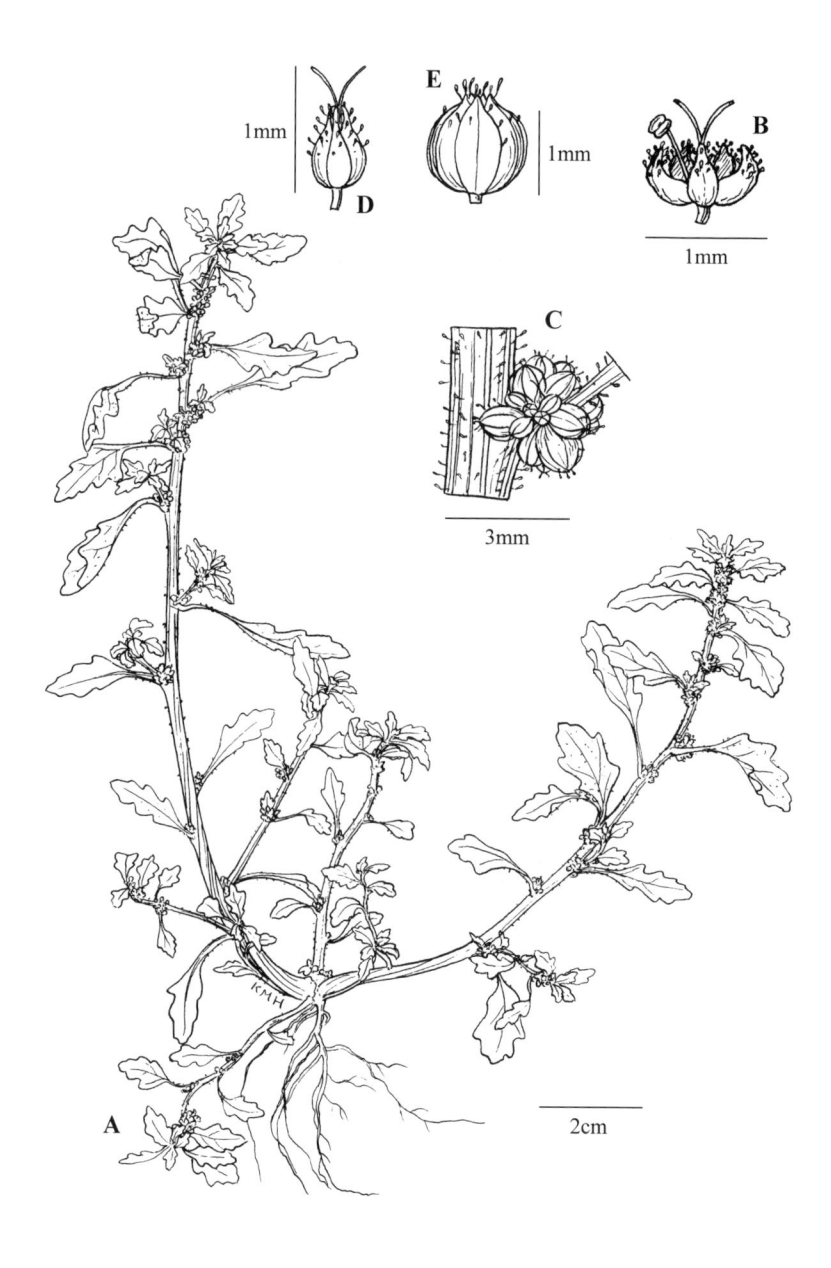

Chenopodium pumilio R. Br. (Clammy Goosefoot). A habit; B flower; C fruits in leaf axil; D fruiting perianth; E fruit. (Horsell, Surrey). STACE 169/138.

45. CHENOPODIACEAE

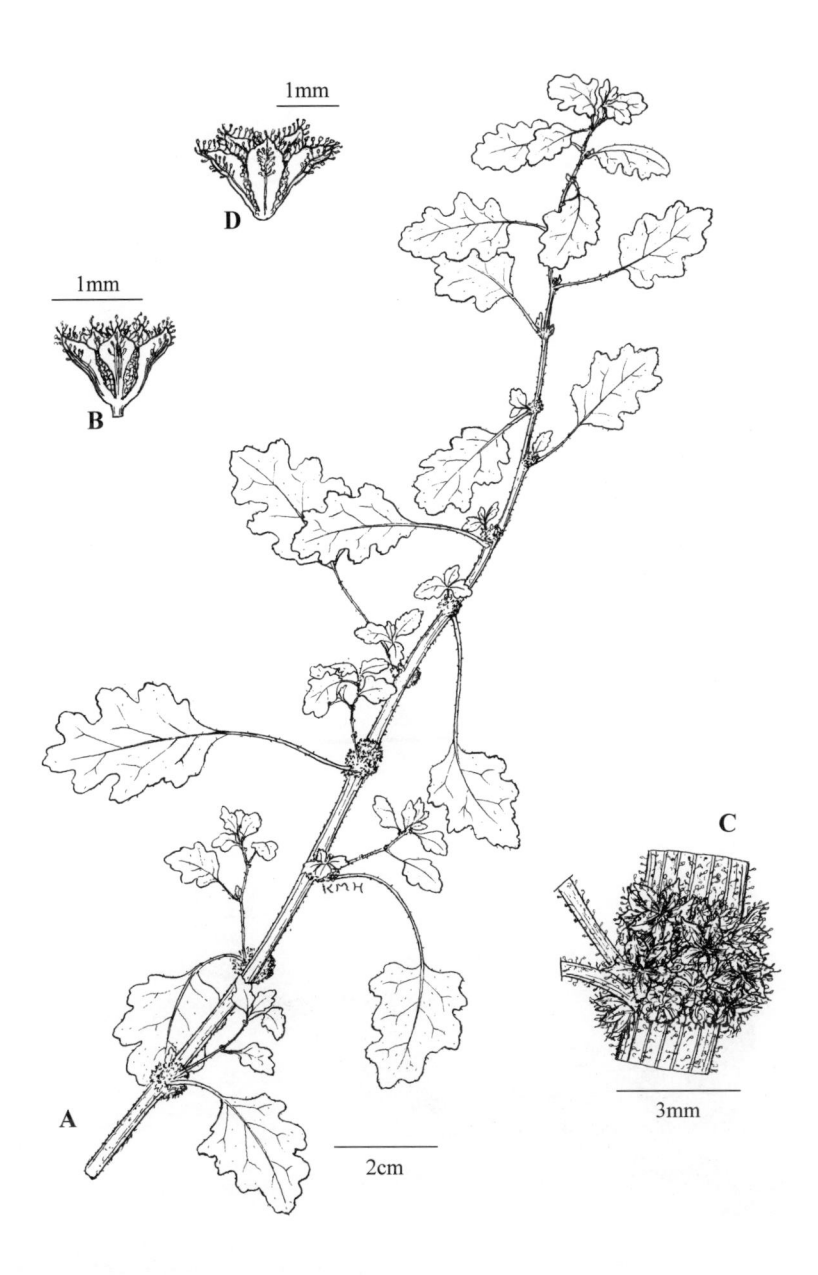

Chenopodium carinatum R. Br. (Keeled Goosefoot). A habit; B flower; C fruits in leaf axil; D fruiting perianth. (Charlton, Worcs). STACE 169/138.

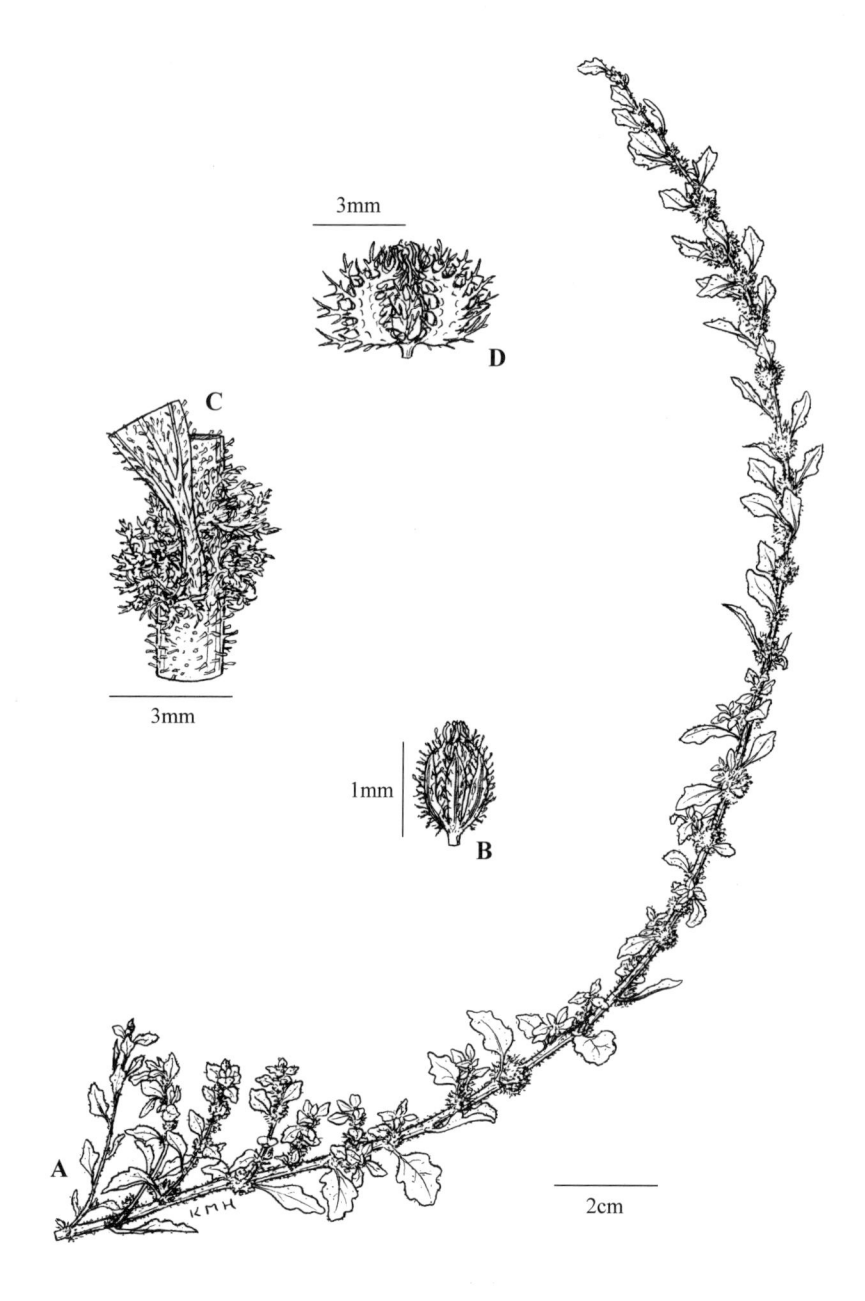

Chenopodium x bontei (F. Muell.) F. Muell. (No English Name). A habit; B flower; C fruits in leaf axil; D fruiting perianth. (Source unknown). STACE 169/138.

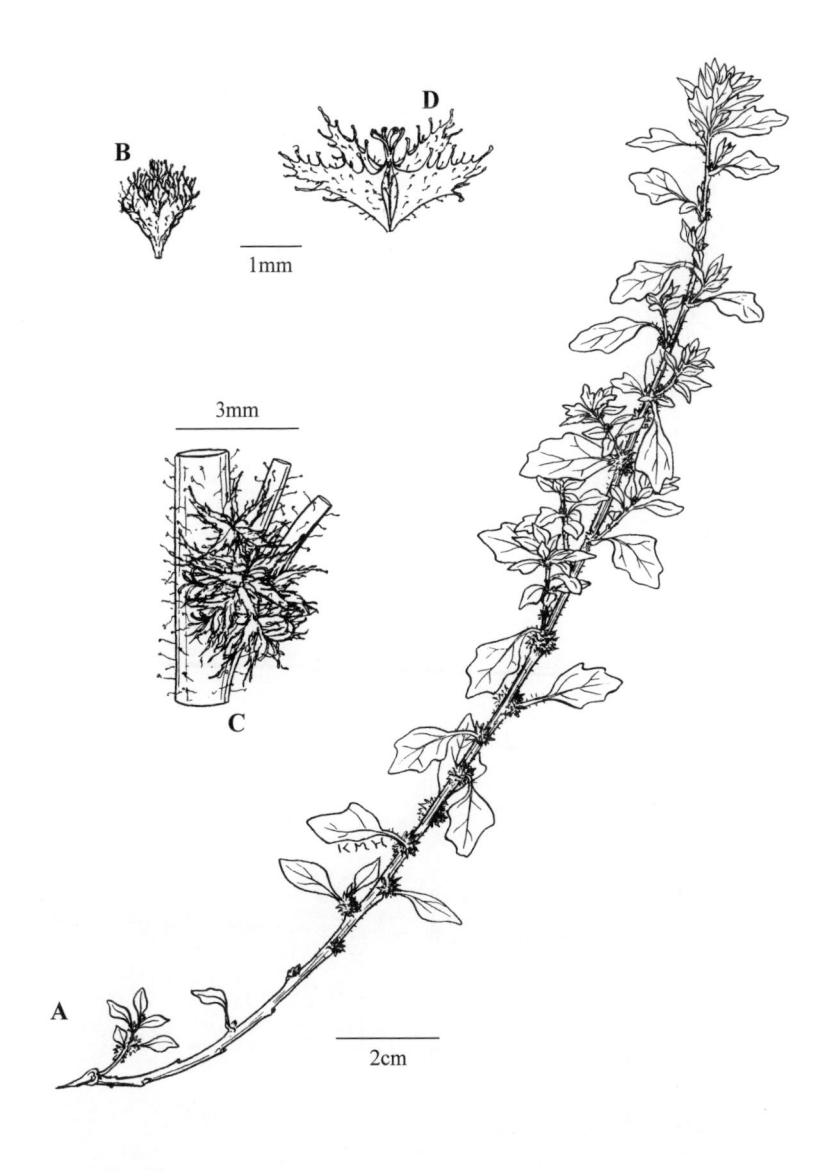

45. CHENOPODIACEAE

Chenopodium cristatum Aellen (Crested Goosefoot). A habit; B flower; C fruits in leaf axil; D fruiting perianth. (Sandwich, E Kent). STACE 169/138.

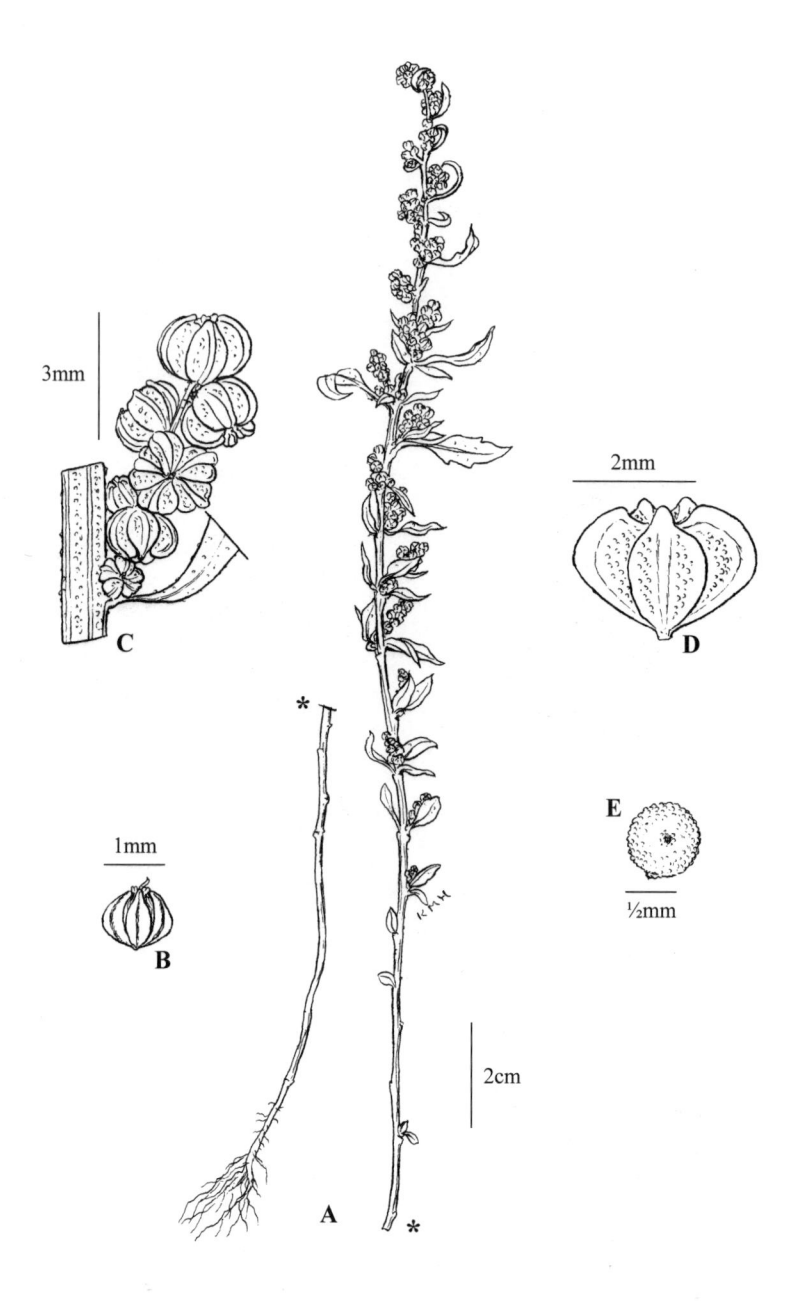

Chenopodium suecicum Murr (Swedish Goosefoot). A habit; B flower; C cluster of fruits; D fruiting perianth; E seed. (*Hort.* Ware, Herts ex RBG, Kew seed). STACE 174/140.

45. CHENOPODIACEAE

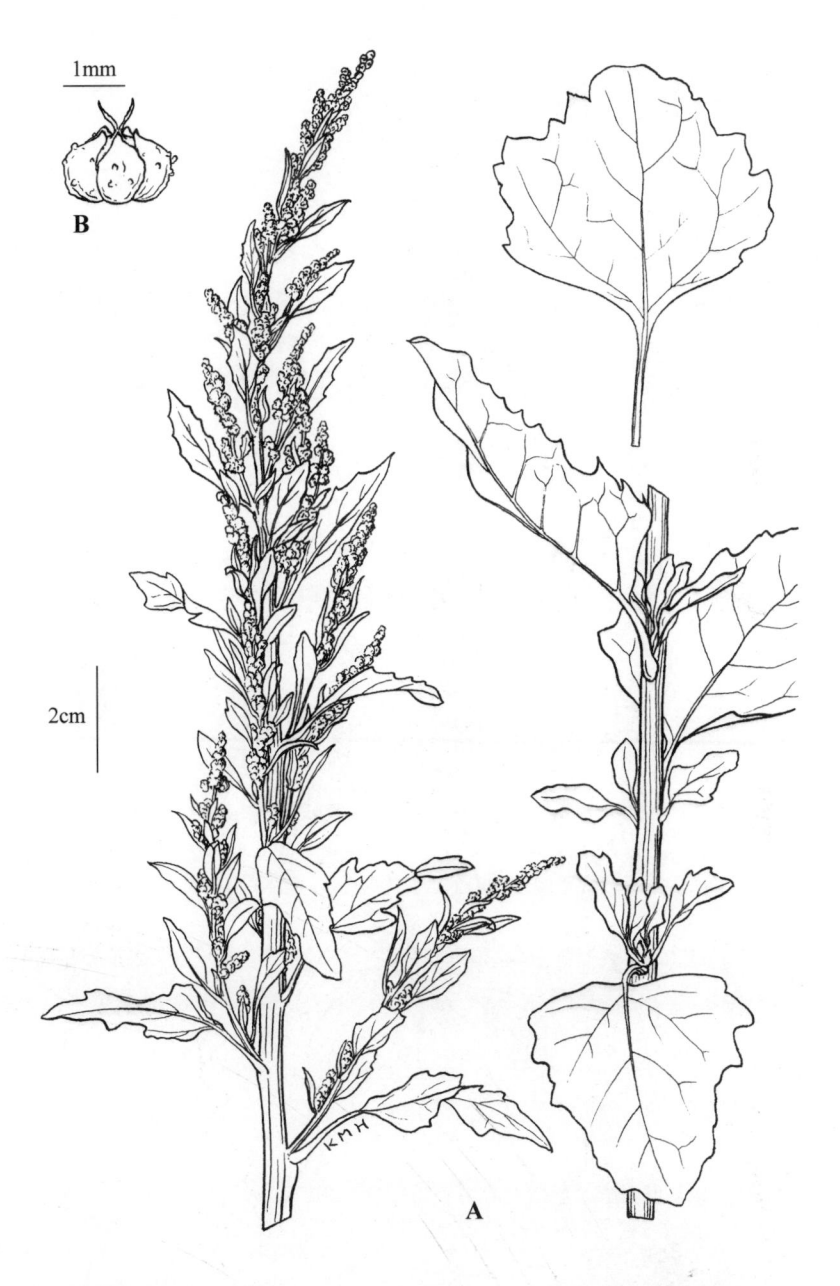

Chenopodium probstii Aellen (Probst's Goosefoot). A habit; B flower. (Source unknown). STACE 174/140.

54

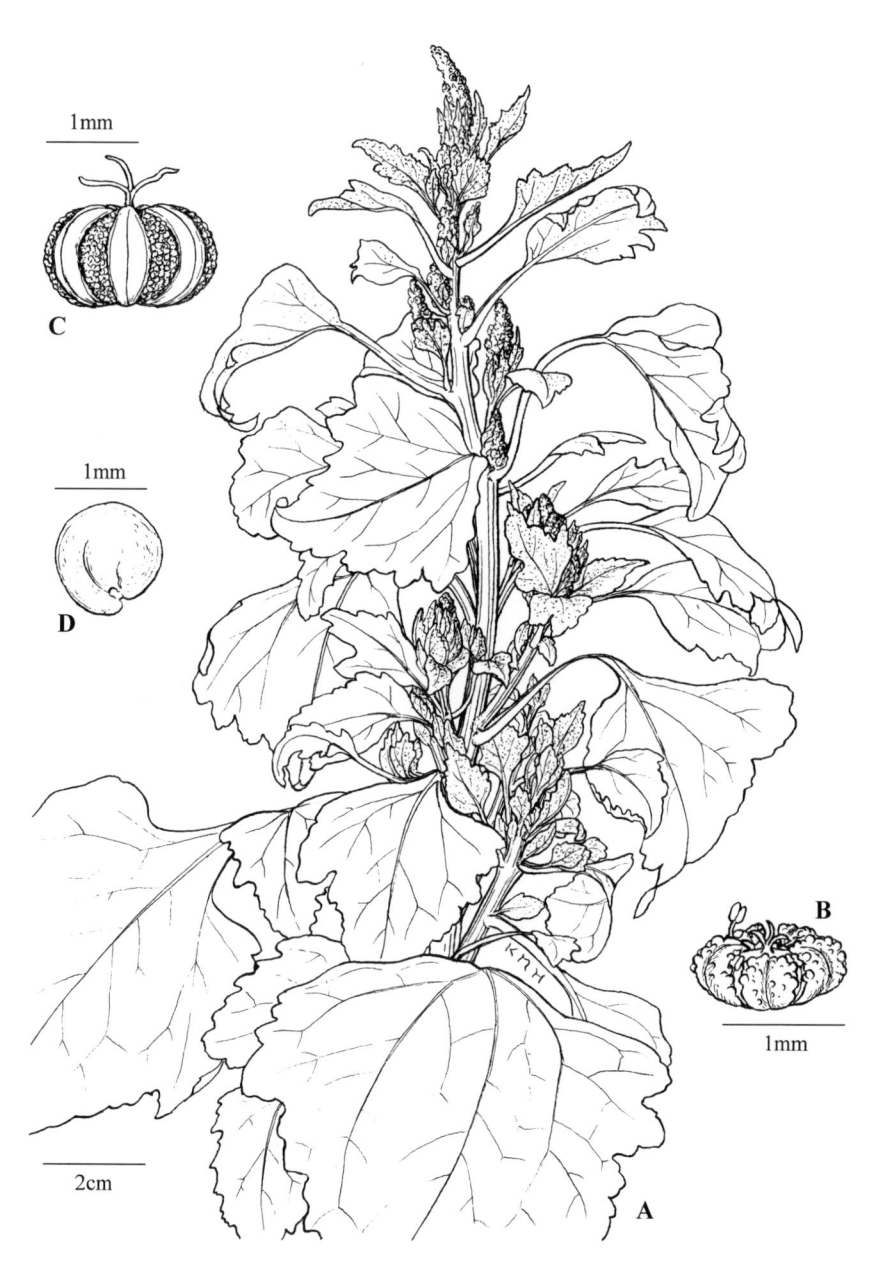

45. CHENOPODIACEAE

Chenopodium giganteum D. Don (Tree Spinach). A habit; B flower; C fruiting perianth; D seed. (Otley, MW Yorks). STACE 174/140.

45. CHENOPODIACEAE

Chenopodium schraderianum Schult. (No English name). A habit; B cluster of flowers; C male flower; D bisexual flower. (Blackmoor, N Hants). STACE —/—.

Corispermum leptopterum (Asch.) Iljin (Bugseed). A habit; B flower; C fruit. (*Hort*. ex Emscote, Warks). STACE —/142.

45. CHENOPODIACEAE

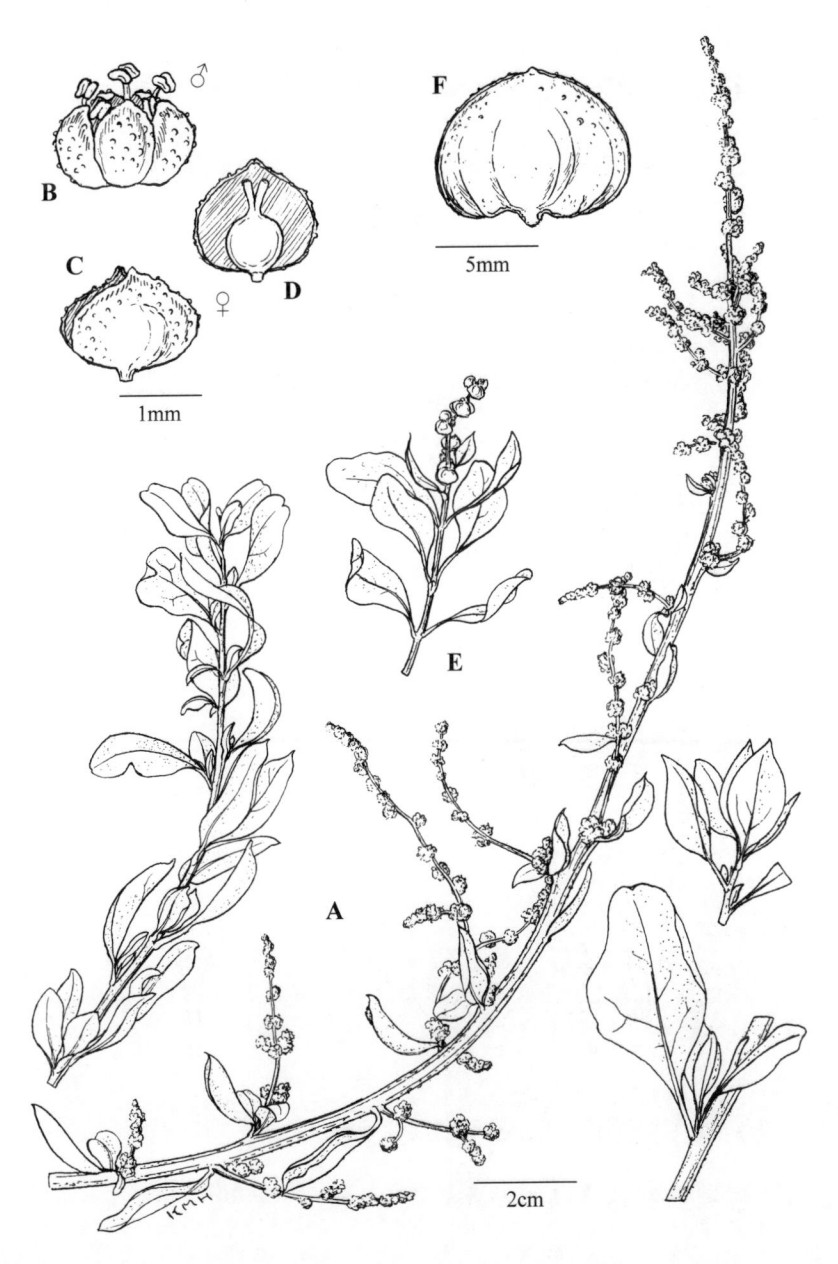

Atriplex halimus L. (Shrubby Orache). A barren and flowering branches; B male flower; C female flower; D female flower with one bracteole removed; E fruiting branchlet; F fruit. (Jersey). STACE 179/146.

45. CHENOPODIACEAE

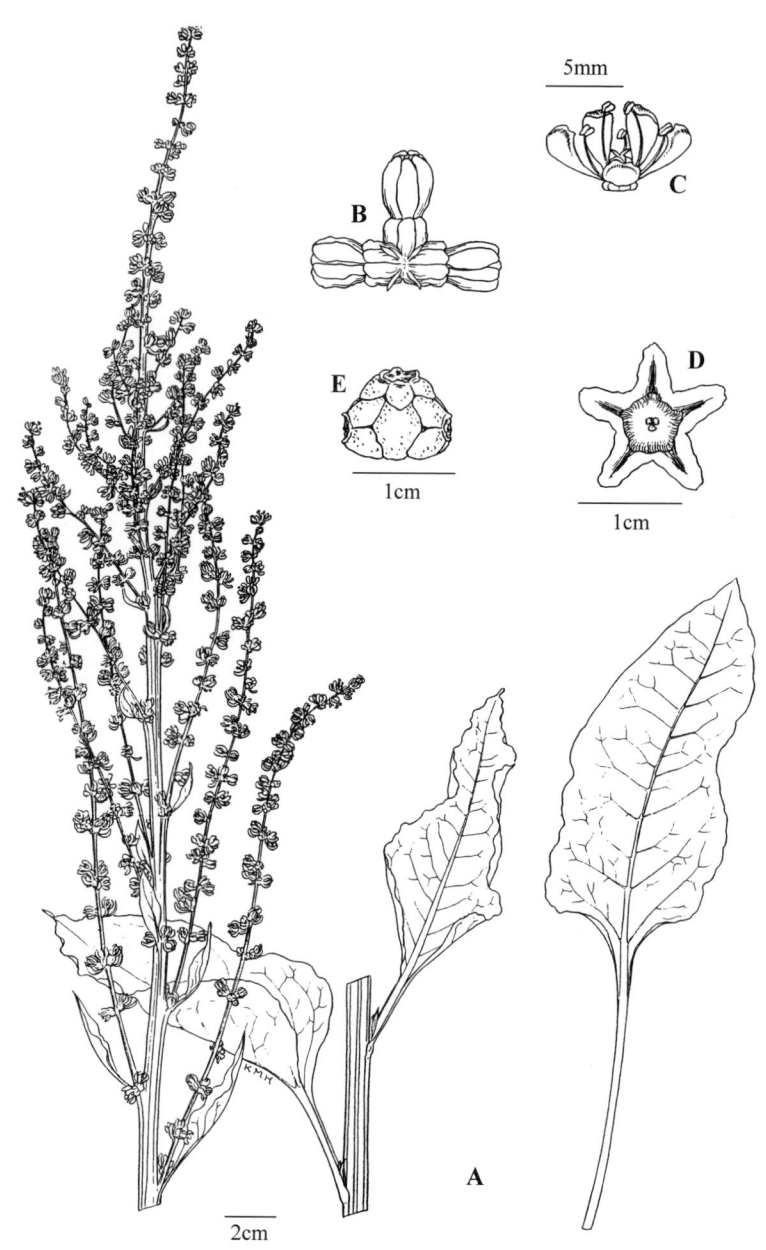

Beta trigyna Waldst. & Kit. (Caucasian Beet). A habit and basal leaf; B cluster of flowers; C flower with one tepal removed; D young fruit surrounded by perianth; E cluster of fruits. (Gog Magog, Cambs). STACE 180/147.

45. CHENOPODIACEAE

Salsola kali L. ssp. **ruthenica** (Iljin) Soó (Spineless Saltwort). A young shoot and flowering stem; B flower; C fruit. (Source unknown). STACE 185/150.

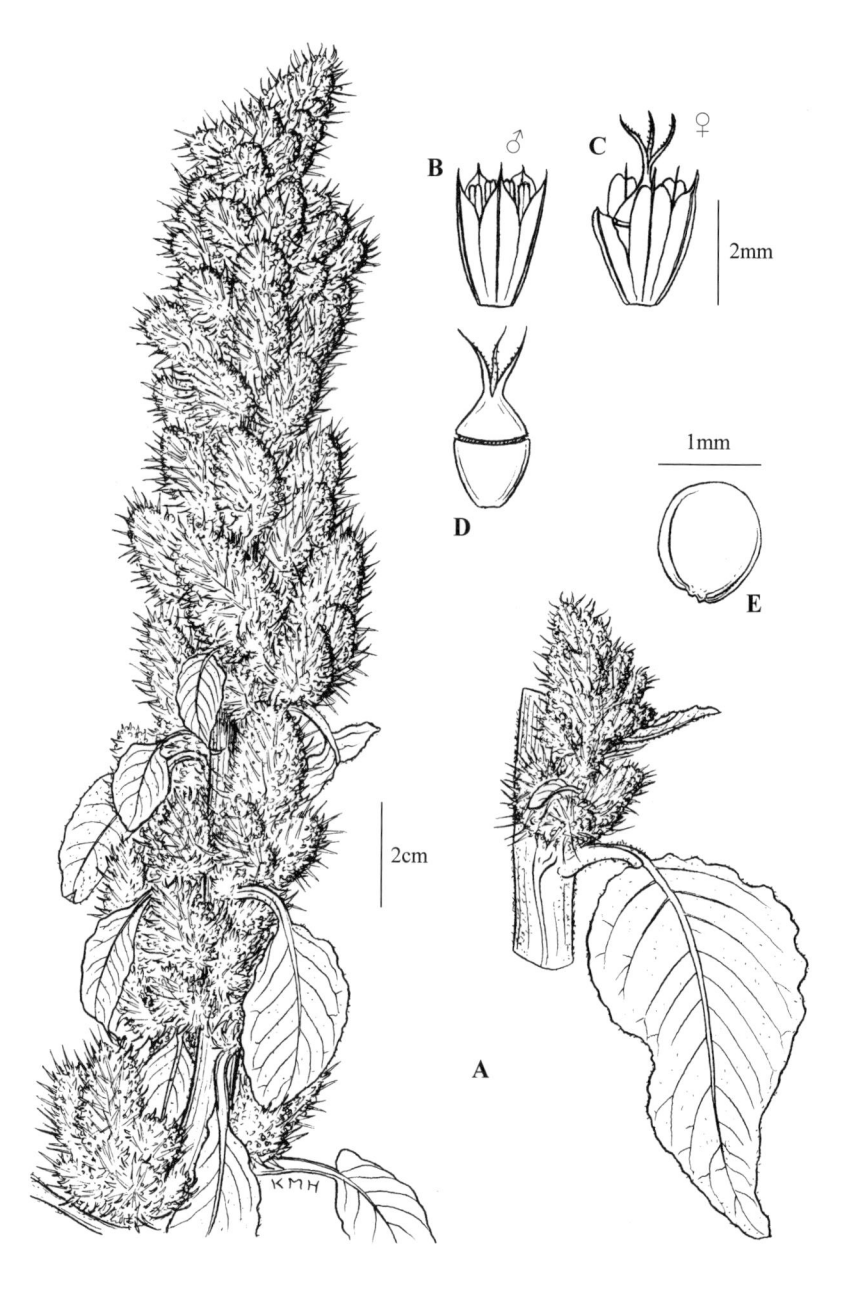

Amaranthus retroflexus L. (Common Amaranth). A habit; B male flower; C female flower; D fruit; E seed. (Blackmoor, N Hants). STACE 187/152.

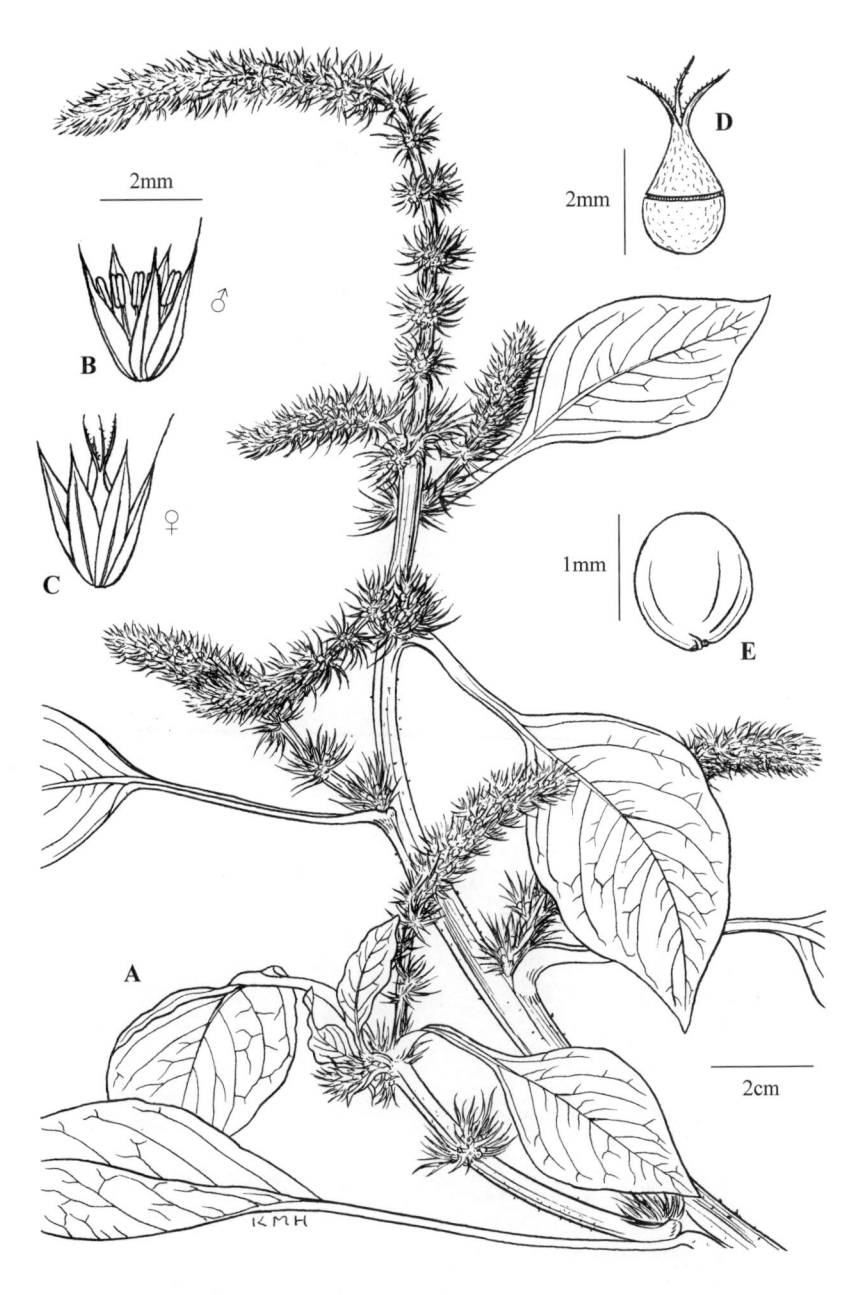

46. AMARANTHACEAE

Amaranthus hybridus L. (Green Amaranth). A habit; B male flower; C female flower; D fruit; E seed. (Croydon, Surrey). STACE 187/152.

Amaranthus deflexus L. (Perennial Pigweed). A habit; B male flower; C female flower; D fruit; E seed. (Newton Abbot, S Devon). STACE 189/154.

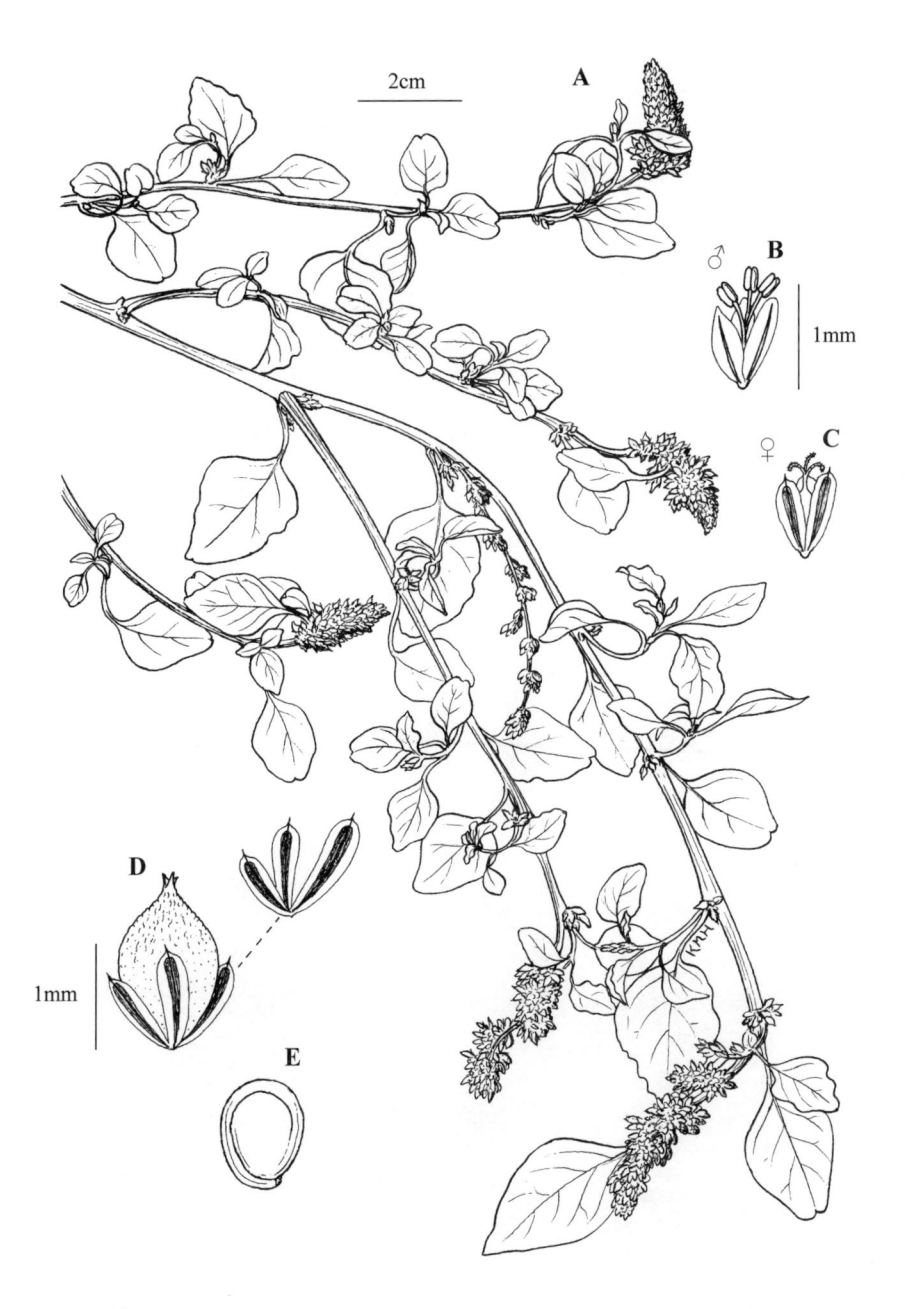

46. AMARANTHACEAE

Amaranthus blitum L. (Guernsey Pigweed). A habit; B male flower; C female flower; D fruit; E seed. (Samares, Jersey). STACE 187/154.

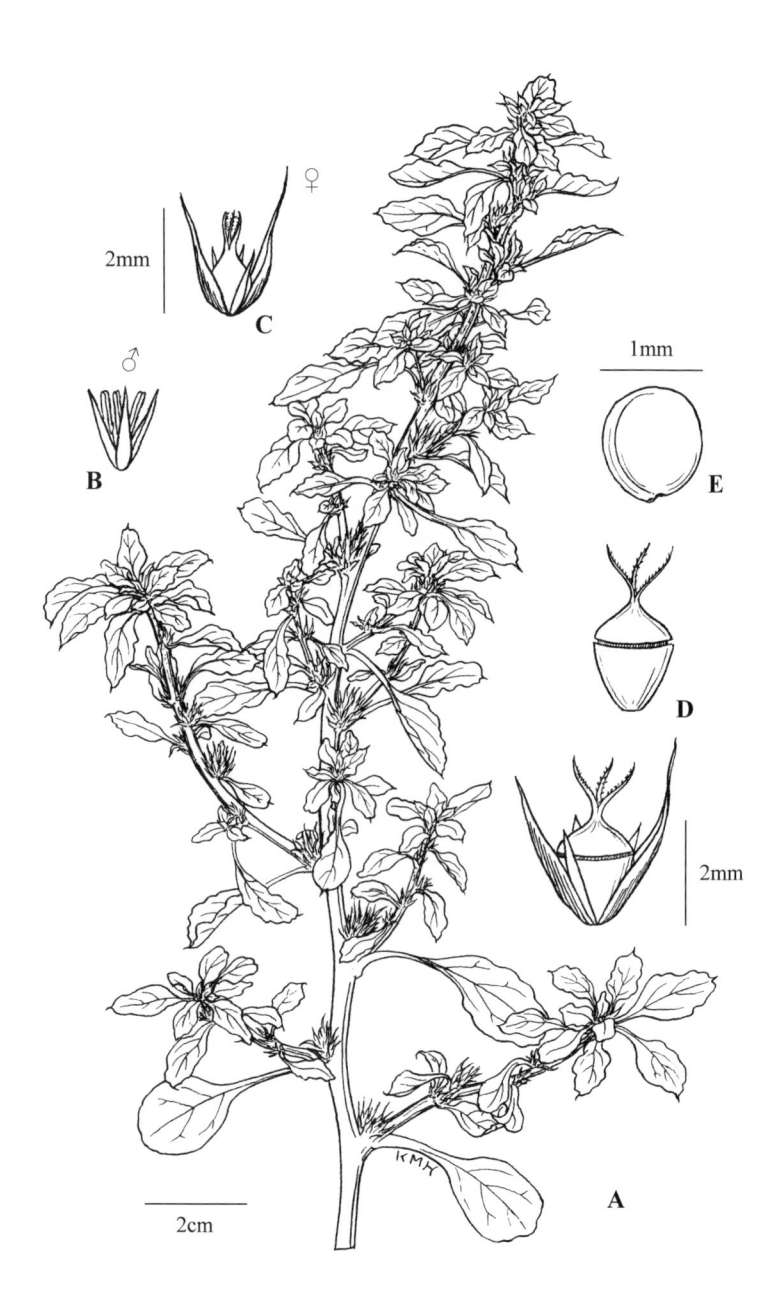

46. AMARANTHACEAE

Amaranthus albus L. (White Pigweed). A habit; B male flower; C female flower; D fruit; E seed. (Blackmoor, N Hants). STACE 189/154.

46. AMARANTHACEAE

Amaranthus thunbergii Moq. (Thunberg's Pigweed). A habit; B blotched leaf of variant; C male flower; D female flower; E fruit; F seed. (Blackmoor, N Hants). STACE 189/154.

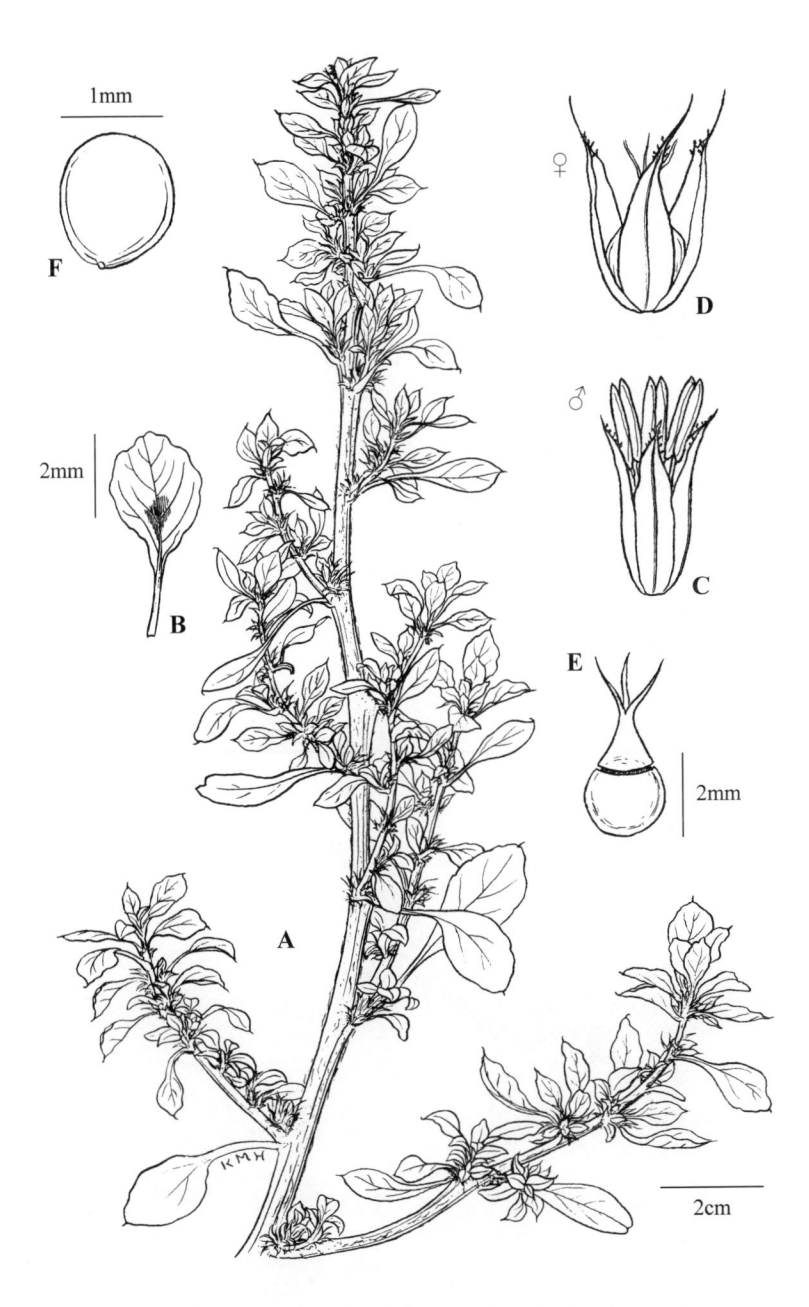

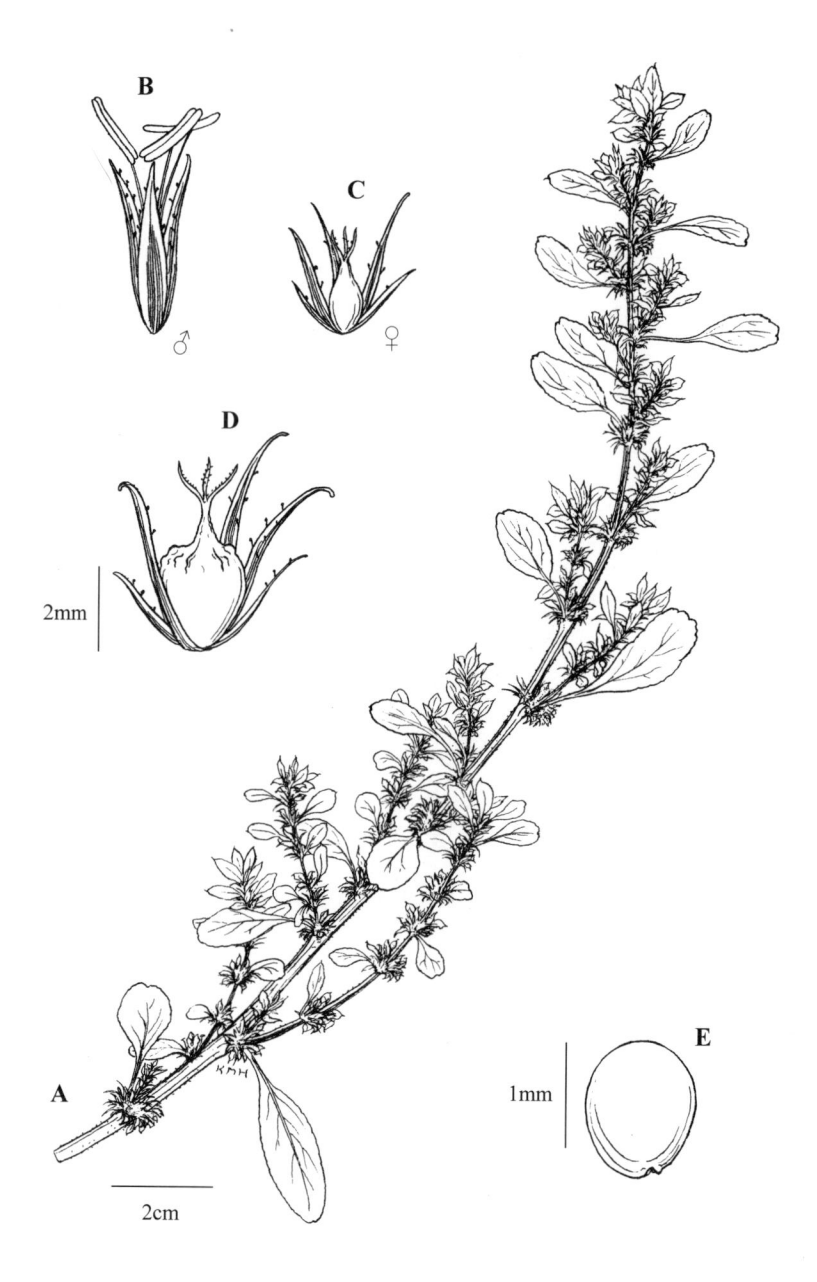

Amaranthus capensis Thell. ssp. **uncinatus** (Thell.) Brenan (Cape Pigweed). A habit; B male flower; C female flower; D fruit; E seed. (Blackmoor, N Hants). STACE 190/155.

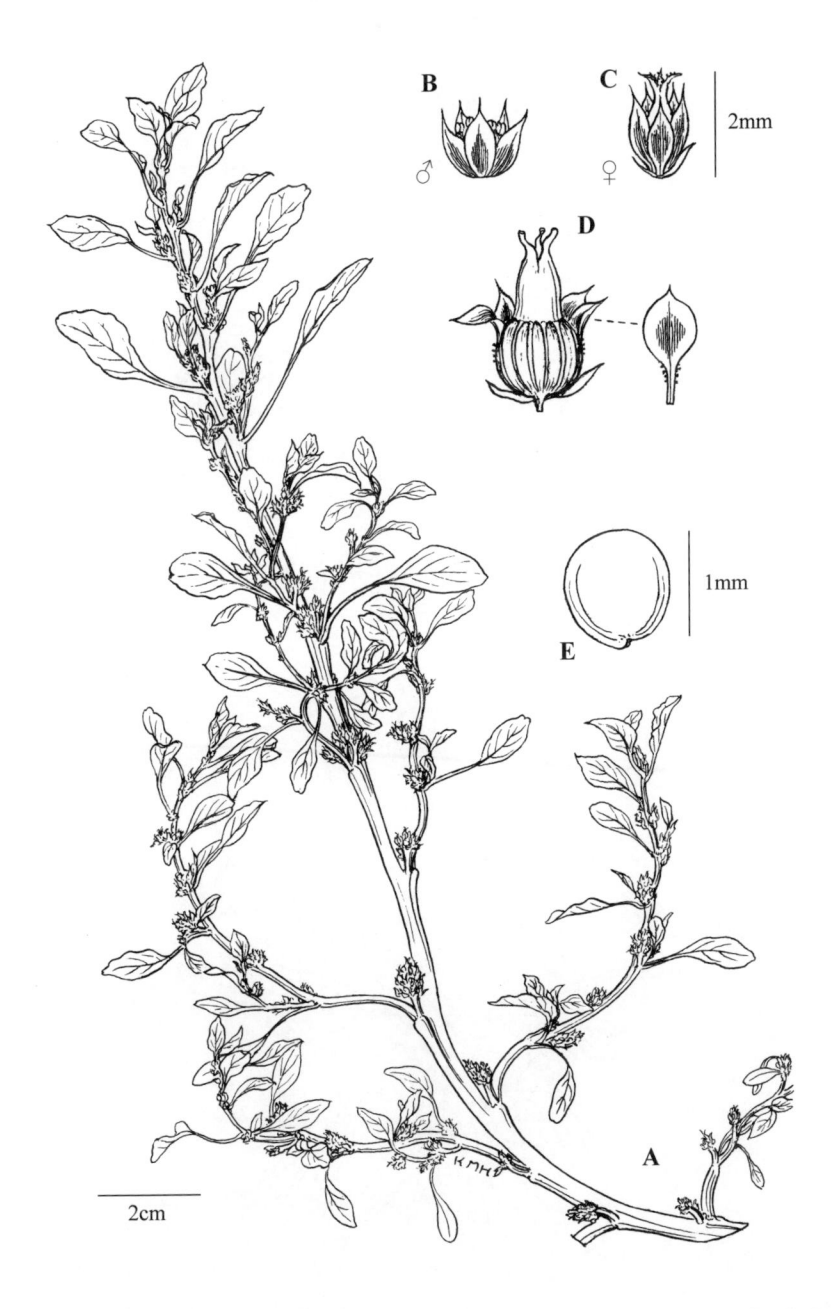

46. AMARANTHACEAE 67

Amaranthus mitchellii Benth. (Mitchell's Amaranth). A habit; B male flower; C female flower; D fruit; E seed. (Blackmoor, N Hants). STACE —/—.

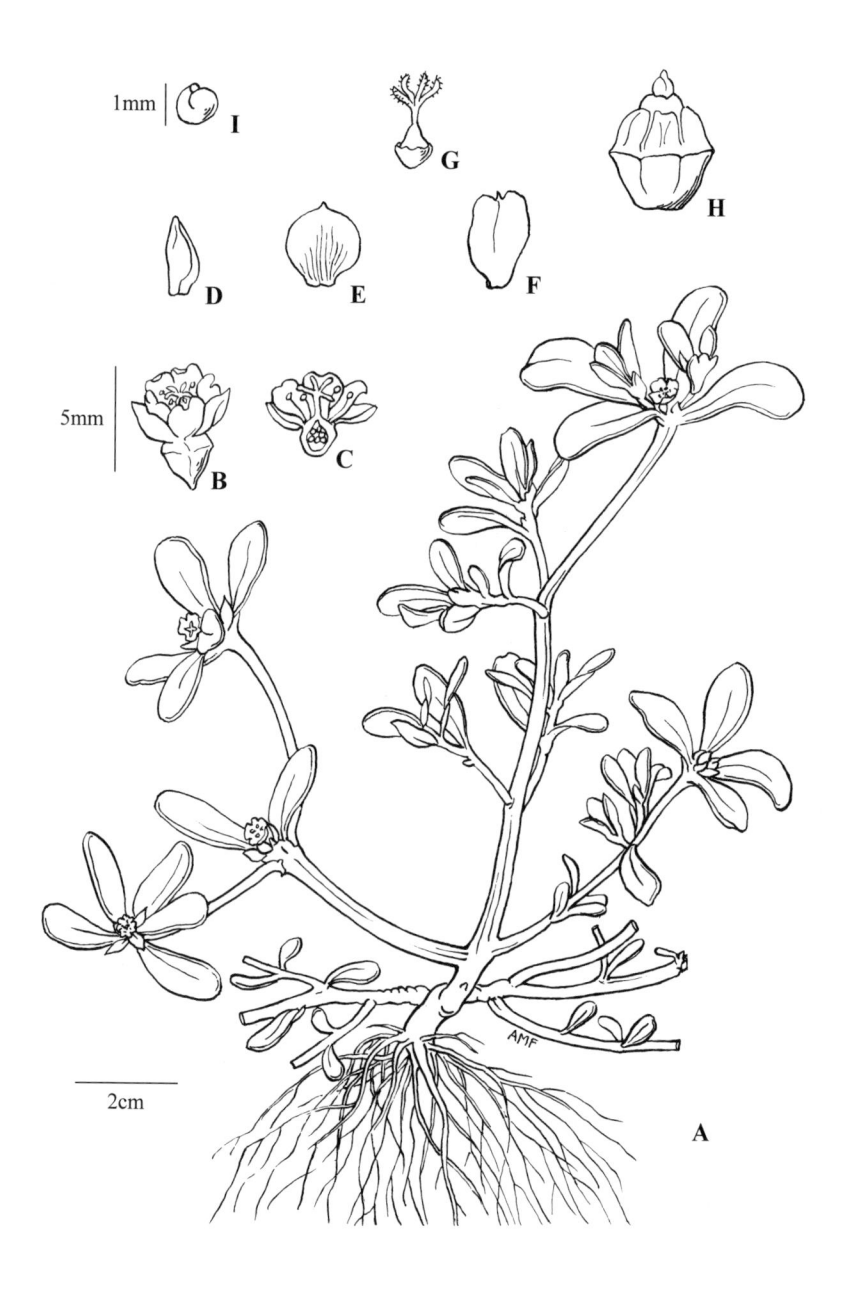

Portulaca oleracea L. (Common Purslane). A habit; B flower; C LS of flower; D sepal; E sepal flattened out; F petal; G gynoecium; H capsule; I seed. (Source unknown). STACE 191/155.

48. CARYOPHYLLACEAE

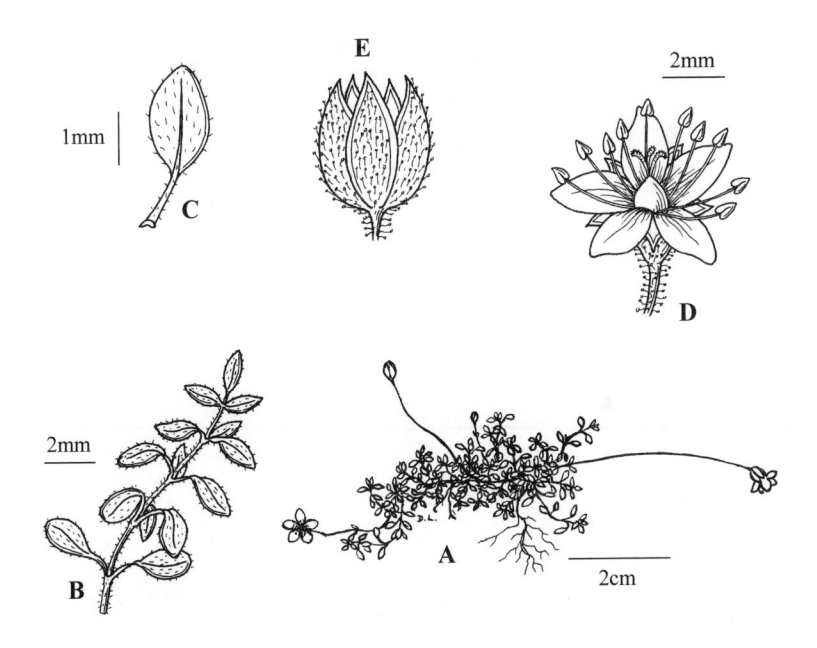

Arenaria balearica L. (Mossy Sandwort). A habit; B leafy shoot; C leaf; D flower; E fruiting calyx. (Balerno, Midlothian). STACE 196/160.

48. CARYOPHYLLACEAE

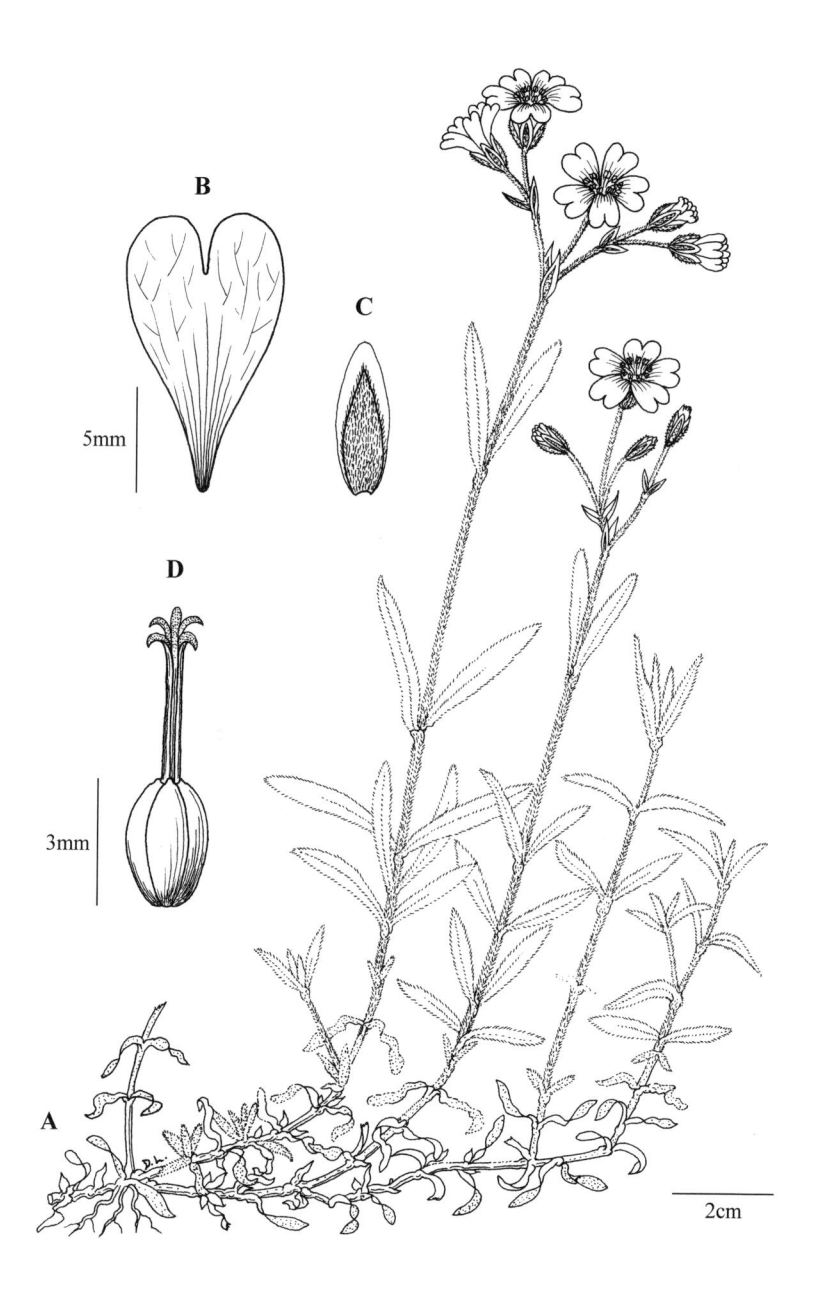

Cerastium tomentosum L. (Snow-in-summer). A habit; B petal; C sepal; D gynoecium. (Source unknown). STACE 202/164.

48. CARYOPHYLLACEAE 71

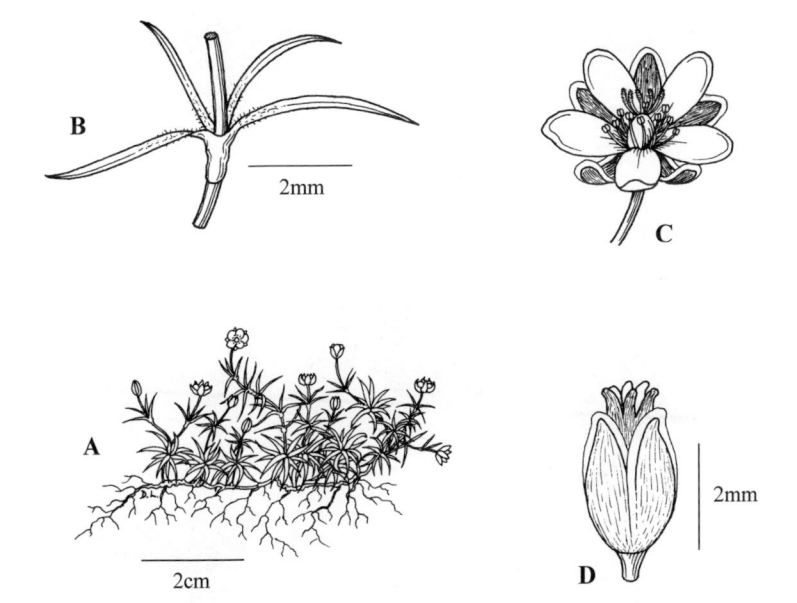

Sagina subulata (Sw.) C. Presl 'Aurea' (as var. *glabrata* Gillot) (Heath Pearlwort cultivar). A habit; B leaves at one node; C flower; D capsule. (Barnes, Surrey). STACE 206/168.

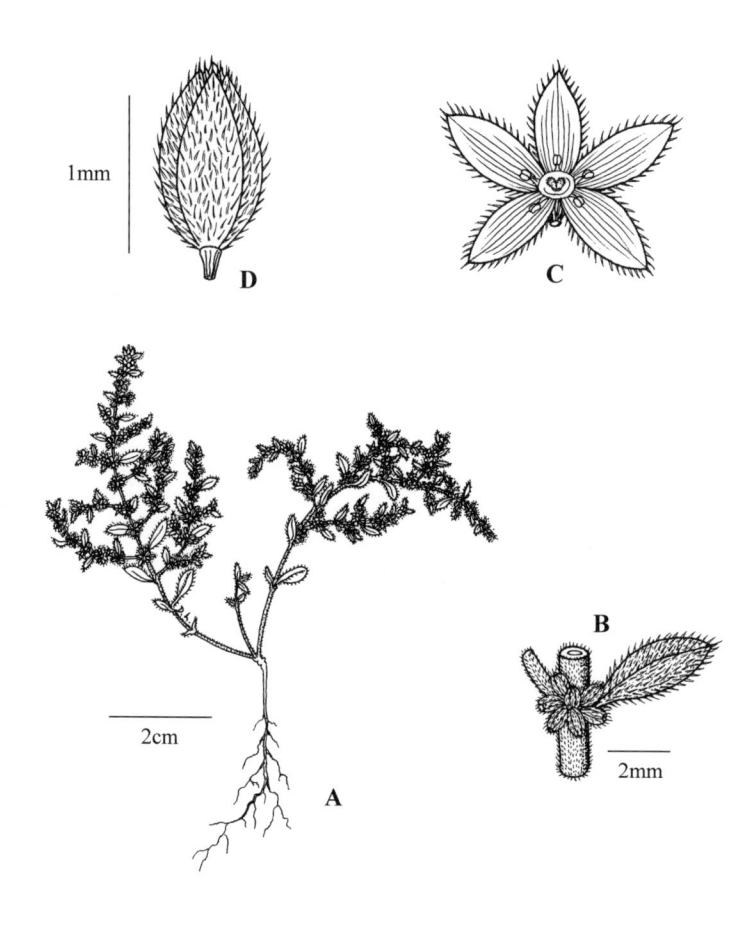

Herniaria hirsuta L. (Hairy Rupturewort). A habit; B cluster of flowers at node; C flower opened out; D fruiting calyx. (Burton-on-Trent, Staffs). STACE 209/170.

48. CARYOPHYLLACEAE

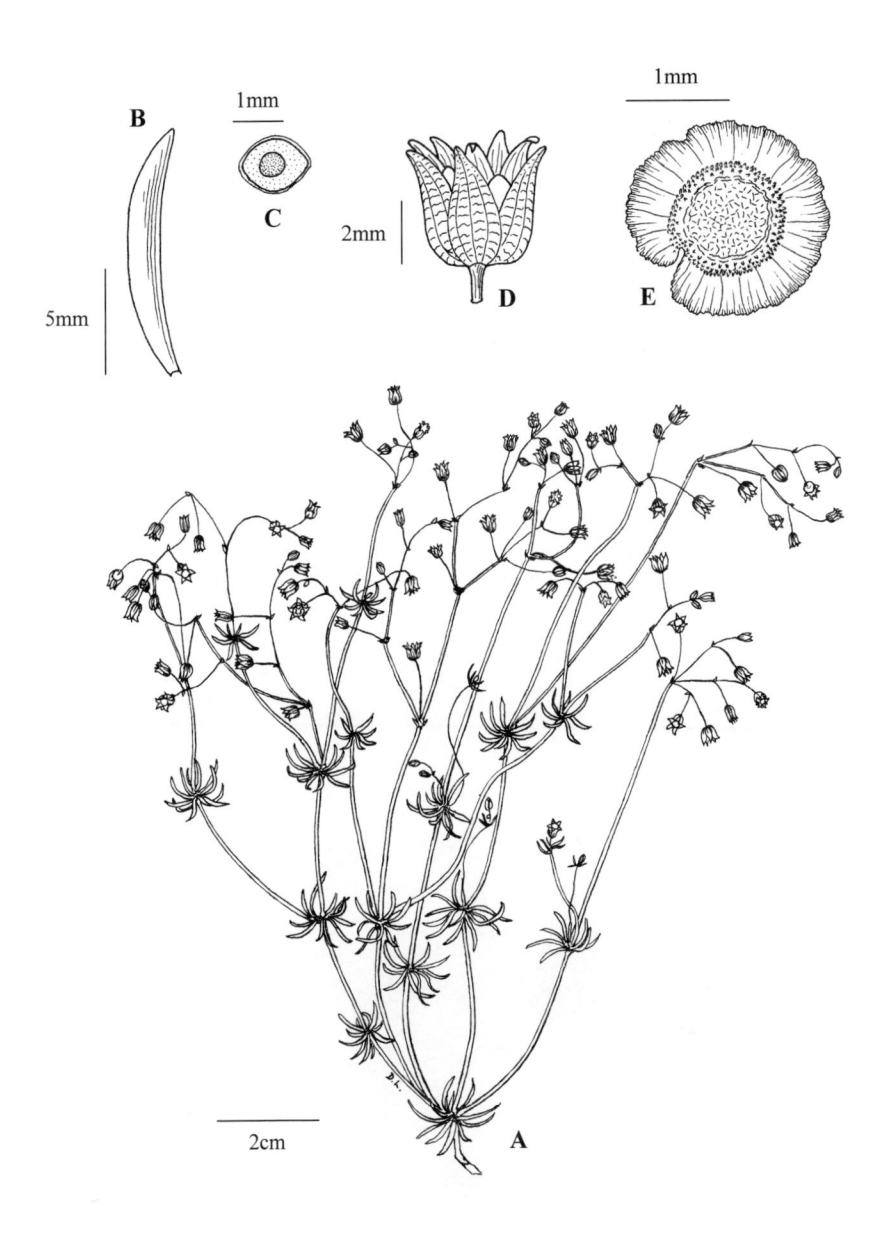

Spergula morisonii Boreau (Pearlwort Spurrey). A habit; B leaf; C TS of leaf; D capsule; E seed. (Broadwater Forest, E Sussex). STACE 210/171.

Silene dichotoma Ehrh. (Forked Catchfly). A habit; B petal; C gynoecium. (*Hort*. Sevenoaks, W Kent). STACE 216/175.

48. CARYOPHYLLACEAE

Silene fimbriata (Adams ex F. Weber & D. Mohr) Sims (Fringed Catchfly). A habit; B petal; C calyx opened out to show young fruit. (Ardrishaig, Kintyre). STACE 216/175.

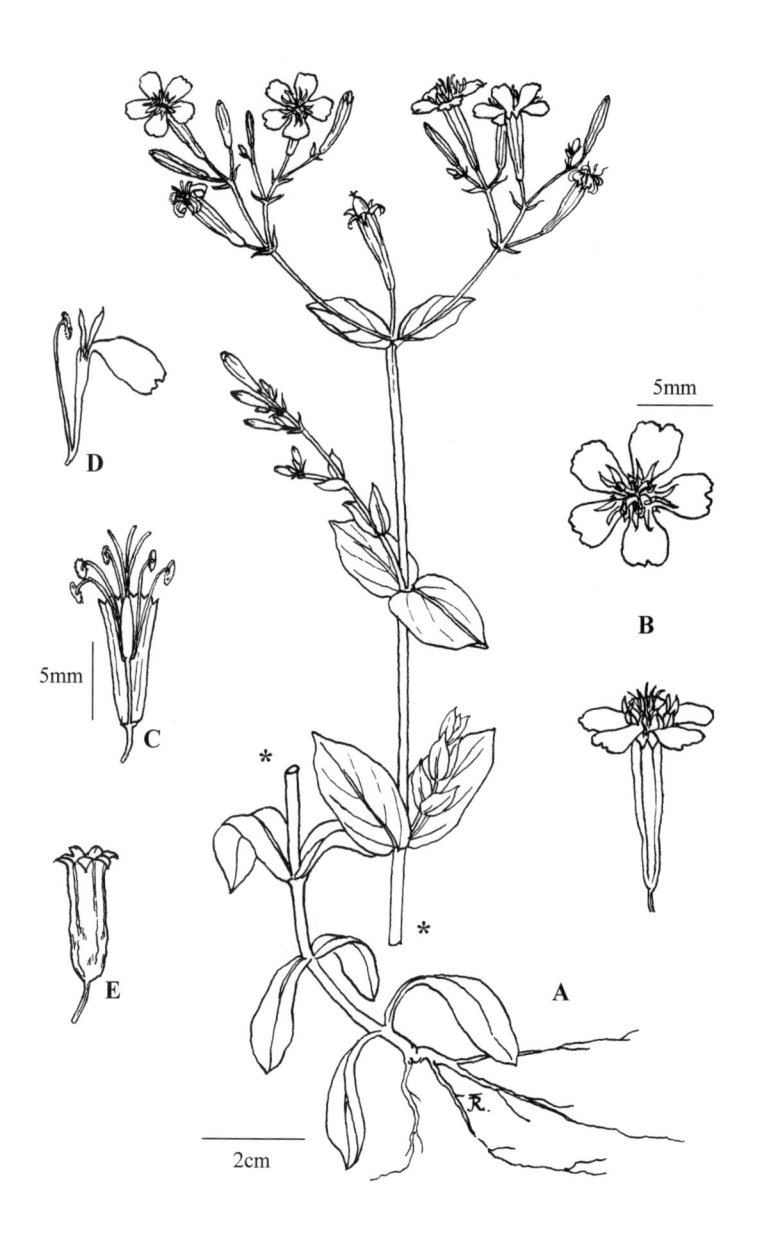

Silene armeria L. (Sweet-William Catchfly). A habit; B flower; C LS of flower minus petals; D petal and stamen; E fruit. (Lancs). STACE 217/176.

48. CARYOPHYLLACEAE

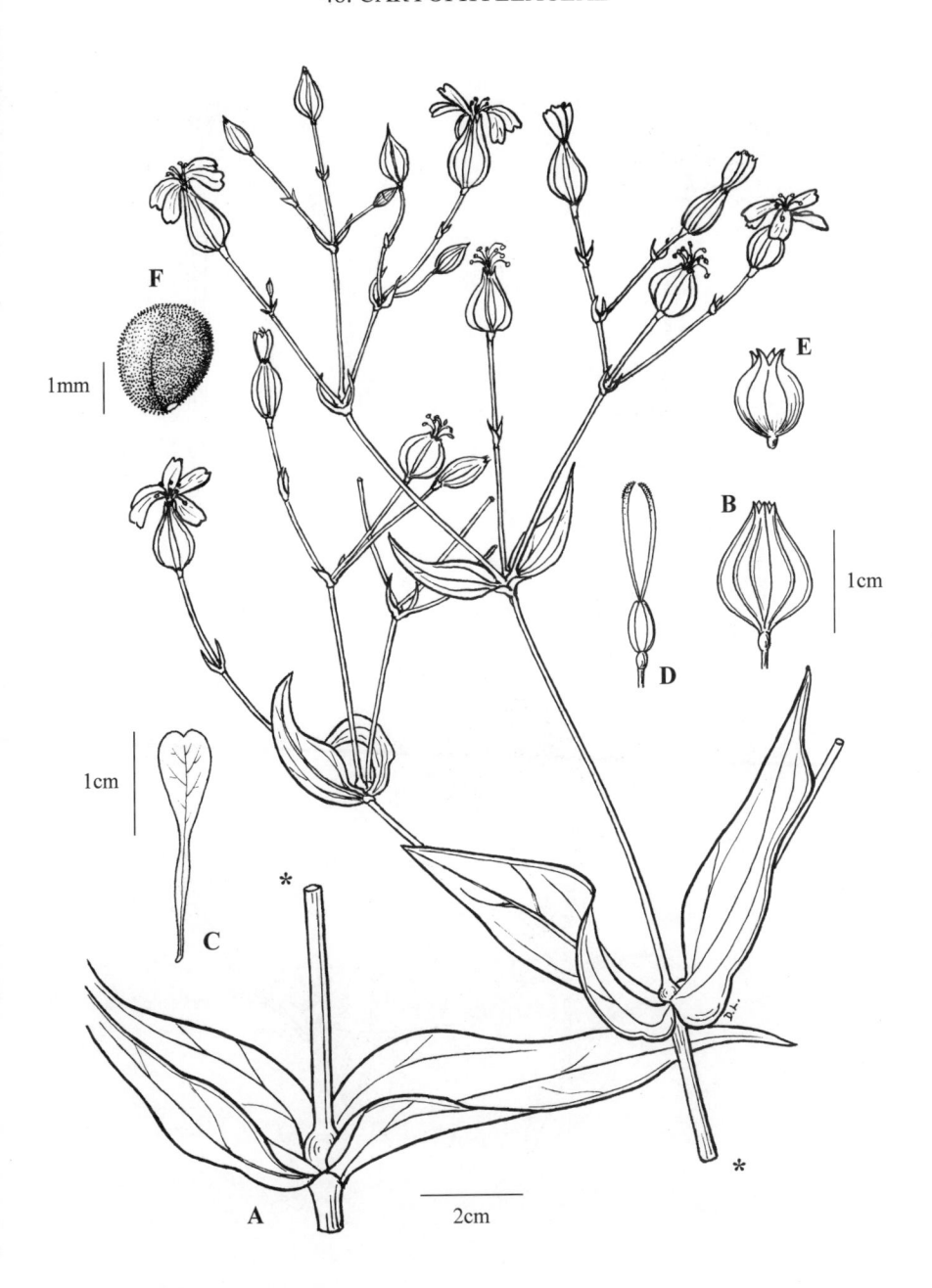

Vaccaria hispanica (Mill.) Rauschert (Cowherb). A habit; B calyx; C petal; D gynoecium; E capsule; F seed. (Esher, Surrey). STACE 219/178.

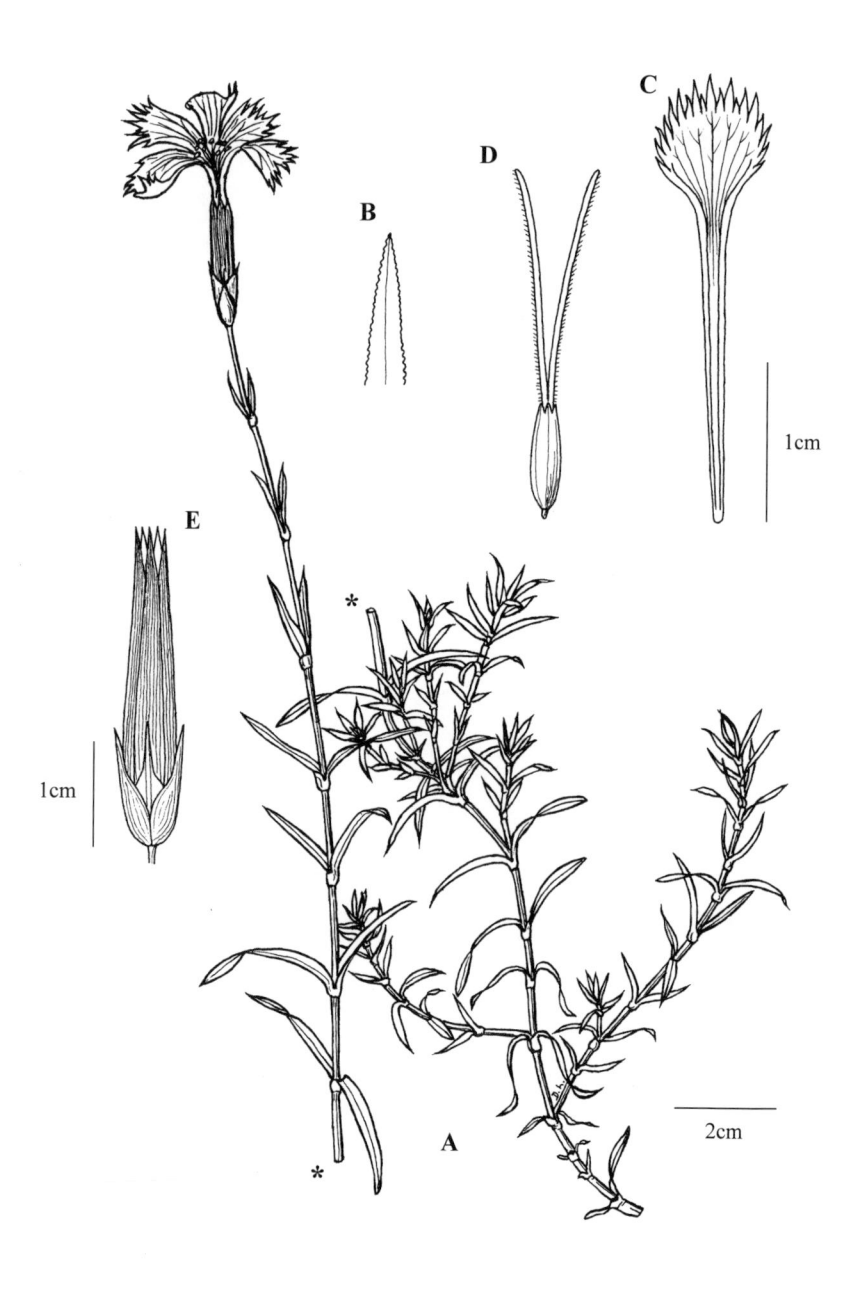

Dianthus gallicus Pers. (Jersey Pink). A habit; B leaf apex; C petal; D gynoecium; E fruiting calyx and epicalyx. (St Ouens Bay, Jersey). STACE 221/180.

48. CARYOPHYLLACEAE

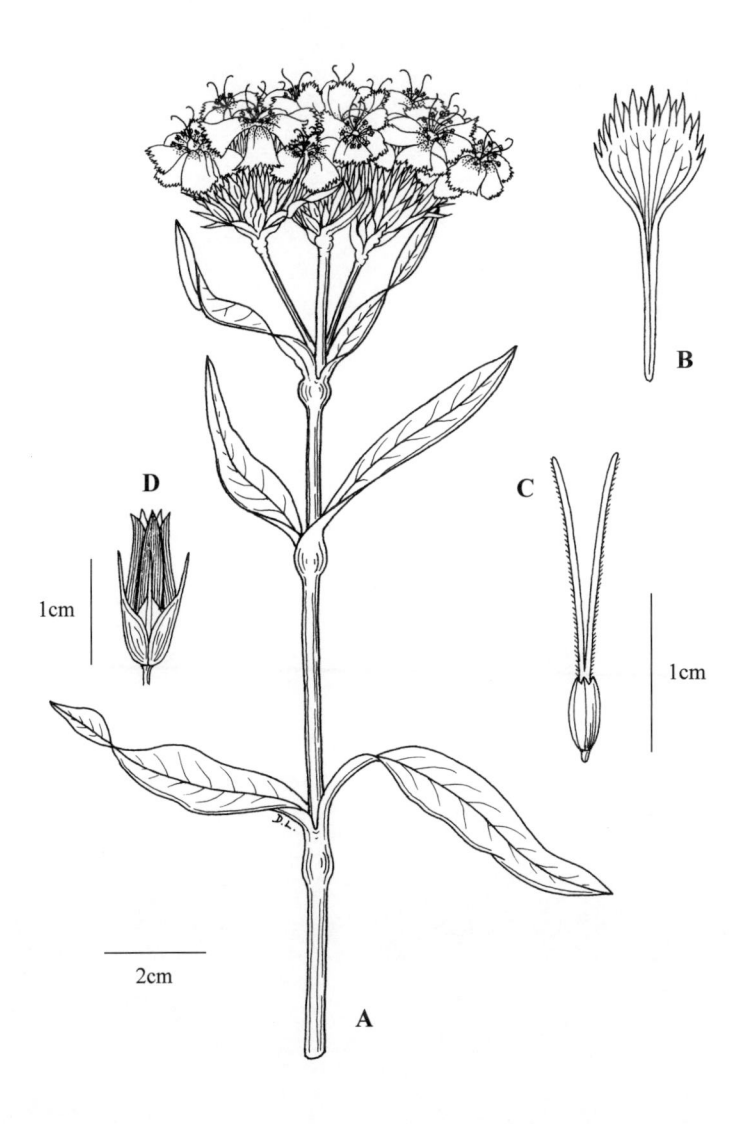

Dianthus barbatus L. (Sweet-William). A flowering stem; B petal; C gynoecium; D fruiting calyx and epicalyx. (Hanicksham, W Kent). STACE 222/180.

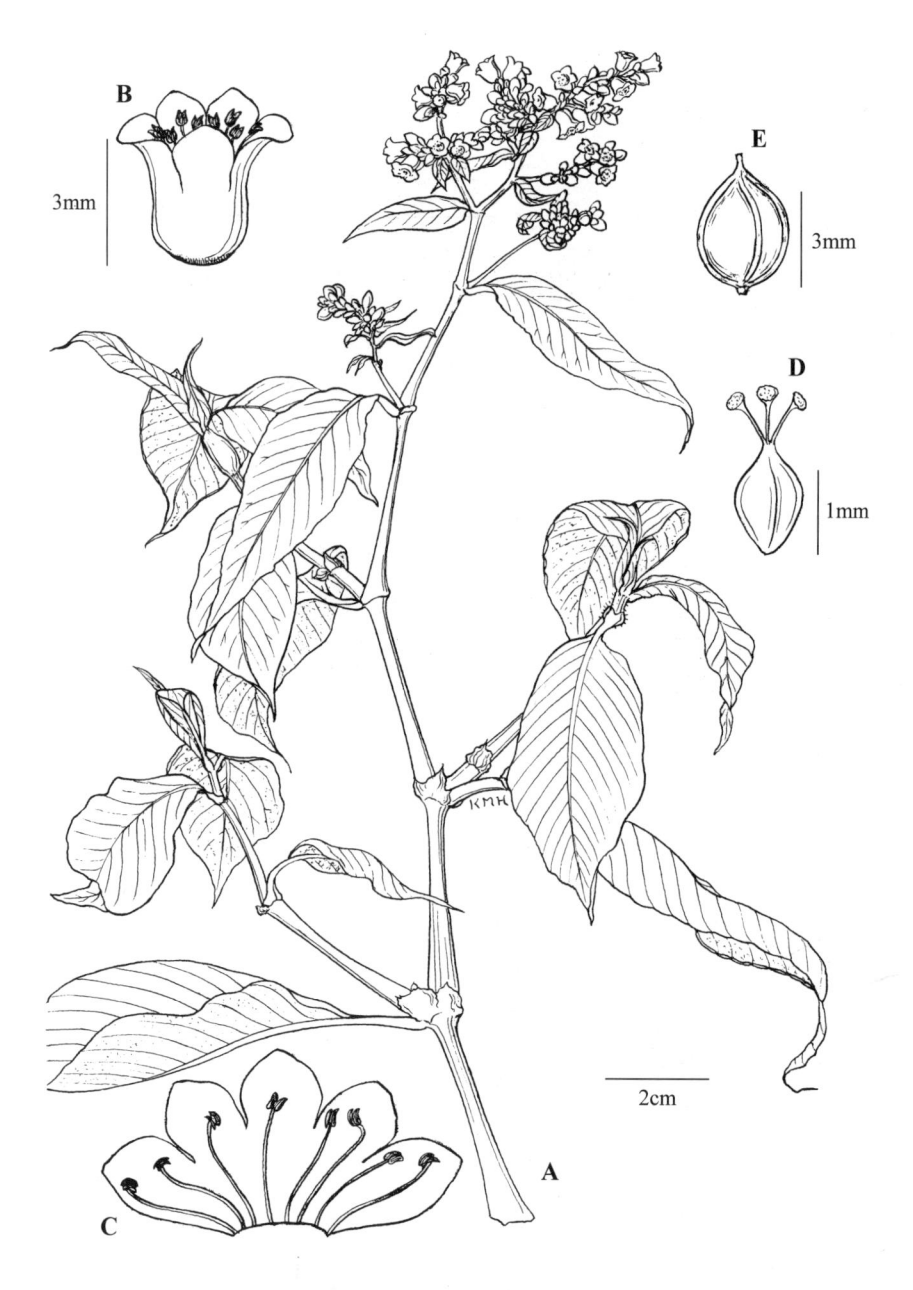

Persicaria campanulata (Hook. f.) Ronse Decr. (Lesser Knotweed). A habit; B flower; C corolla opened out; D gynoecium; E achene. (Carradale, Kintyre). STACE 225/182.

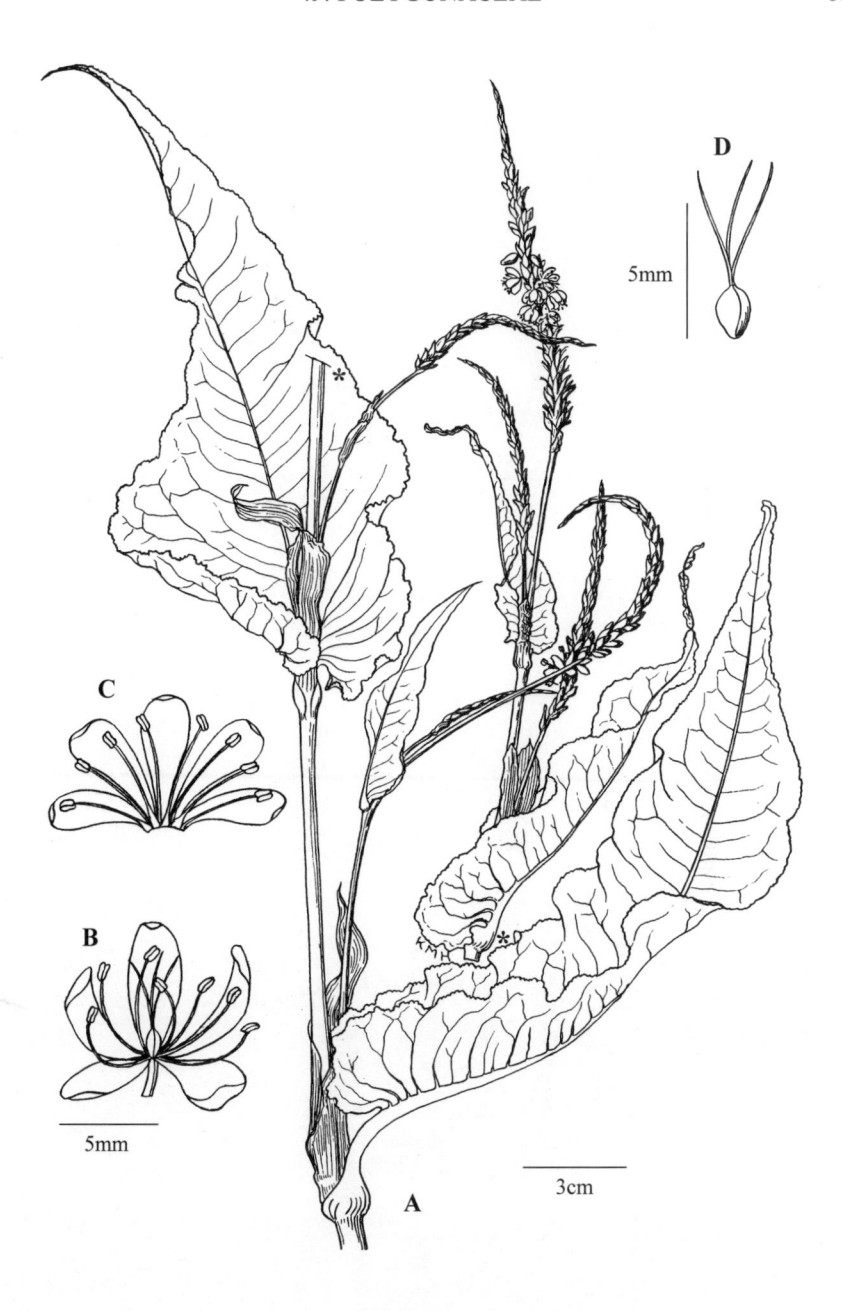

49. POLYGONACEAE

Persicaria amplexicaulis (D. Don) Ronse Decr. (Red Bistort). A habit; B flower; C corolla opened out; D gynoecium. (Kilkenny, Co Kilkenny). STACE 225/183.

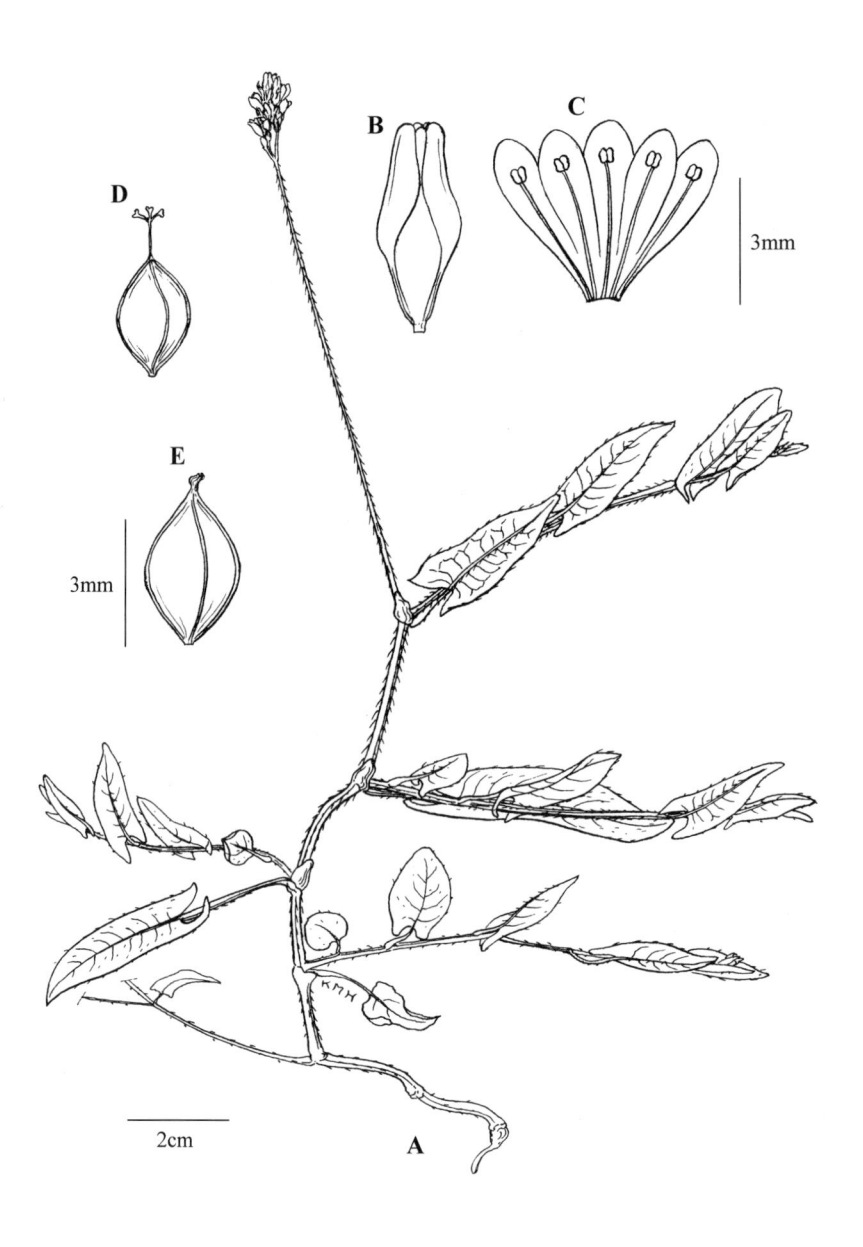

Persicaria sagittata (L.) H. Gross ex Nakai (American Tear-thumb). A habit; B flower; C corolla opened out; D gynoecium; E achene. (Waterville, S Kerry). STACE 227/184.

49. POLYGONACEAE

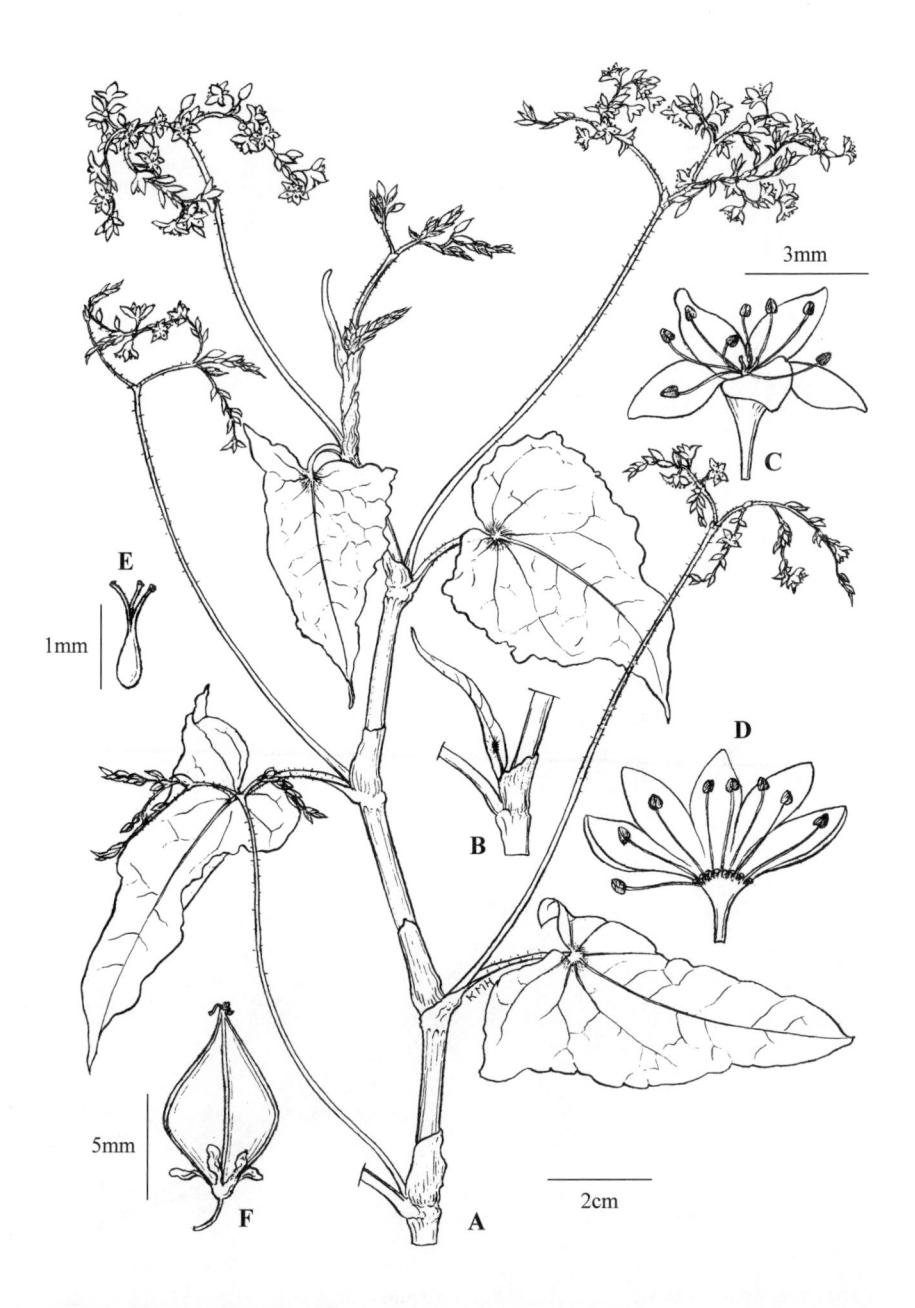

Fagopyrum dibotrys (D. Don) H. Hara (Tall Buckwheat). A habit; B stem node with fused stipules; C flower; D corolla opened out; E gynoecium; F achene. (Dale, Pembs). STACE 228/185.

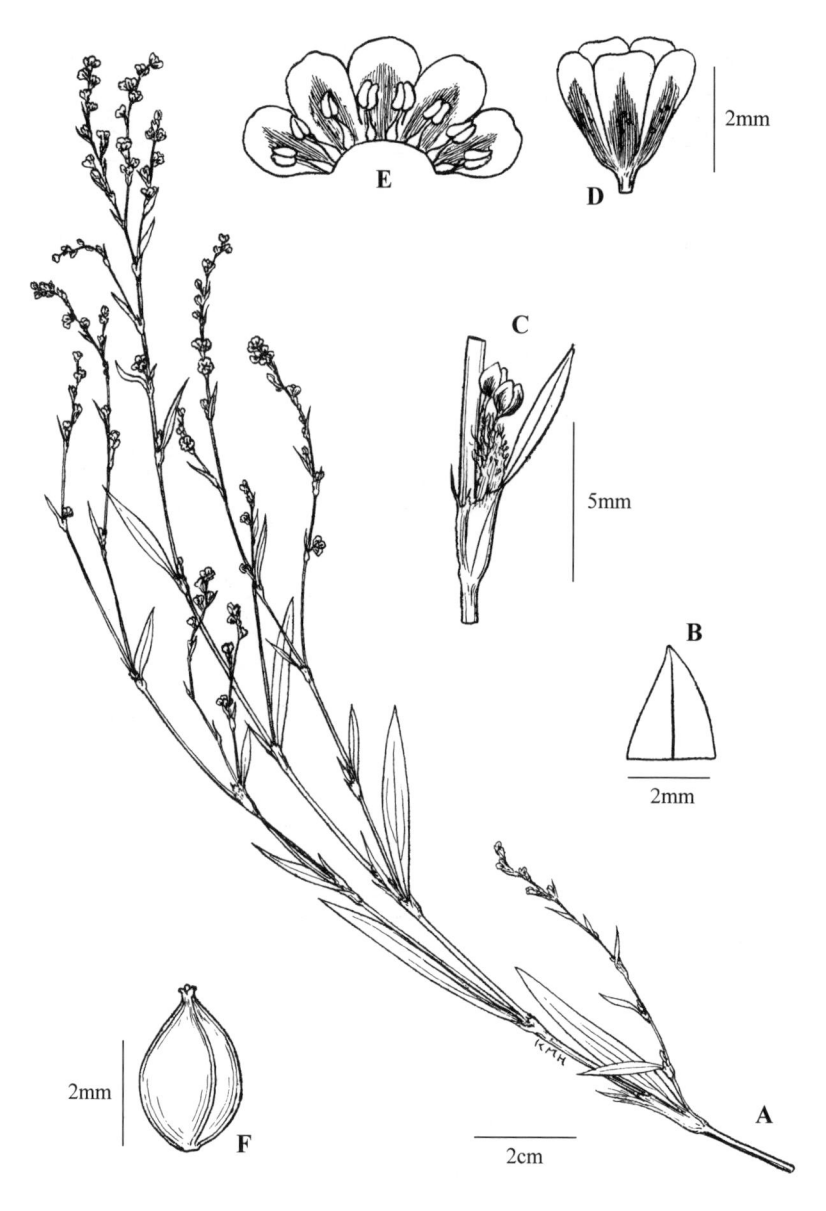

Polygonum arenarium Waldst. & Kit. ssp. **pulchellum** (Lois.) Thell. (Lesser Red-knotgrass). A habit; B apex of leaf; C axillary inflorescence; D flower; E corolla opened out; F achene. (Ringwood, S Hants). STACE 230/186.

49. POLYGONACEAE

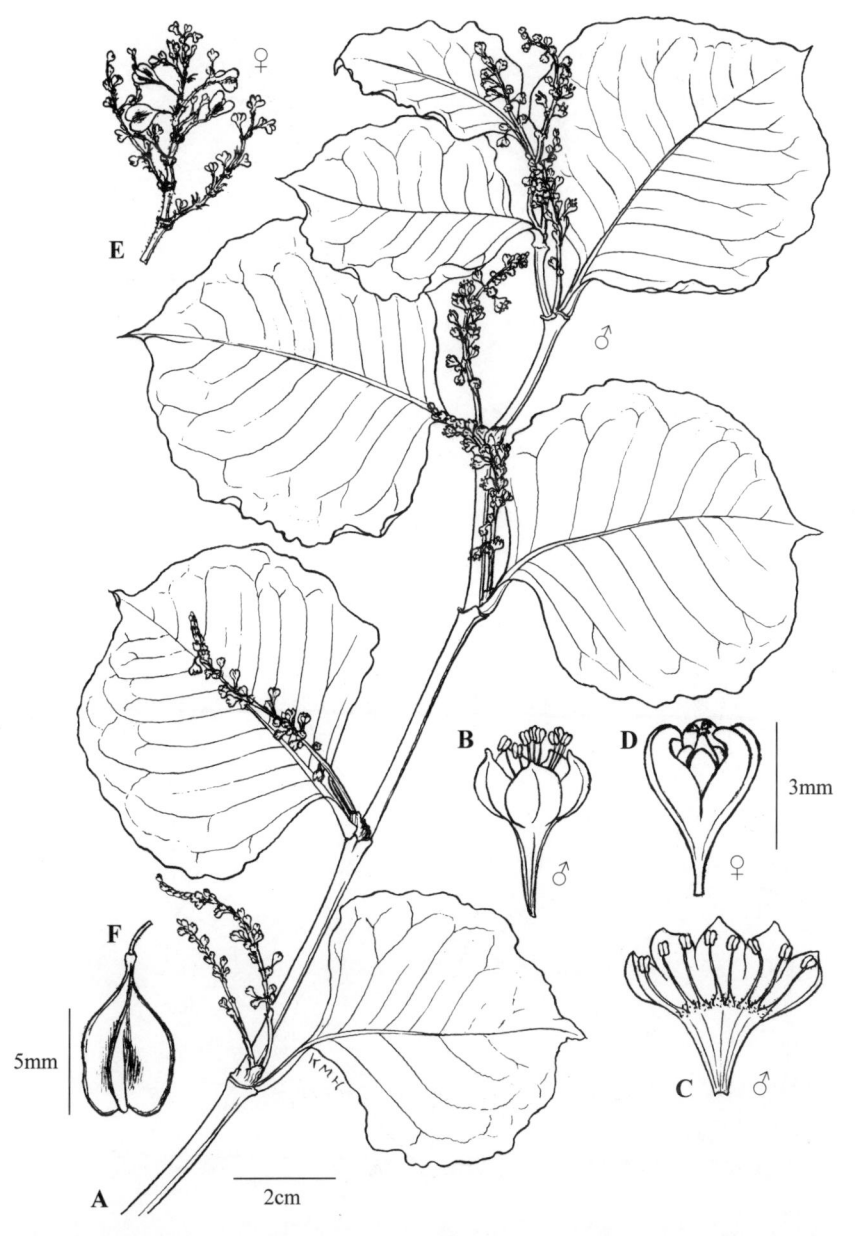

Fallopia japonica (Houtt.) Ronse Decr. var. **compacta** (Hook. f.) J.P. Bailey (Japanese Knotweed variety). A flowering branch; B male flower; C male flower opened out; D female flower; E fruiting racemes; F fruit. (Source unknown). STACE 231/187.

86   49. POLYGONACEAE

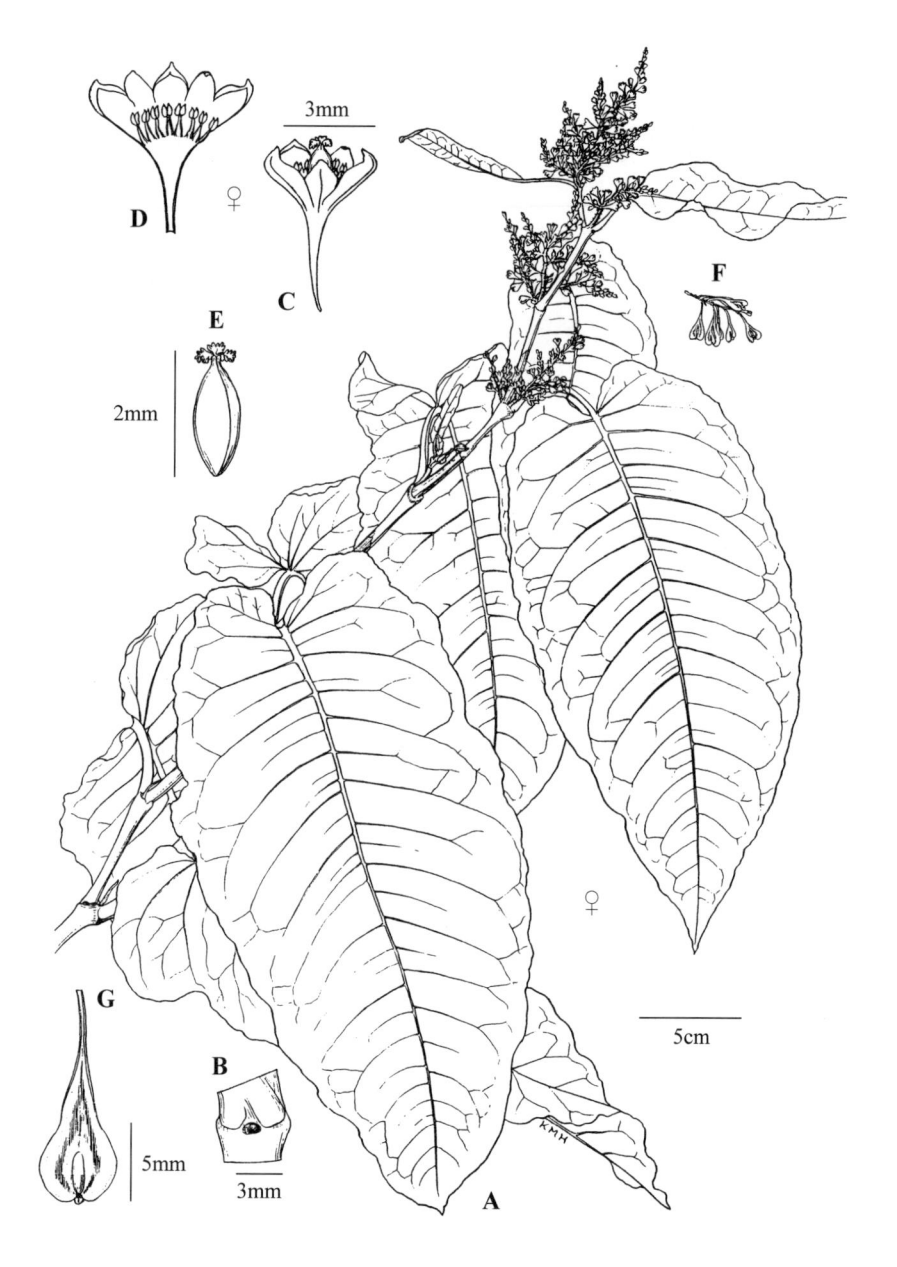

Fallopia sachalinensis (F. Schmidt ex Maxim.) Ronse Decr. (Giant Knotweed). A flowering branch; B gland at stem node; C female flower; D corolla opened out; E gynoecium; F fruiting raceme; G fruit. (Caterham, Surrey). STACE 231/187.

49. POLYGONACEAE

Fallopia baldschuanica (Regel) Holub (Russian-vine). A non-flowering and fertile branches; B flower; C flower with part of corolla cut away; D fruiting raceme; E fruit; F achene. (Knowl Hill, Berks). STACE 231/187.

49. POLYGONACEAE

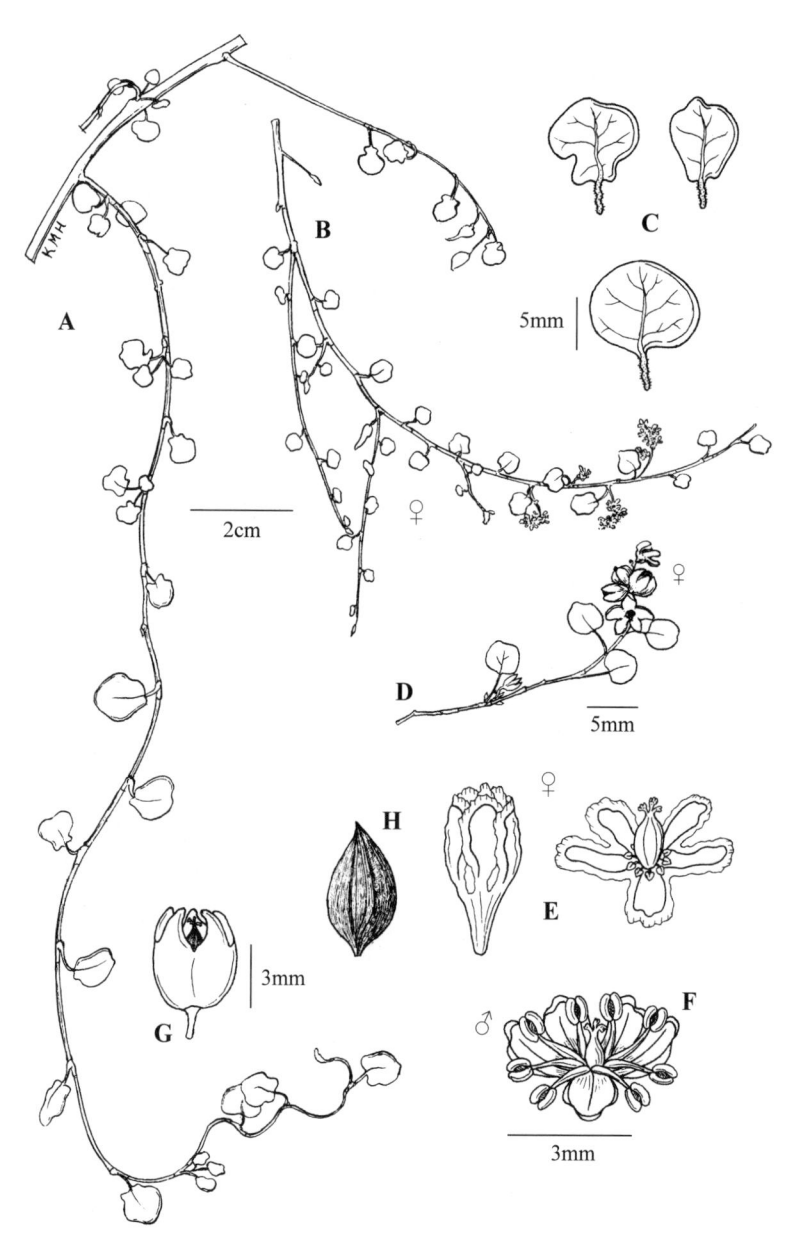

Muehlenbeckia complexa (A. Cunn.) Meisn. (Wireplant). A non-flowering branch; B female plant branch; C leaves; D female inflorescence; E female flower and opened out; F male flower; G fruit with succulent tepals; H achene. (A-E, G-H Le Gouffre, Guernsey; F North Island, New Zealand). STACE 232/187.

49. POLYGONACEAE

Rumex salicifolius Weinm. ssp. **triangulivalvis** Danser (Willow-leaved Dock). A habit; B young stem showing fused stipules; C flower at anthesis; D flower after anthesis; E fruiting panicle; F fruit; G TS of fruit. (Cardiff, Glam). STACE 237/191.

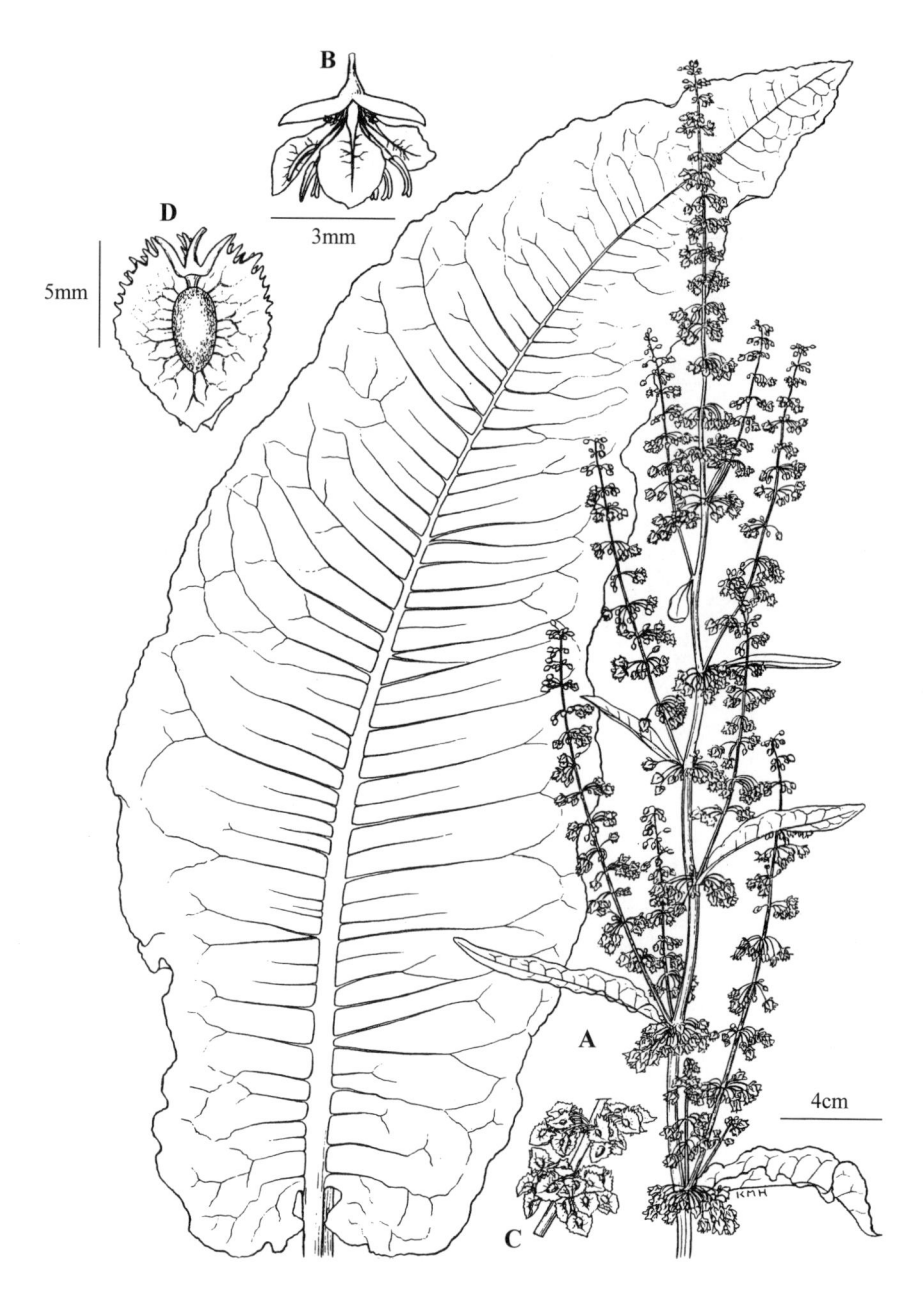

Rumex cristatus DC. (Greek Dock). A habit and basal leaf; B flower; C two whorls of fruits; D fruit. (Reading, Berks). STACE 238/192.

49. POLYGONACEAE

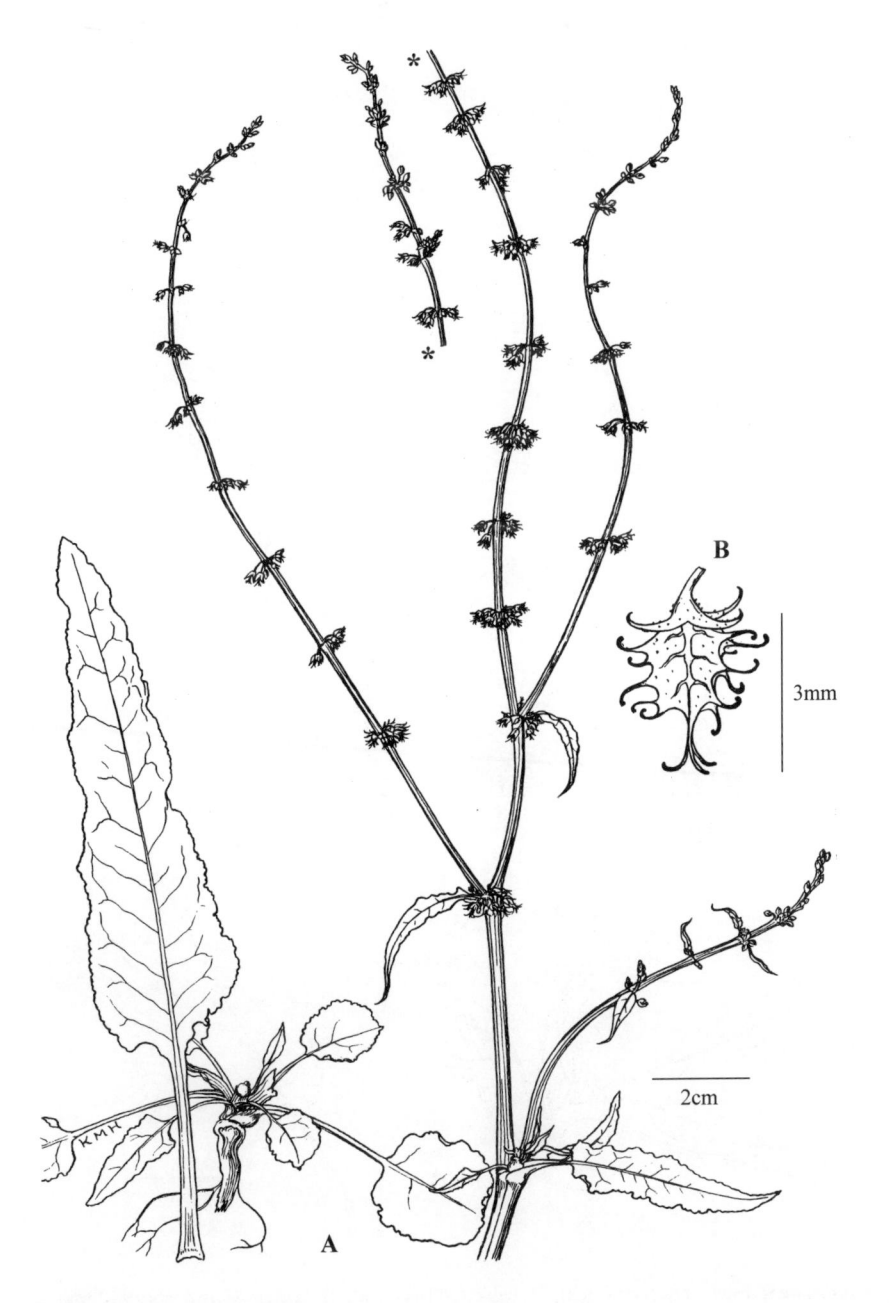

Rumex brownii Campd. (Hooked Dock). A habit and base of plant; B fruit. (Source unknown). STACE 240/195.

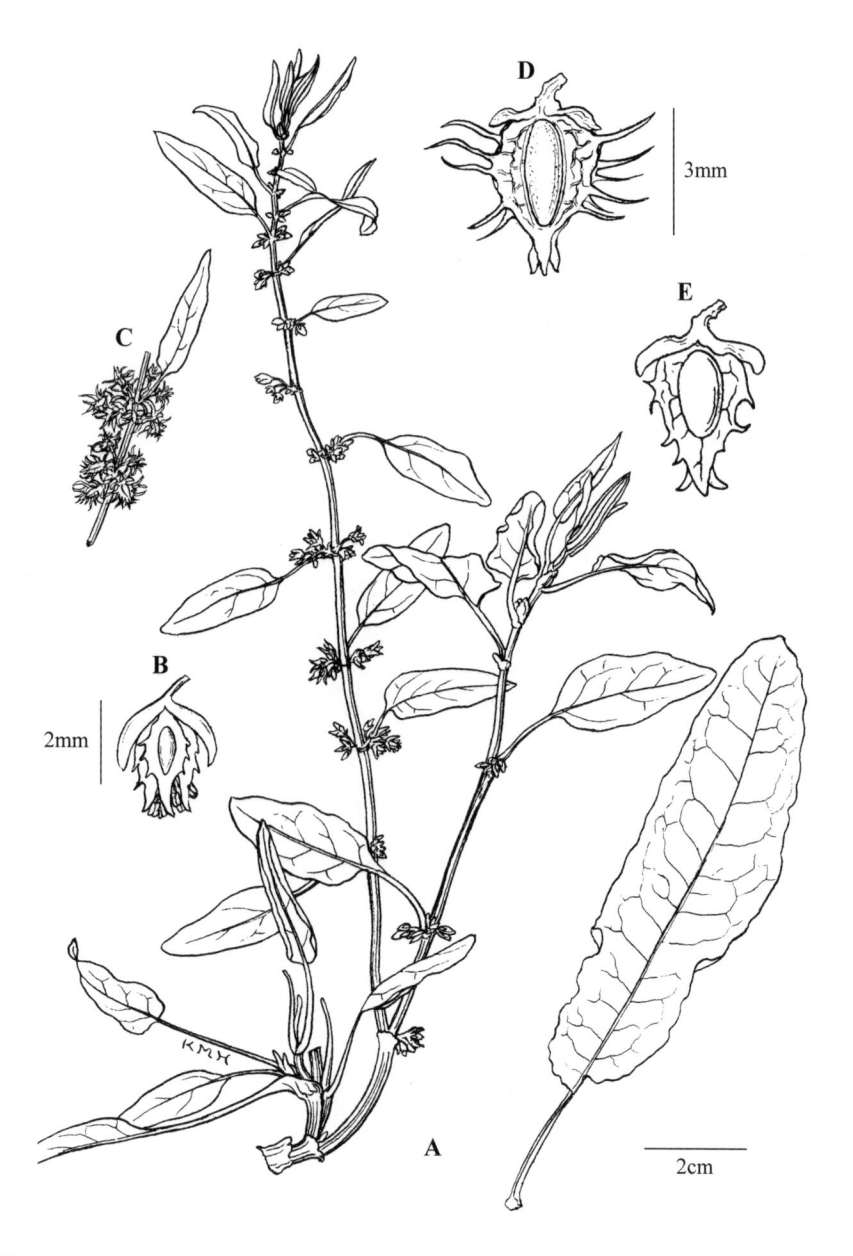

Rumex dentatus L. (Aegean Dock). A habit and basal leaf; B flower; C two whorls of fruits; D fruit of ssp. **halacsyi** (Rech.) Rech. f.; E fruit of ssp. **klotzschianus** (Meissner) Rech. f. (Sources unknown). STACE 242/195.

53. CLUSIACEAE

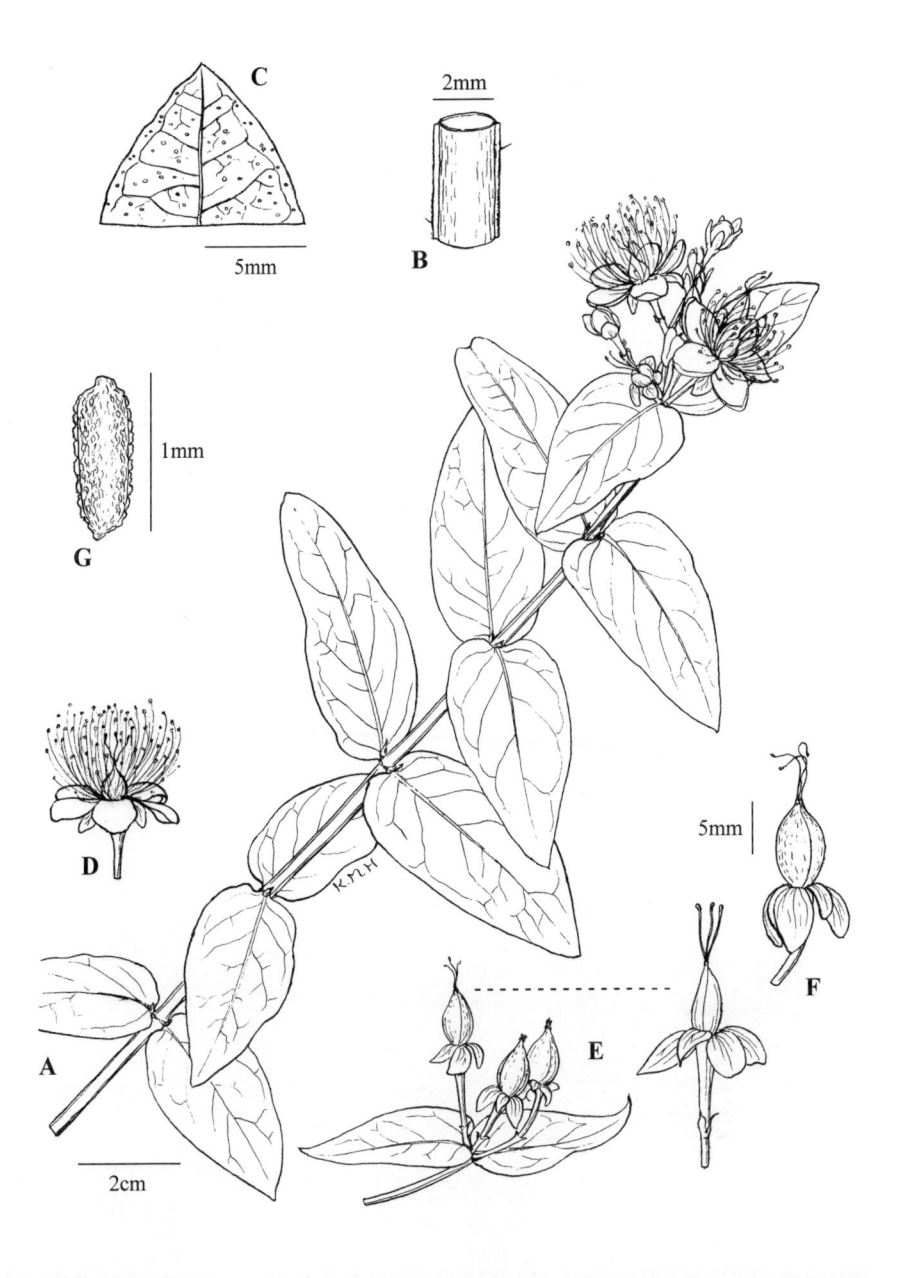

Hypericum x inodorum Mill. (Tall Tutsan). A flowering branch; B part of stem; C apex of leaf; D flower; E cluster of fruits; F fruit; G seed. (*Hort*. Teddington, Middlesex). STACE 254/207.

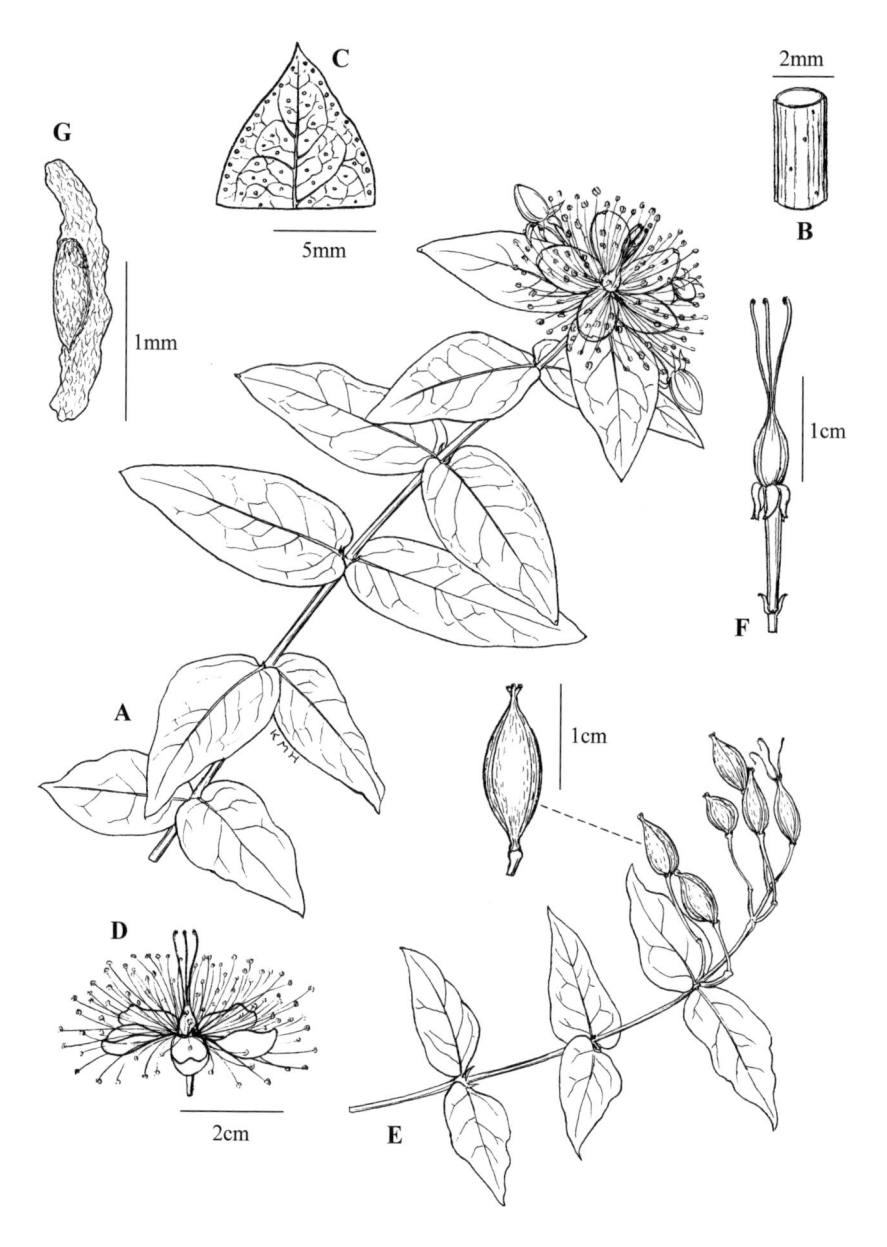

Hypericum hircinum L. (Stinking Tutsan). A flowering branch; B part of stem; C apex of leaf; D flower; E cluster of fruits; F fruit; G winged seed. (*Hort*. Marlborough House, Middlesex). STACE 255/207.

55. MALVACEAE

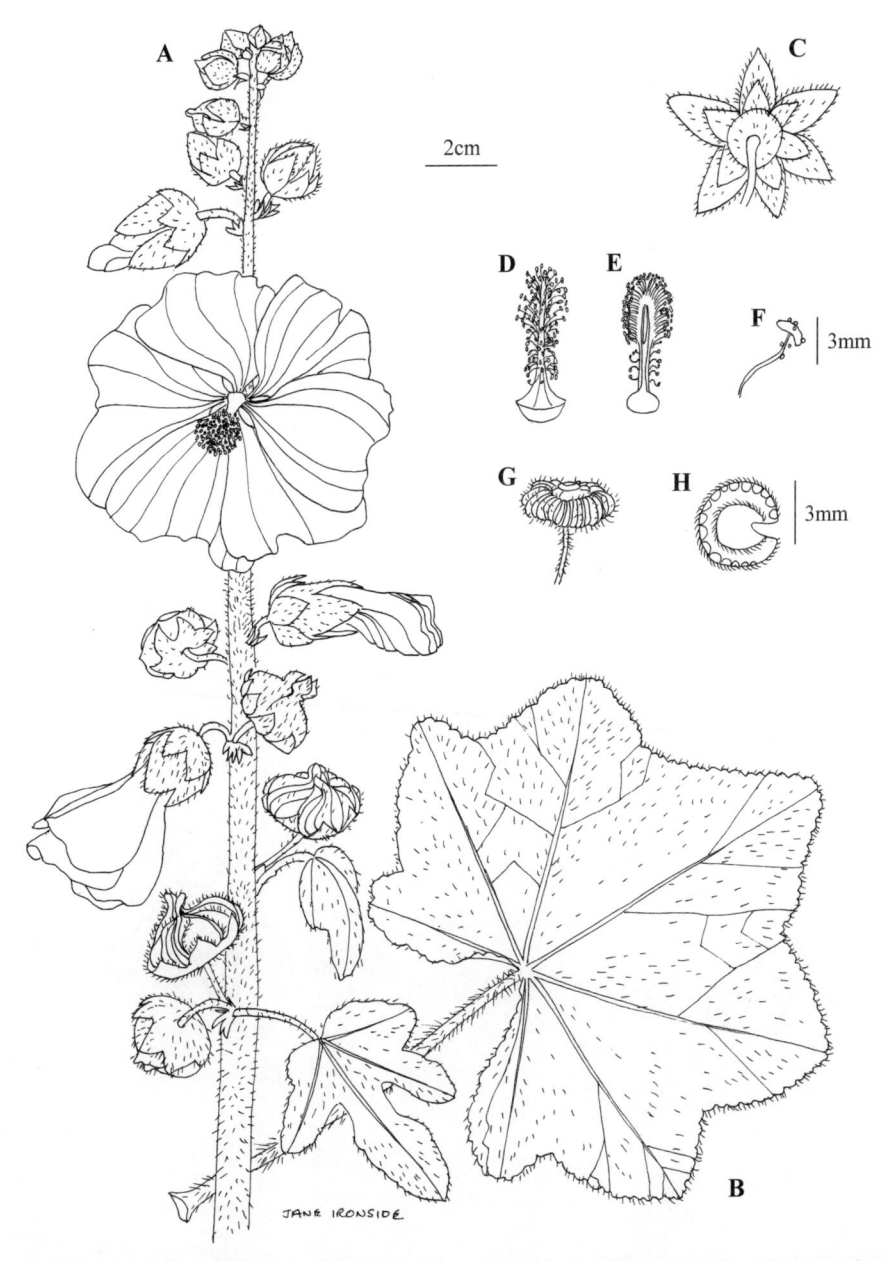

Alcea rosea L. (Hollyhock). A flowering stem; B mature leaf; C calyx and epicalyx; D gynoecium; E LS of gynoecium; F anther and filament; G schizocarp; H mericarp. (Kettlestone Parish, W Norfolk). STACE 263/215.

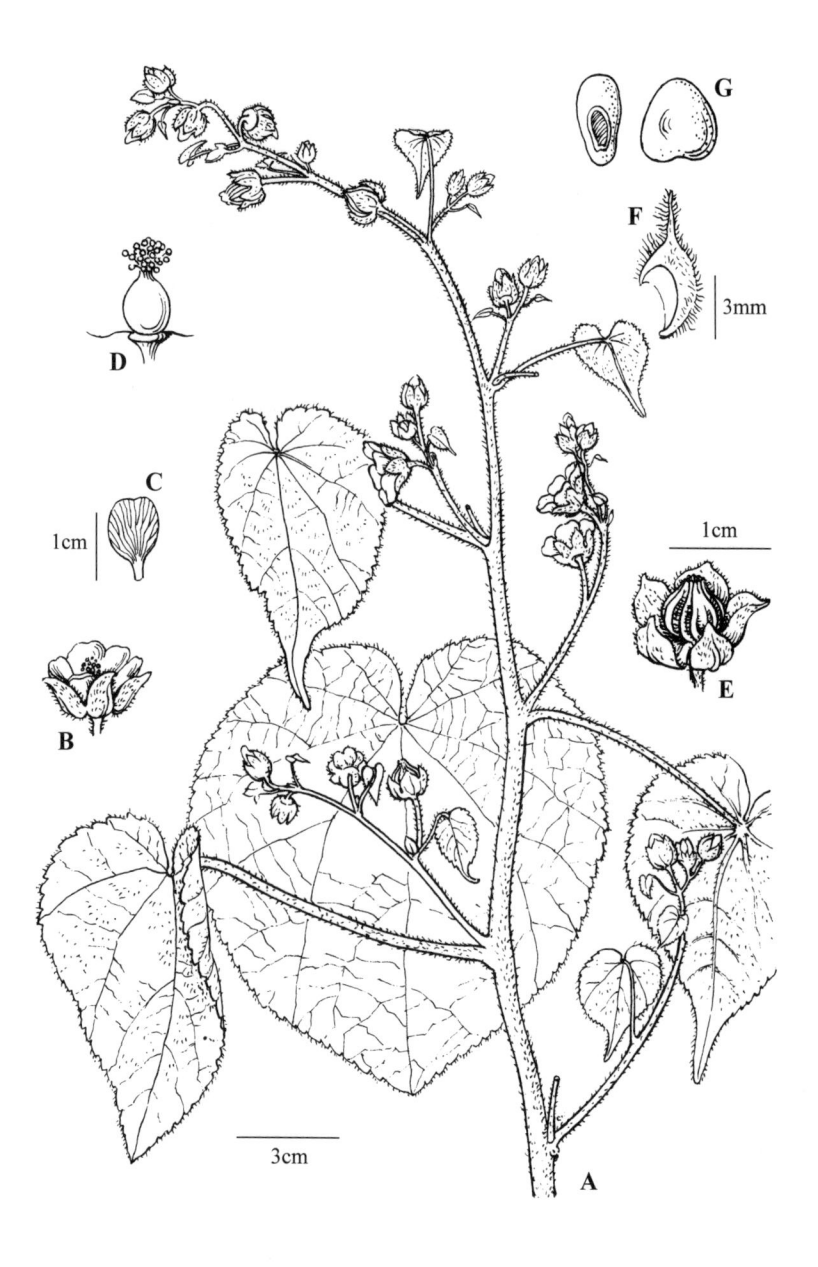

Abutilon theophrasti Medik. (Velvetleaf). A habit; B flower; C petal; D flower minus sepals and petals; E schizocarp; F mericarp minus seed; G seed. (Hoddesdon, Herts). STACE 264/215.

55. MALVACEAE

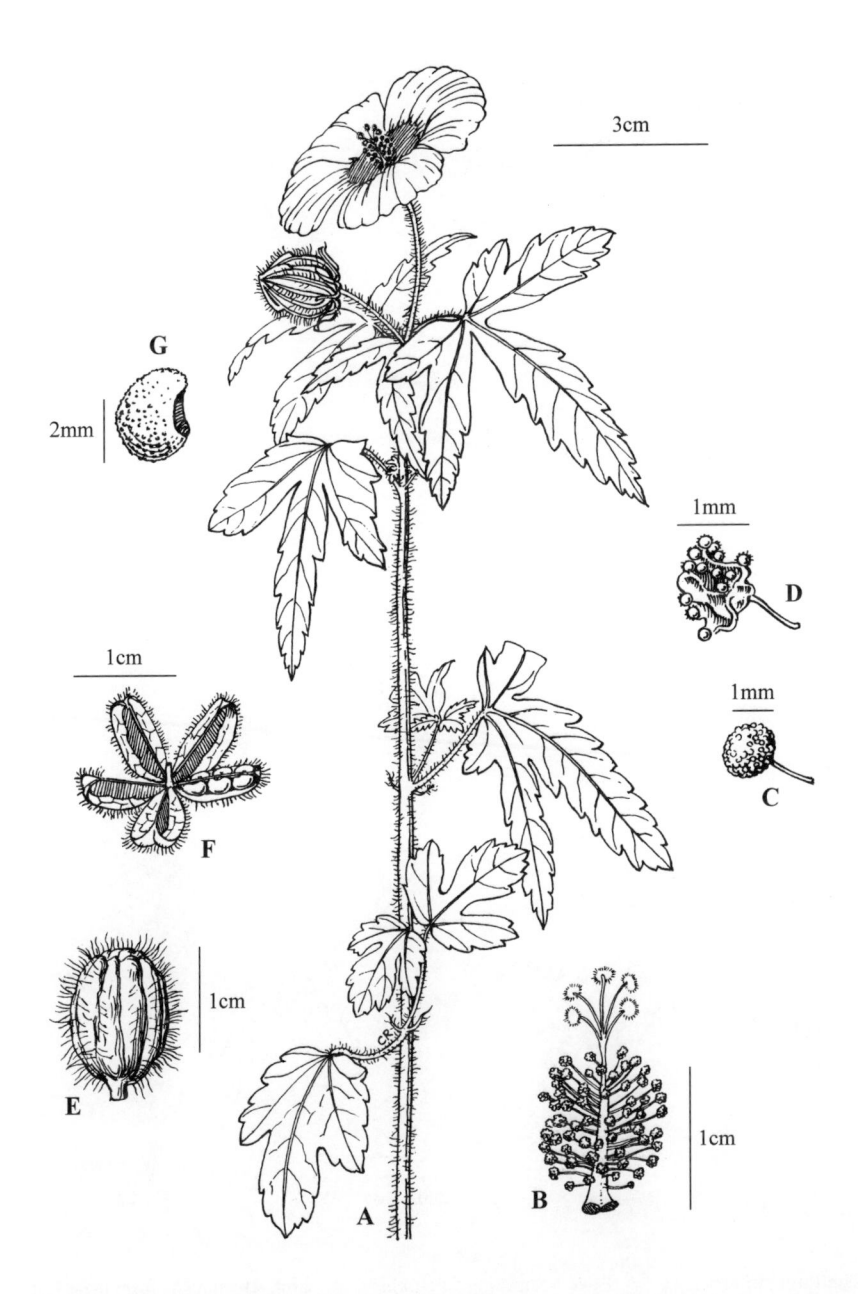

Hibiscus trionum L. (Bladder Ketmia). A habit; B central parts of flower; C anther; D stigma with pollen grains; E capsule; F dehisced capsule; G seed. (Source unknown). STACE 264/216.

98   62. CUCURBITACEAE

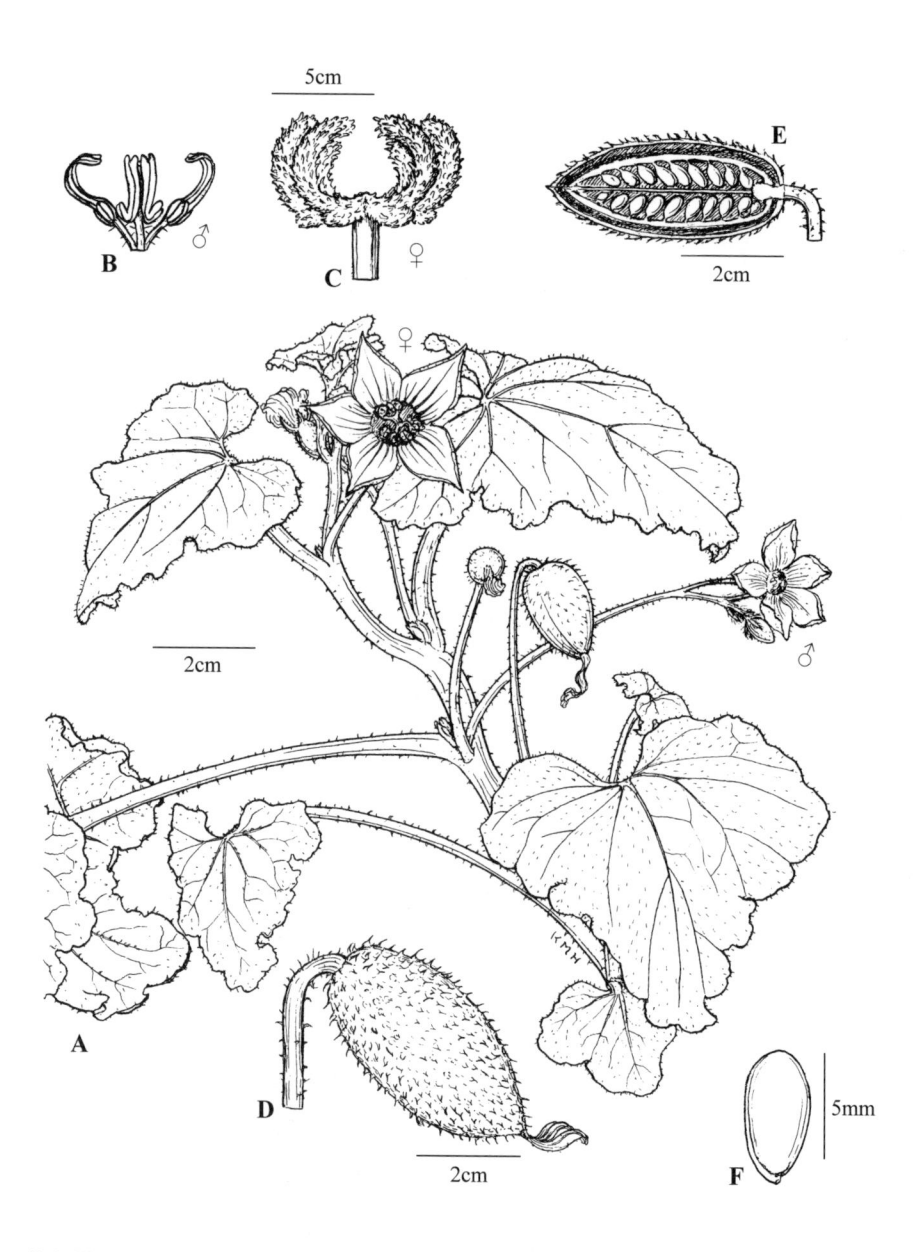

Ecballium elaterium (L.) A. Rich. (Squirting Cucumber). A habit; B stamens from male flower; C style and stigma from female flower; D fruit; E LS of fruit; F seed. (Channel Islands). STACE 275/224.

63. SALICACEAE

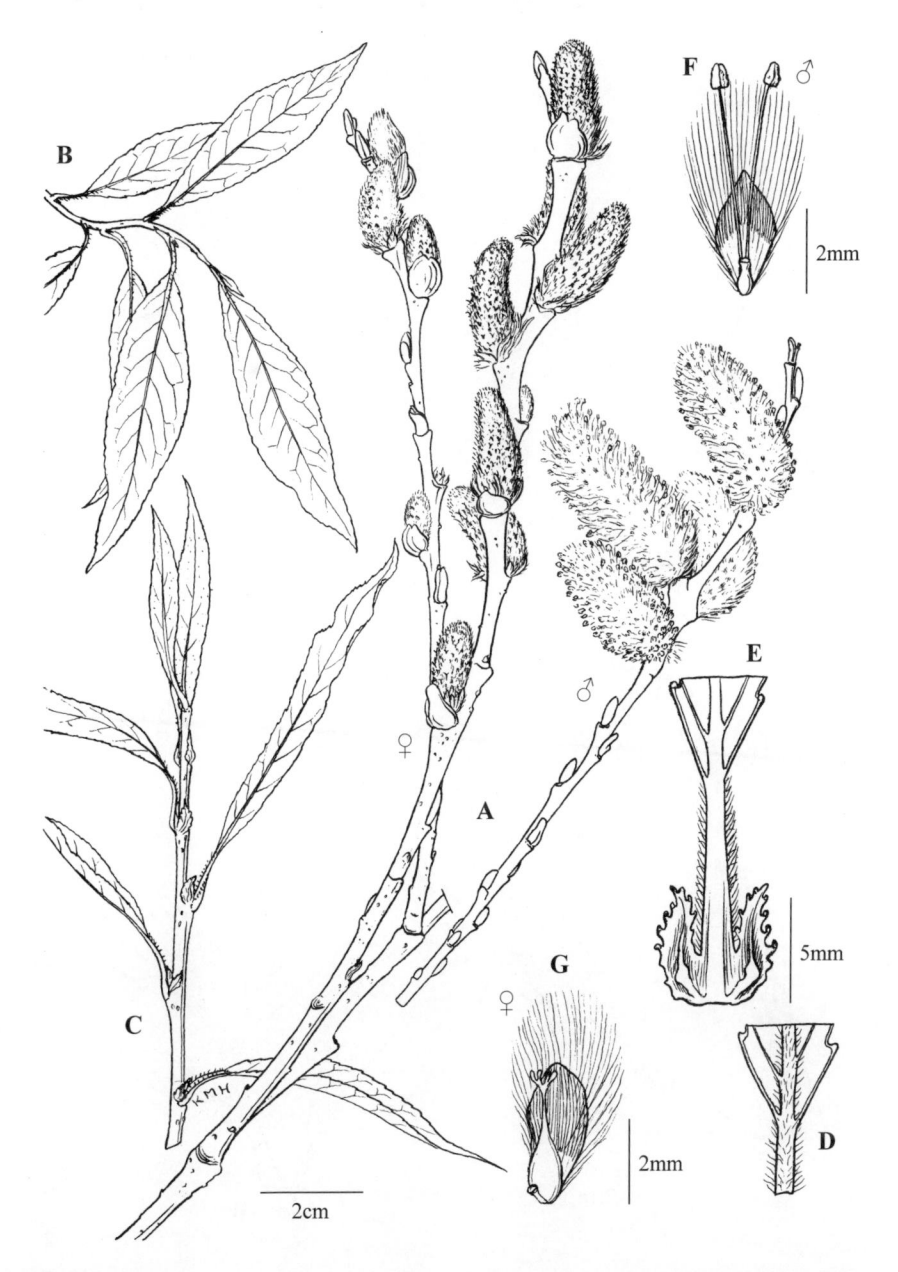

Salix daphnoides Vill. (European Violet-willow). A flowering branches; B branch in summer; C branch in summer of ssp. **acutifolia** (Willd.) Zahn; D upper surface of leaf base; E lower surface of leaf base and stipules; F male flower; G female flower. (Bradley, Derbys). STACE 289/236.

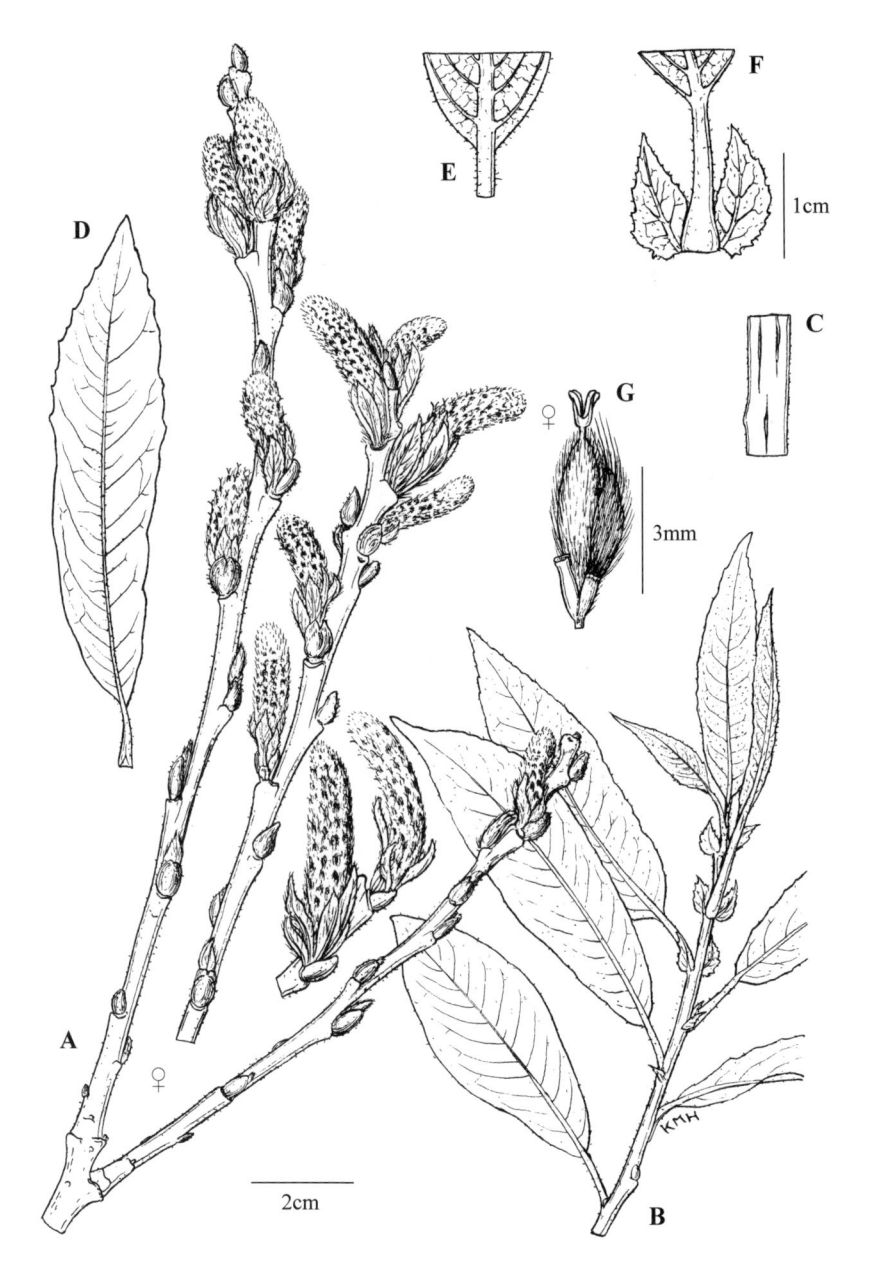

Salix × calodendron Wimm. (Holme Willow). A flowering branch; B branch in summer; C striations under bark; D leaf; E base of leaf; F stipules; G female flower, bract and nectary. (Lunan Bay, Angus). STACE 292/237.

64. BRASSICACEAE

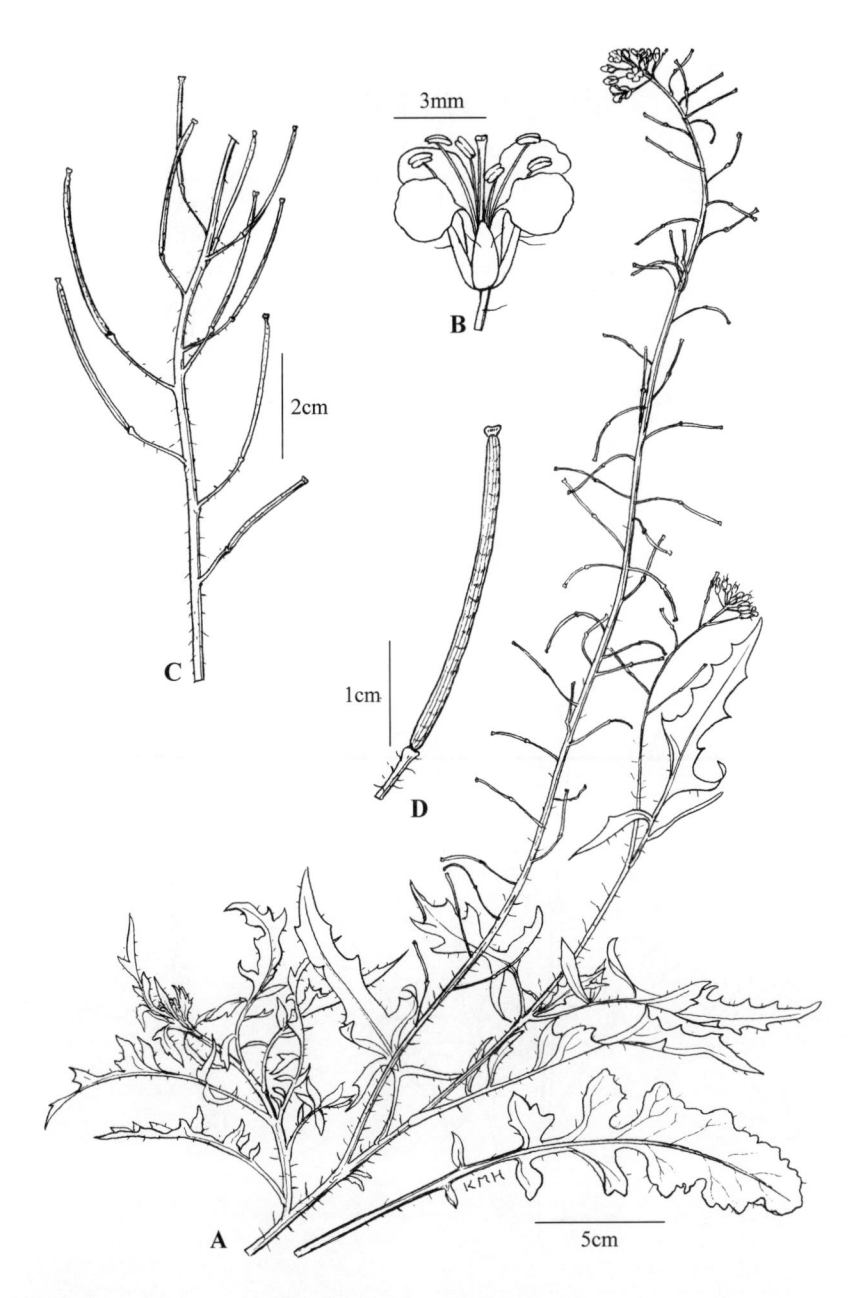

Sisymbrium loeselii L. (False London-rocket). A habit and lower leaf; B flower; C part of fruiting stem; D fruit. (Bedford Square, Middlesex). STACE 304/250.

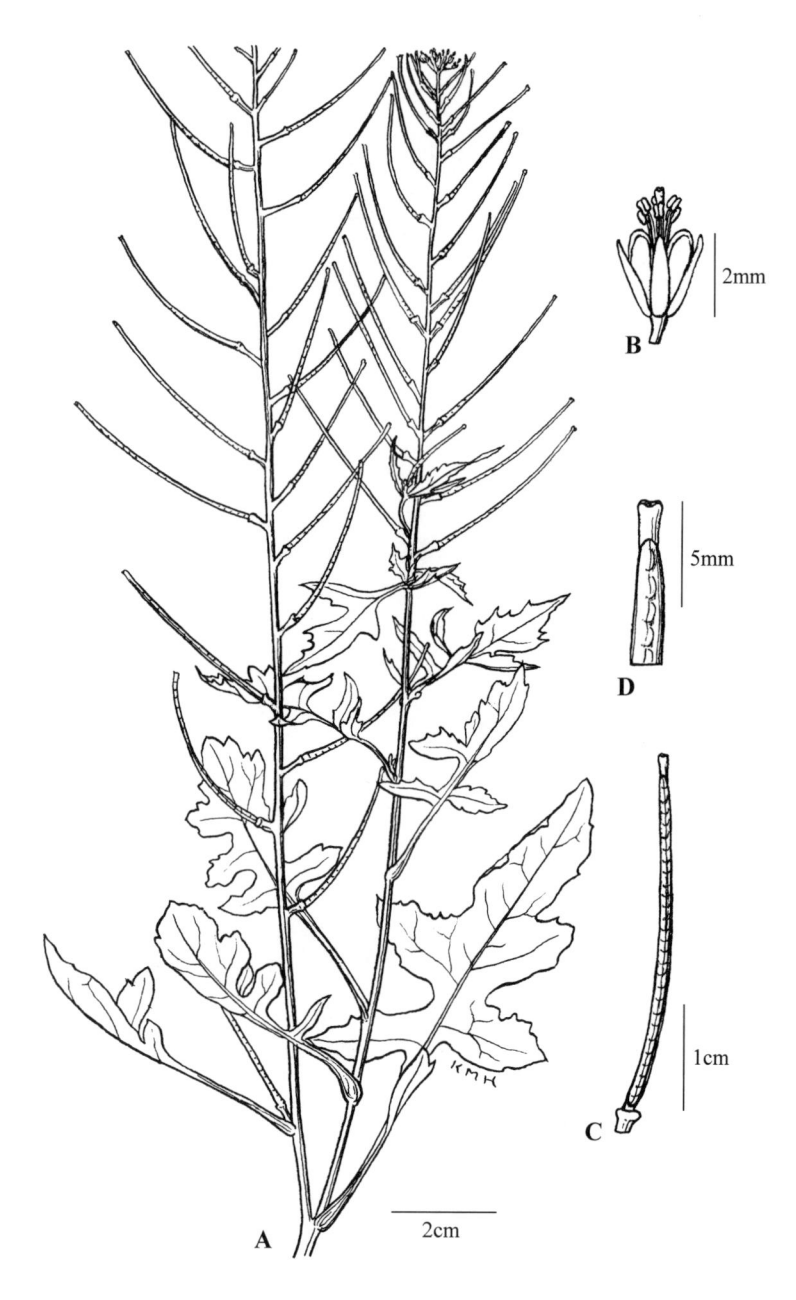

Sisymbrium erysimoides Desf. (French Rocket). A habit; B flower; C fruit; D apex of fruit. (Source unknown). STACE 305/250.

64. BRASSICACEAE

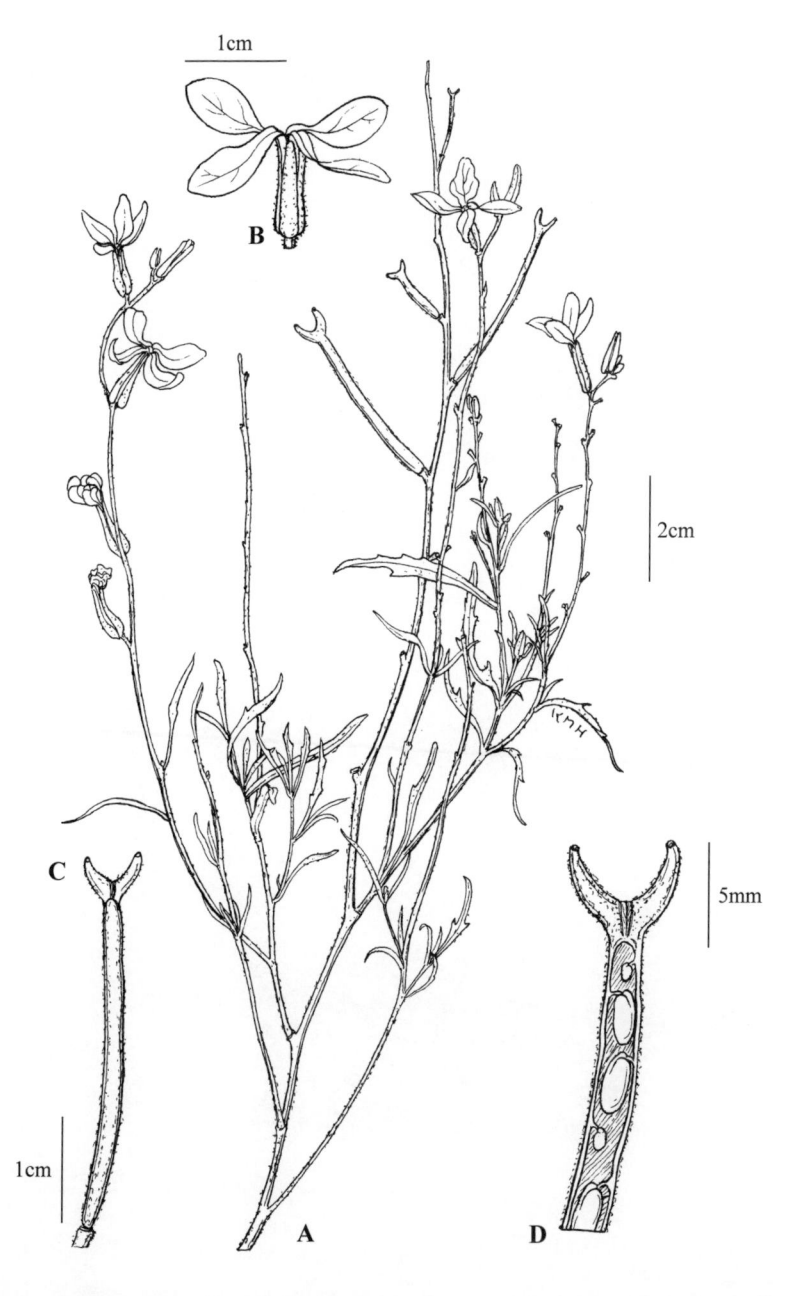

Matthiola longipetala (Vent.) DC. ssp. **bicornis** (Sibth. & Sm.) P.W. Ball (Night-scented Stock). A habit; B flower; C fruit; D LS of part of fruit. (Bucks). STACE 310/255.

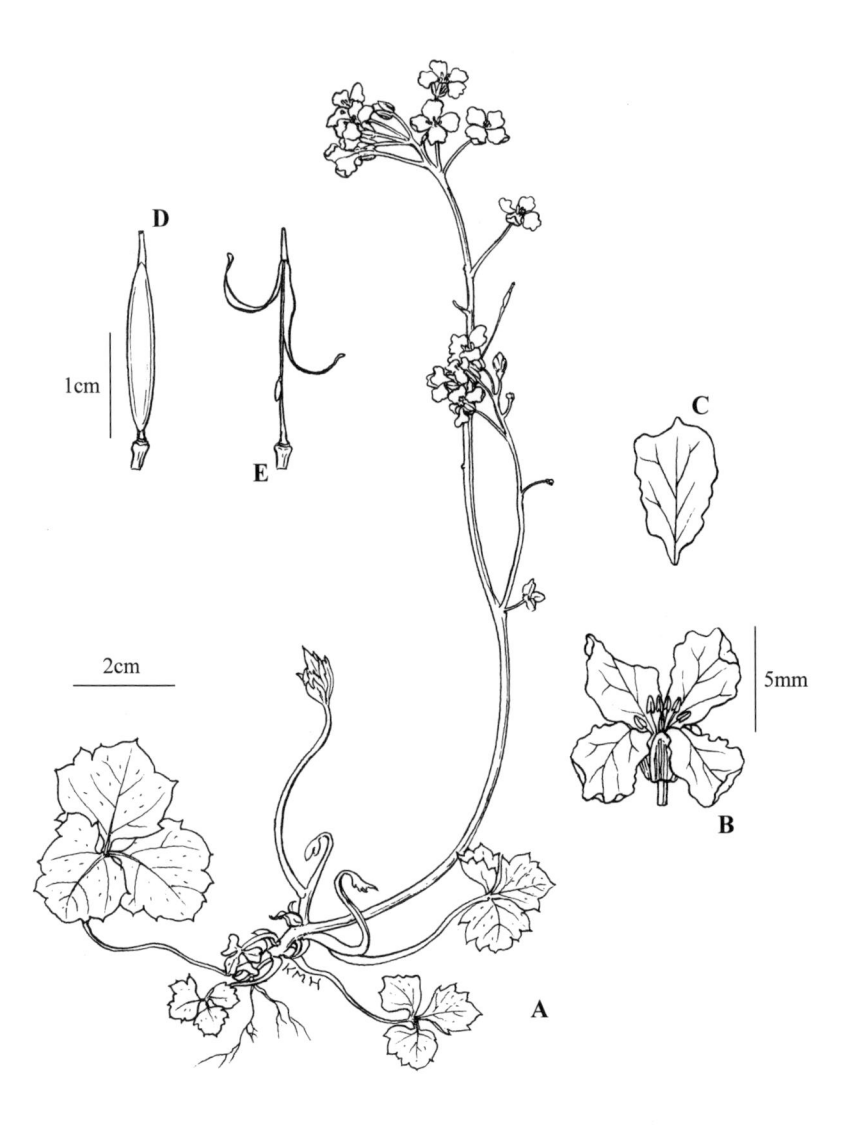

Cardamine trifolia L. (Trefoil Cress). A habit; B flower; C petal; D fruit; E fruit with valves open. (Trenstishoe, N Devon). STACE 315/258.

64. BRASSICACEAE

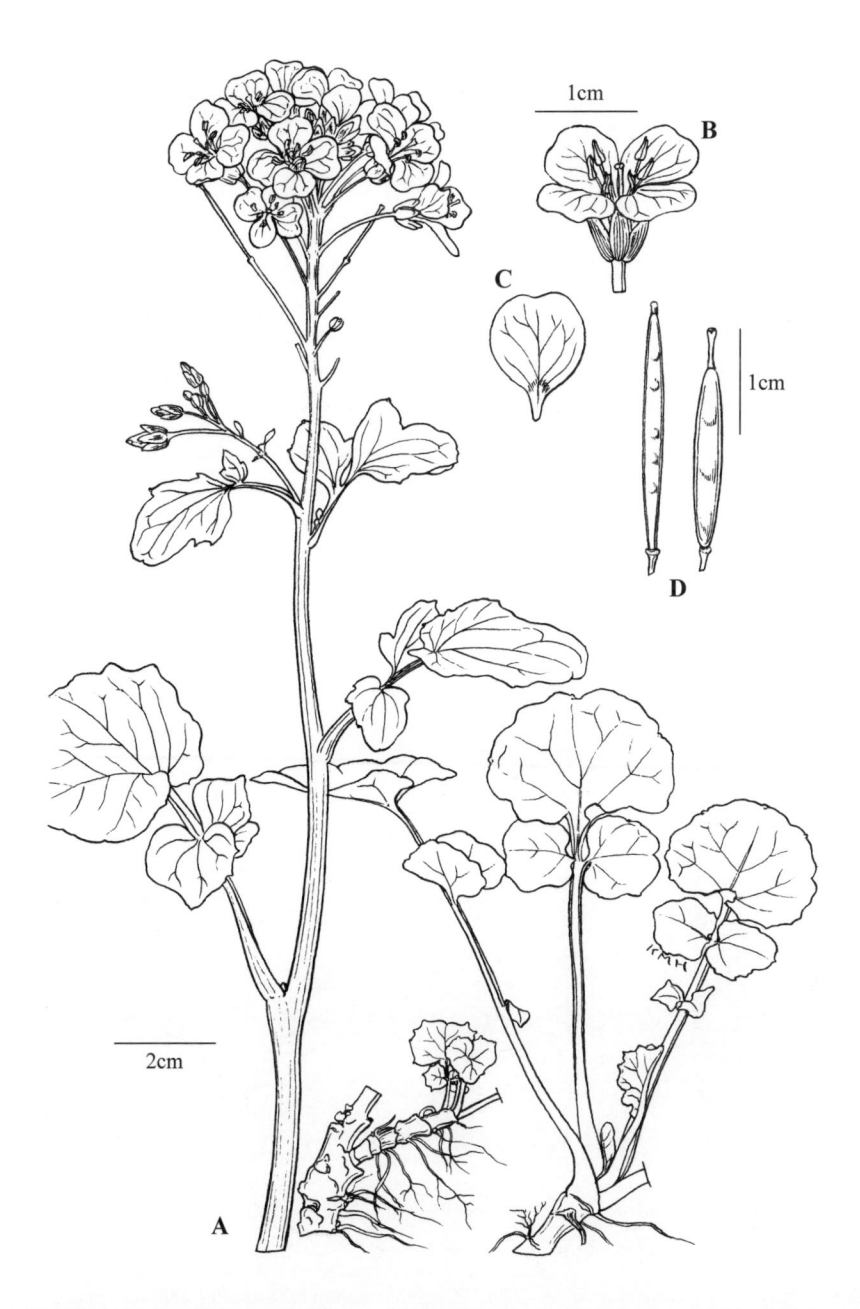

Cardamine raphanifolia Pourr. (Greater Cuckooflower). A flowering stem and basal leaves; B flower; C petal; D fruit. (Ambleside, Westmorland). STACE 315/259.

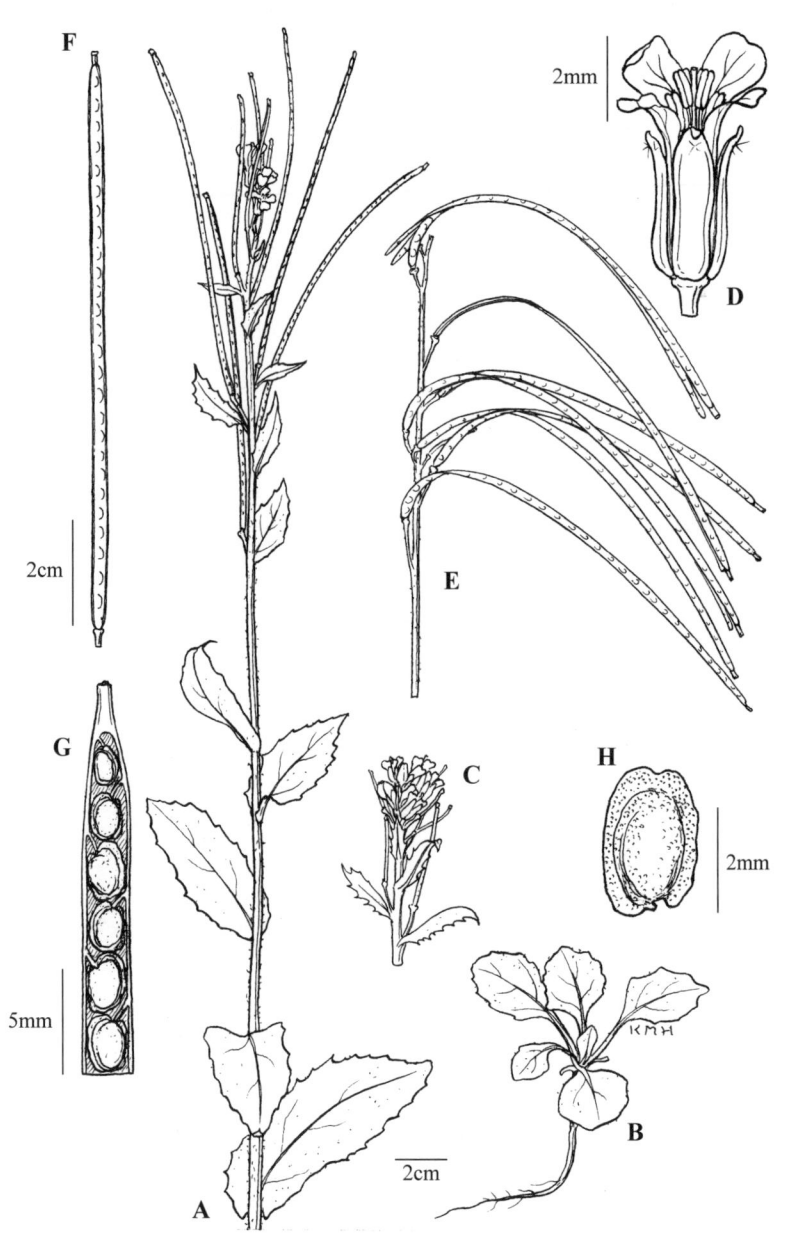

Arabis turrita L. (Tower Cress). A habit; B rosette; C top of flowering stem with young fruits; D flower; E cluster of fruits; F fruit; G LS of part of fruit; H seed. (Source unknown). STACE 317/260.

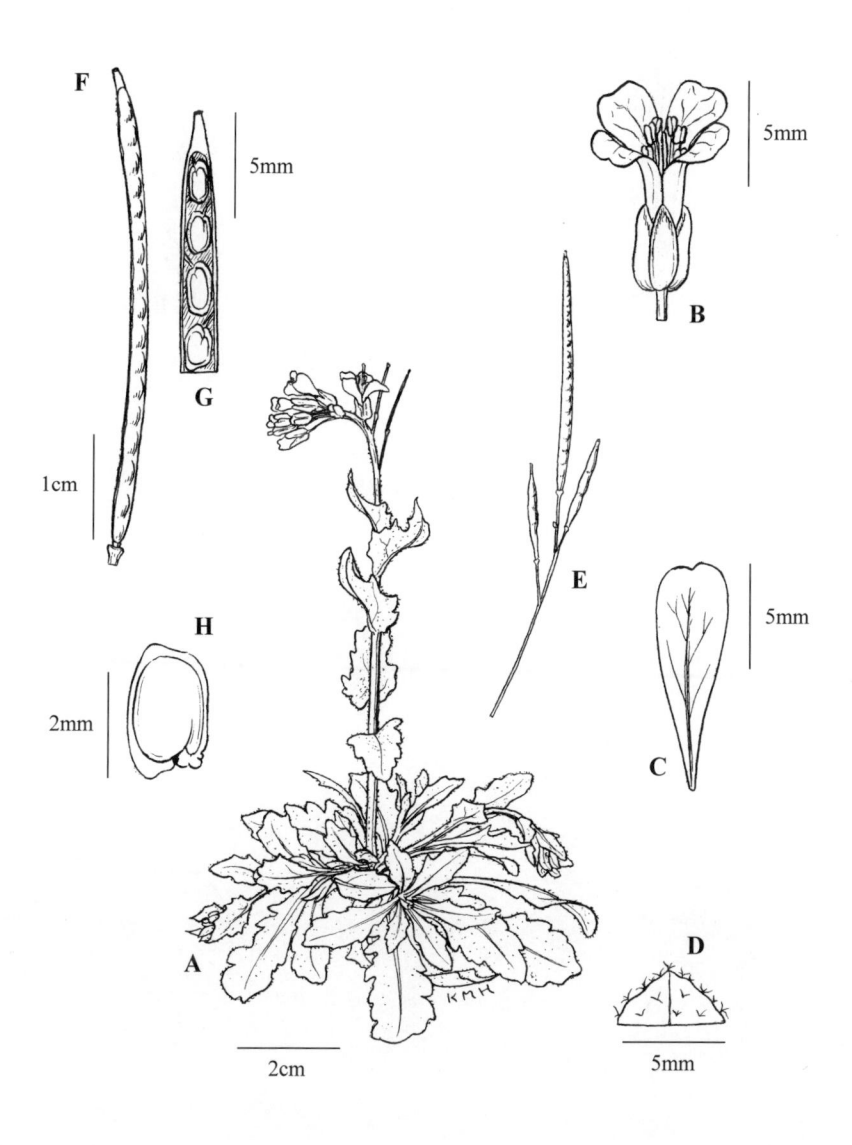

64. BRASSICACEAE

Arabis collina Ten. (Rosy Cress). A habit; B flower; C petal; D apex of leaf; E cluster of fruits; F fruit; G LS of part of fruit; H seed. (Source unknown). STACE 319/262.

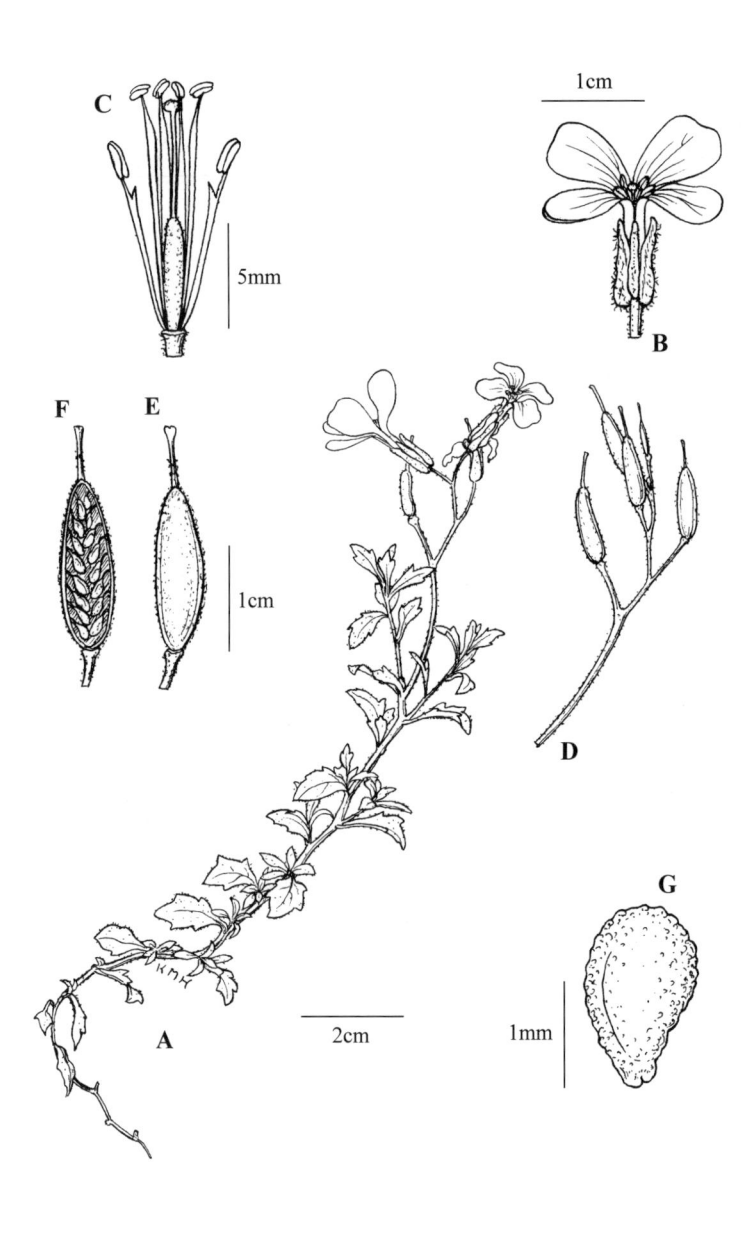

Aubrieta deltoidea (L.) DC. (Aubretia). A habit; B flower; C flower with petals and sepals removed; D cluster of fruits; E fruit; F LS of fruit; G seed. (Source unknown). STACE 319/262.

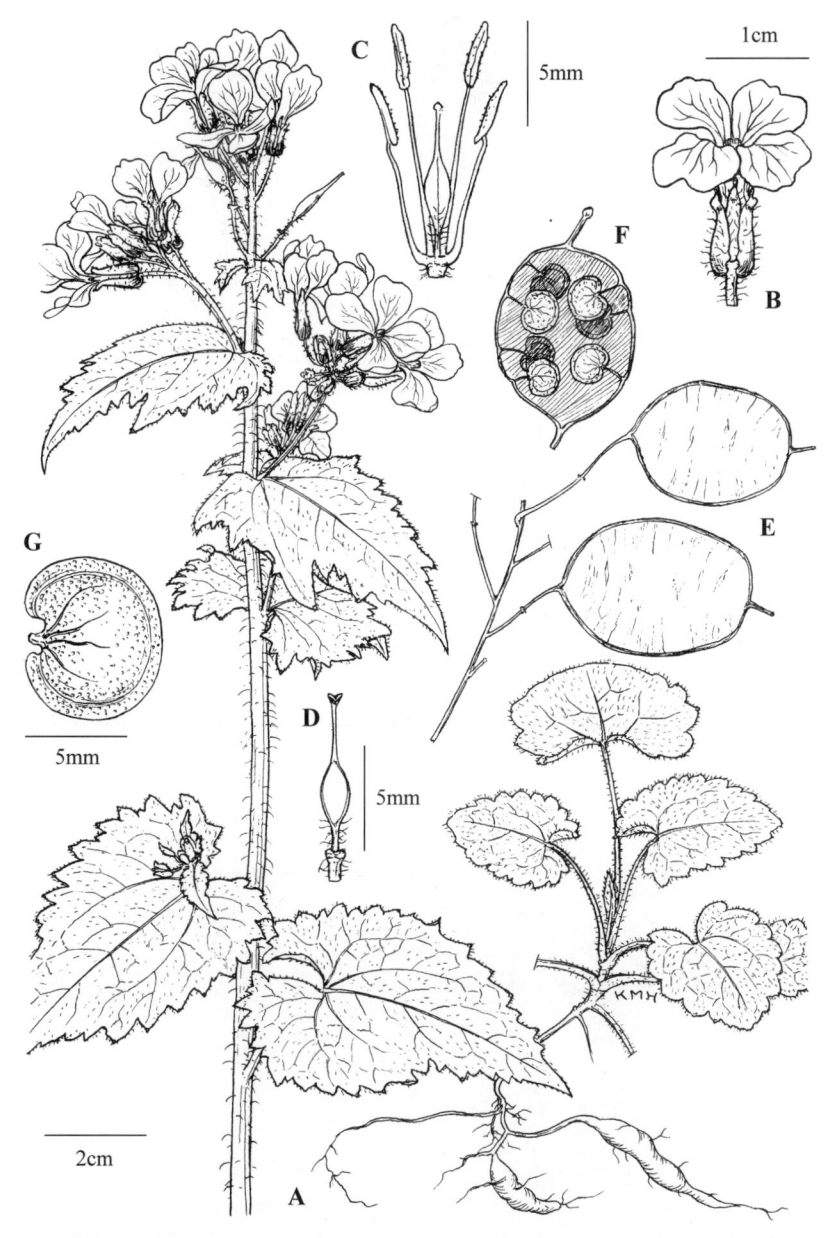

Lunaria annua L. (Honesty). A flowering stem and rosette; B flower; C flower minus petals and sepals; D gynoecium; E fruits; F LS of fruit showing seeds; G seed. (*Hort*. Ashbourne, Derbys). STACE 320/262.

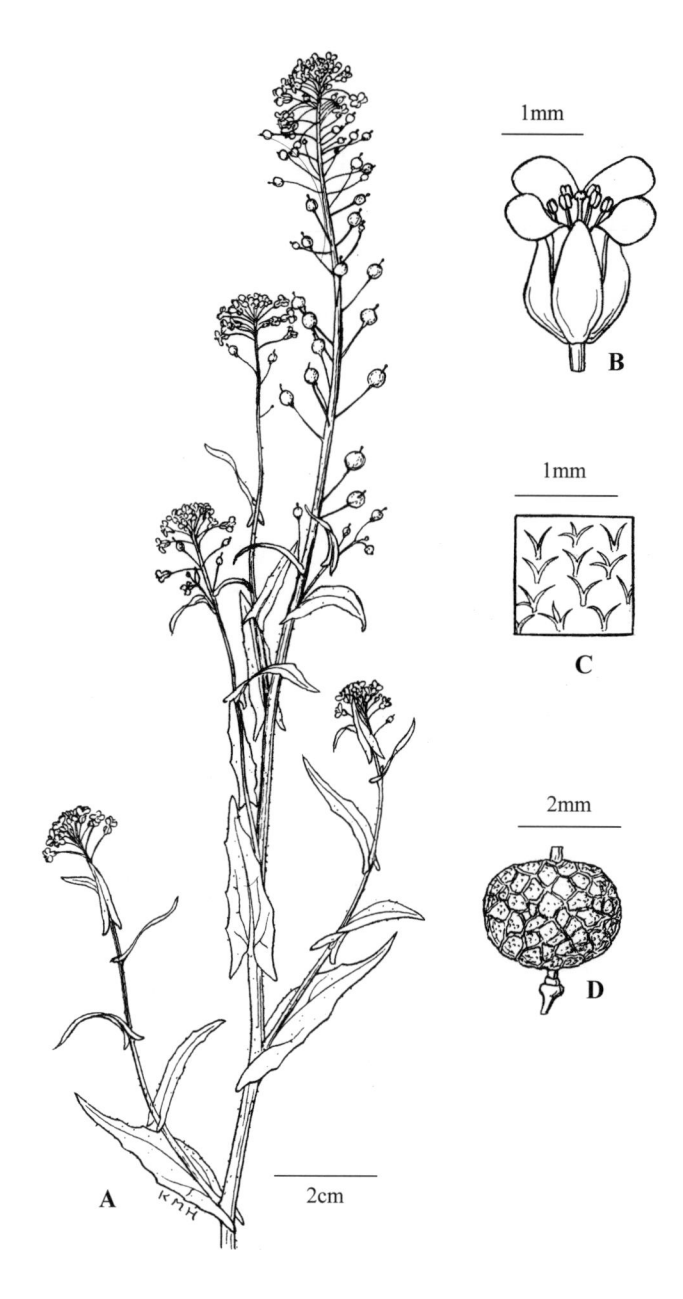

Neslia paniculata (L.) Desv. (Ball Mustard). A habit; B flower; C close-up of part of stem; D fruit. (Harrietsham, E Kent). STACE 325/267.

64. BRASSICACEAE

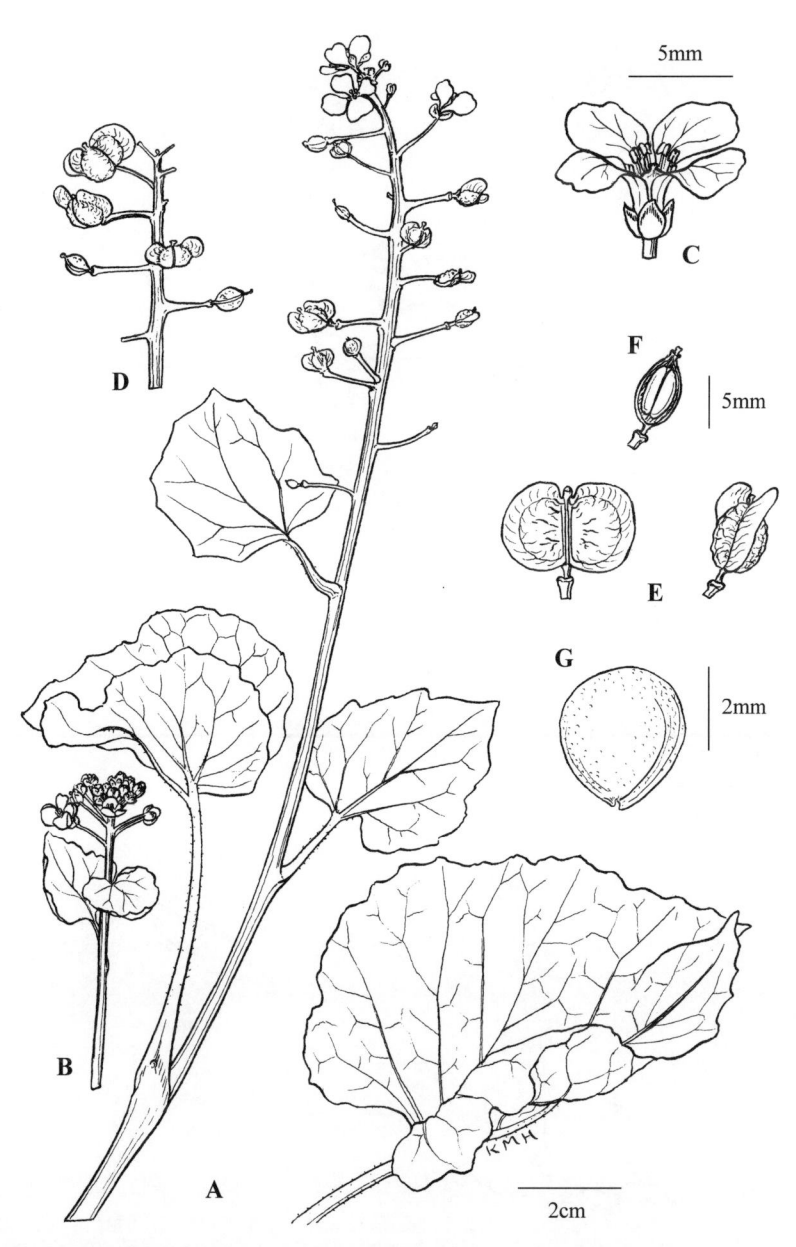

Thlaspi macrophyllum Hoffm. (Caucasian Penny-cress). A habit and basal leaf; B apex of flowering stem; C flower; D part of stem with fruits; E two views of fruit; F LS of fruit; G seed. (Bedford Square, Middlesex). STACE 329/270.

64. BRASSICACEAE

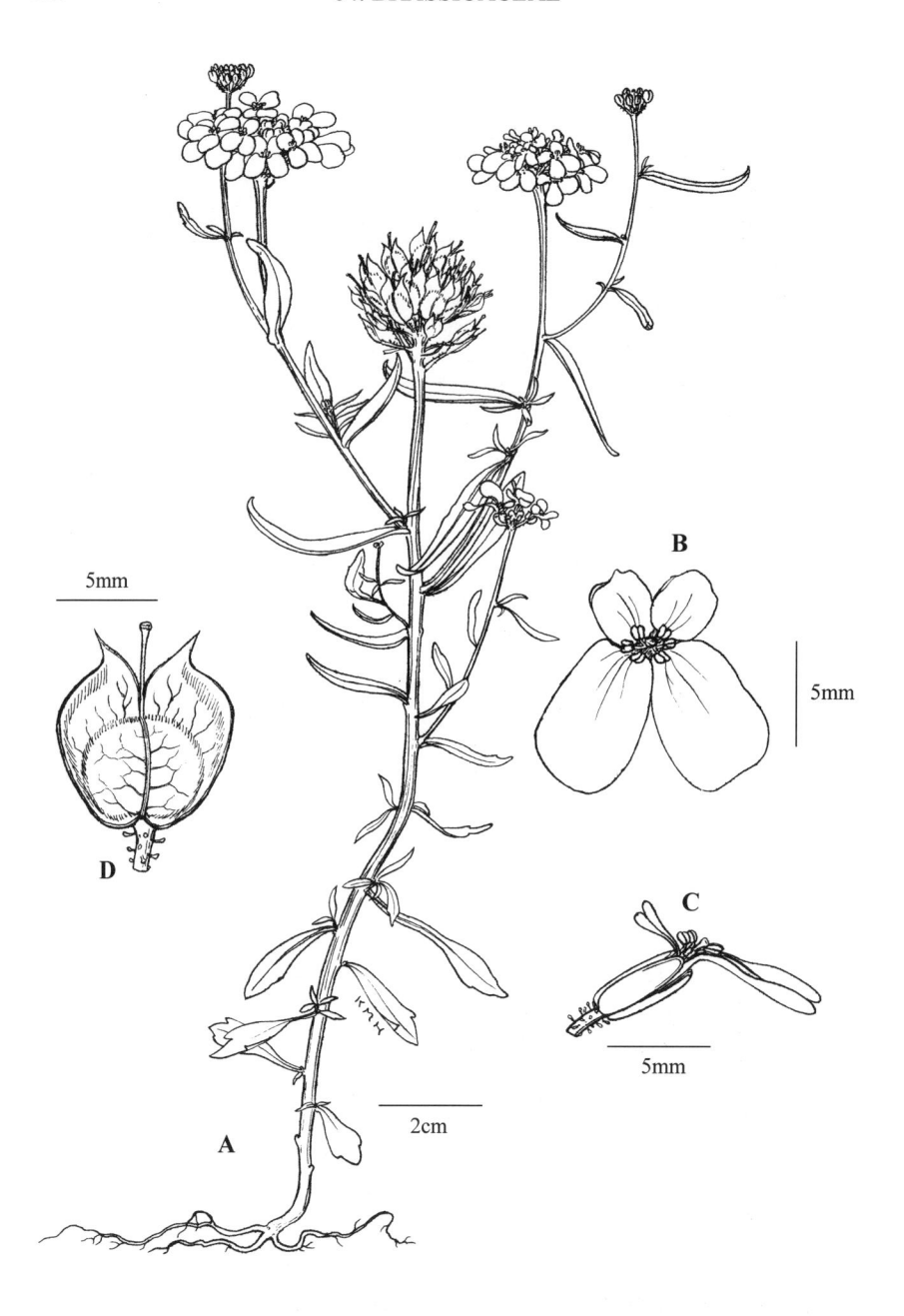

Iberis umbellata L. (Garden Candytuft). A habit; B flower; C lateral view of flower; D fruit. (*Hort*. Ashbourne, Derbys). STACE 329/271.

64. BRASSICACEAE

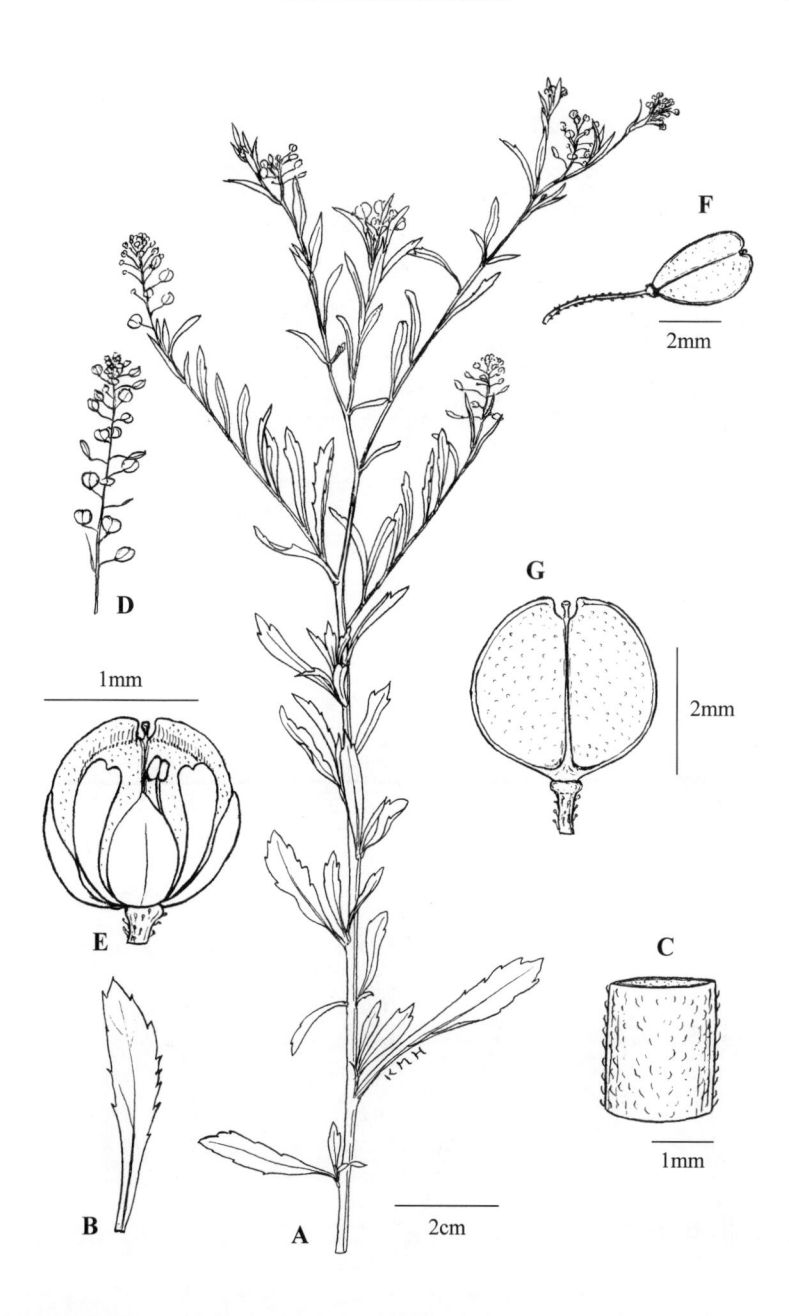

Lepidium virginicum L. (Least Pepperwort). A habit; B leaf; C part of stem; D apex of fruiting stem; E flower; F young fruit; G fruit. (Source unknown). STACE 331/272.

64. BRASSICACEAE

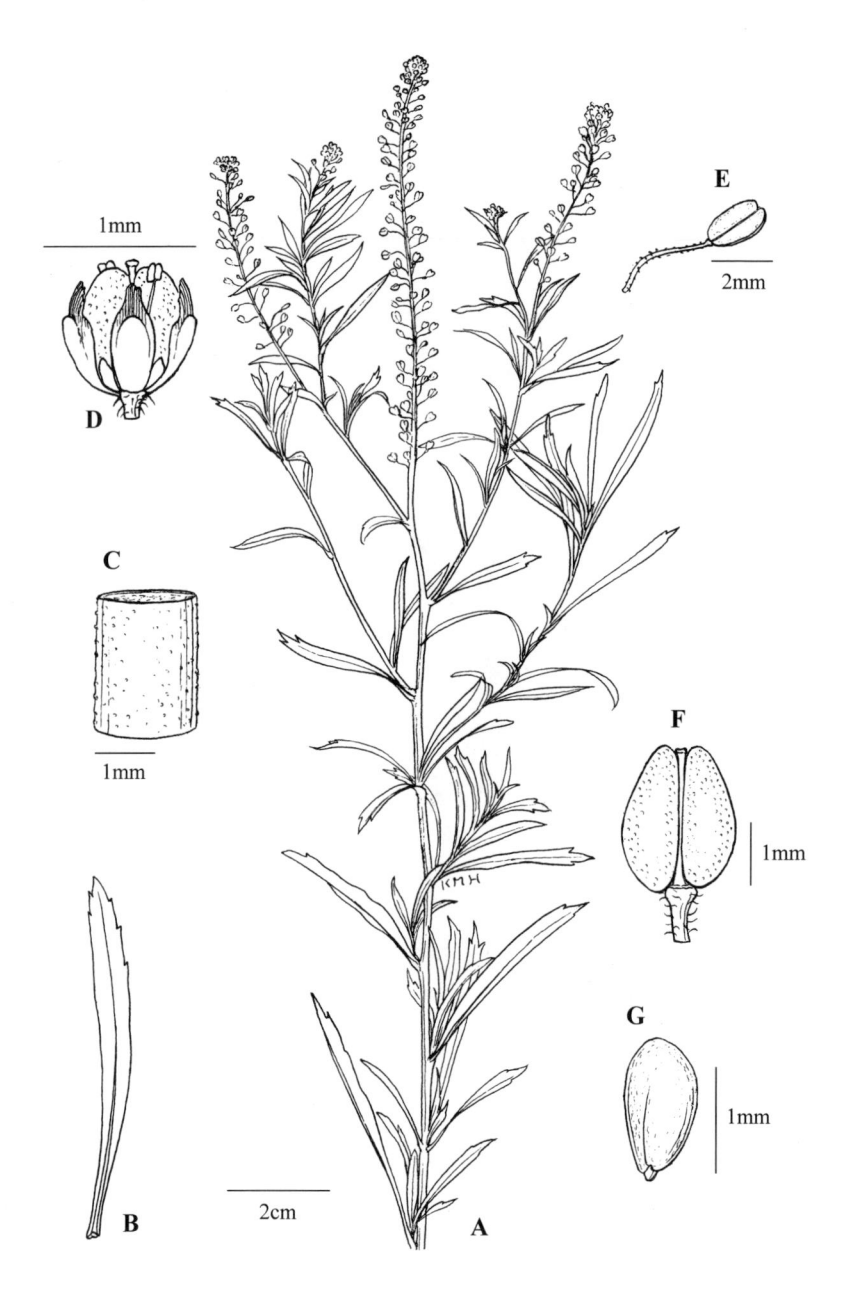

Lepidium hyssopifolium Desv. (African Pepperwort). A habit; B leaf; C part of stem; D flower; E young fruit; F fruit; G seed. (Source unknown). STACE 331/272.

64. BRASSICACEAE

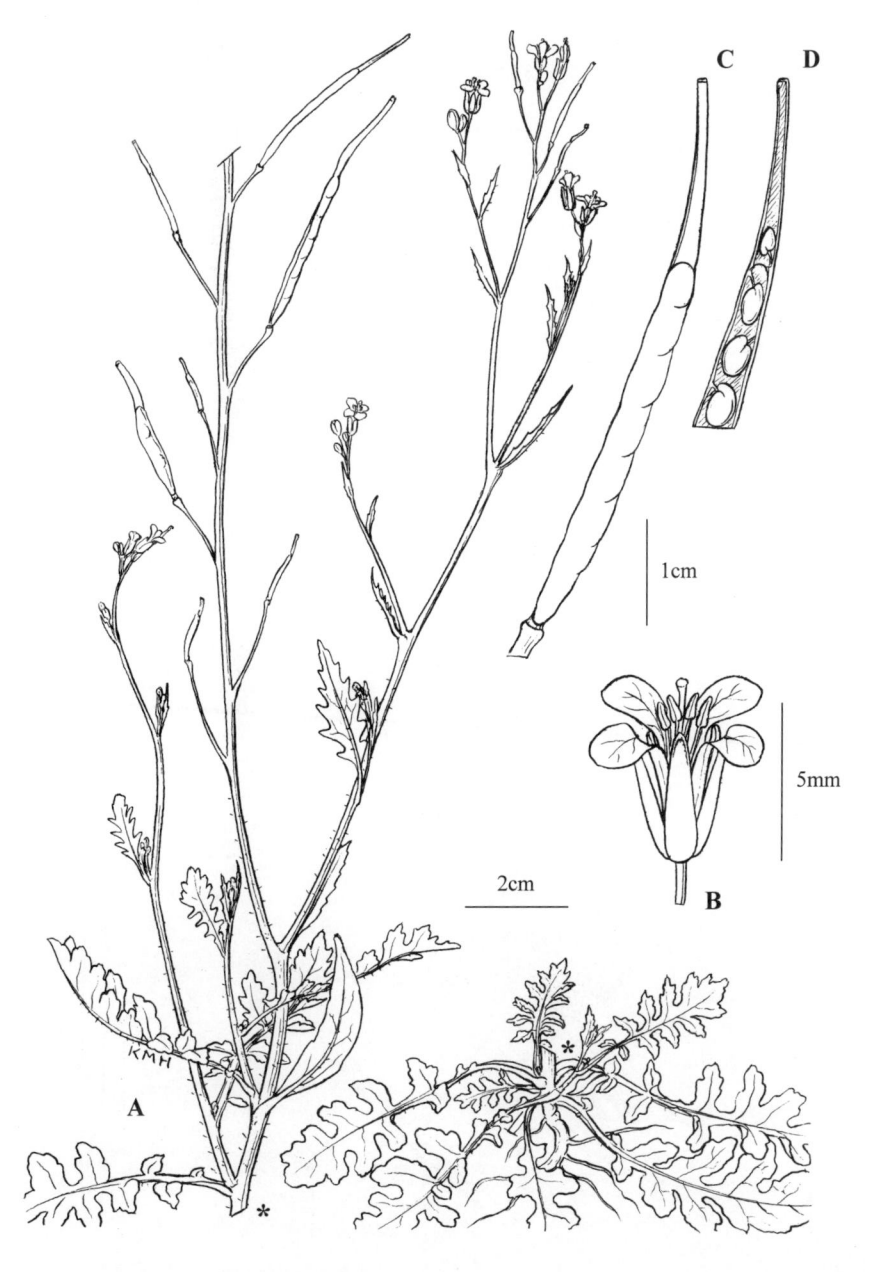

Brassica tournefortii Gouan (Pale Cabbage). A habit; B flower; C fruit; D LS of part of fruit. (Stretford Tip, Lancs). STACE 336/276.

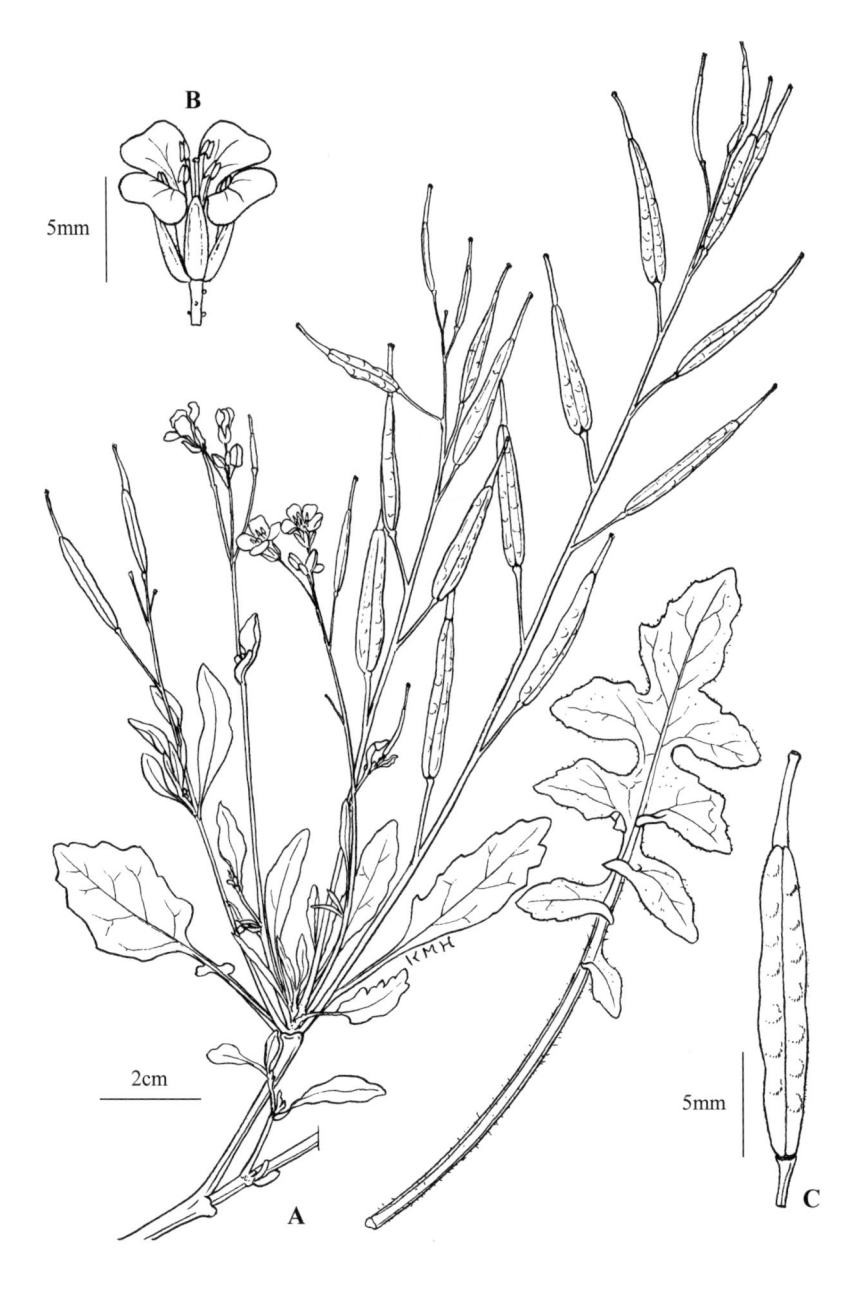

Brassica juncea (L.) Czern. (Chinese Mustard). A habit and basal leaf; B flower; C fruit. (Source unknown). STACE 336/276.

64. BRASSICACEAE 117

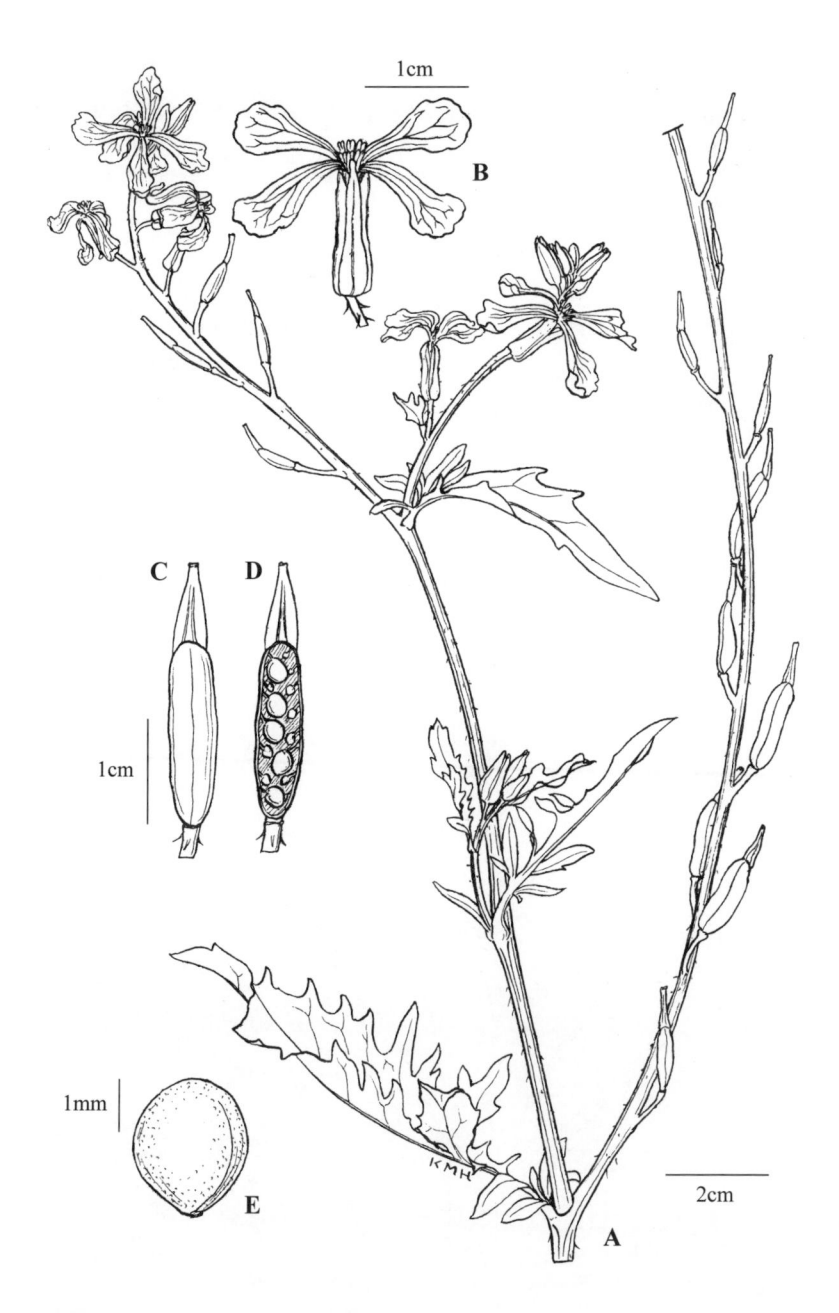

Eruca vesicaria (L.) Cav. ssp. **sativa** (Mill.) Thell. (Garden Rocket). A habit; B flower; C fruit; D LS of fruit; E seed. (York, MW Yorks). STACE 337/277.

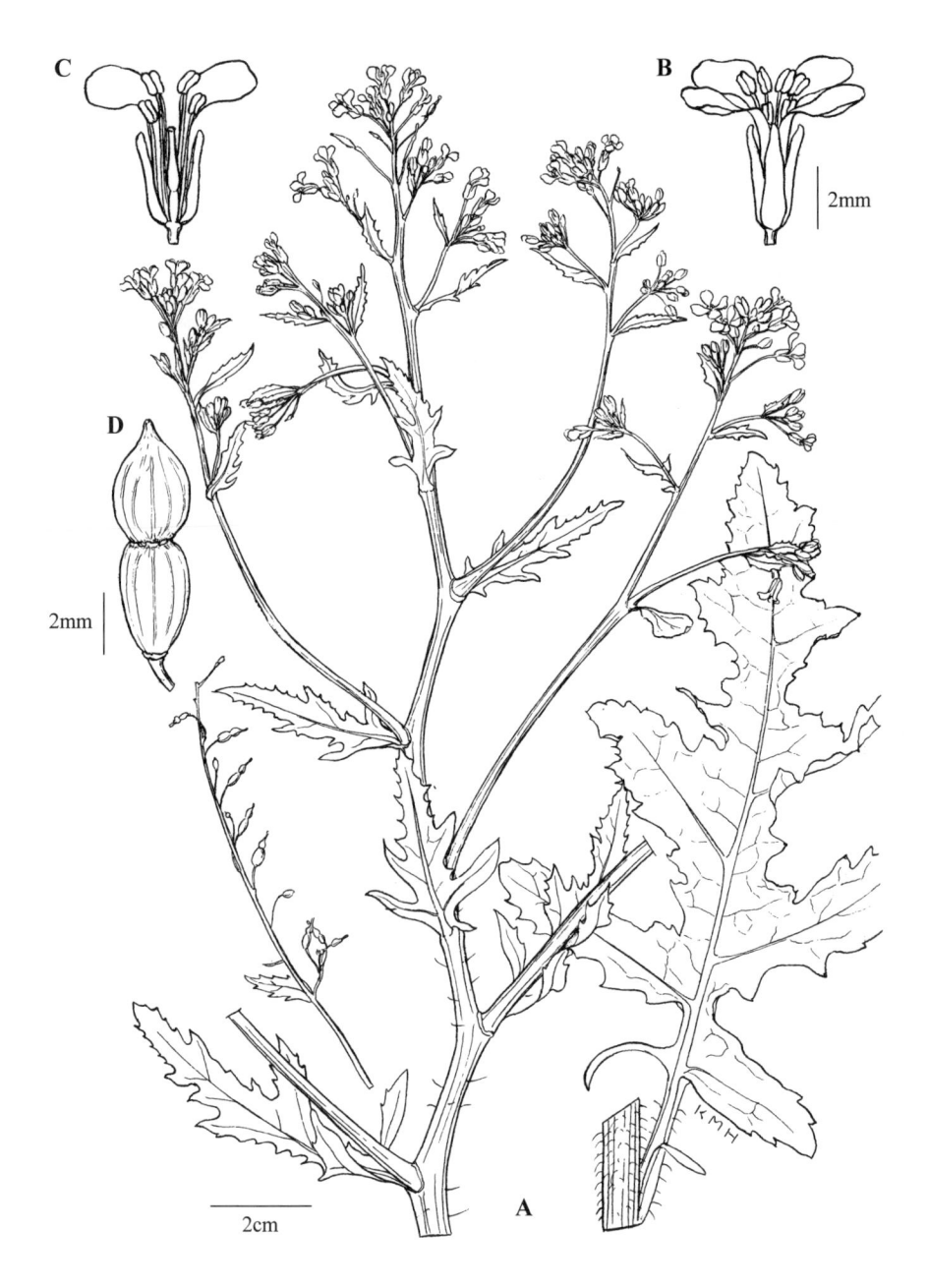

Rapistrum perenne (L.) All. (Steppe Cabbage). A habit and lower leaf; B flower; C flower with some sepals and petals removed; D fruit. (Breckland). STACE 341/280.

65. RESEDACEAE 119

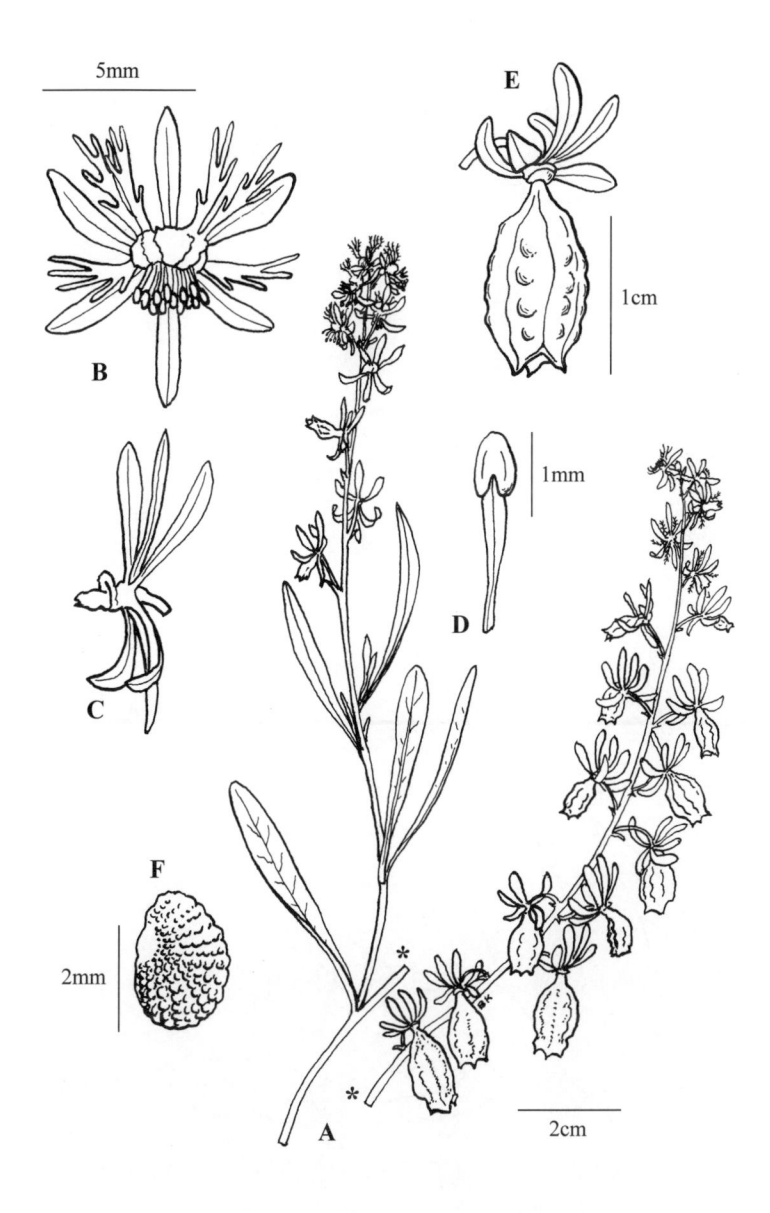

Reseda phyteuma L. (Corn Mignonette). A flowering and fruiting stems; B flower; C calyx and gynoecium; D stamen; E fruit; F seed. (Trottiscliffe, W Kent). STACE 344/282.

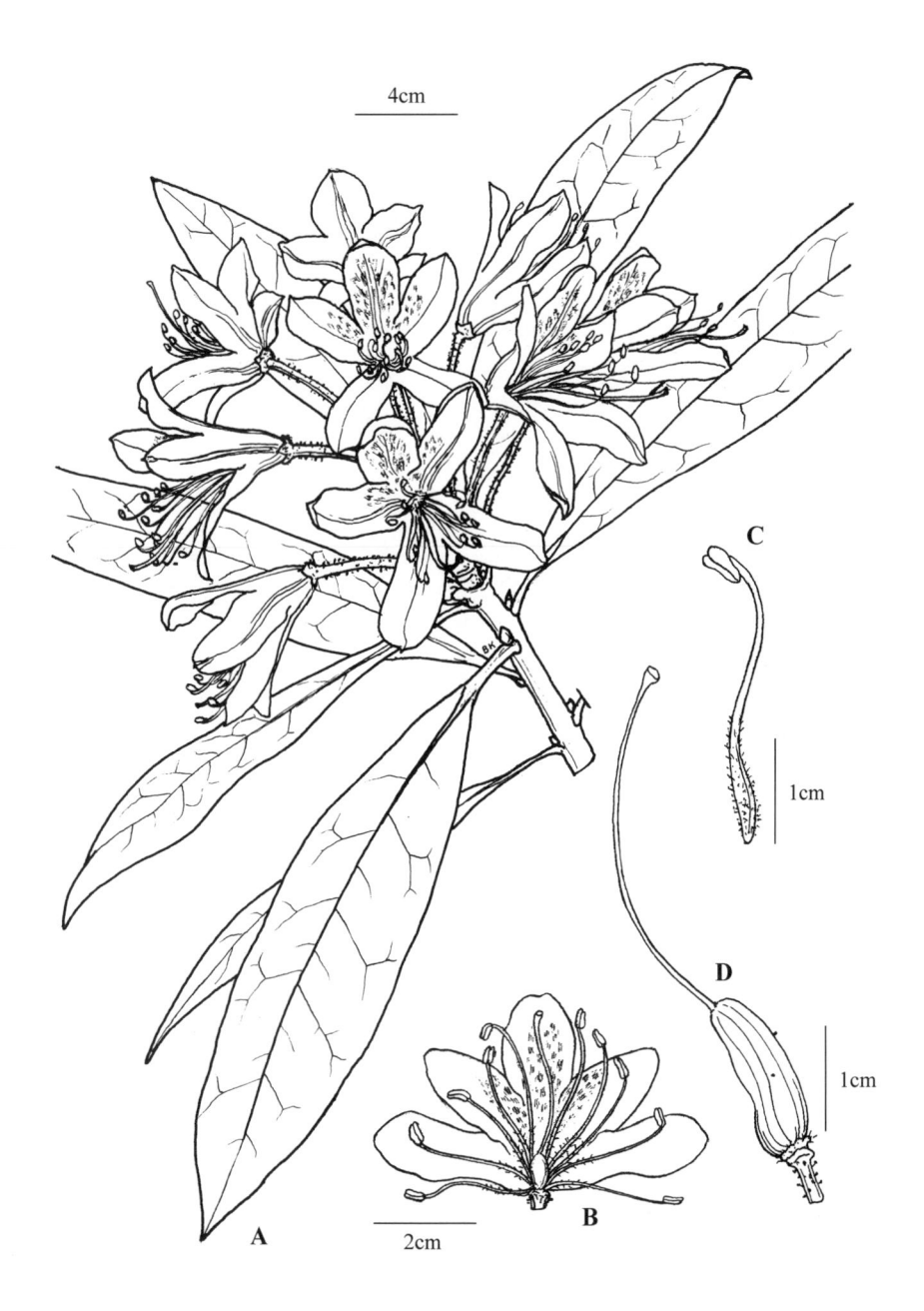

Rhododendron ponticum L. (Rhododendron). A flowering branch; B flower opened out; C stamen; D fruit. (Source unknown). STACE 347/285.

67. ERICACEAE

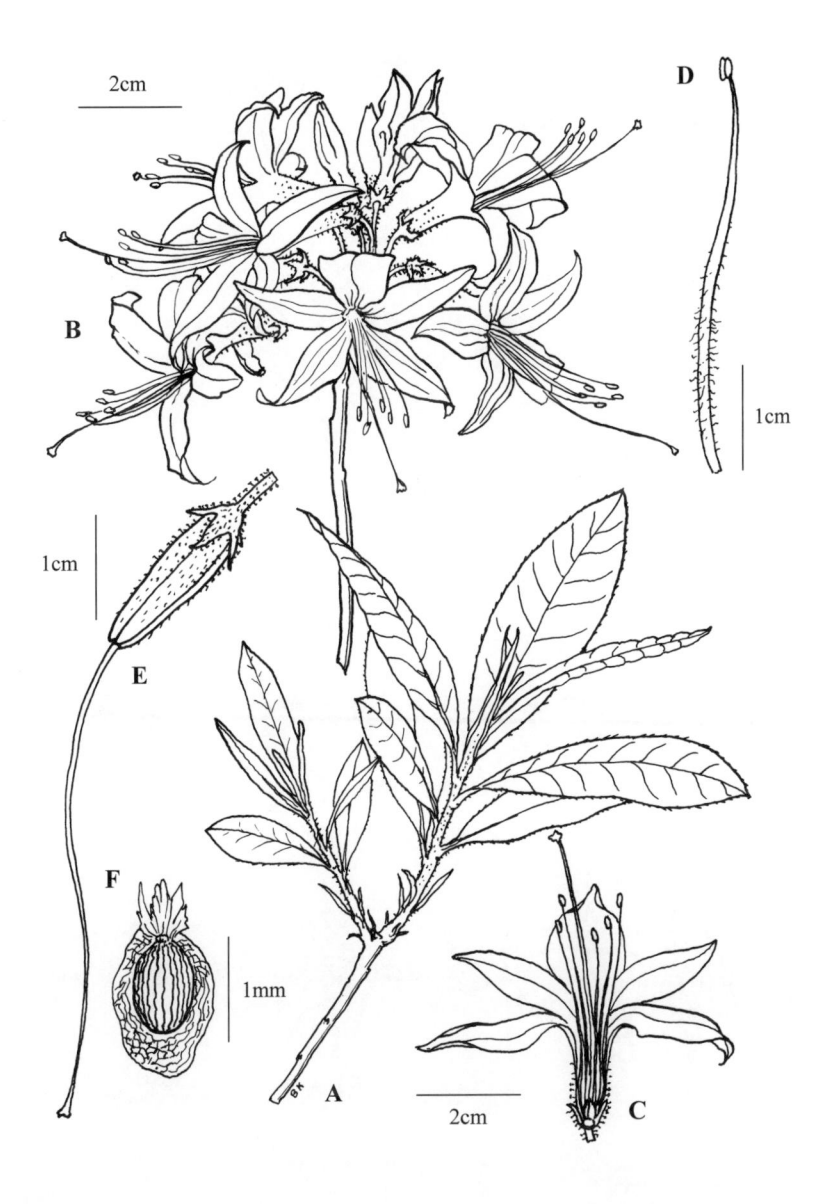

Rhododendron luteum Sweet (Yellow Azalea). A leafy branch in summer; B leafless flowering branch; C flower with corolla opened out; D stamen; E fruit; F seed. (Source unknown). STACE 347/285.

122   67. ERICACEAE

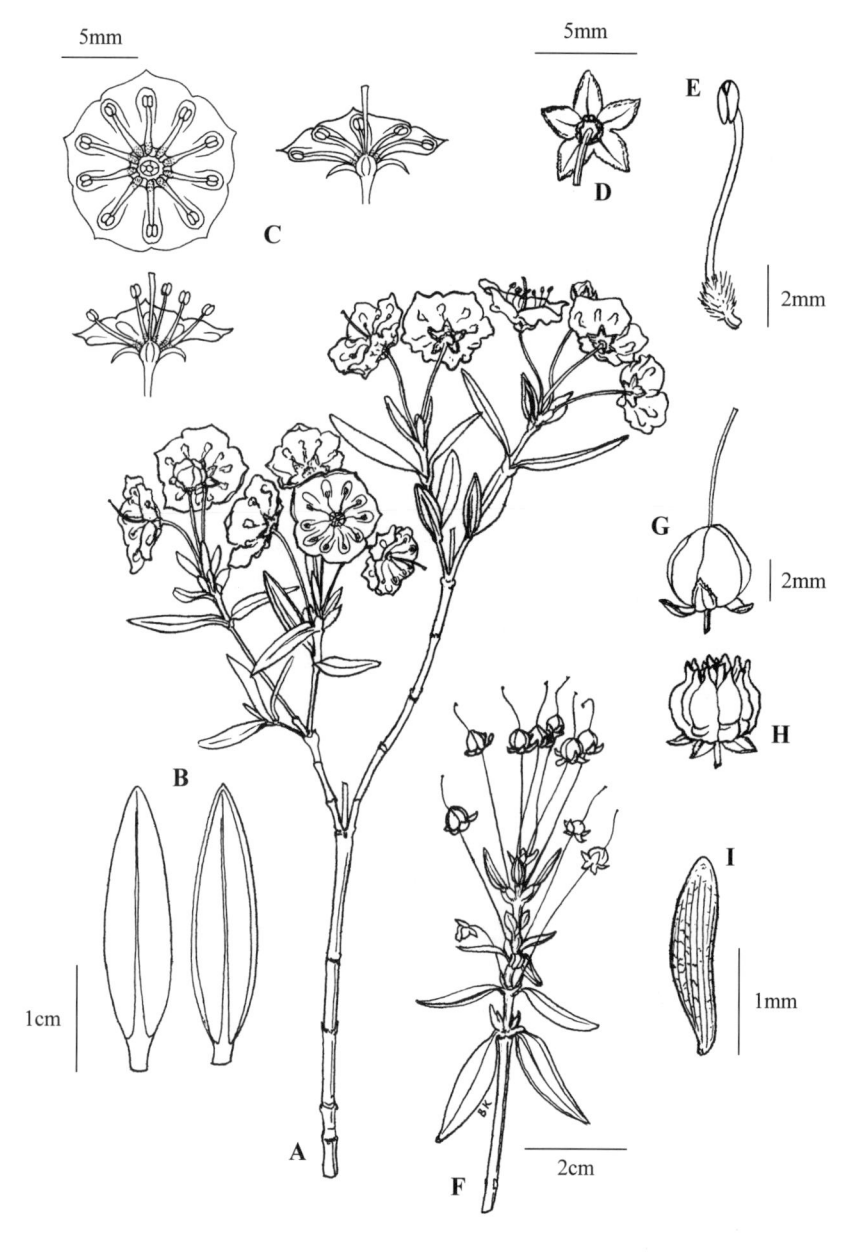

Kalmia polifolia Wangenh. (Bog-laurel). A flowering branch; B leaf; C flower; D calyx; E stamen; F fruiting branch; G gynoecium; H fruit; I seed. (Chobham Common, Surrey). STACE 347/286.

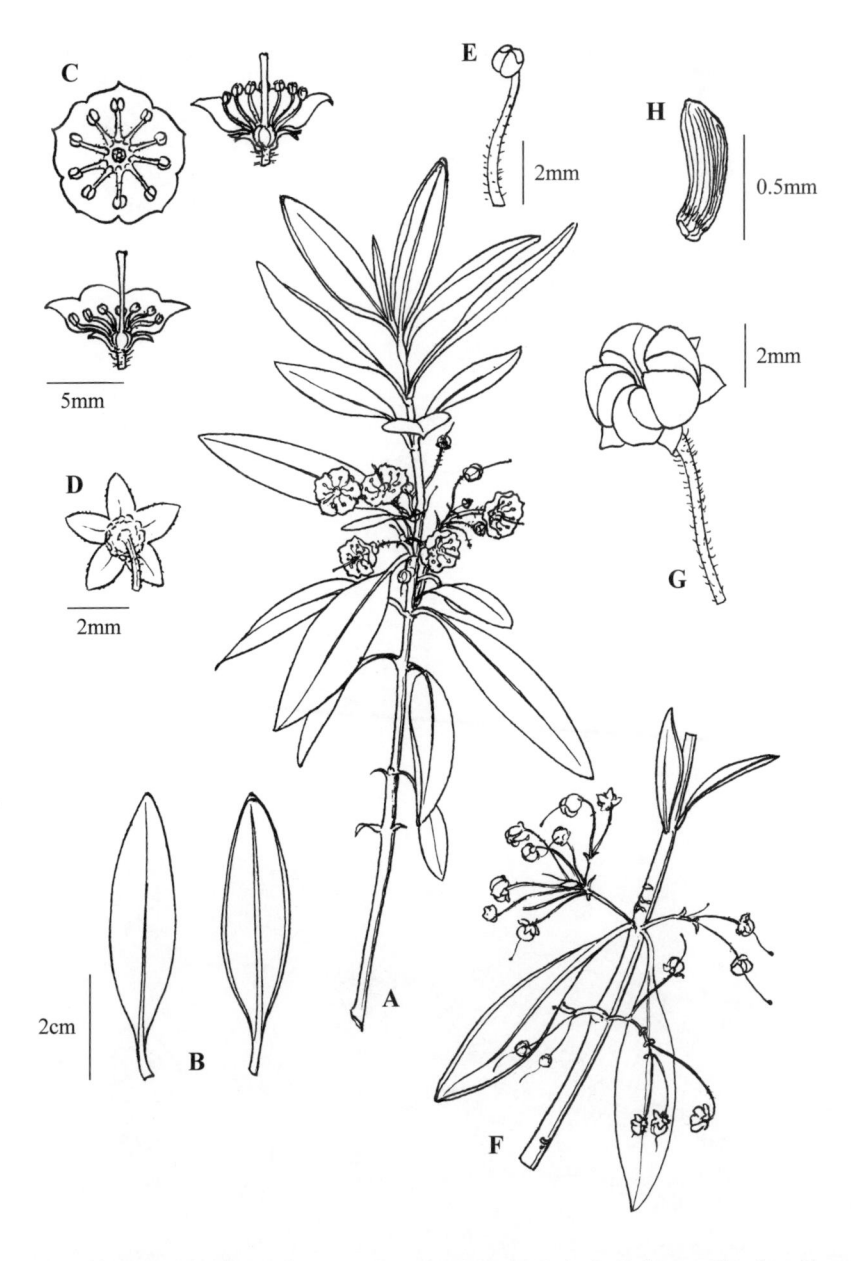

67. ERICACEAE

Kalmia angustifolia L. (Sheep-laurel). A flowering branch; B leaf; C flower; D calyx; E stamen; F part of fruiting branch; G fruit; H seed. (Source unknown). STACE 349/286.

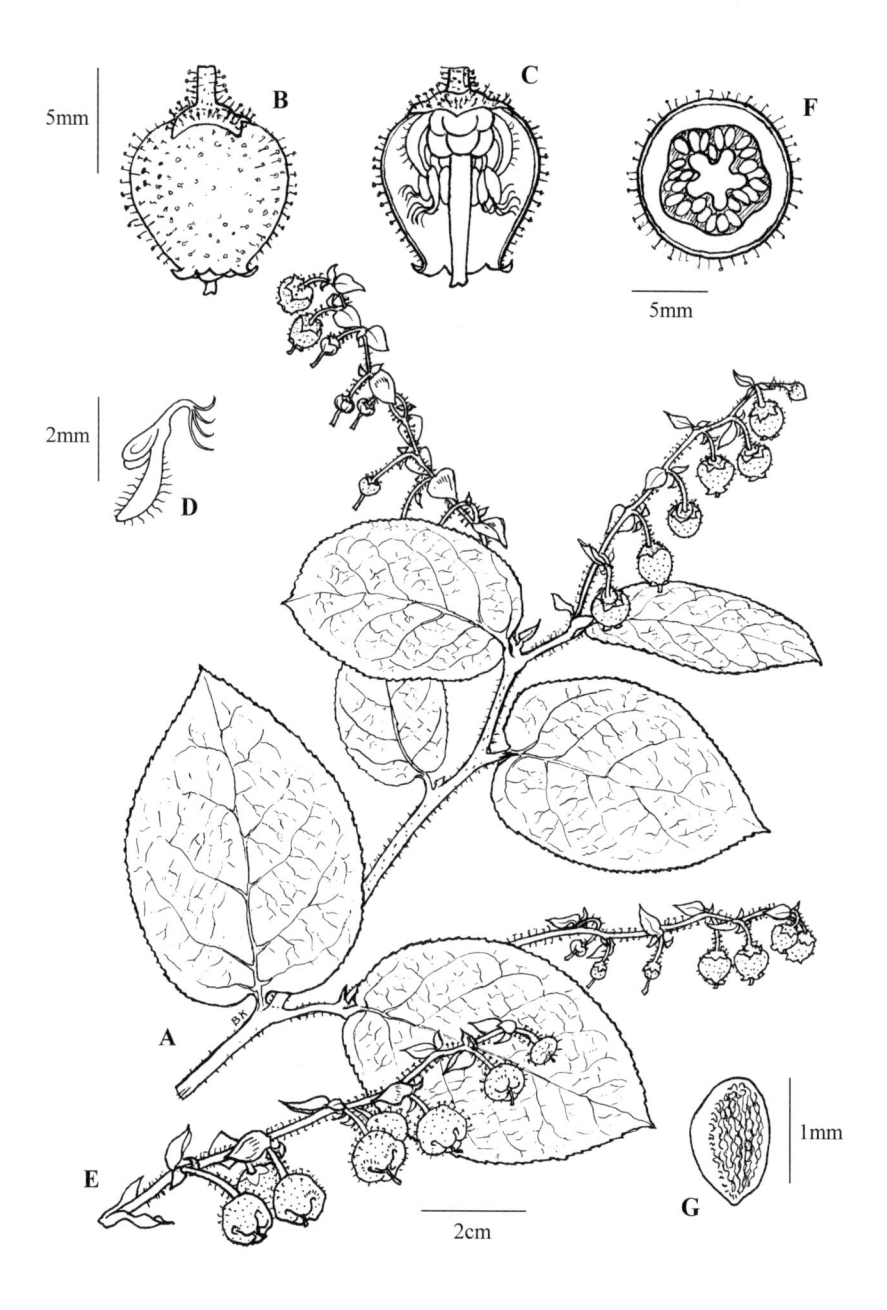

Gaultheria shallon Pursh (Shallon). A flowering branch; B flower; C flower with part of corolla cut away; D stamen; E fruiting branch; F TS of fruit; G seed. (Source unknown). STACE 350/287.

67. ERICACEAE

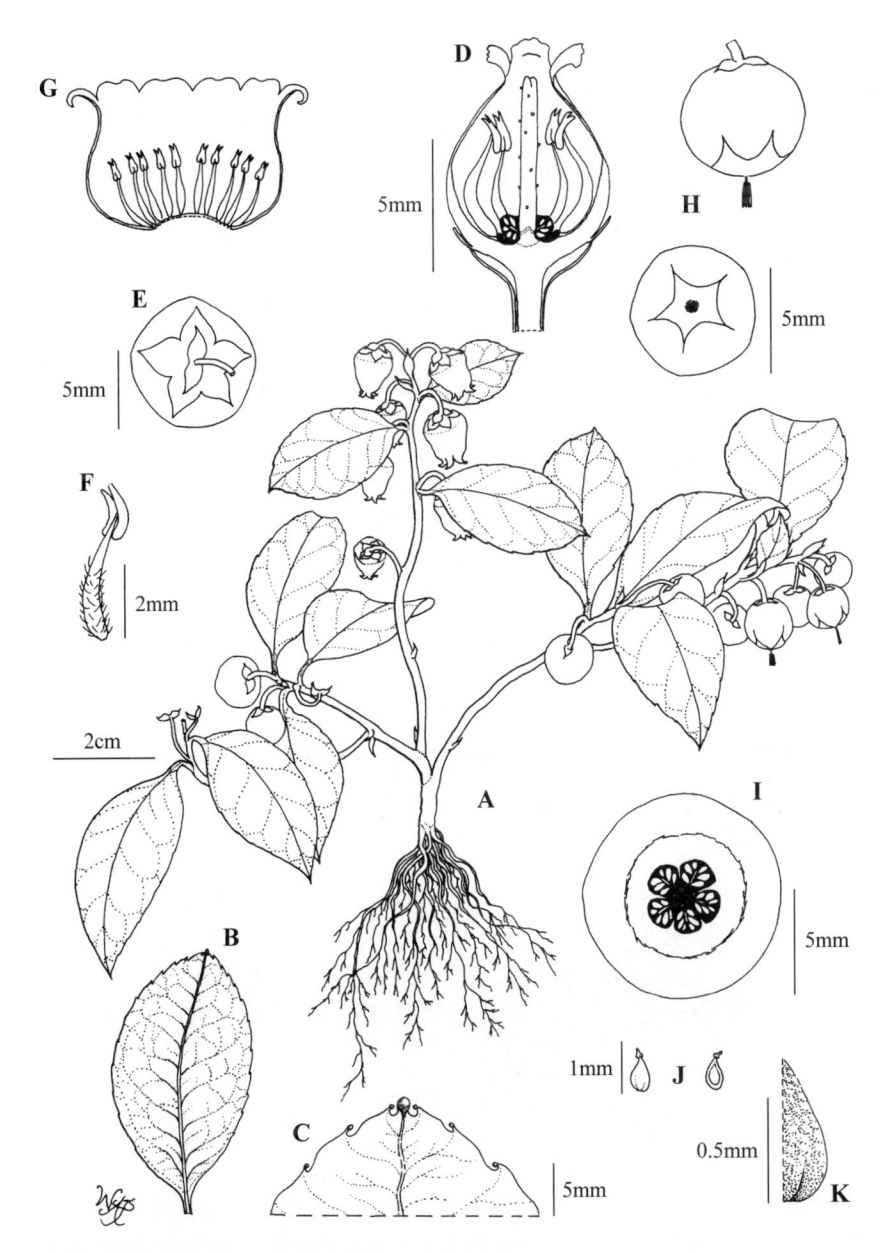

Gaultheria procumbens L. (Checkerberry). A habit; B leaf; C leaf tip; D LS of flower; E flower showing 2 bracts and 5 sepals; F stamen; G corolla opened out; H side and top views of swollen, fleshy calyx; I TS of fruit; J seed and LS of seed; K surface of seed. (*Hort.* Horsham, W Sussex). STACE 350/287.

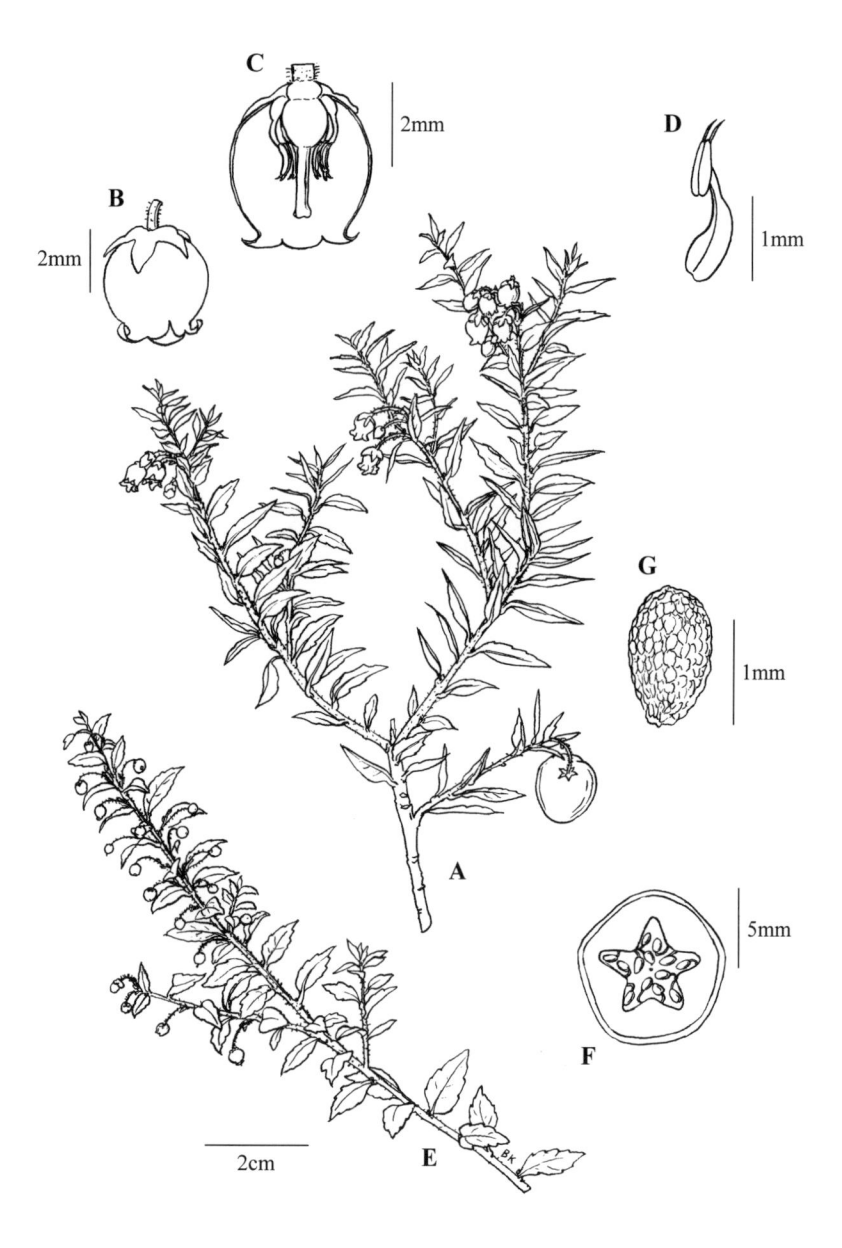

Gaultheria mucronata (L. f.) Hook. & Arn. (Prickly Heath). A flowering branch with one persisting berry; B flower; C flower with part of corolla cut away; D stamen; E branch with young fruit; F TS of fruit; G seed. (Hengistbury Head, S Hants). STACE 350/287.

67. ERICACEAE

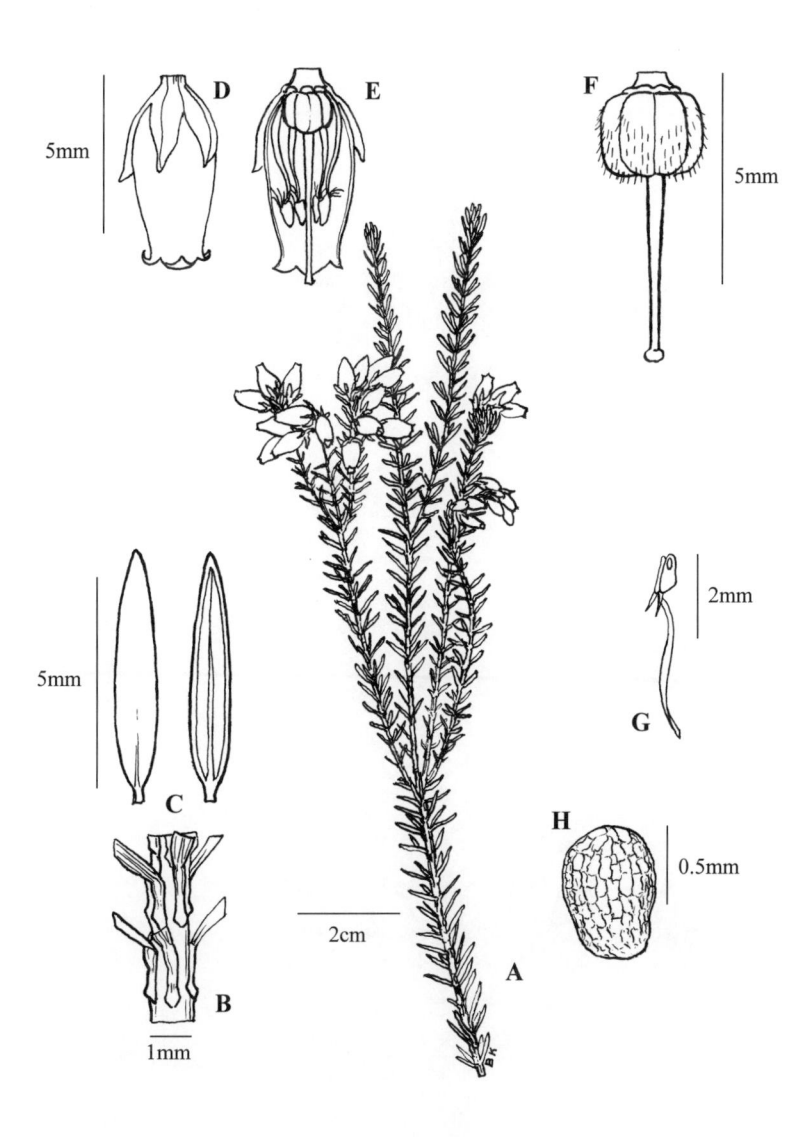

Erica terminalis Salisb. (Corsican Heath). A flowering branch; B stem and leaf bases; C two views of leaf; D flower; E LS of flower; F gynoecium; G stamen; H seed. (Magilligan, Londonderry). STACE 353/289.

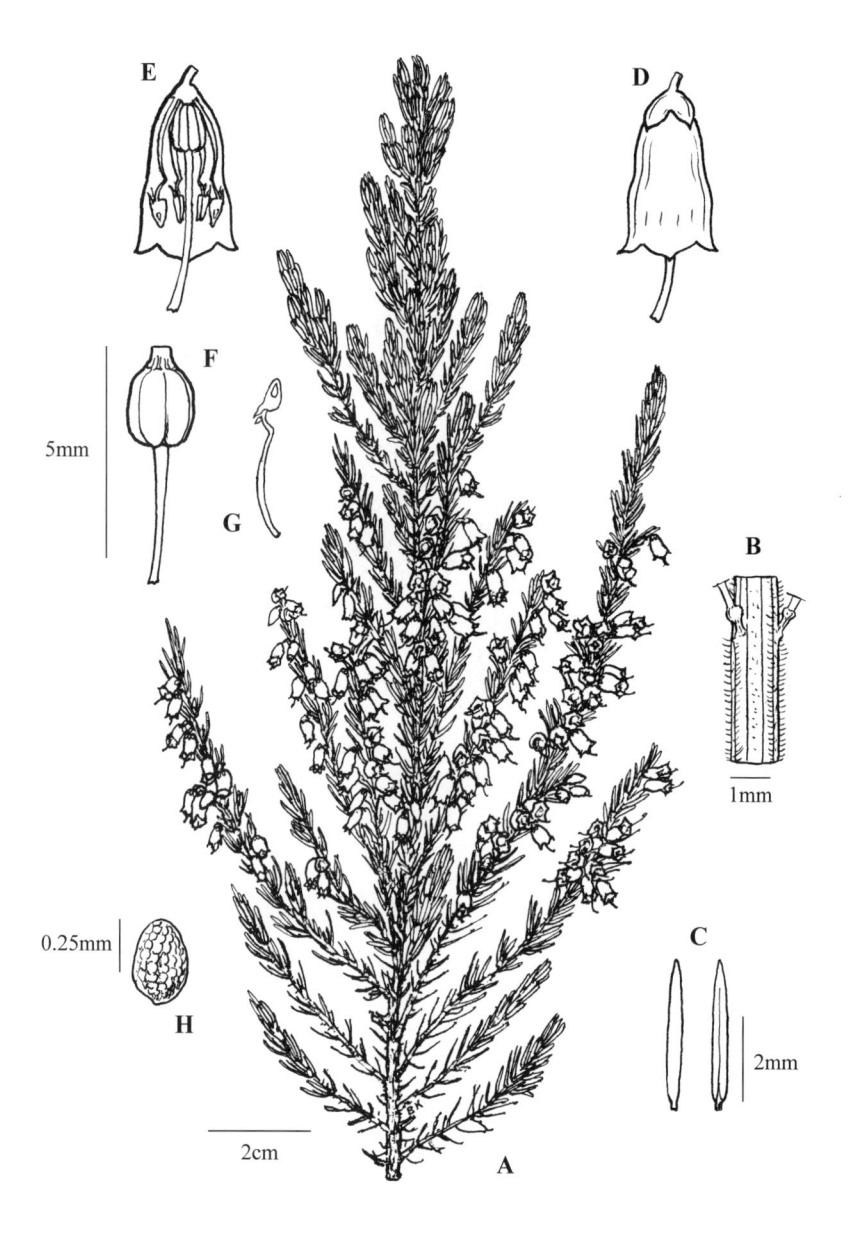

67. ERICACEAE

Erica lusitanica Rudolphi (Portuguese Heath). A flowering branch; B stem and leaf bases; C two views of leaf; D flower; E LS of flower; F gynoecium; G stamen; H seed. (Source unknown). STACE 353/291.

67. ERICACEAE

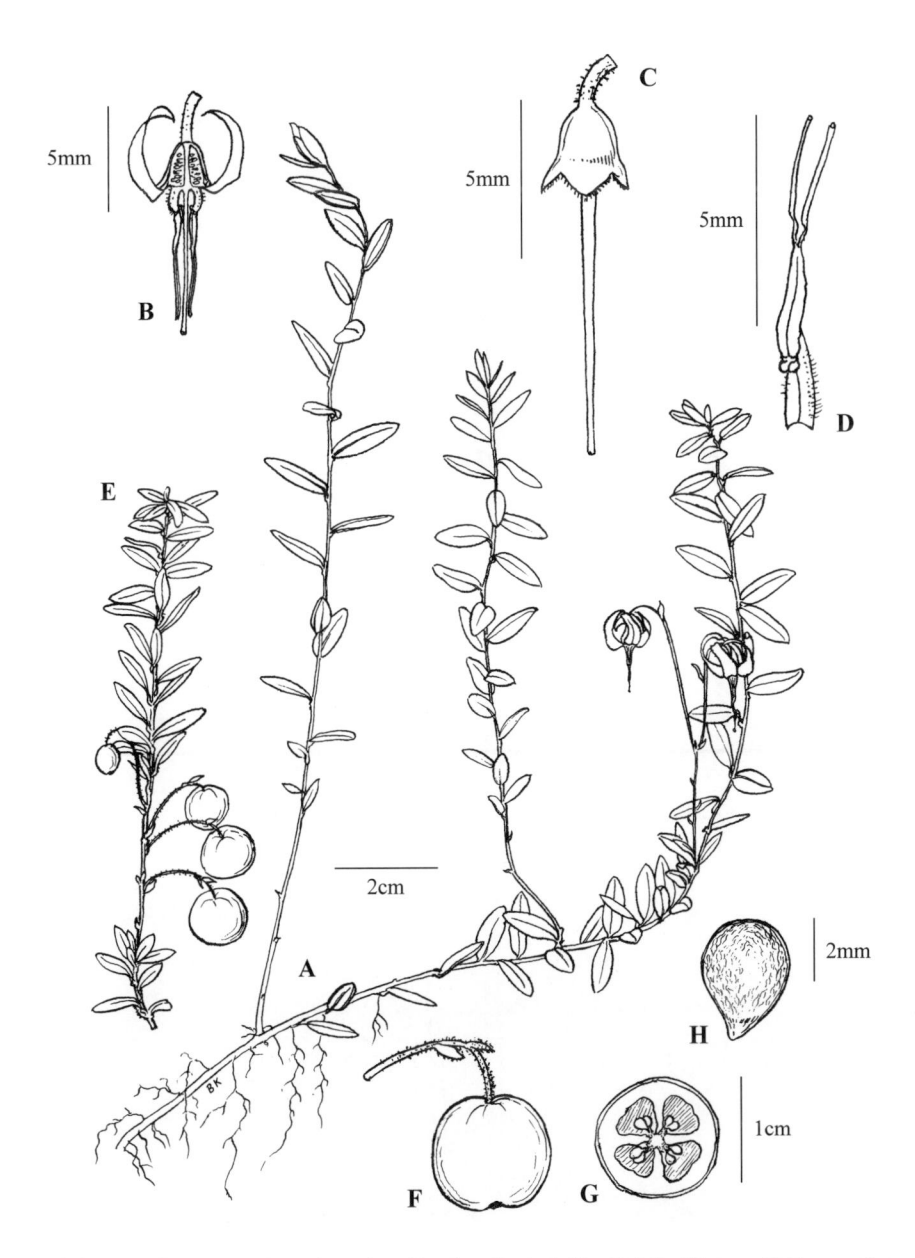

Vaccinium macrocarpon Aiton (American Cranberry). A habit; B LS of flower; C flower minus corolla; D stamen; E fruiting stem; F fruit; G TS of fruit; H seed. (Puttenham Common, Surrey). STACE 356/292.

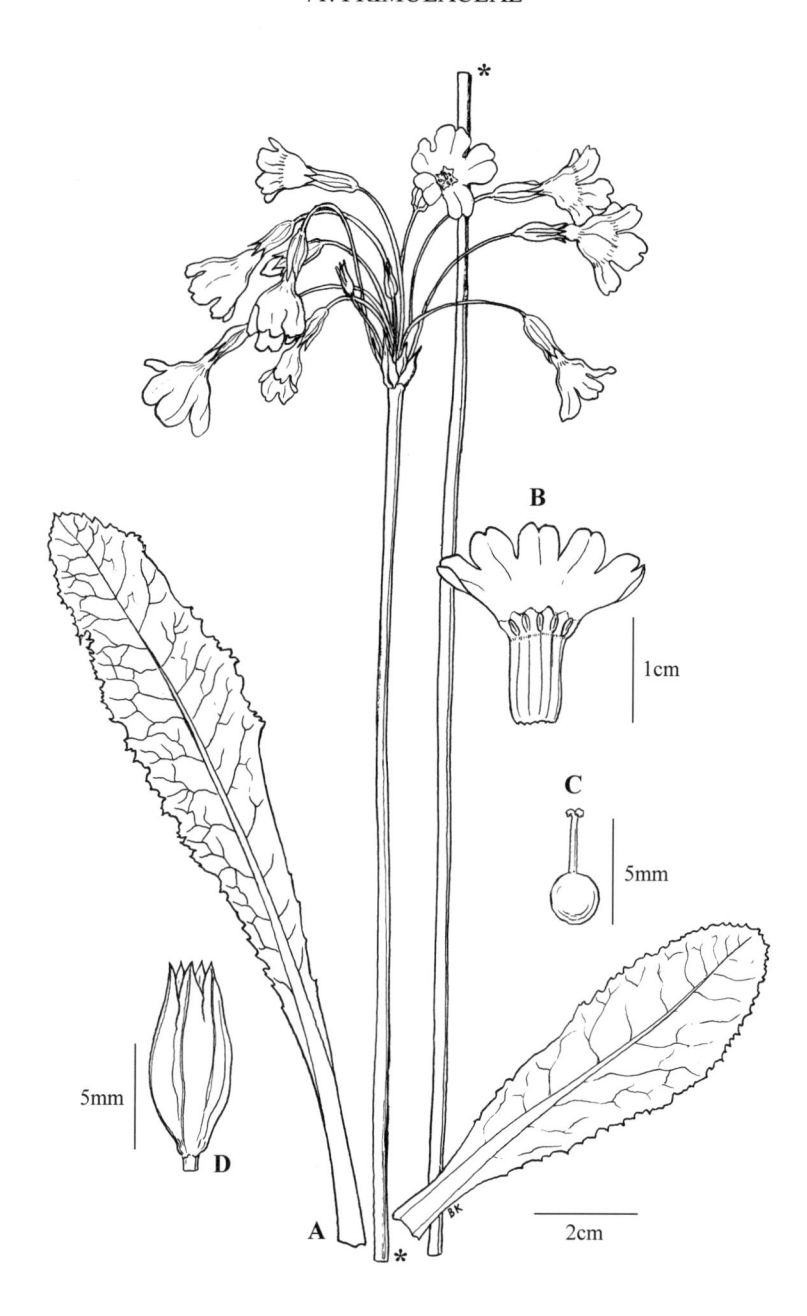

Primula sikkimensis Hook. f. (Sikkim Cowslip). A flowering stem and basal leaves; B corolla opened out; C gynoecium; D fruiting calyx. (Snowdon, Caerns). STACE 364/297.

71. PRIMULACEAE 131

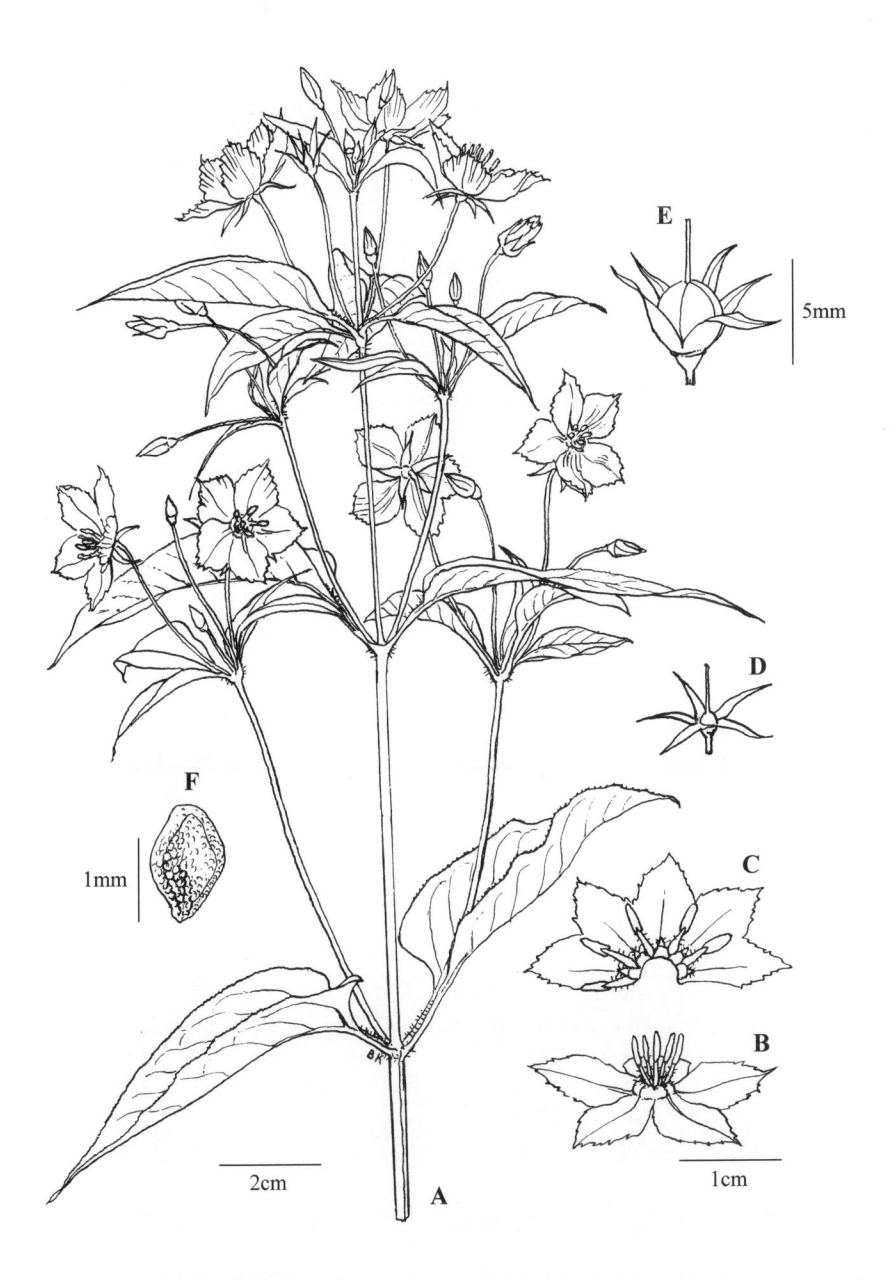

Lysimachia ciliata L. (Fringed Loosestrife). A habit; B flower; C corolla opened out; D flower minus corolla; E capsule; F seed. (Henley-on-Thames, Oxon). STACE 366/300.

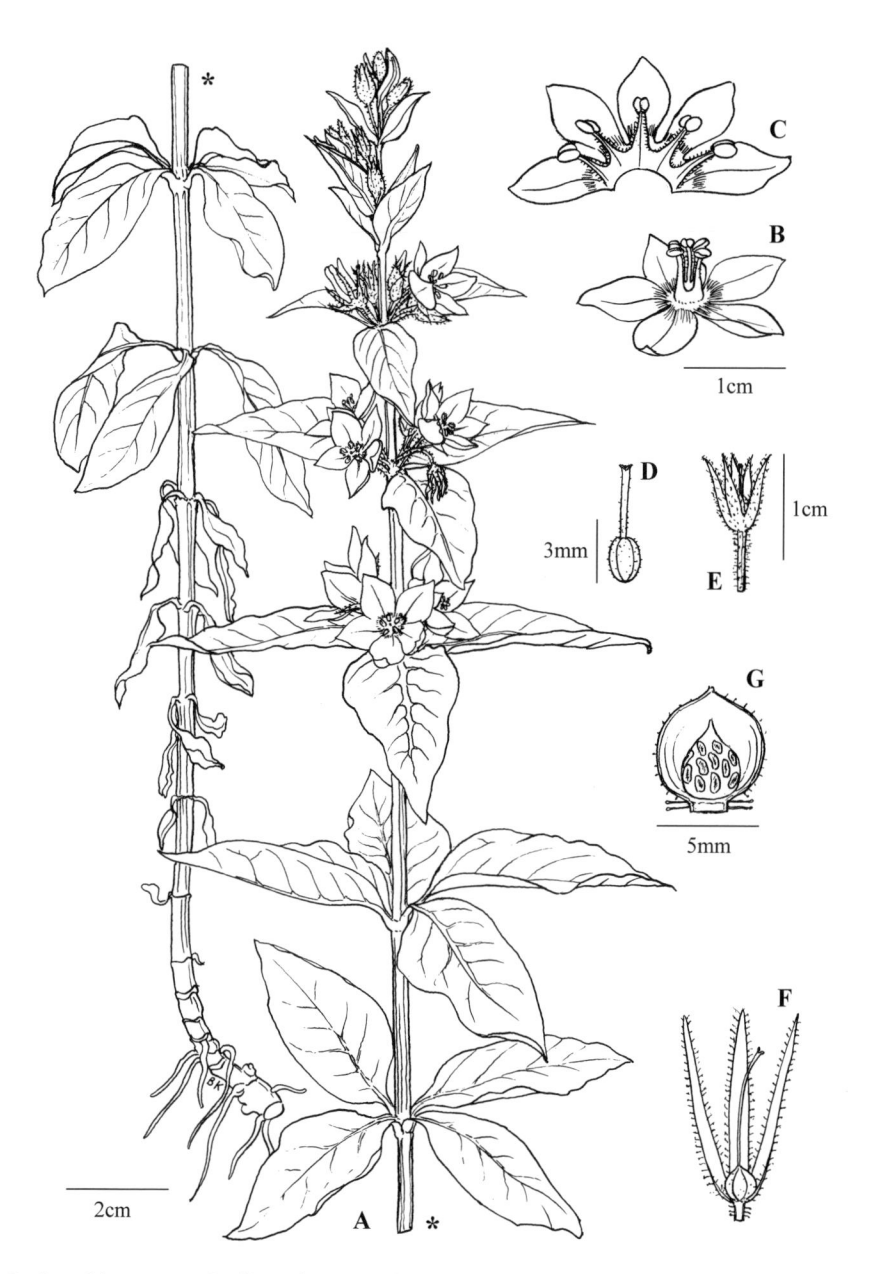

71. PRIMULACEAE

Lysimachia punctata L. (Dotted Loosestrife). A habit; B flower; C corolla opened out; D gynoecium; E flower minus corolla; F sterile capsule; G LS of capsule. (Fulwood, W Lancs). STACE 366/300.

71. PRIMULACEAE 133

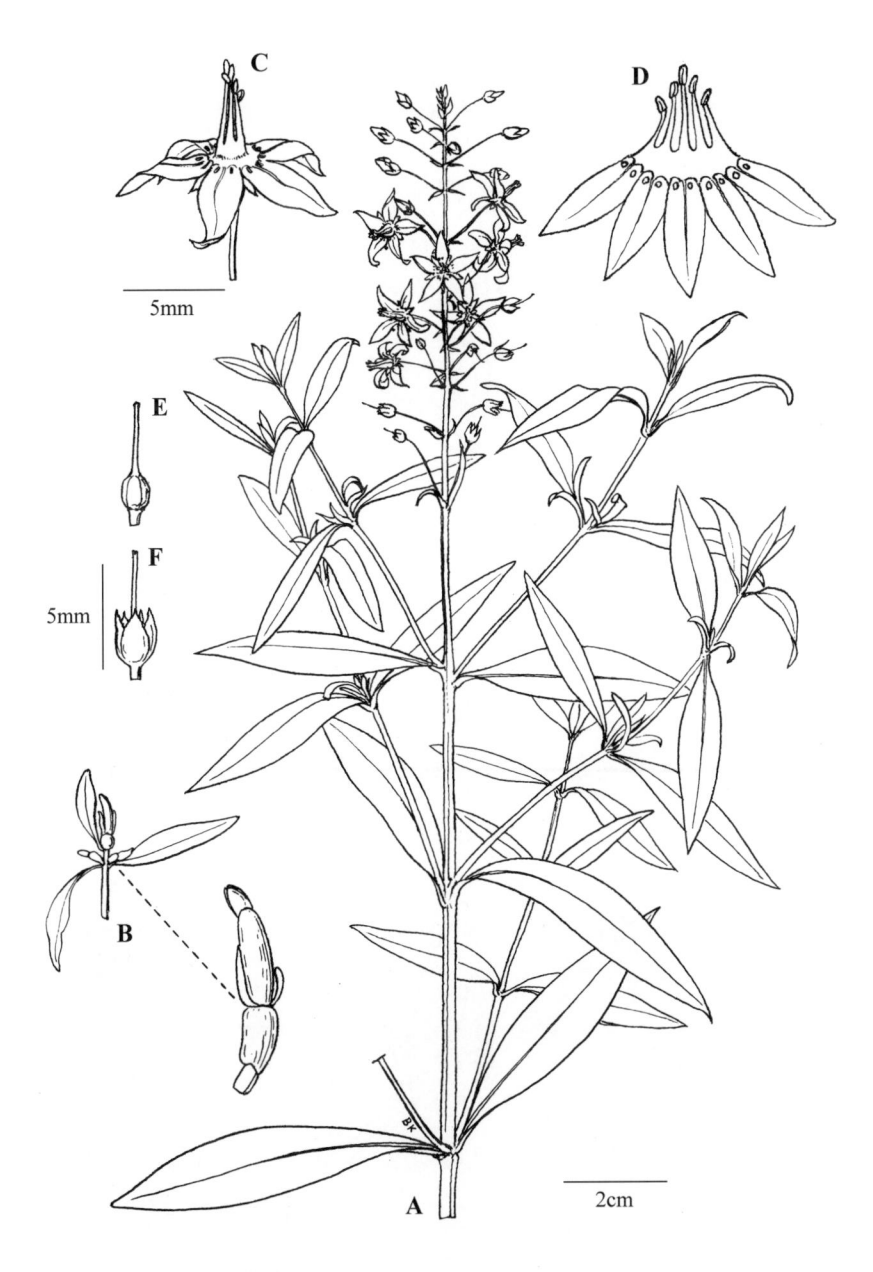

Lysimachia terrestris (L.) Britton, Sterns & Poggenb. (Lake Loosestrife). A habit; B bulbils in leaf axils; C flower; D corolla opened out; E gynoecium; F flower minus corolla. (Lake Windermere, Westmorland). STACE 366/300.

Pittosporum crassifolium Banks & Sol. ex A. Cunn. (Karo). A habit; B flower; C LS of flower; D unripe fruit; E fruit with seeds exposed. (St Mary's, Scilly). STACE 369/302.

74. GROSSULARIACEAE

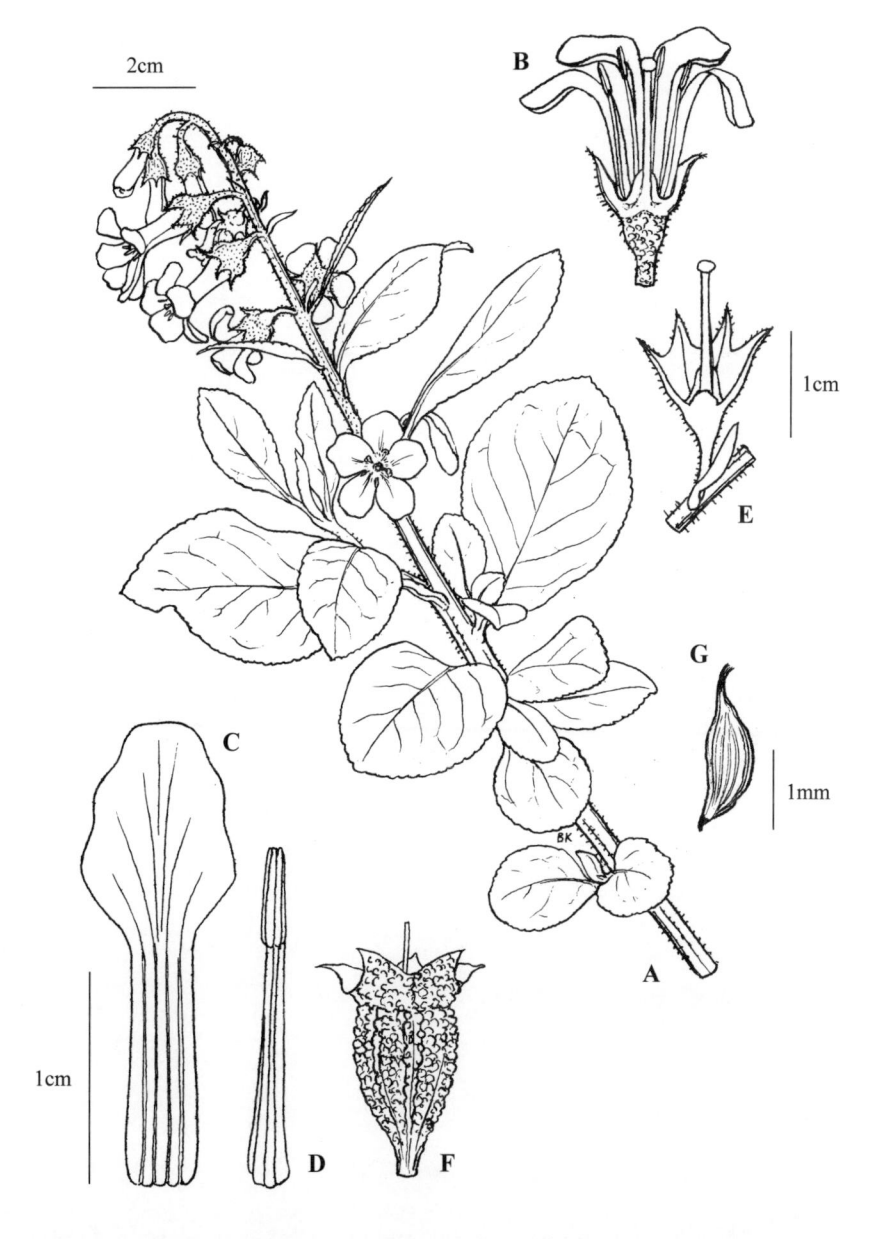

Escallonia macrantha Hook. & Arn. (Escallonia). A flowering branch; B LS of flower; C petal; D stamen; E gynoecium within part of calyx; F capsule; G seed. (Source unknown). STACE 372/305.

75. CRASSULACEAE

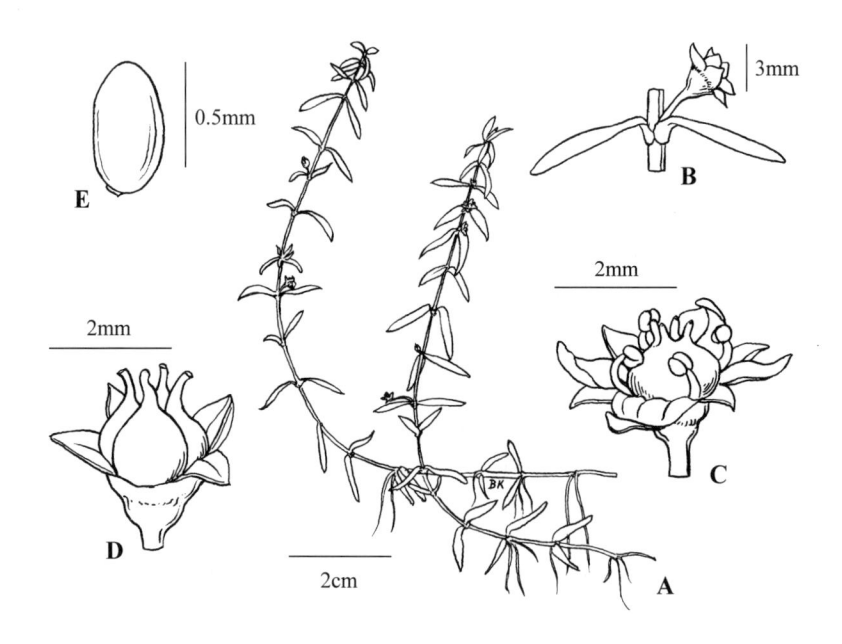

Crassula helmsii (Kirk) Cockayne (New Zealand Pigmyweed). A habit; B part of inflorescence; C flower; D fruit; E seed. (Farnham, Surrey). STACE 375/307.

75. CRASSULACEAE

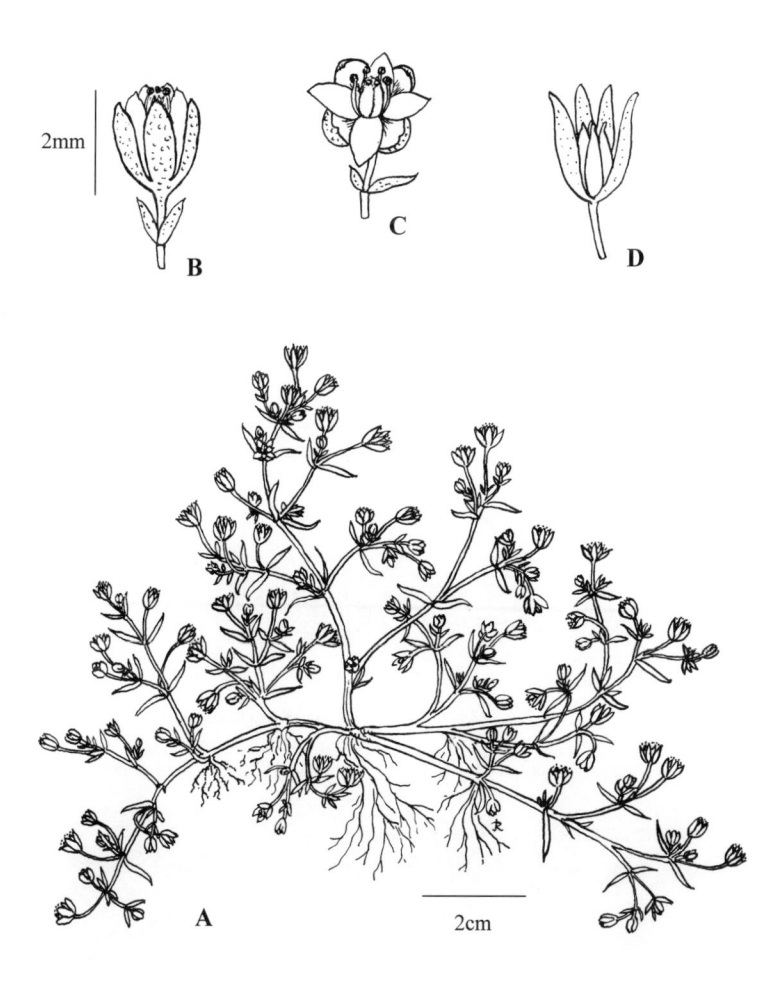

Crassula decumbens Thunb. (Scilly Pigmyweed). A habit; B flower; C flower opened out; D fruit with calyx opened out. (St Mary's, Isles of Scilly). STACE 375/308.

Sedum praealtum A. DC. (Greater Mexican-stonecrop). A habit; B flower. (Jersey). STACE 378/310.

75. CRASSULACEAE

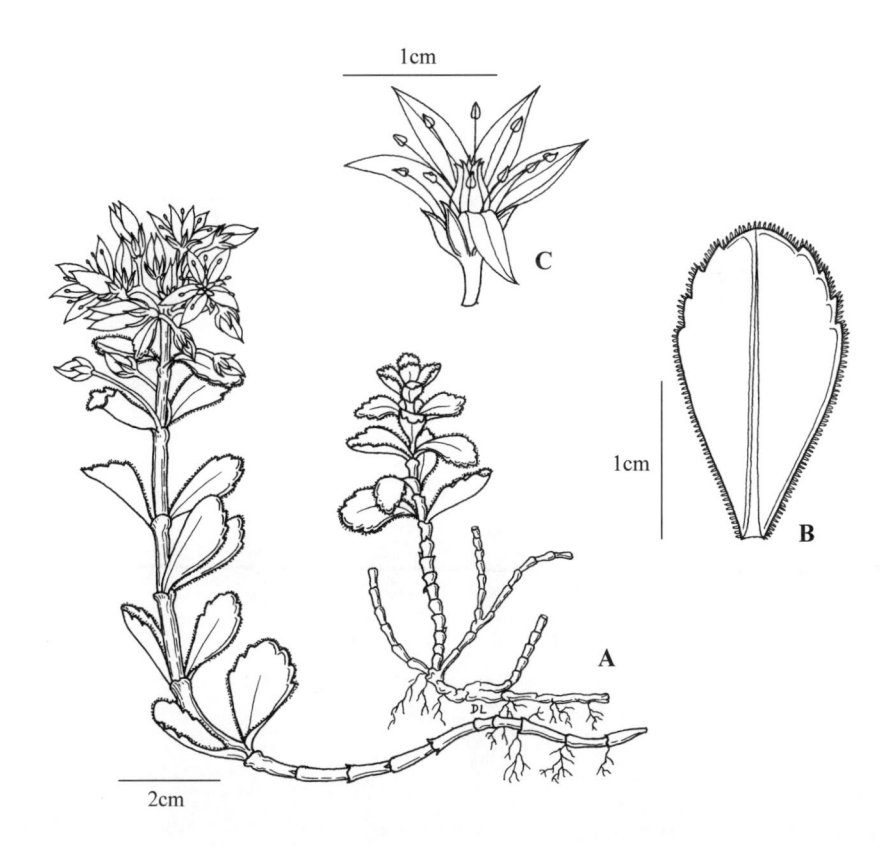

Sedum spurium M. Bieb. (Caucasian-stonecrop). A habit; B leaf; C flower. (Source unknown). STACE 380/312.

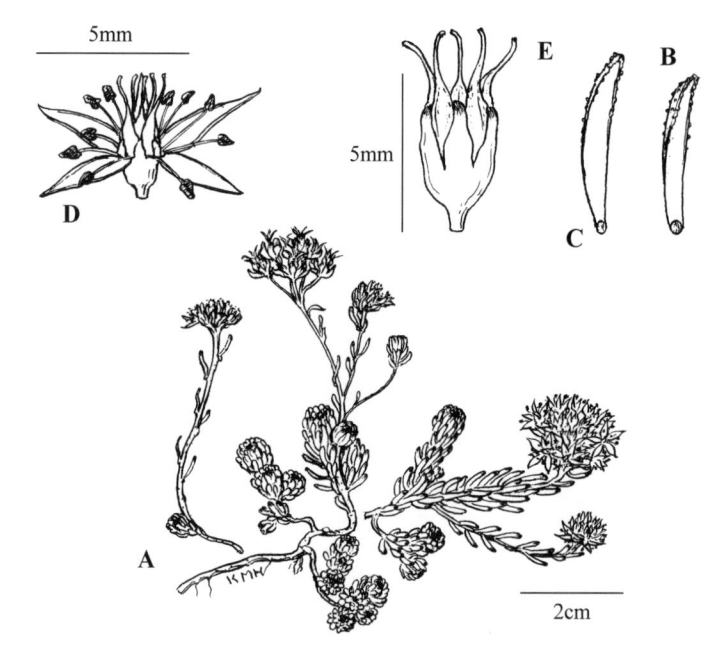

Sedum lydium Boiss. (Least Stonecrop). A habit; B rosette leaf; C stem leaf; D flower; E fruit. (*Hort.* Sevenoaks, W Kent). STACE 381/313.

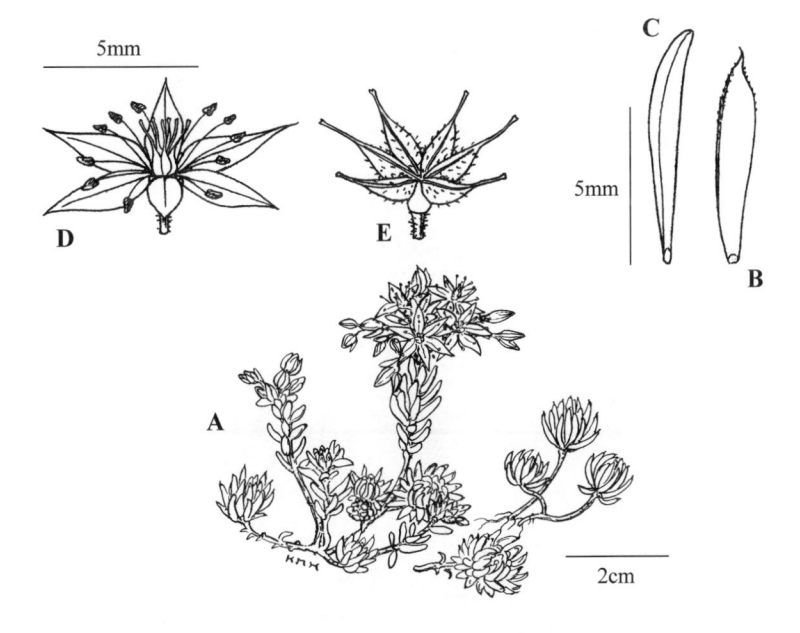

75. CRASSULACEAE 141

Sedum hispanicum L. *sensu lato* (Spanish Stonecrop). A habit; B rosette leaf; C stem leaf; D flower; E fruit. (*Hort*. Sevenoaks, W Kent). STACE 381/313.

76. SAXIFRAGACEAE

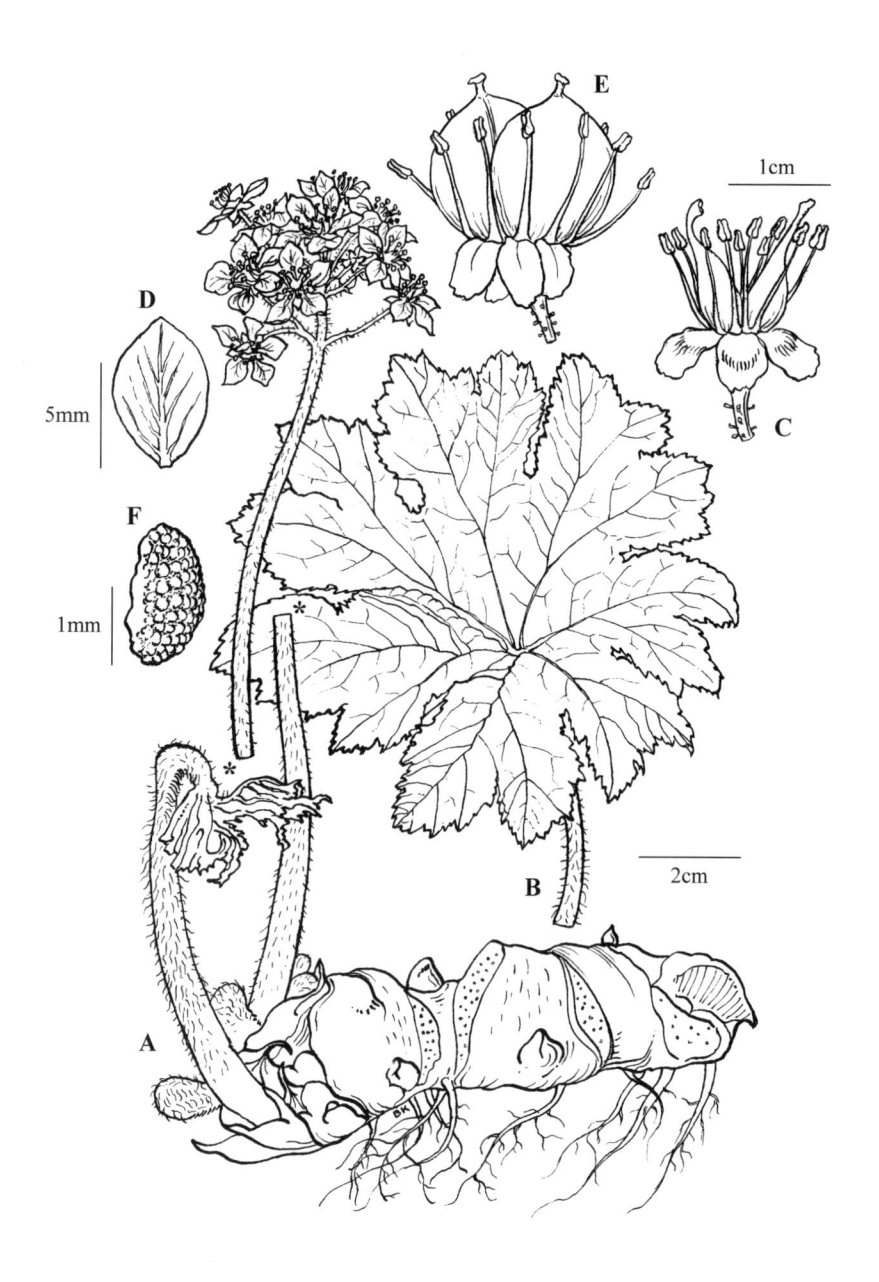

Darmera peltata (Torr. ex Benth.) Voss ex Post & Kuntze (Indian-rhubarb). A habit; B summer leaf; C flower; D petal; E fruit; F seed. (Ilam, Staffs). STACE 385/317.

76. SAXIFRAGACEAE 143

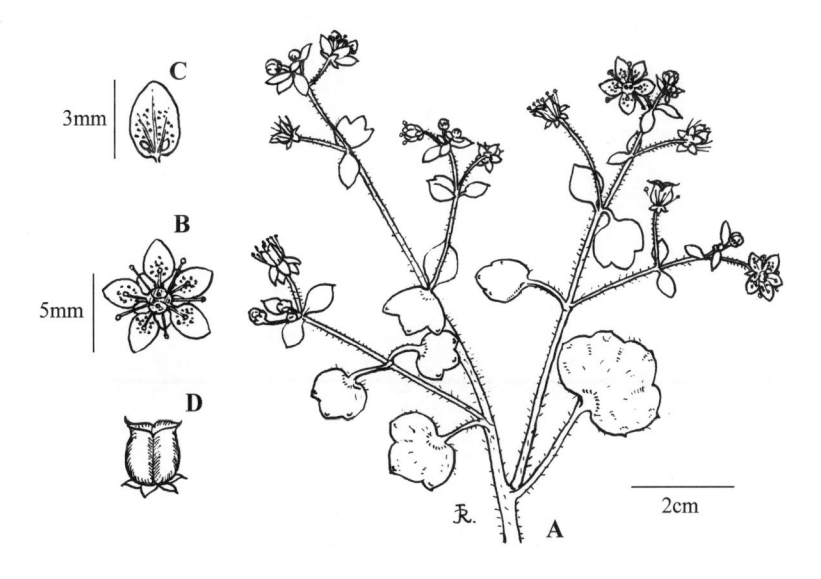

Saxifraga cymbalaria L. var. **huetiana** (Boiss.) Engl. & Irmsch. (Celandine Saxifrage). A habit; B flower; C petal; D fruit. (*Hort.* Ashbourne, Derbys). STACE 387/318.

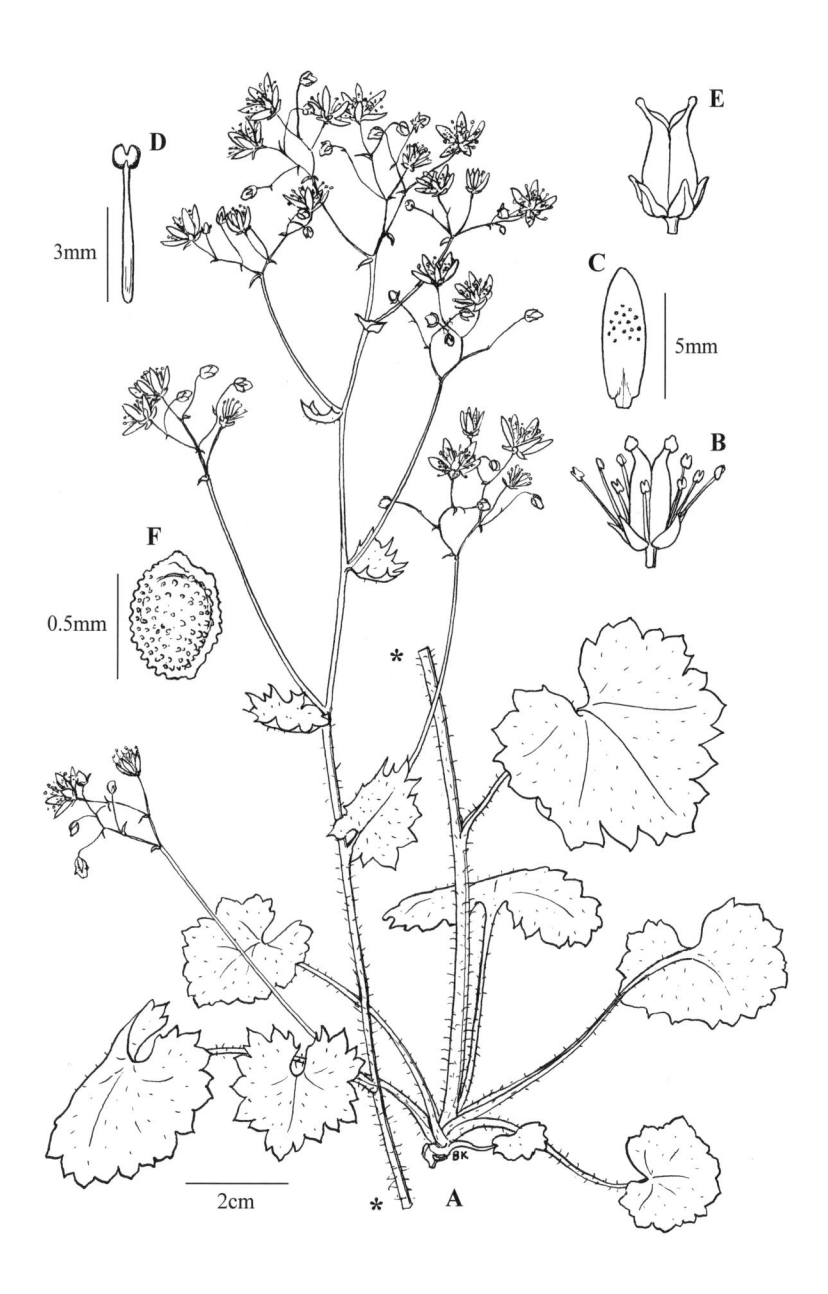

Saxifraga rotundifolia L. (Round-leaved Saxifrage). A habit; B flower minus petals; C petal; D stamen; E fruit; F seed. (Source unknown). STACE 387/318.

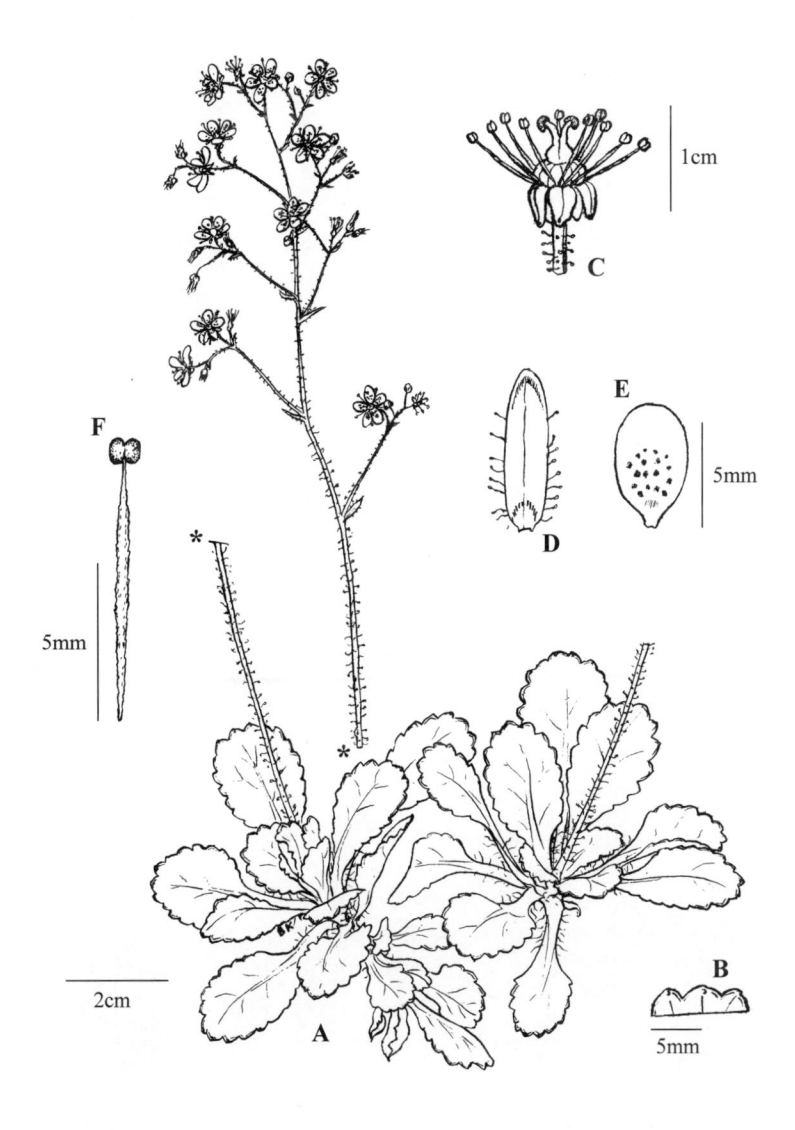

76. SAXIFRAGACEAE 145

Saxifraga × urbium D.A. Webb (Londonpride). A habit; B apex of leaf; C flower minus petals; D bract; E petal; F stamen. (Source unknown). STACE 388/319.

76. SAXIFRAGACEAE

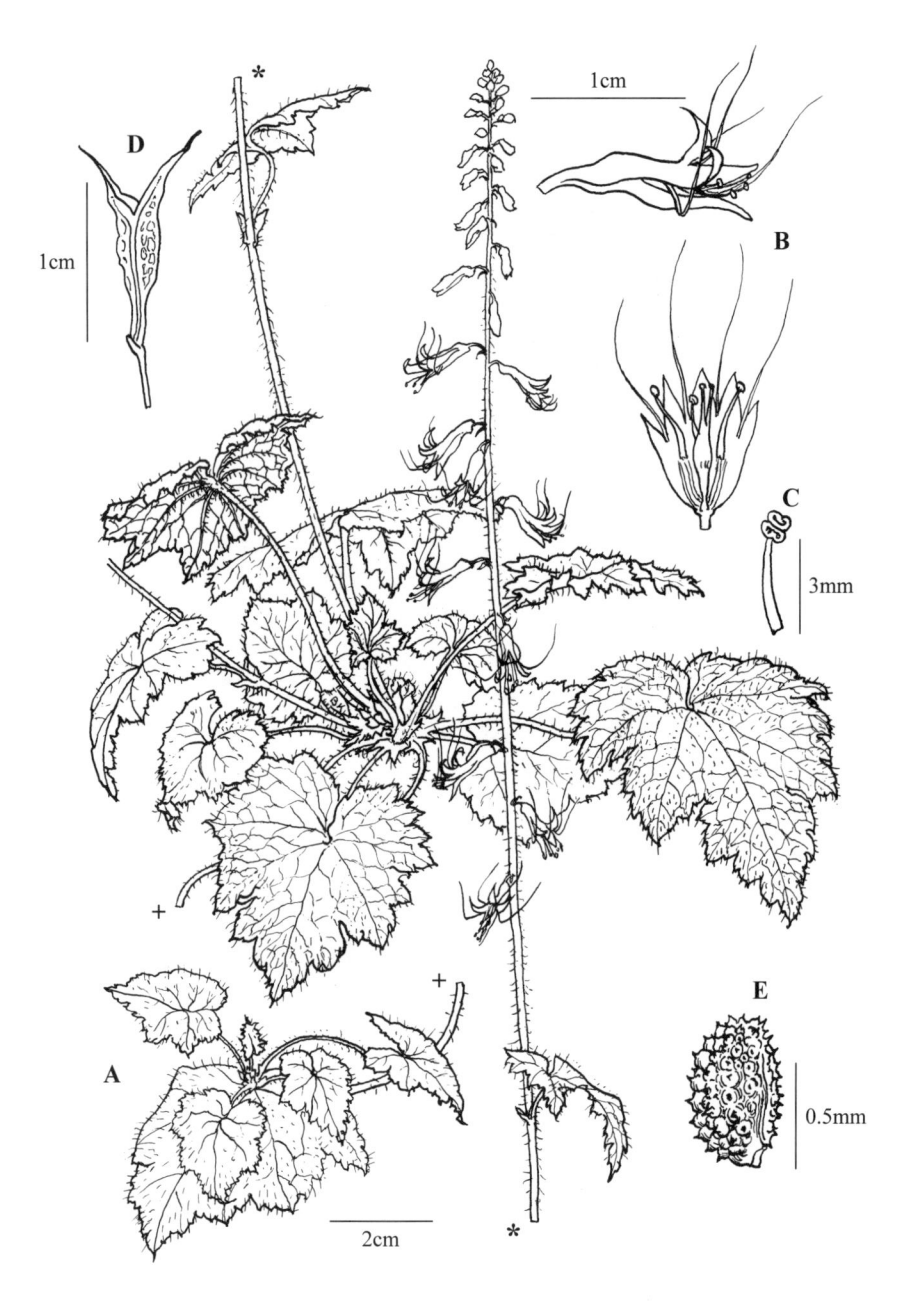

Tolmiea menziesii (Pursh) Torr. & A. Gray (Pick-a-back-plant). A habit; B flower; C stamen; D fruit; E seed. (Source unknown). STACE 390/321.

76. SAXIFRAGACEAE

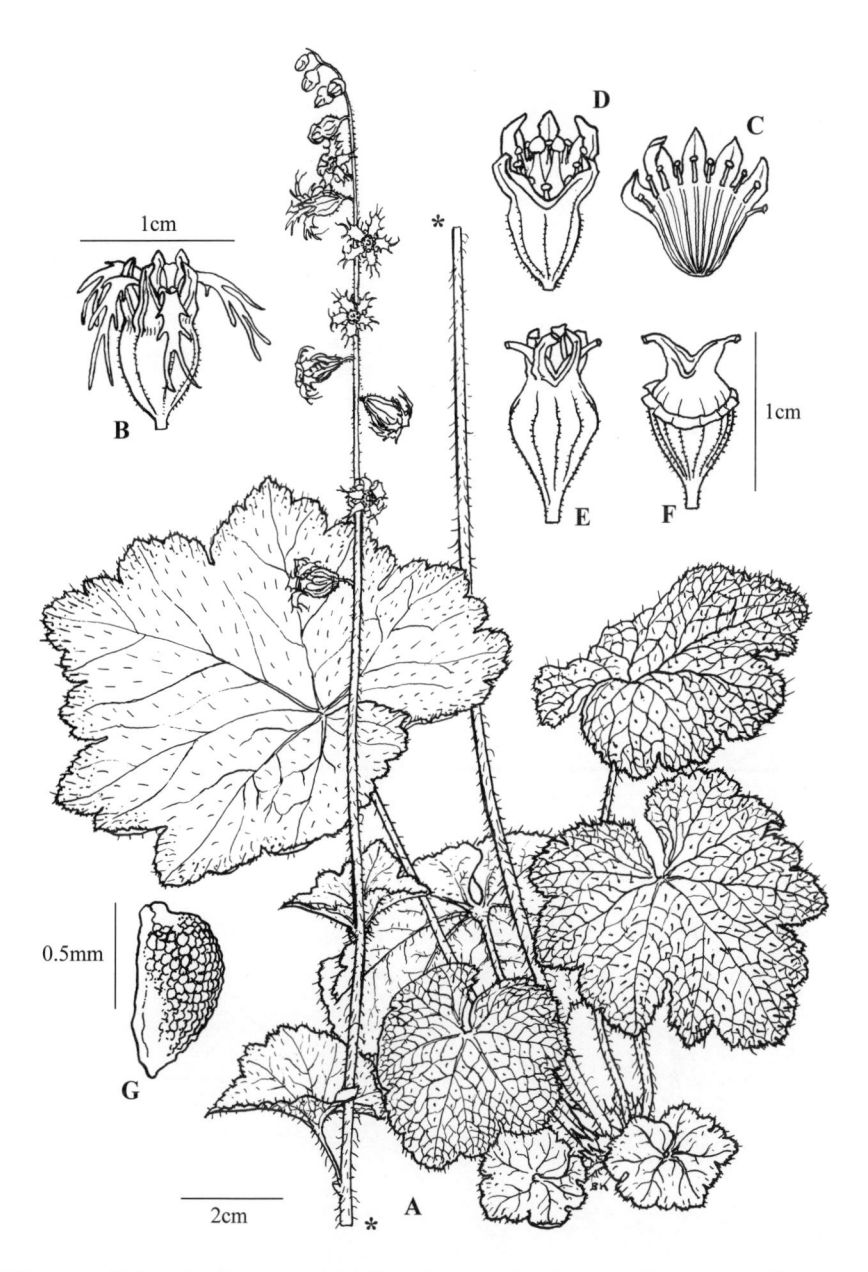

Tellima grandiflora (Pursh) Douglas ex Lindl. (Fringecups). A habit; B flower; C corolla opened out; D flower minus petals; E fruiting perianth; F fruit; G seed. (Source unknown).
STACE 391/321.

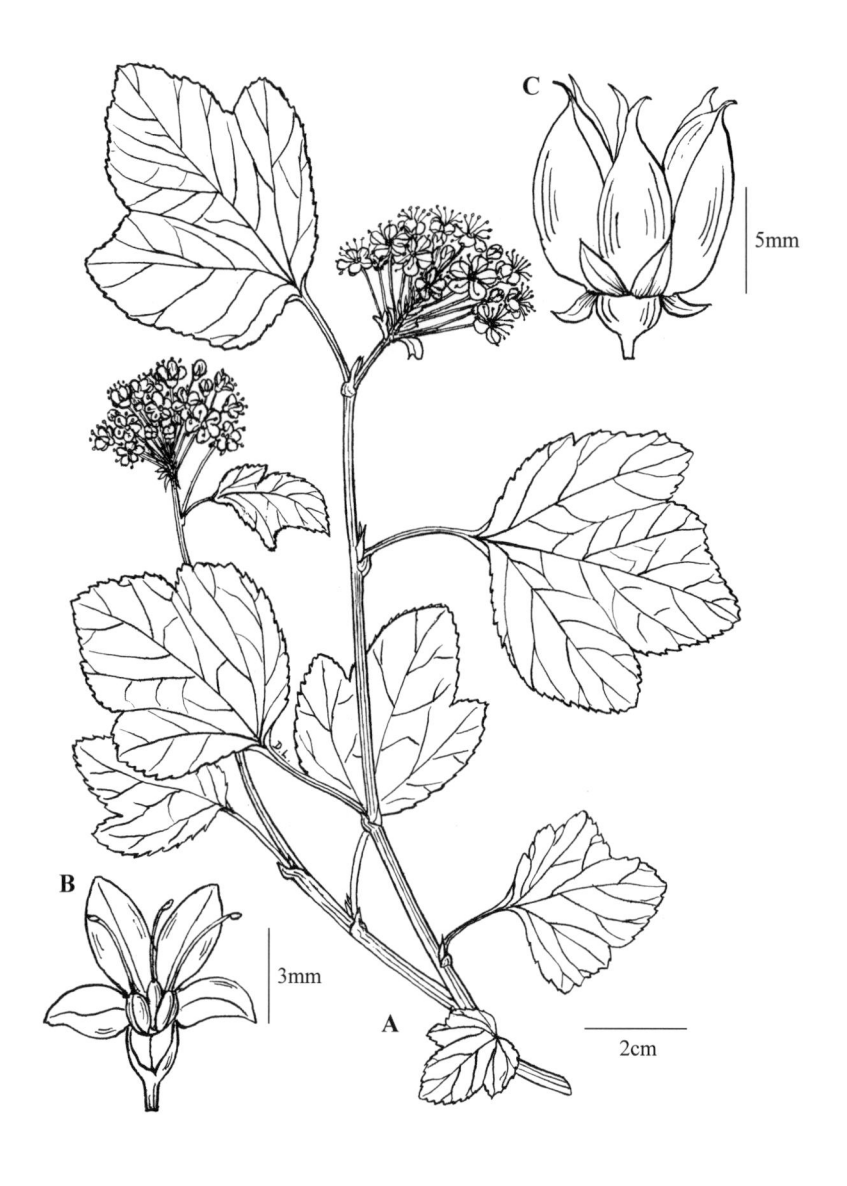

Physocarpus opulifolius (L.) Maxim. (Ninebark). A flowering branch; B flower; C fruit. (Conon Bridge, E Ross). STACE 396/325.

77. ROSACEAE

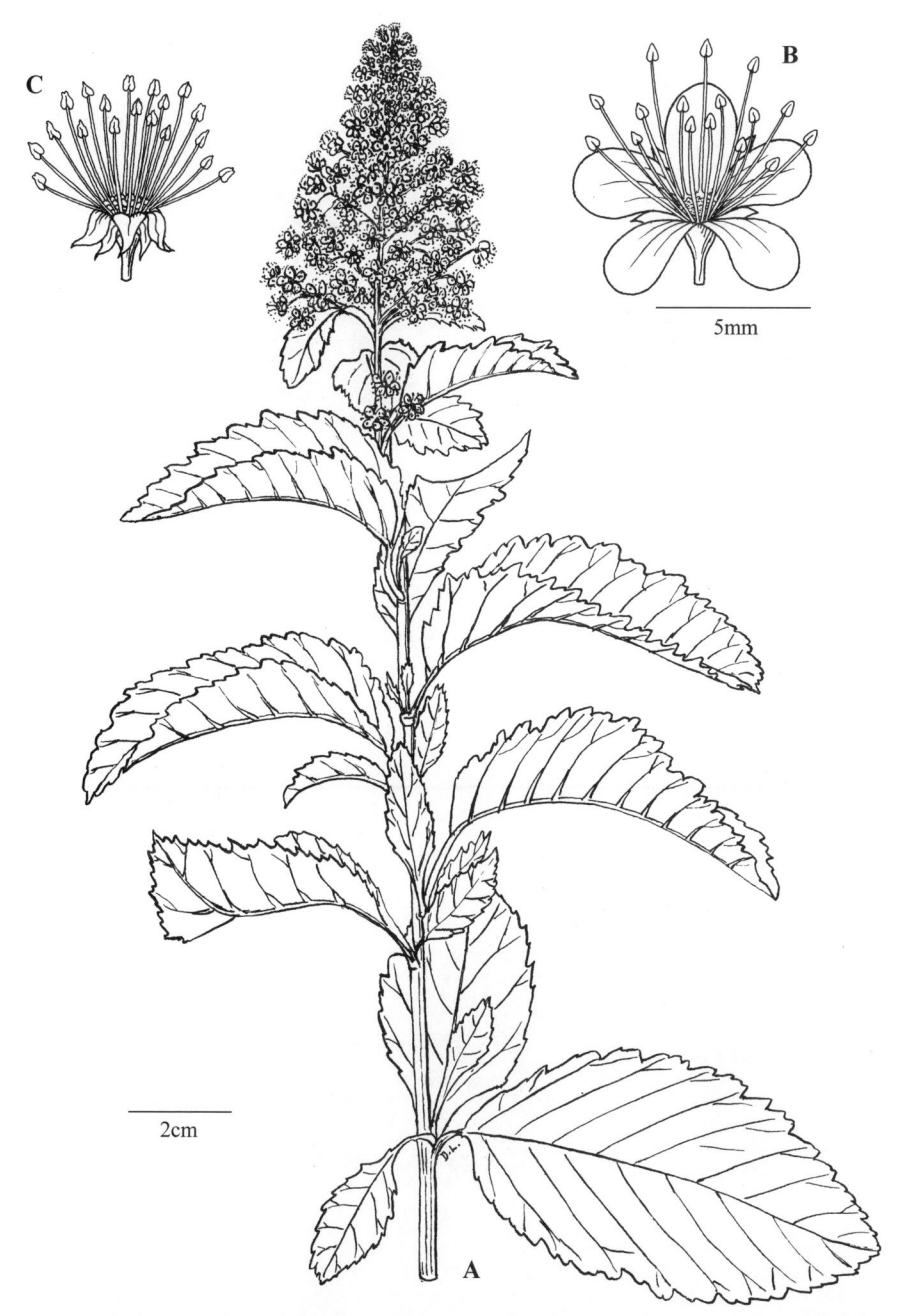

Spiraea alba Du Roi (Pale Bridewort, variant). A flowering branch; B flower; C flower after anthesis. (Crowborough Common, E Sussex). STACE 399/327.

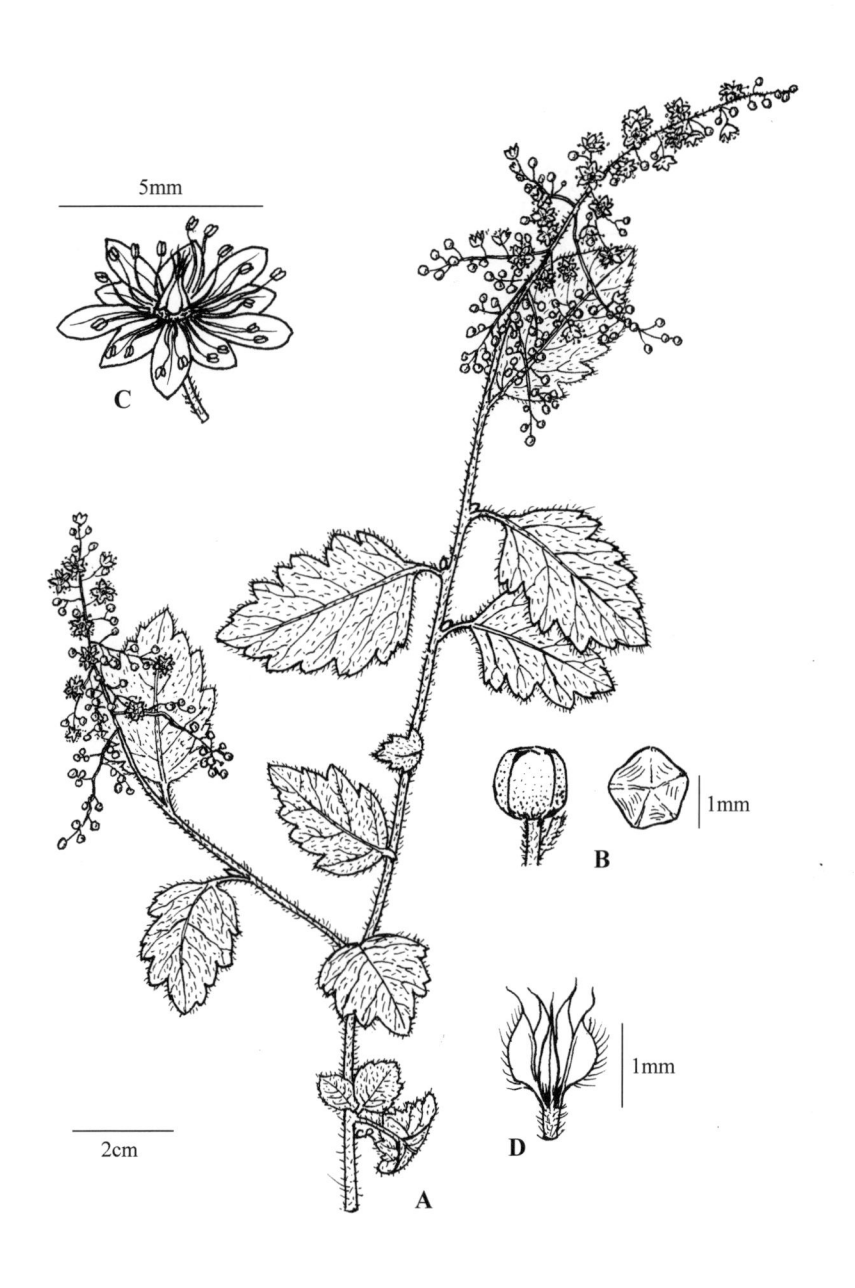

Holodiscus discolor (Pursh) Maxim. (Oceanspray). A flowering branch; B flower bud; C flower; D fruit. (Source unknown). STACE 400/329.

77. ROSACEAE

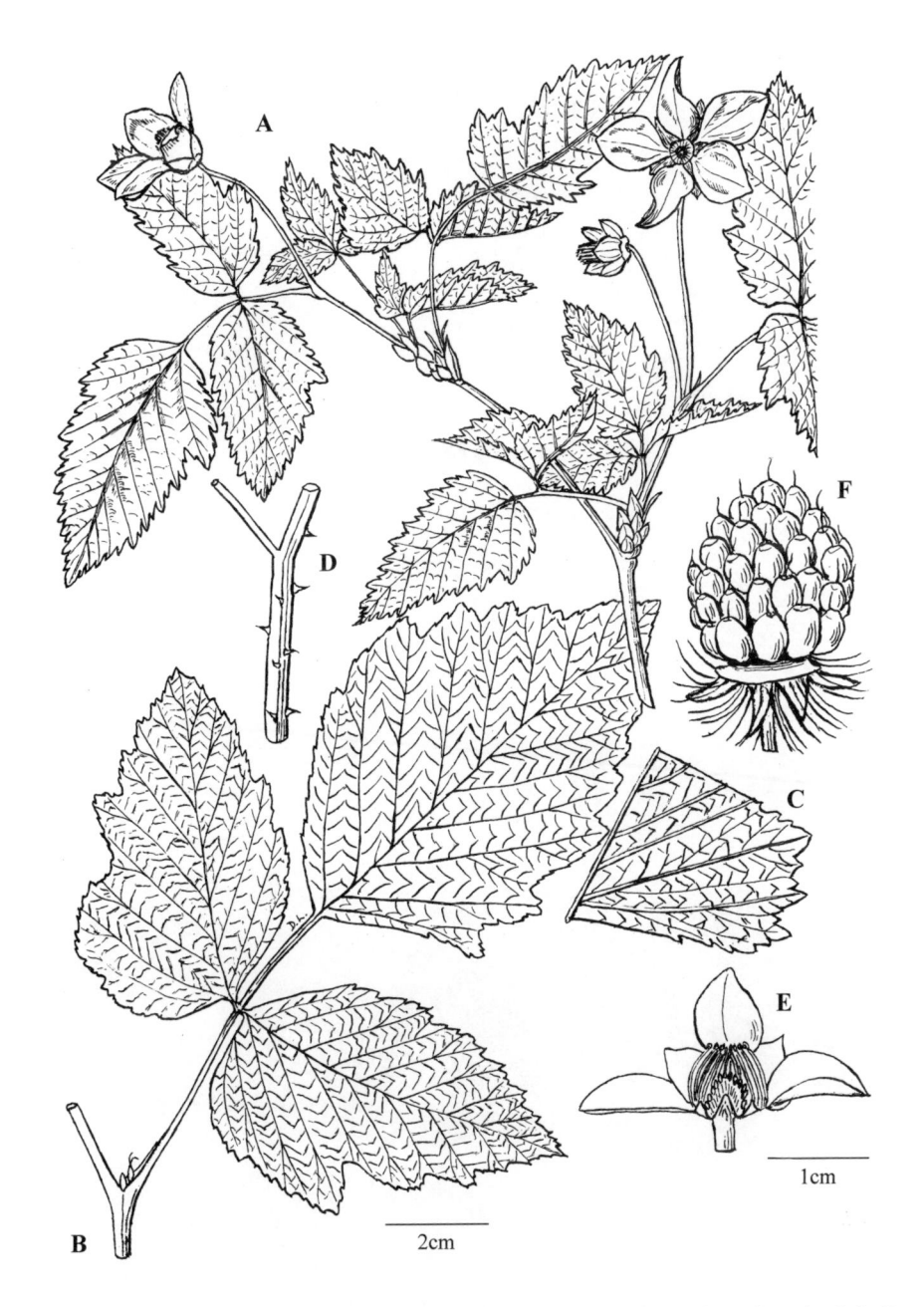

Rubus spectabilis Pursh (Salmonberry). A flowering branch; B leaf from lower branch; C leaflet margin; D lower part of stem; E LS of flower; F fruit. (Sandling, E Kent). STACE 406/334.

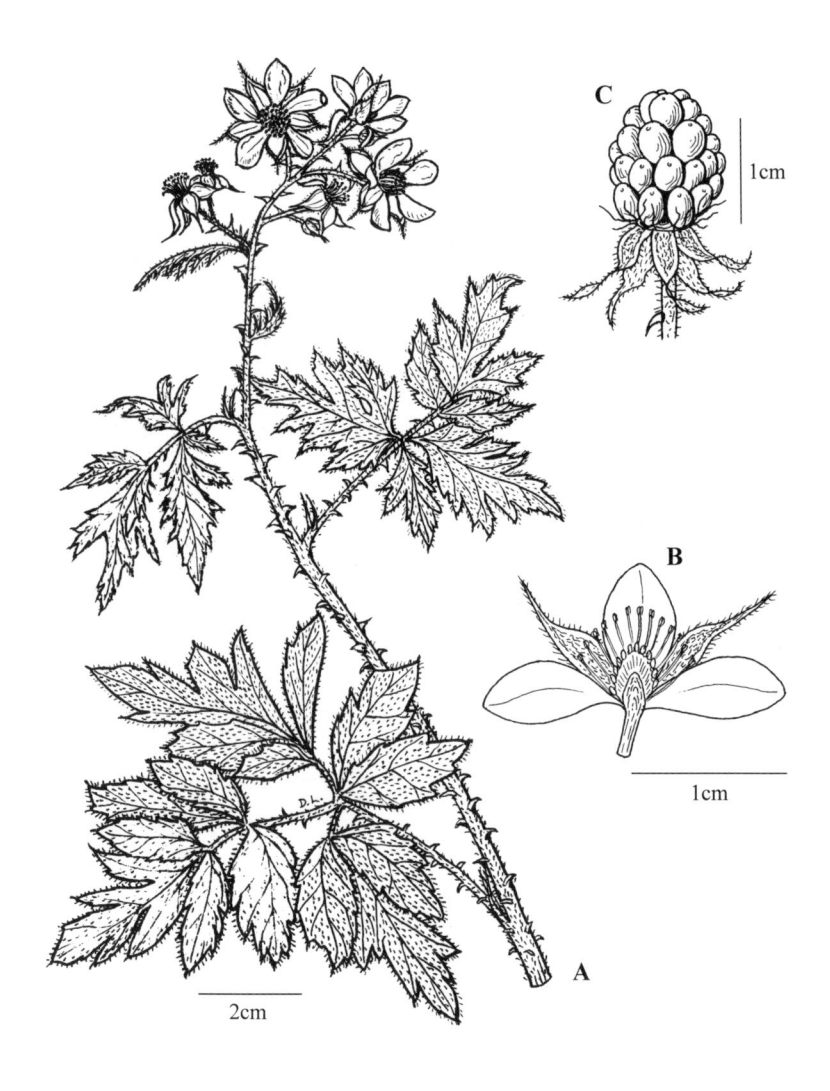

Rubus laciniatus Willd. (Bramble microspecies). A flowering branch; B LS of flower; C fruit. (Channel Islands). STACE 407/335.

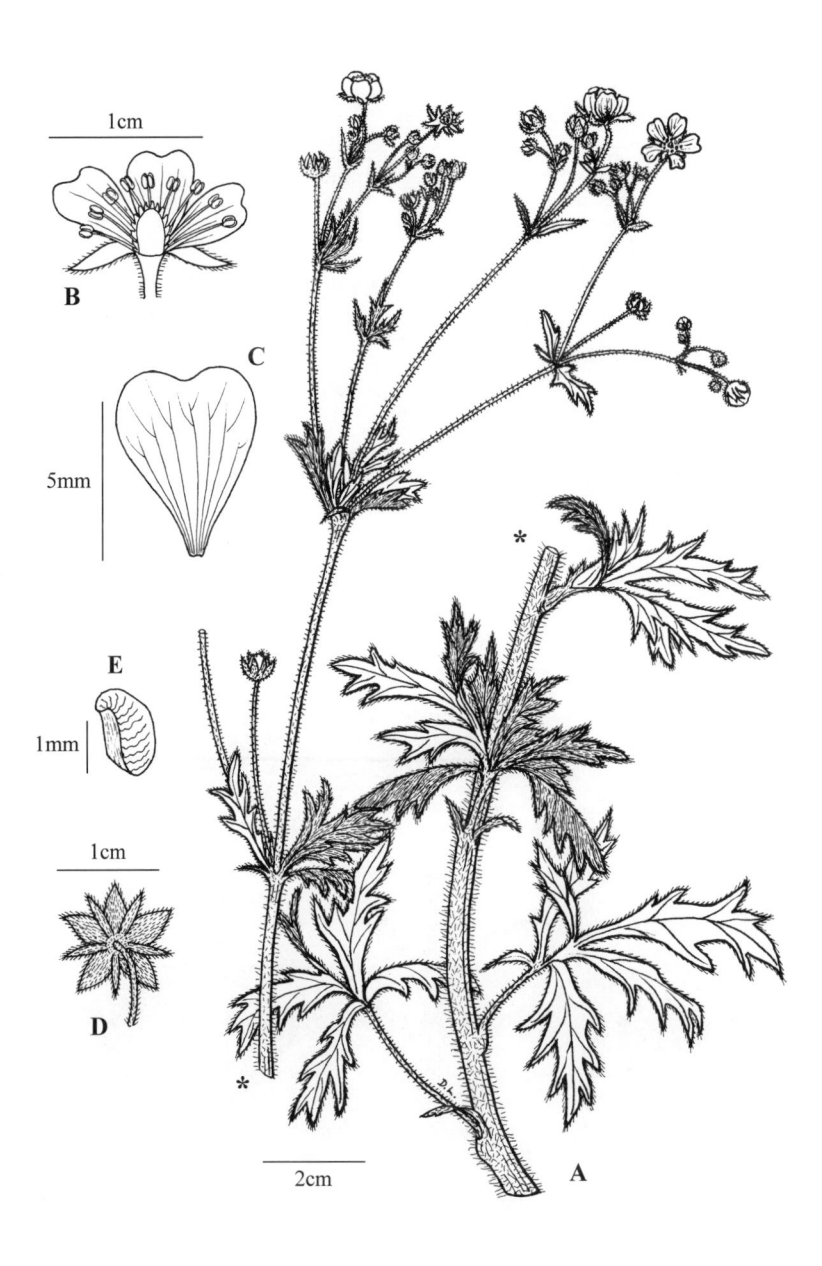

Potentilla inclinata Vill. (Grey Cinquefoil). A habit; B LS of flower; C petal; D calyx and epicalyx; E achene. (Reigate Heath, Surrey). STACE 411/342.

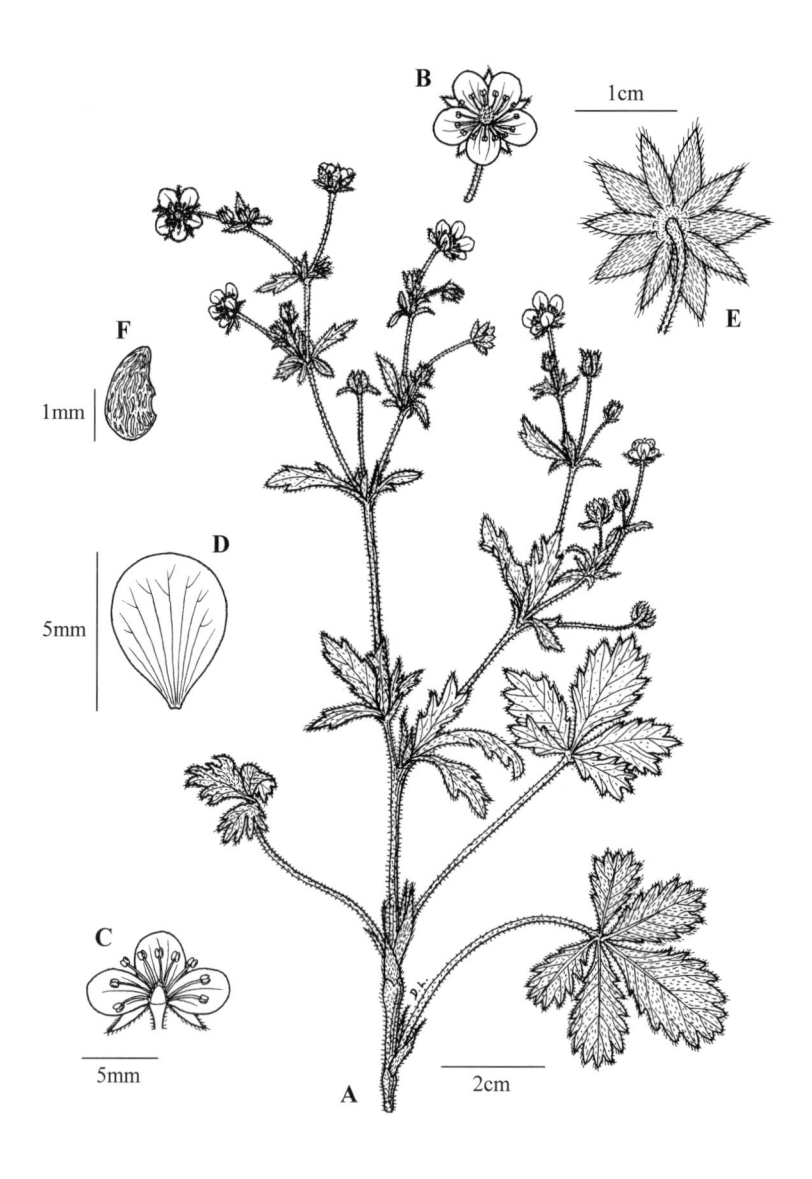

Potentilla intermedia L. (Russian Cinquefoil). A habit; B flower; C LS of flower; D petal; E inflated calyx and epicalyx in fruit; F achene. (Riverhead, W Kent). STACE 411/342.

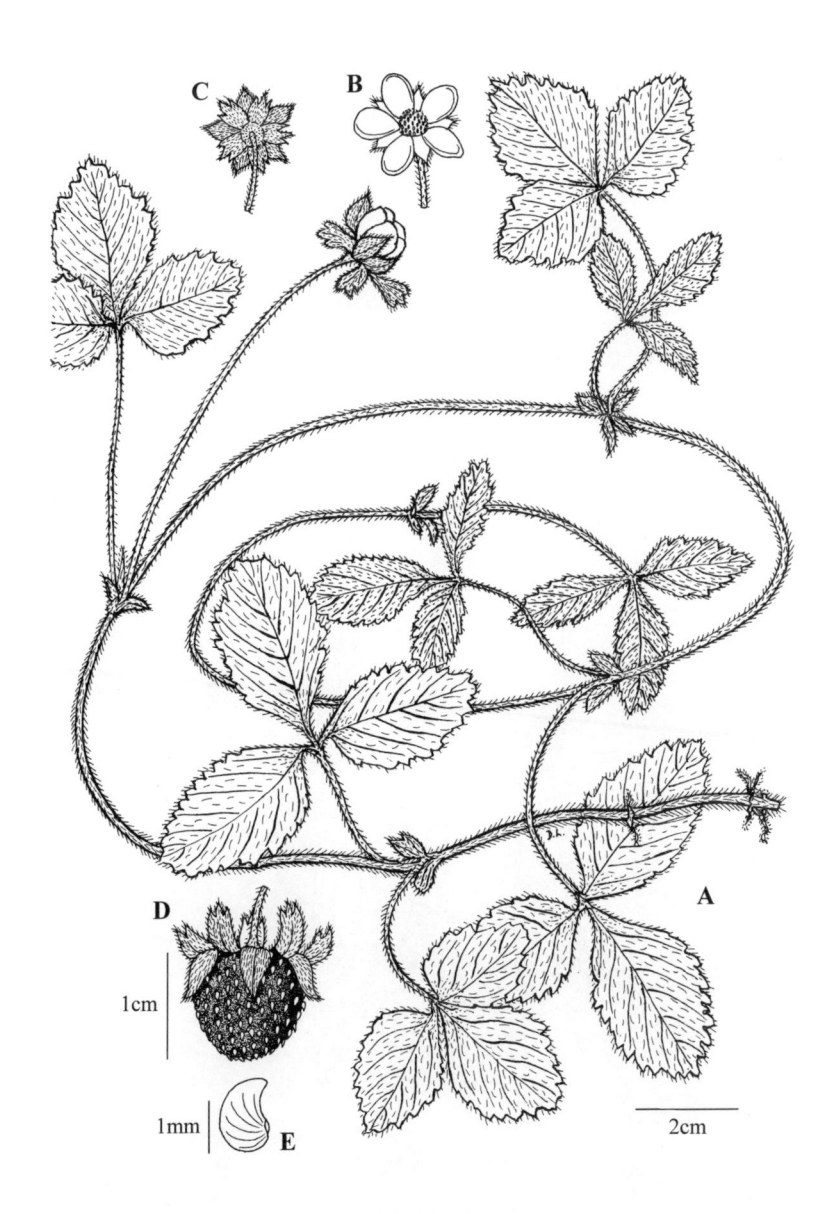

Duchesnea indica (Jacks.) Focke (Yellow-flowered Strawberry). A habit; B flower; C calyx and epicalyx; D fruit; E achene. (Source unknown). STACE 415/346.

156 77. ROSACEAE

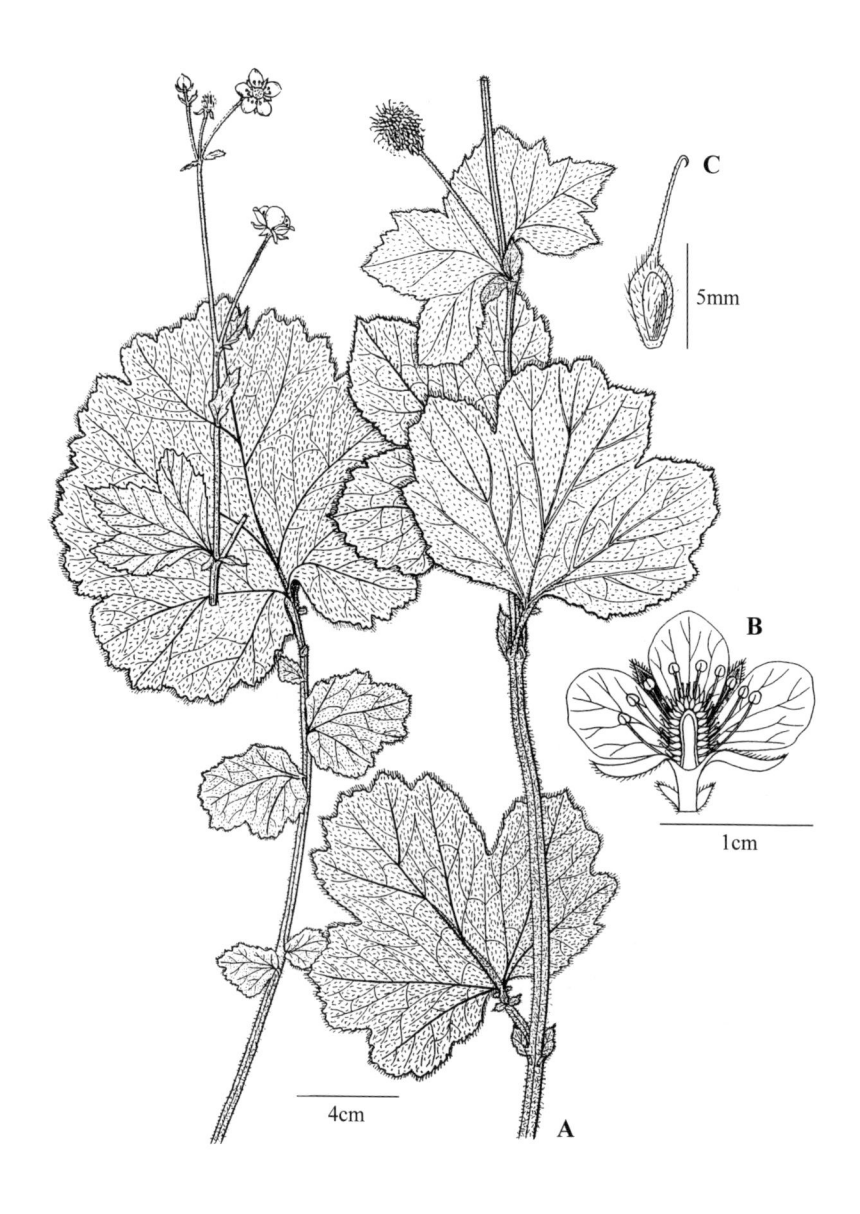

Geum macrophyllum Willd. (Large-leaved Avens). A habit and basal leaf; B LS of flower; C achene. (Kinnordy, Angus). STACE 416/347.

77. ROSACEAE

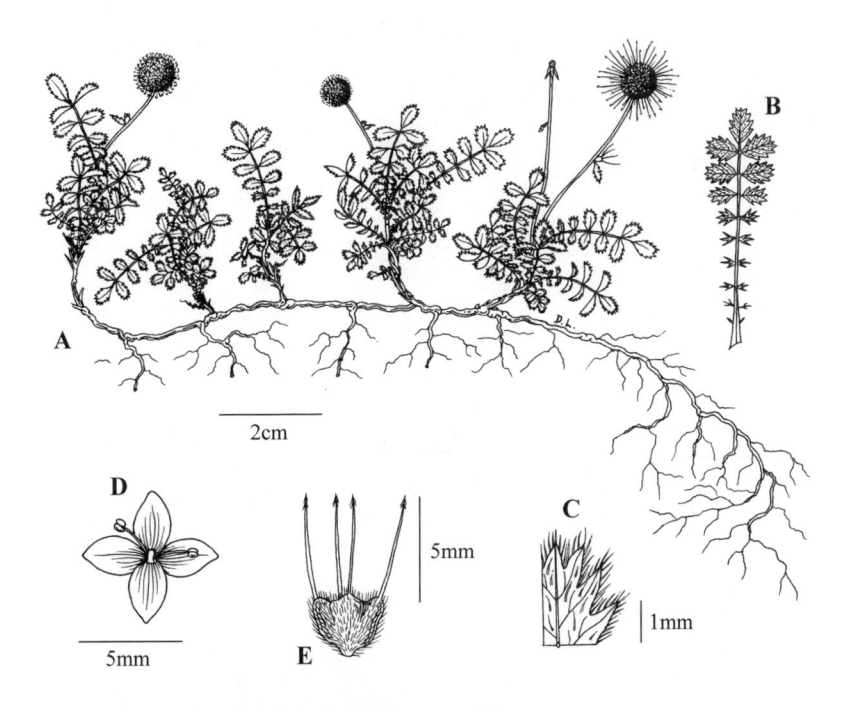

Acaena anserinifolia (J.R. & G. Forst.) Druce (Bronze Pirri-pirri-bur, variant). A habit; B leaf; C upper surface of part of leaflet; D flower; E fruit. (*Hort*. Cambridge Botanic Garden, as **A. pusilla** (Bitter) Allan). STACE 420/350.

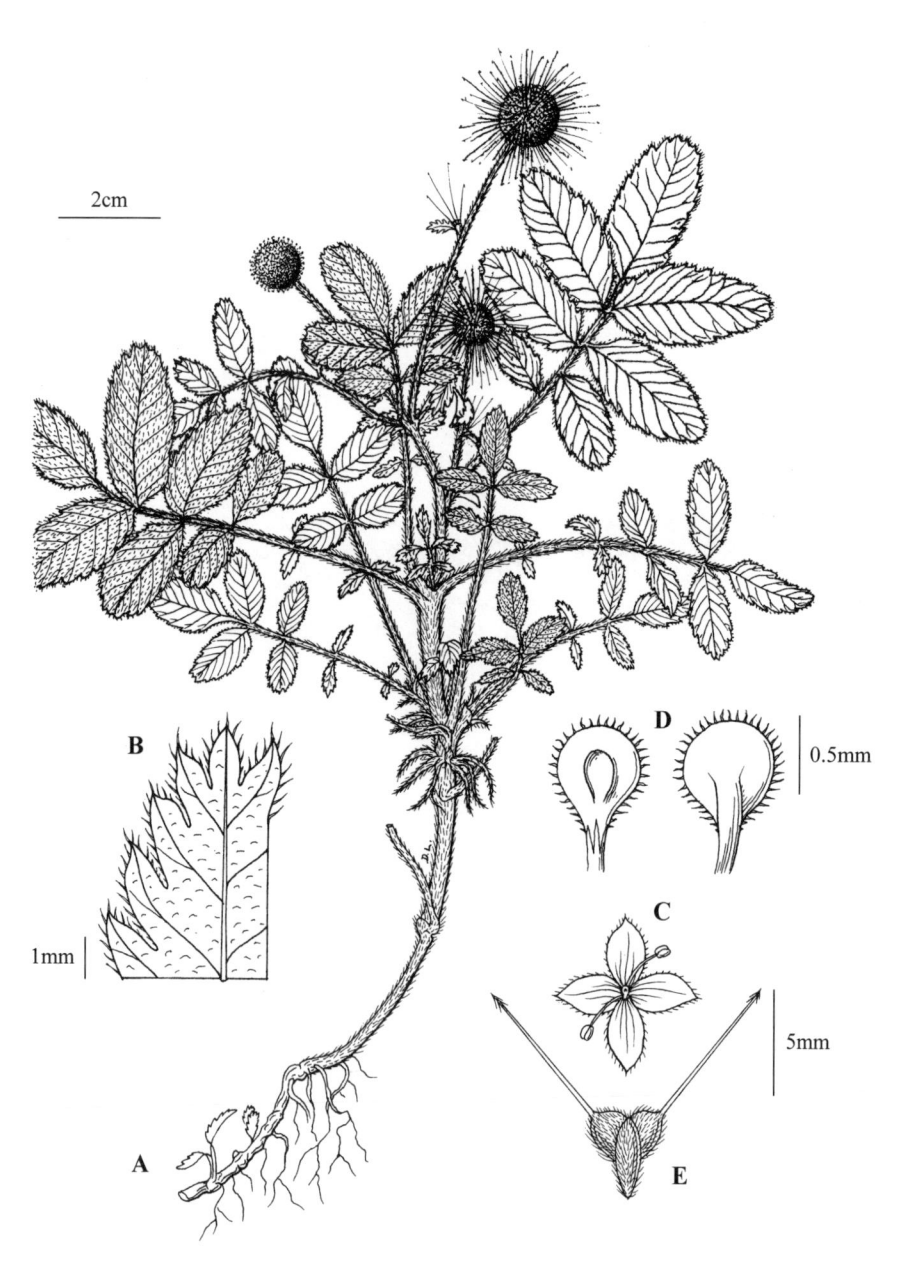

Acaena ovalifolia Ruiz & Pav. (Two-spined Acaena). A habit; B upper surface of part of leaflet; C flower; D stigma; E fruit. (Bunclody, Co Wexford). STACE 420/350.

77. ROSACEAE

Acaena microphylla Hook. f. (New Zealand Burr). A habit; B leaf; C upper surface of part of leaflet; D young flower; E flower; F fruit. (*Hort*. Cambridge Botanic Garden). STACE —/—.

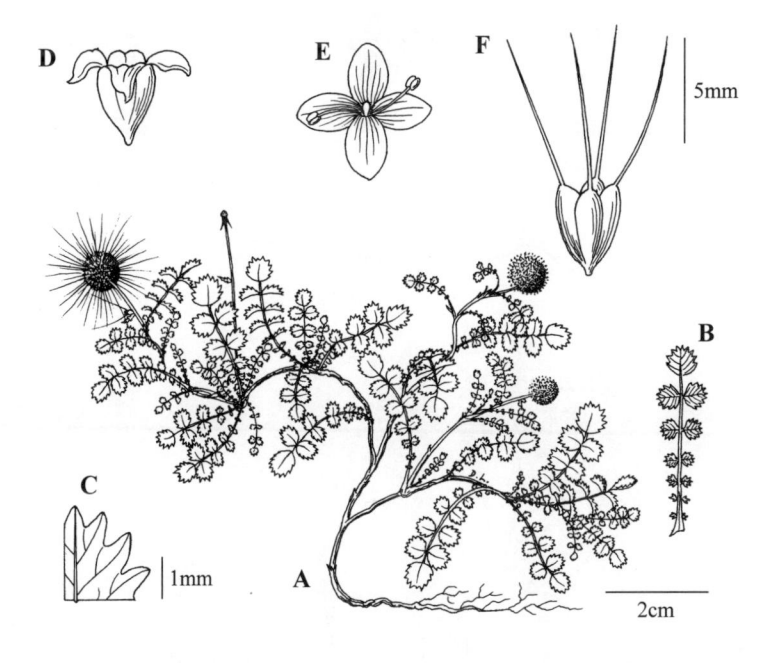

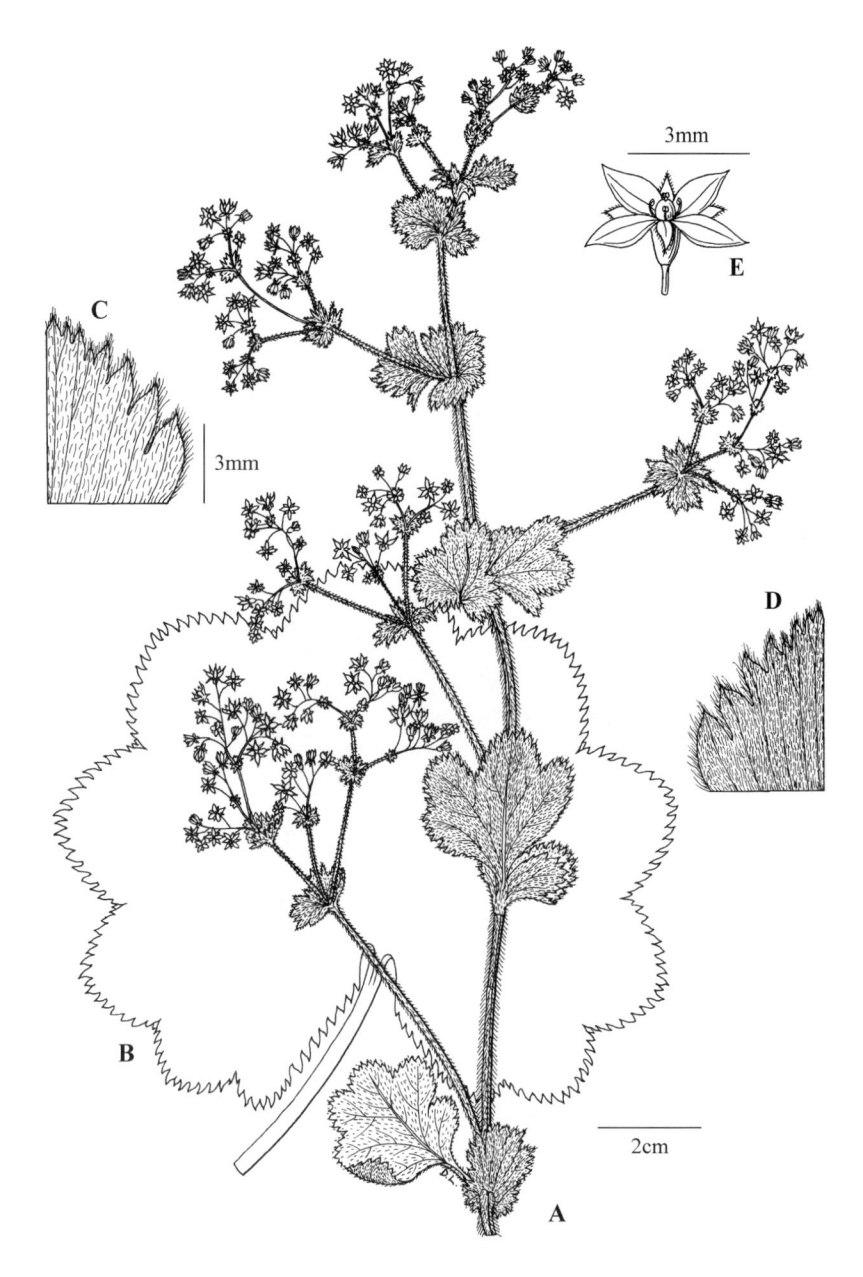

Alchemilla tytthantha Juz. (Lady's-mantle segregate). A habit; B outline of basal leaf; C upperside of leaf margin; D underside of leaf margin; E flower. (Bowhill, Selkirks). STACE 423/352.

77. ROSACEAE 161

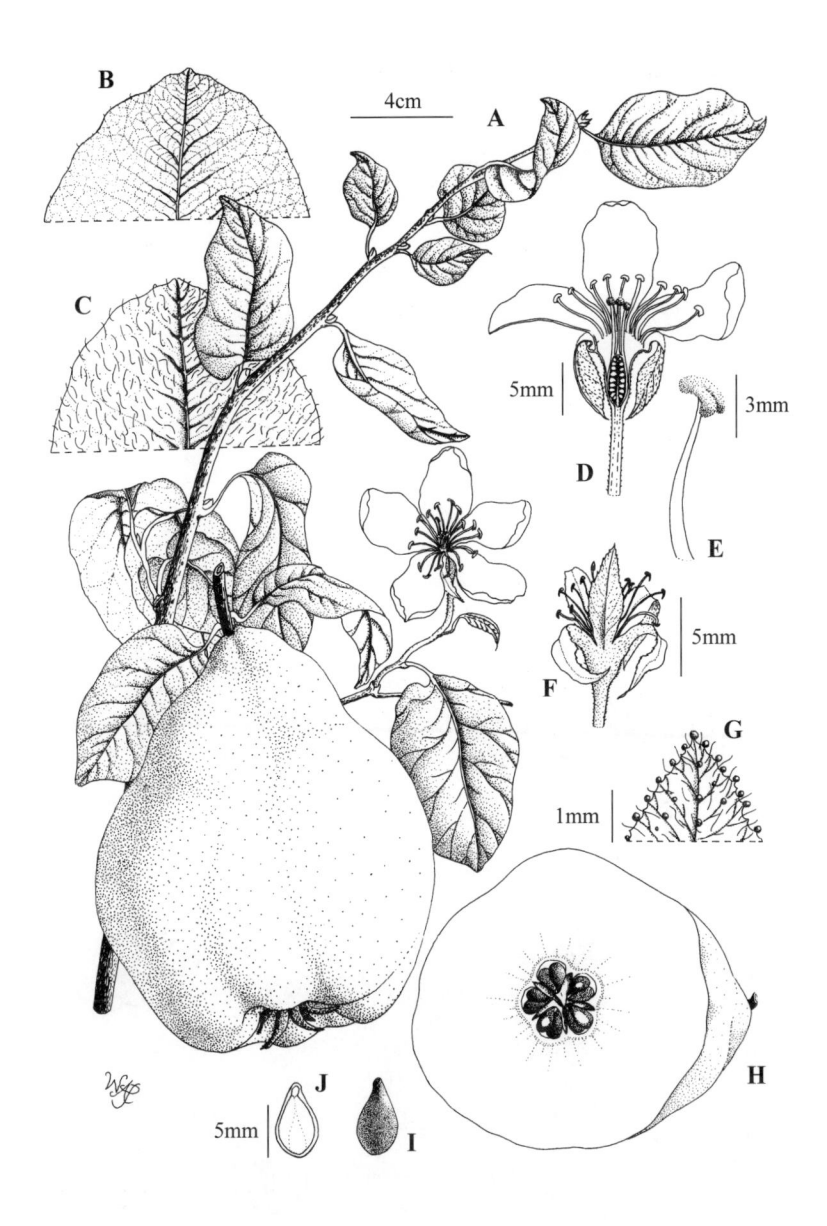

Cydonia oblonga Mill. (Quince). A flowering stem; B upper surface of leaf tip; C lower surface of leaf tip; D LS of flower; E stamen; F flower minus petals; G tip of sepal; H fruit and TS of fruit; I seed; J LS of seed. (*Hort*. Horsham, W Sussex). STACE 442/368.

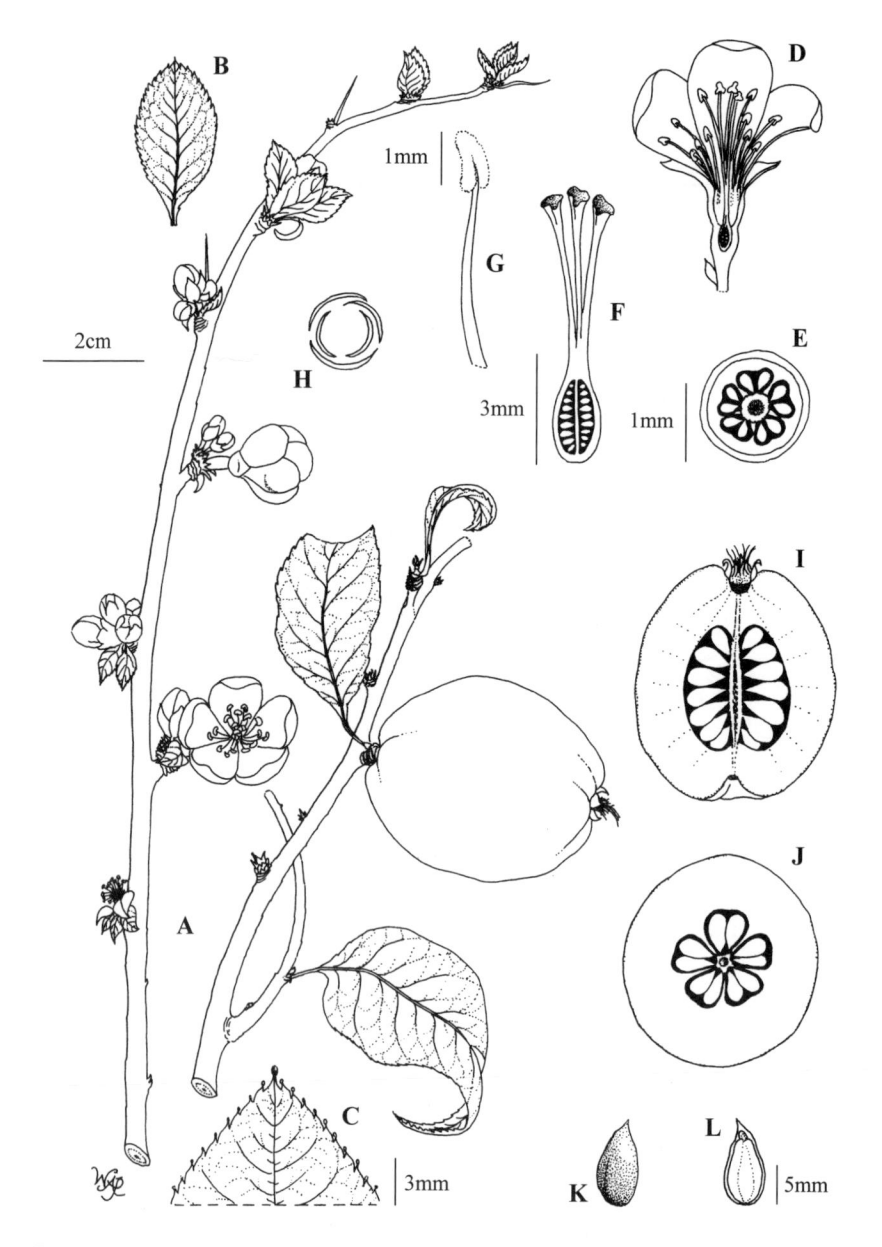

Chaenomeles × superba (Frahm) Rehder (Hybrid Quince). A flowering and fruiting branches; B upper surface of leaf; C leaf tip; D LS of flower; E TS of ovary; F LS of gynoecium; G stamen; H TS of flower bud; I LS of fruit; J TS of fruit; K seed; L LS of seed. (*Hort.* Horsham, W Sussex). STACE —/368.

77. ROSACEAE

Amelanchier lamarckii F.G. Schroed. (Juneberry). A flowering branch; B LS of flower; C fruiting raceme; D LS of fruit. (Dorset). STACE 451/377.

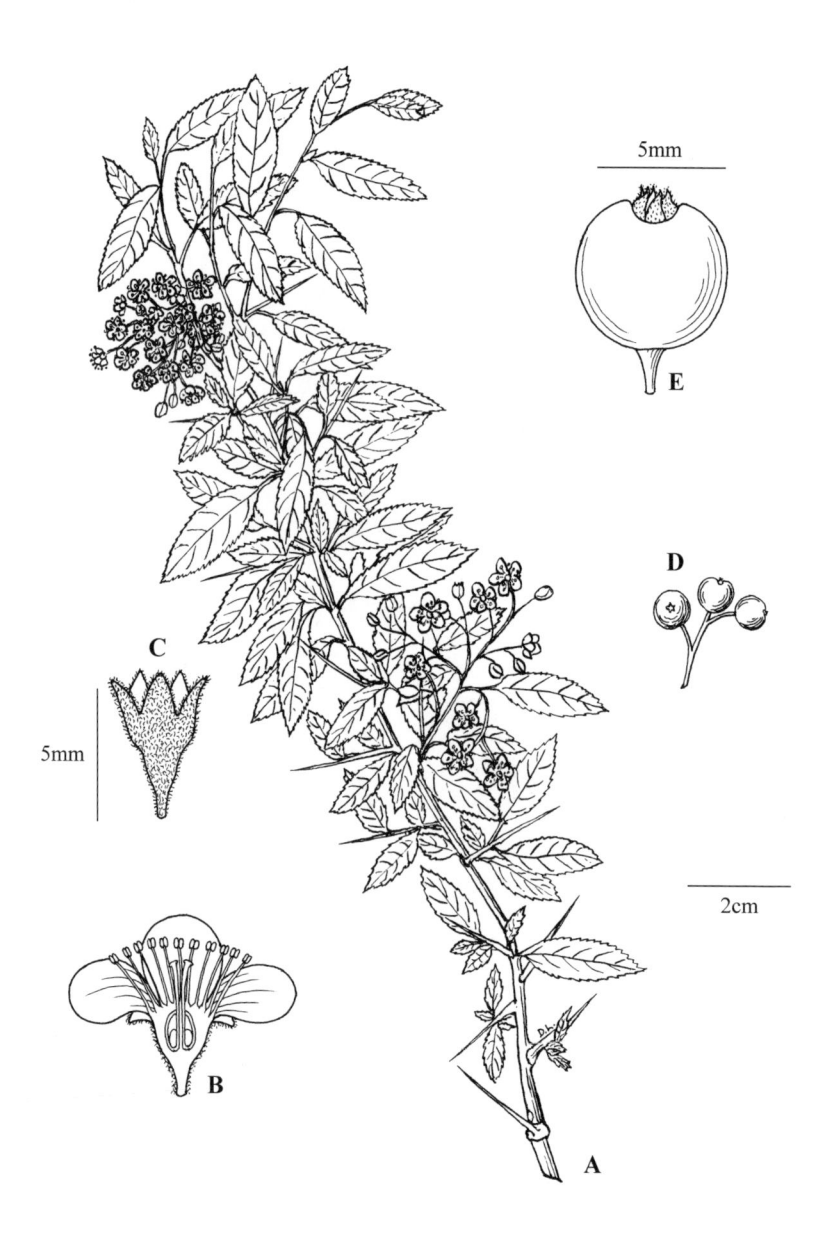

Pyracantha coccinea M. Roem. (Firethorn). A flowering branch; B LS of flower; C hypanthium and sepals; D corymb of fruits; E fruit. (Banstead Downs, Surrey). STACE 463/395.

79. FABACEAE 165

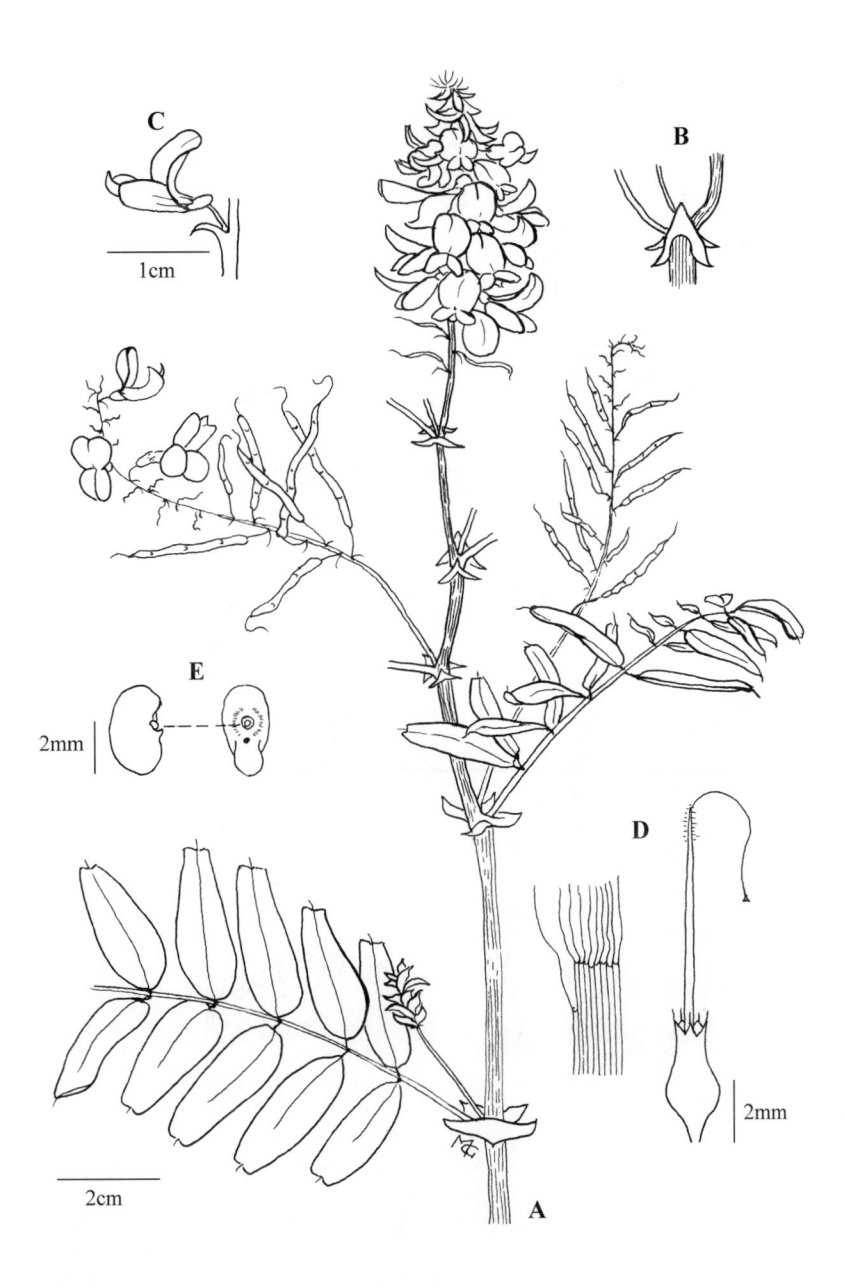

Galega officinalis L. (Goat's-rue). A habit; B stipules at stem node; C flower; D staminal sheath and gynoecium; E seed. (*Hort.* Bexley, W Kent). STACE 474/402.

79. FABACEAE

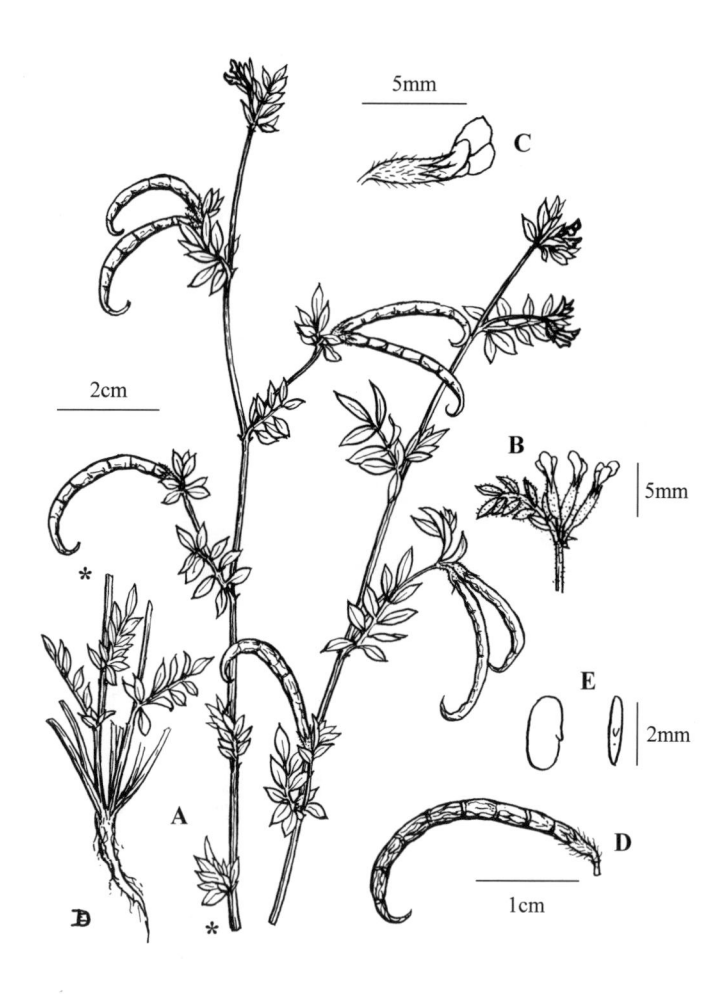

Ornithopus compressus L. (Yellow Serradella). A habit; B cluster of flowers; C flower; D fruit; E seed. (Source unknown). STACE 479/407.

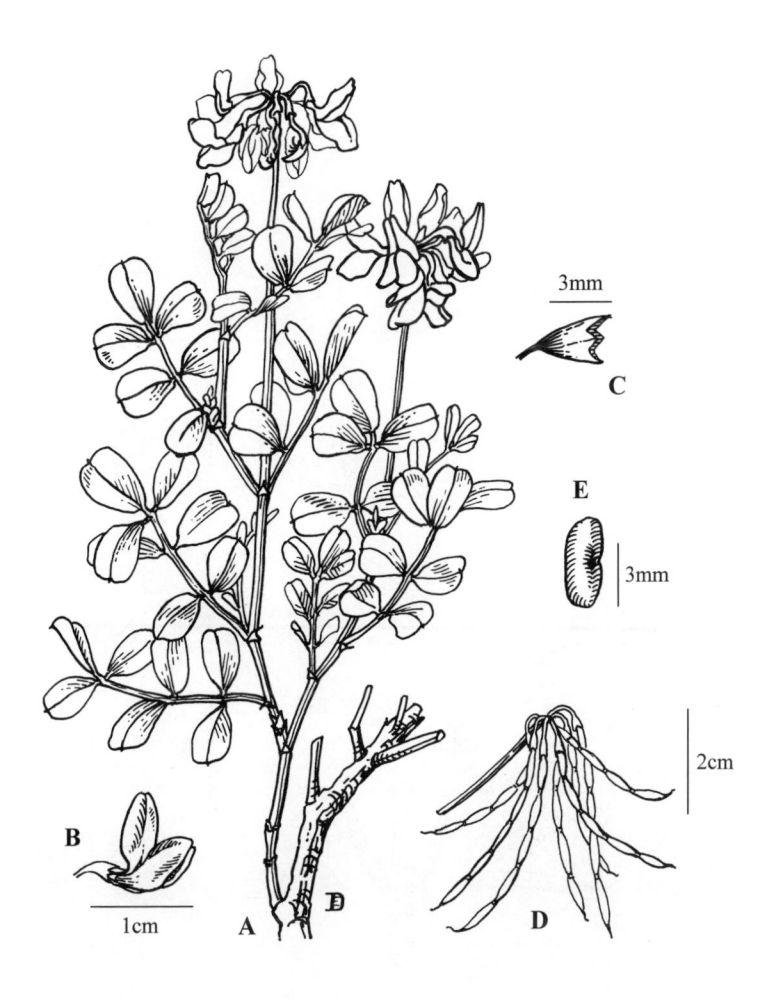

Coronilla valentina L. ssp. **glauca** (L.) Battand. (Shrubby Scorpion-vetch). A flowering branch; B flower; C calyx; D fruits; E seed. (Torquay, S Devon). STACE 480/408.

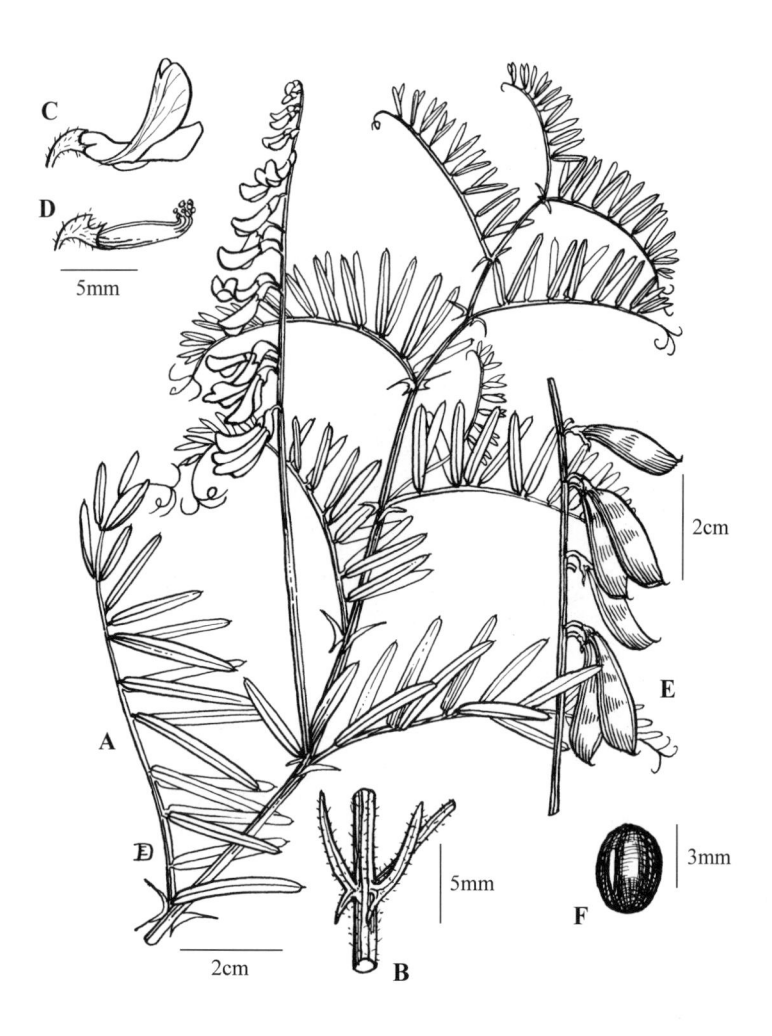

Vicia tenuifolia Roth (Fine-leaved Vetch). A habit; B stipules; C flower; D flower minus petals; E fruits; F seed. (Source unknown). STACE 484/411.

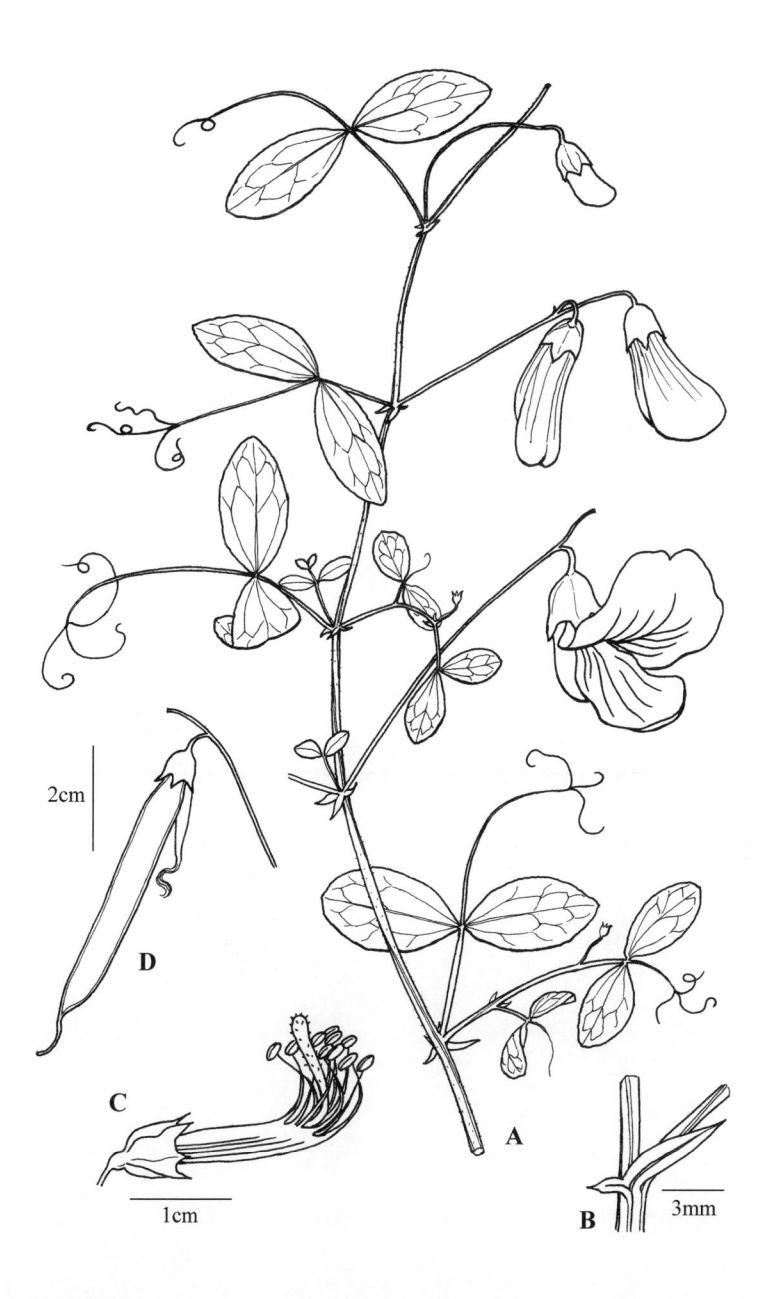

Lathyrus grandiflorus Sm. (Two-flowered Everlasting-pea). A habit; B stipule; C flower minus petals; D fruit. (Betchworth, Surrey). STACE 488/415.

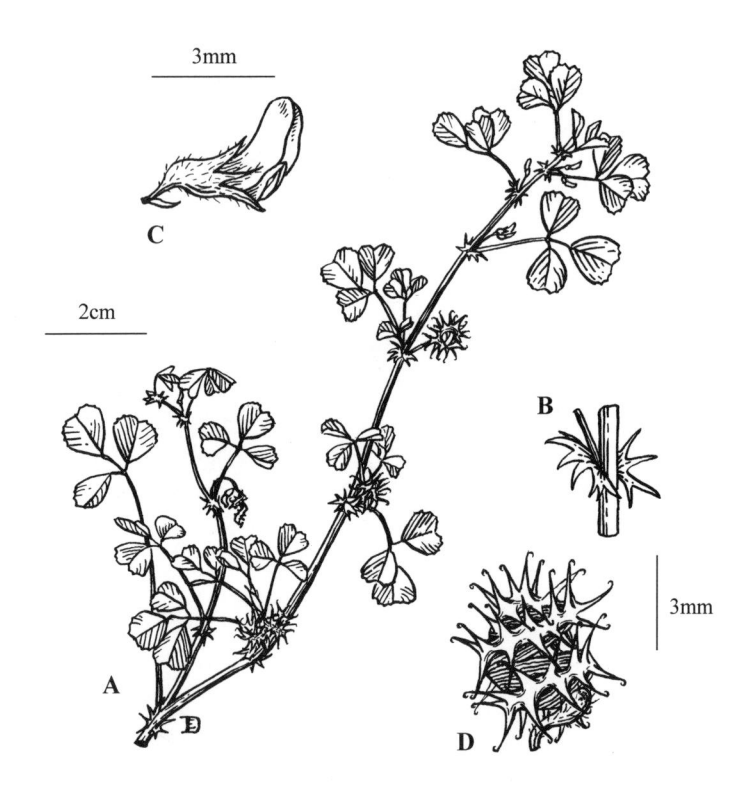

Medicago praecox DC. (Early Medick). A habit; B stipules; C flower; D fruit. (Source unknown). STACE 498/421.

82. GUNNERACEAE

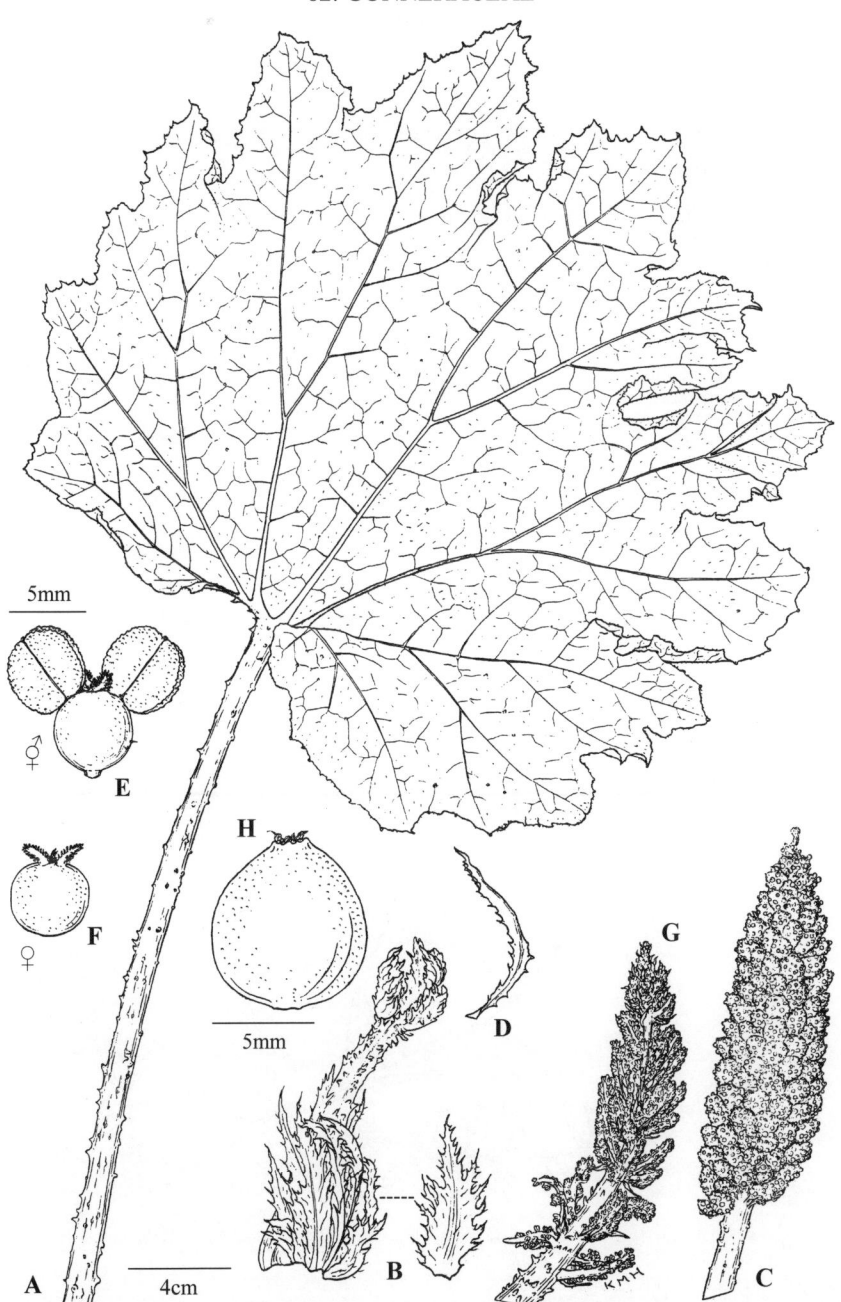

Gunnera tinctoria (Molina) Mirb. (Giant-rhubarb). A leaf; B young leaf surrounded by scale-leaves; C inflorescence; D scale-leaf from scape; E bisexual flower; F female flower; G fruiting panicle; H drupe. (Ballynahinch, W Galway). STACE 519/439.

83. LYTHRACEAE

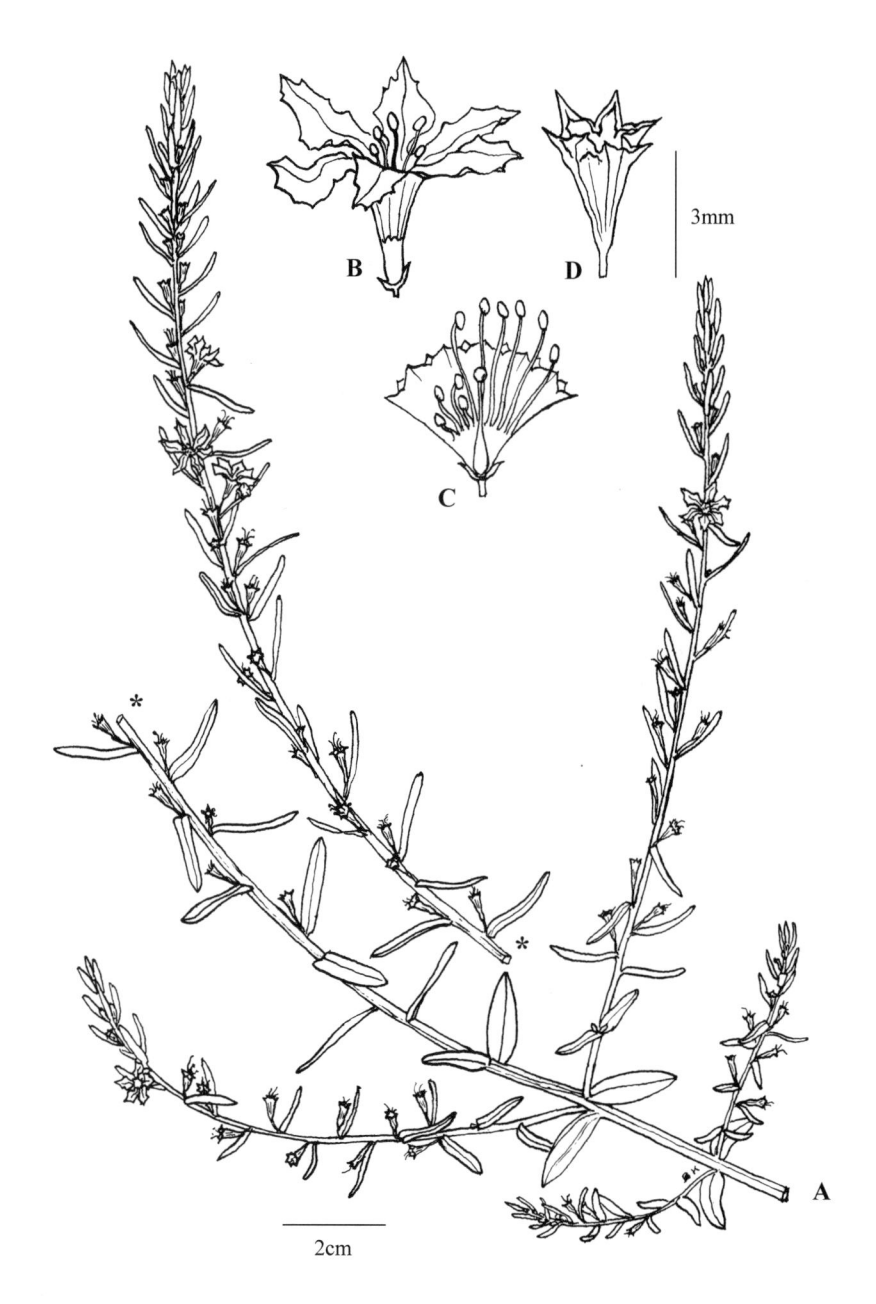

Lythrum junceum Banks & Sol. (False Grass-poly). A habit; B flower; C flower opened out; D calyx. (N Lincs). STACE 520/440.

85. MYRTACEAE

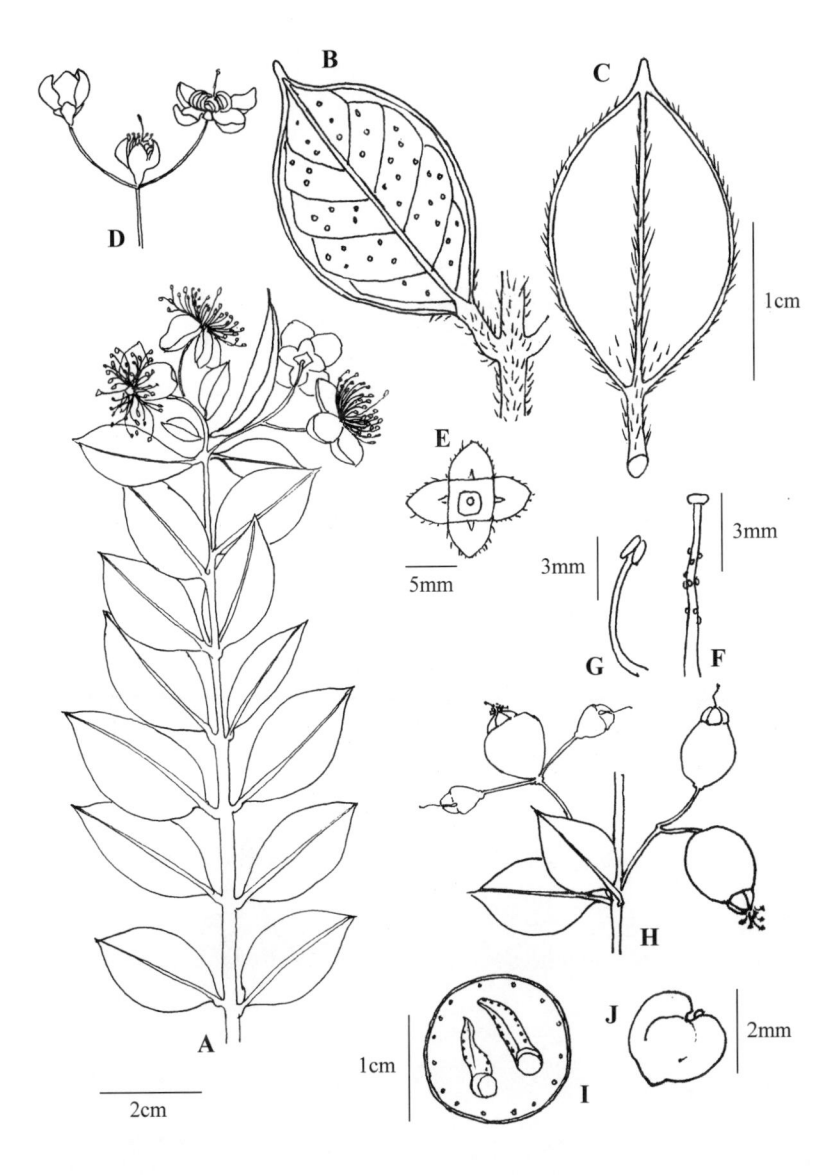

Luma apiculata (DC.) Burret (Chilean Myrtle). A flowering branch; B upperside of leaf; C underside of leaf; D cluster of three flowers (often only one or two); E calyx; F style and stigma; G stamen; H cluster of young fruits; I TS of fruit; J seed. (*Hort.* Stanhoe, W Norfolk). STACE 523/442.

86. ONAGRACEAE

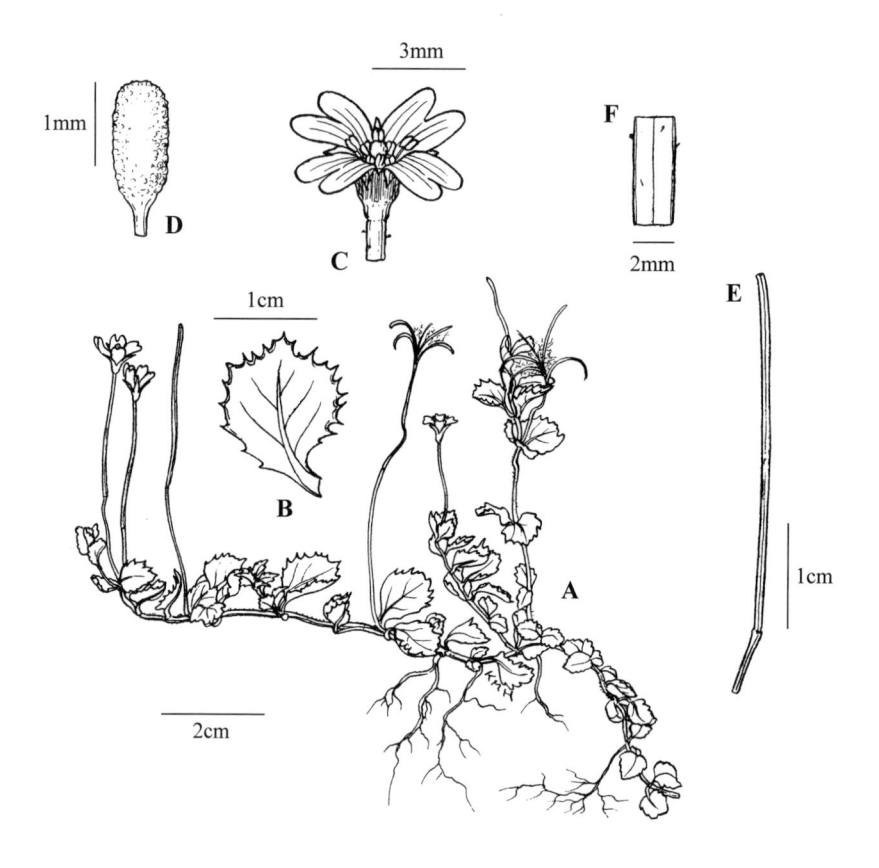

Epilobium pedunculare A. Cunn. (Rockery Willowherb). A habit; B leaf; C flower; D stigma; E capsule; F part of capsule. (*Hort*. ex Leenane, W Galway). STACE 530/448.

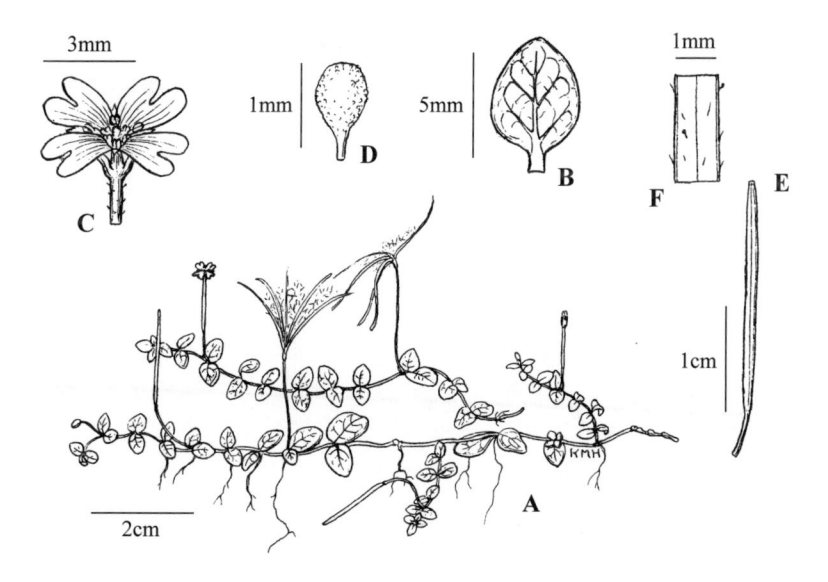

86. ONAGRACEAE

Epilobium komarovianum H. Lév. (Bronzy Willowherb). A habit; B leaf; C flower; D stigma; E capsule; F part of capsule. (Source unknown). STACE 530/448.

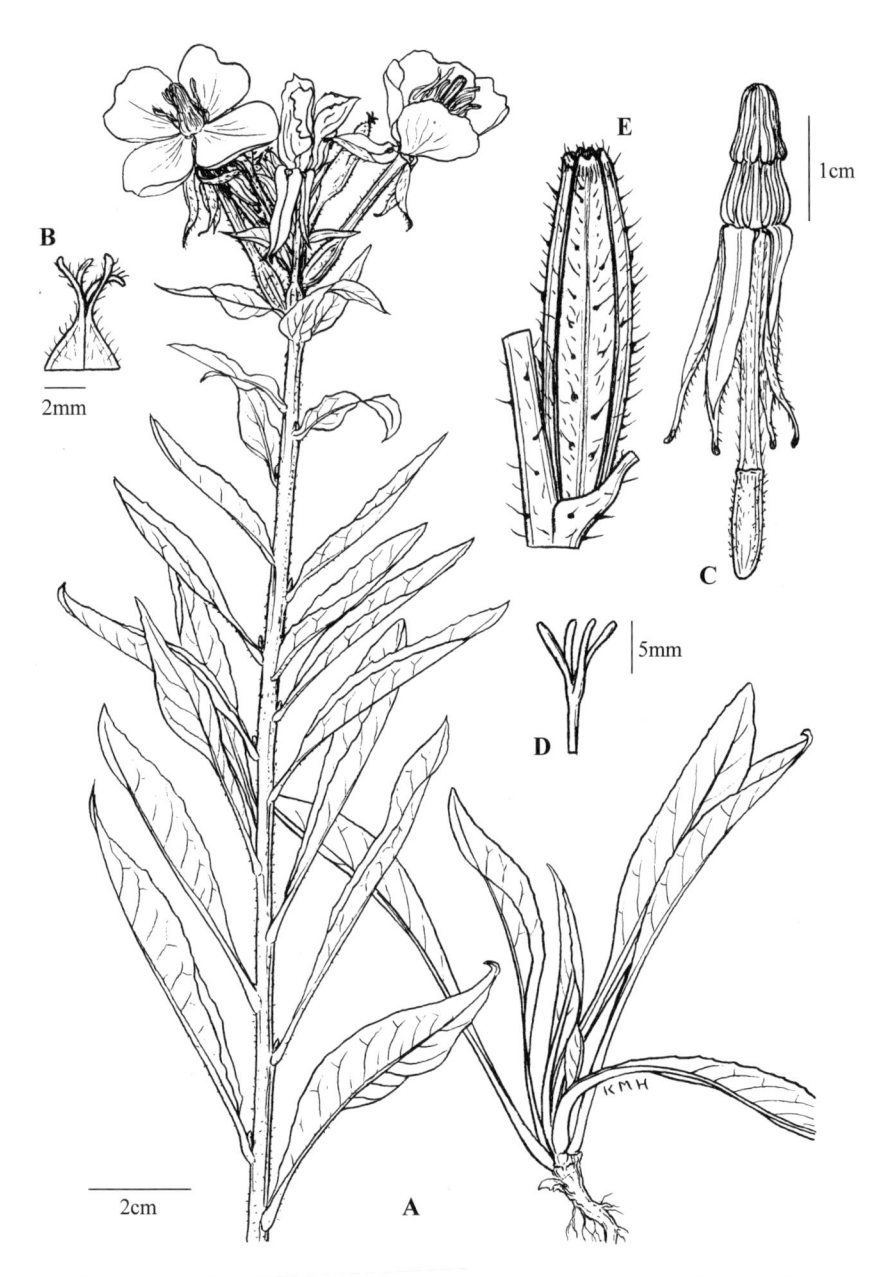

Oenothera rubricaulis Kleb. (No English name). A flowering stem and leaf rosette; B apex of sepals in bud; C flower minus petals; D stigma; E fruit. (Barry Docks, Glam – *det*. K. Rostański, Oct. 1969). STACE 532/449.

86. ONAGRACEAE

Oenothera renneri H. Scholz (Renner's Evening-primrose). A flowering stem and leaf rosette; B apex of sepals in bud; C flower minus petals; D stigma; E fruit. (*Hort.* Ashbourne, Derbys – *det.* K. Rostański, Oct. 1969). STACE 532/451.

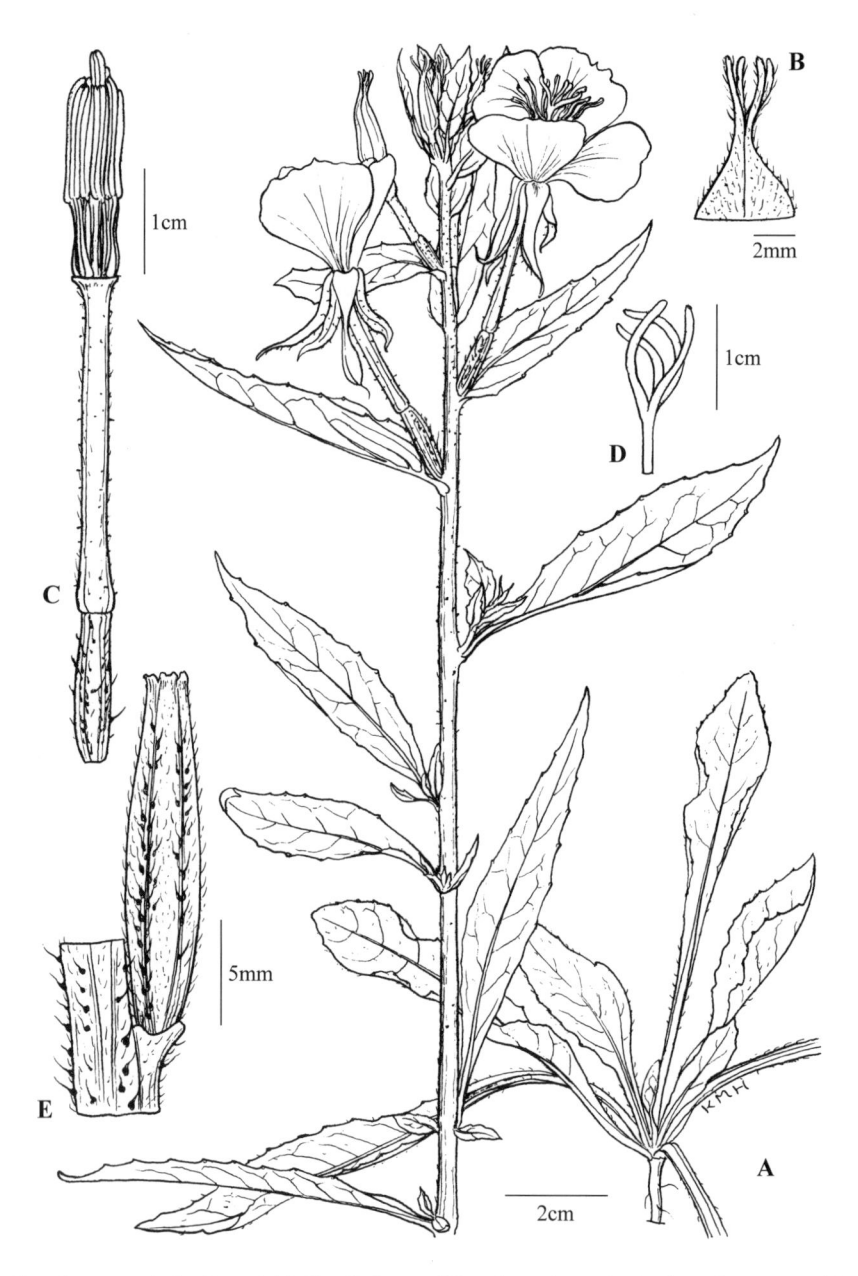

Oenothera cambrica Rostański (Small-flowered Evening-primrose). A flowering stem and leaf rosette; B apex of sepals in bud; C flower minus petals and sepals; D stigma; E fruit. (Pembrey, Carms). STACE 534/451.

86. ONAGRACEAE

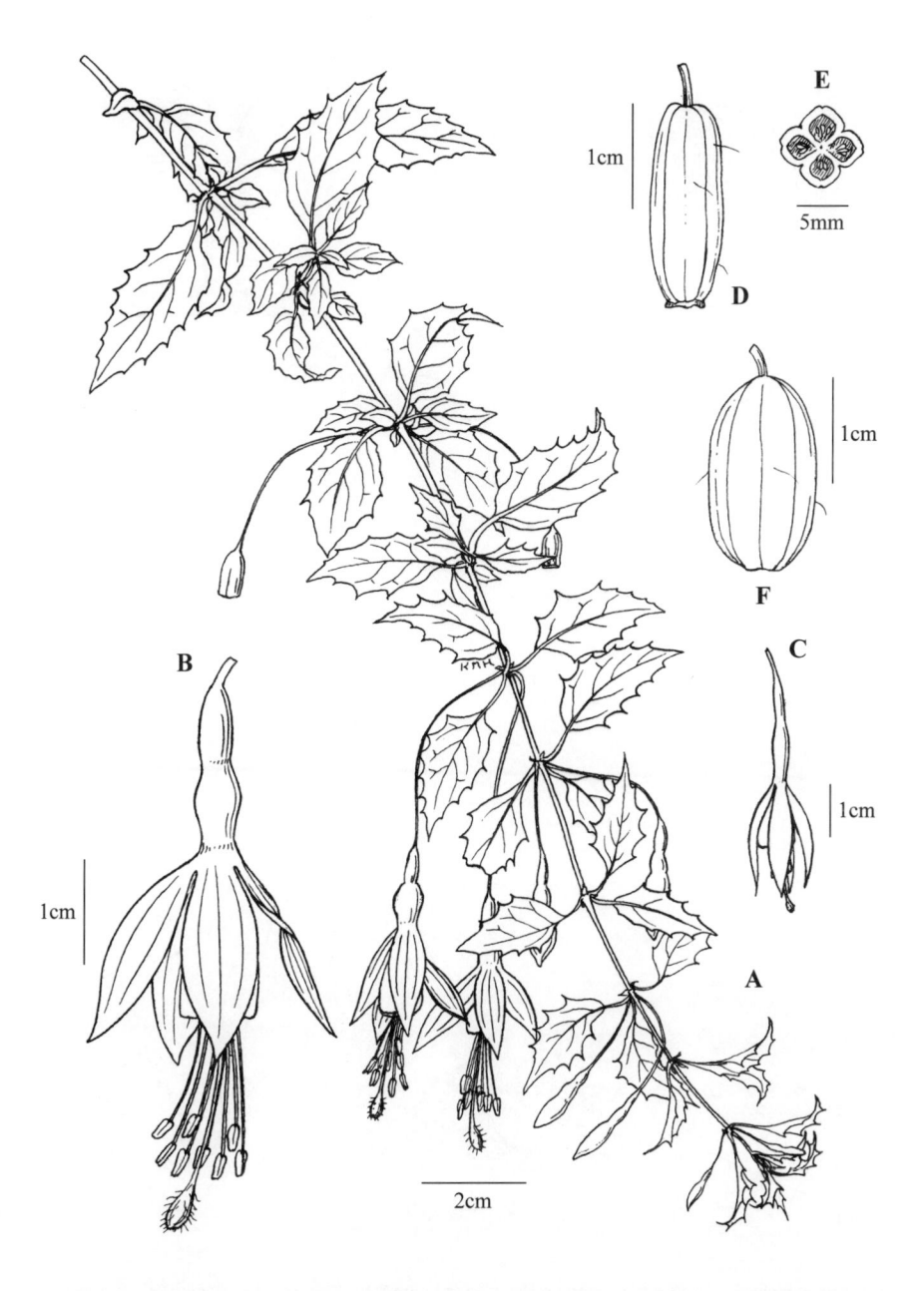

Fuchsia 'Riccartonii' (No English name). A flowering branch; B flower; C flower of **F. magellanica**; D fruit; E TS of fruit; F fruit of **F. magellanica**. (Ireland). STACE 535/452.

Cornus sericea L. (Red-osier Dogwood). A sterile branch; B flowering branch; C flower; D fruiting corymb; E young fruit; F drupe; G stone from inside drupe; H TS of stone with one ovule developed. (Wokingham, Berks). STACE 537/453.

93. EUPHORBIACEAE

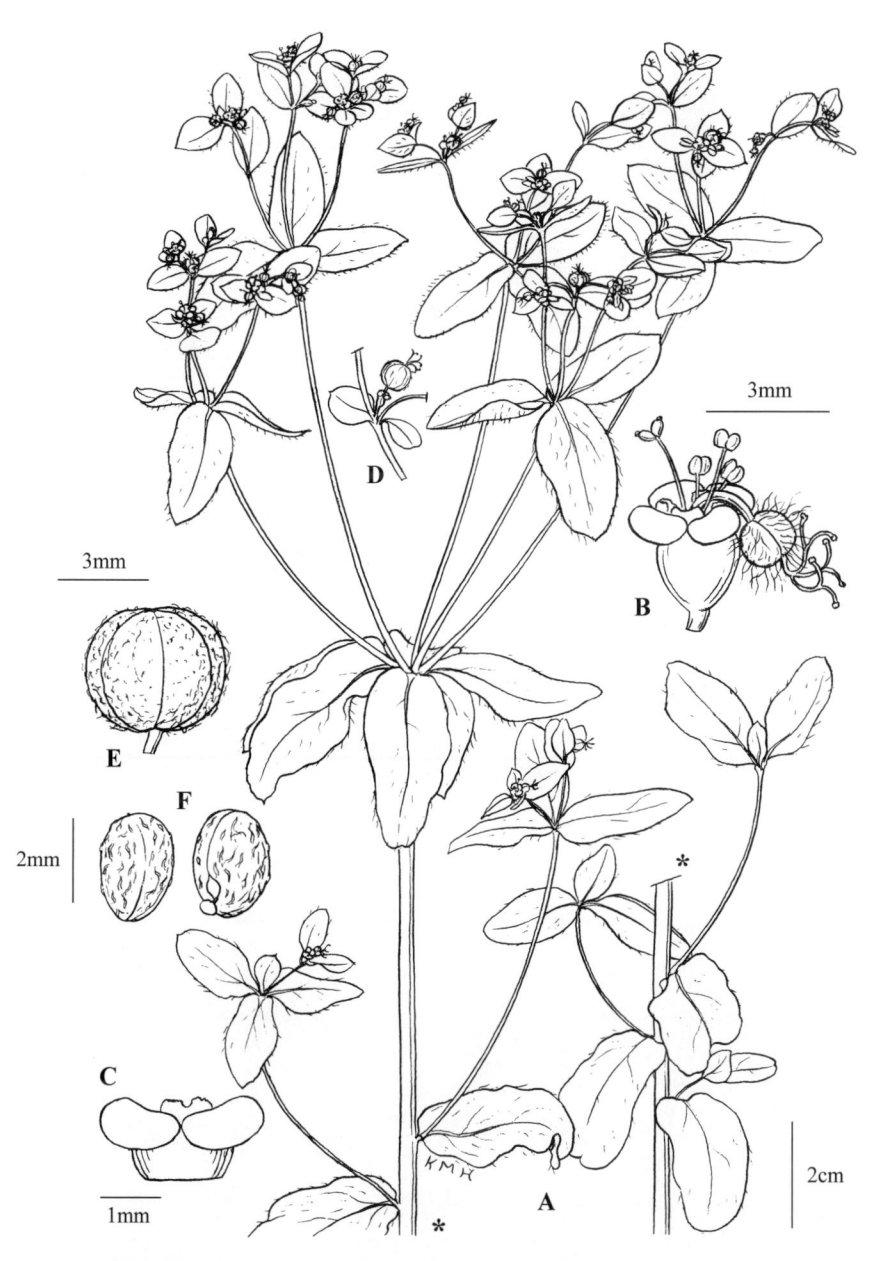

Euphorbia corallioides L. (Coral Spurge). A habit; B cyathium with four male flowers and one female flower; C glands on cyathium; D young fruit in leaf axil; E fruit; F seed. (Slinfold, W Sussex). STACE 545/459.

Parthenocissus inserta (A. Kern.) Fritsch (False Virginia-creeper). A barren branch; B flowering branch; C underside of leaf lobe; D swollen tips of tendrils; E flower; F cluster of fruits; G seed. (Source unknown). STACE 549/464.

100. HIPPOCASTANACEAE

Aesculus hippocastanum L. (Horse-chestnut). A flowering branch; B bisexual flower; C male flower; D unopened fruit. (Source unknown). STACE 553/468.

184　　　　　　　　　　101. ACERACEAE

Acer platanoides L. (Norway Maple). A flowering branch; B summer leaf; C male flower; D female flower; E fruit; F seed. (Source unknown). STACE 556/470.

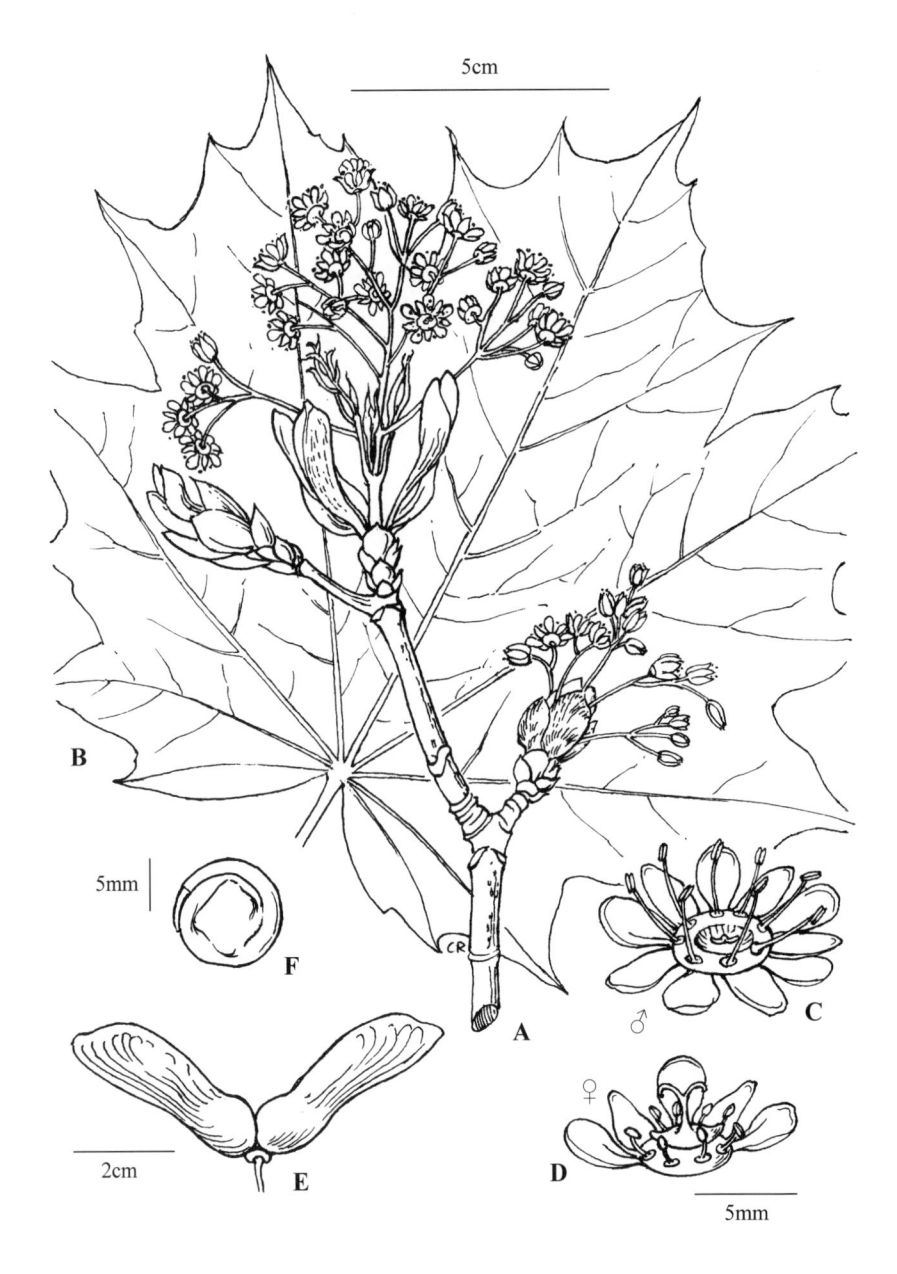

103. SIMAROUBACEAE

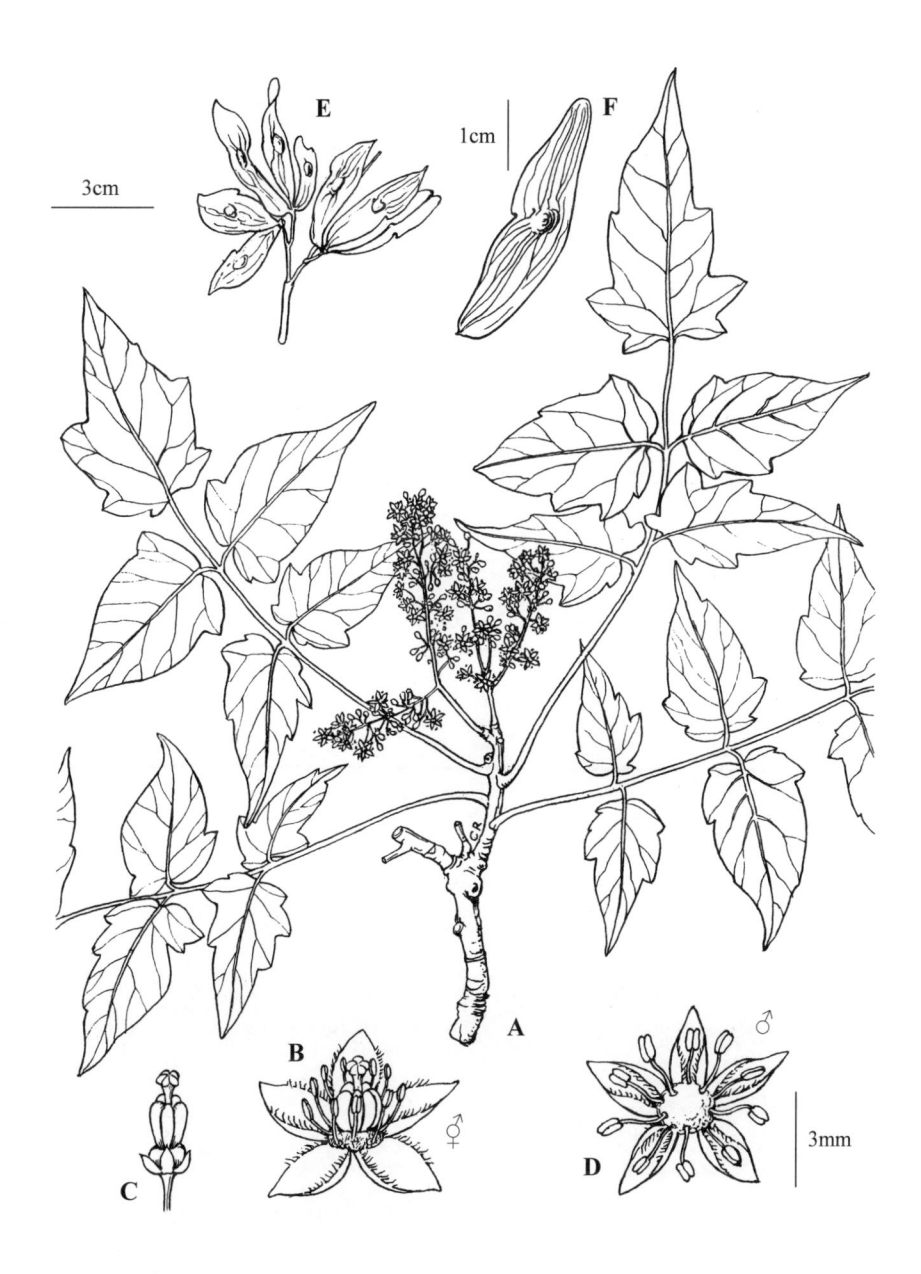

Ailanthus altissima (Mill.) Swingle (Tree-of-heaven). A flowering branch; B bisexual flower; C gynoecium; D male flower; E cluster of fruits; F winged achene. (Source unknown). STACE 557/471.

105. OXALIDACEAE

Oxalis articulata Savigny (Pink-sorrel). A habit; B leaf; C capsule. (Isles of Scilly). STACE 560/475.

105. OXALIDACEAE

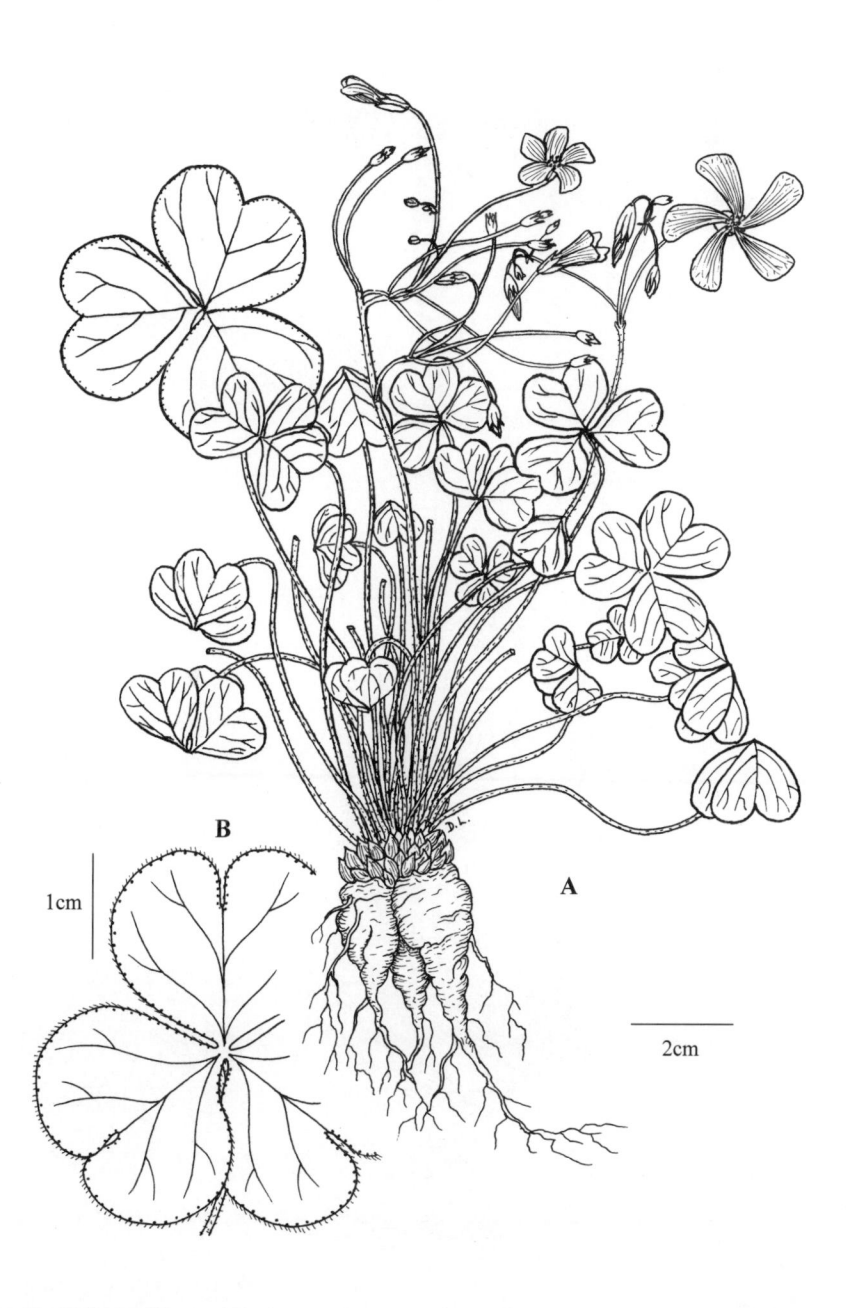

Oxalis debilis Kunth var. **corymbosa** (DC.) Lourt. (Large-flowered Pink-sorrel). A habit; B leaf. (Isle of Man). STACE 562/475.

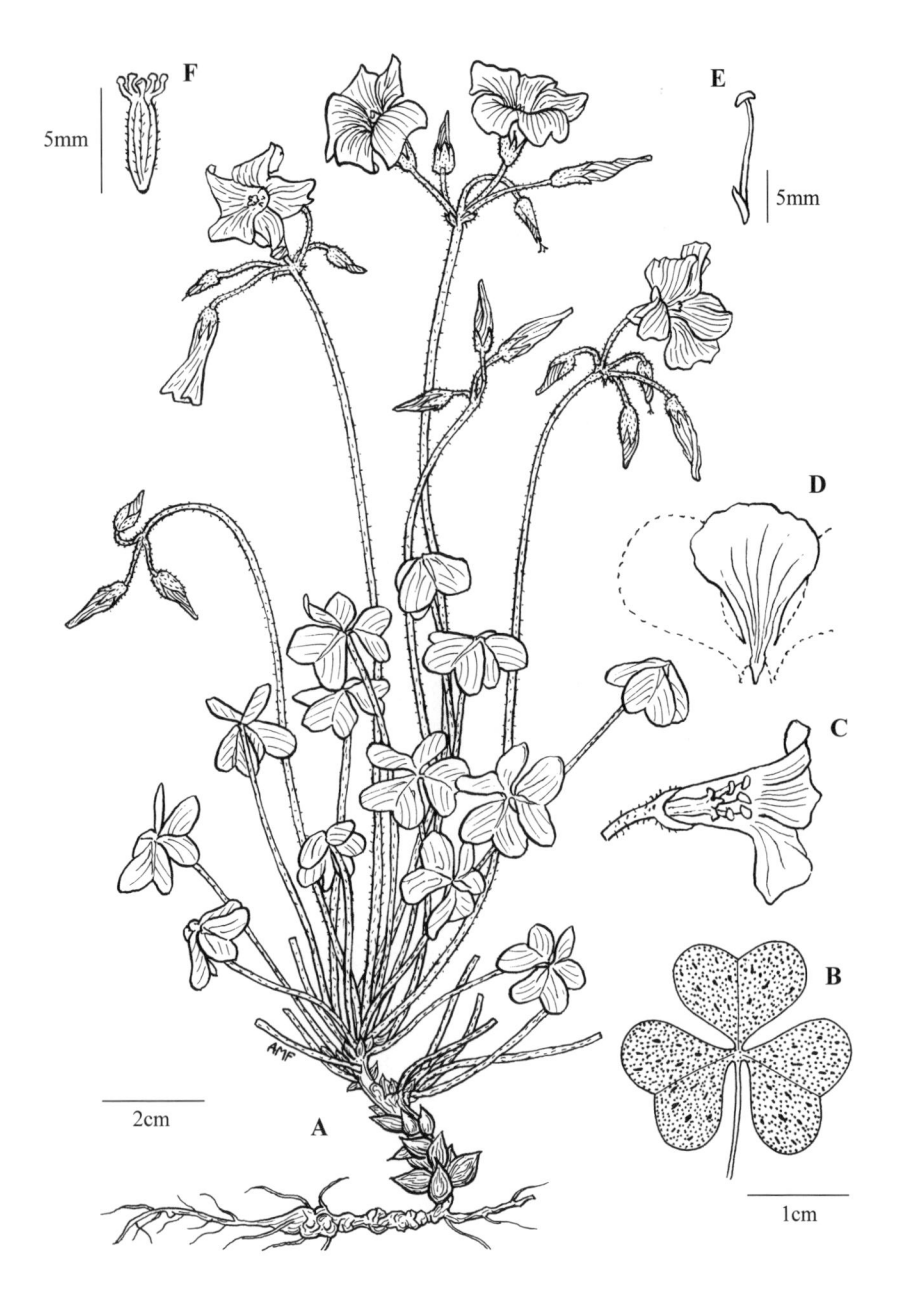

Oxalis pes-caprae L. (Bermuda-buttercup). A habit; B leaf; C LS of flower; D petal; E outer stamen with basal appendage; F immature capsule. (Isles of Scilly). STACE 562/476.

105. OXALIDACEAE   189

Oxalis incarnata L. (Pale Pink-sorrel). A habit; B bulbil; C leaf; D capsule. (Portsmouth, S Hants). STACE 563/476.

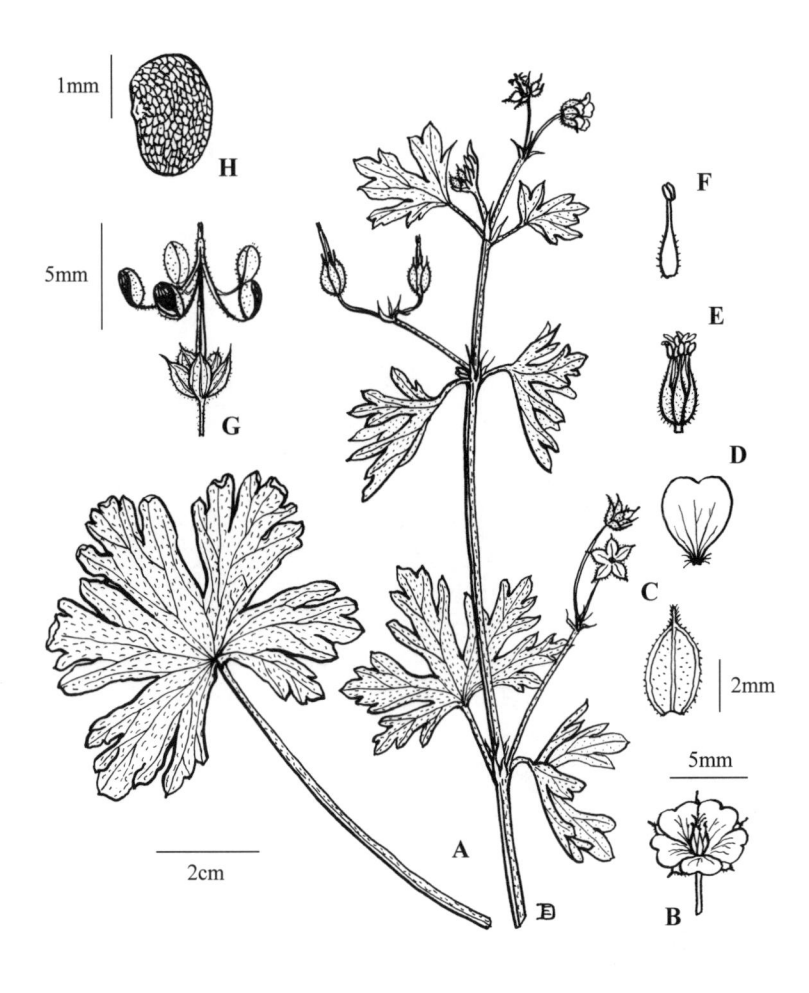

Geranium submolle Steud. (Alderney Crane's-bill). A habit and basal leaf; B flower; C sepal; D petal; E flower minus petals and sepals; F stamen; G fruit dehiscing; H seed. (Guernsey). STACE 568/480.

106. GERANIACEAE

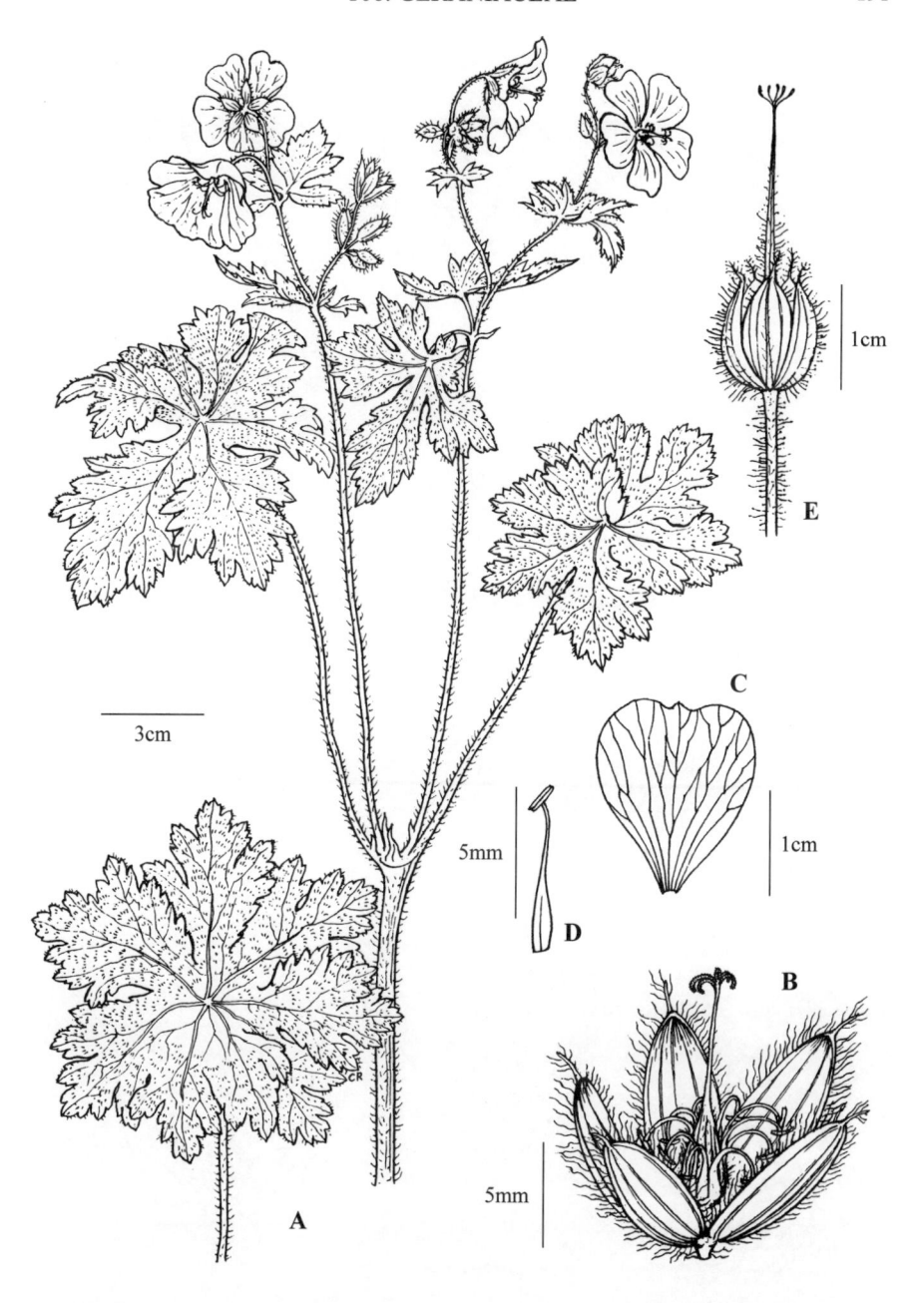

Geranium × magnificum Hyl. (Purple Crane's-bill). A habit and basal leaf; B flower minus petals; C petal; D stamen; E sterile fruit with calyx. (*Hort*. Bromley, W Kent). STACE 568/480.

192   106. GERANIACEAE

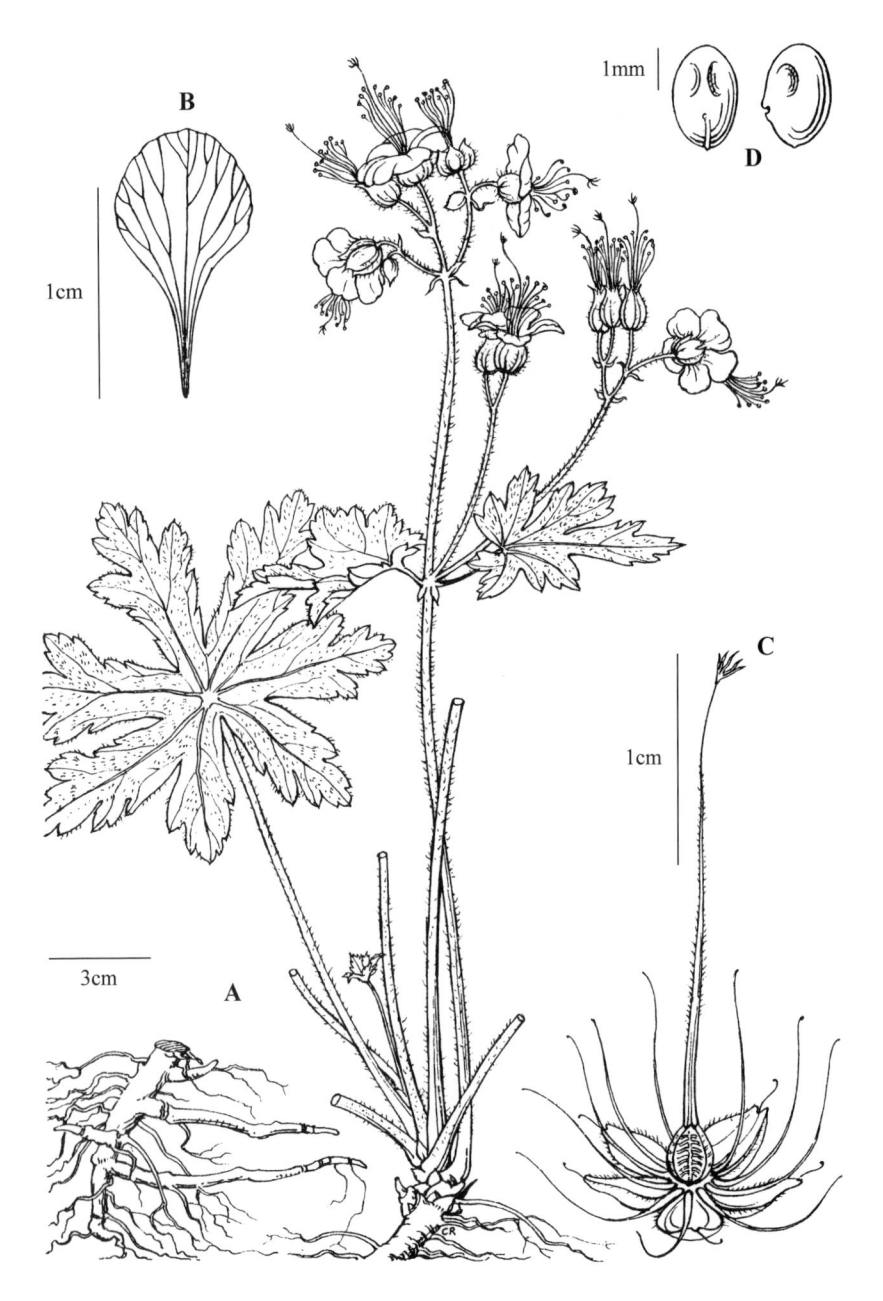

Geranium macrorrhizum L. (Rock Crane's-bill). A habit; B petal; C flower after anthesis, opened out; D seed. (Source unknown). STACE 570/482.

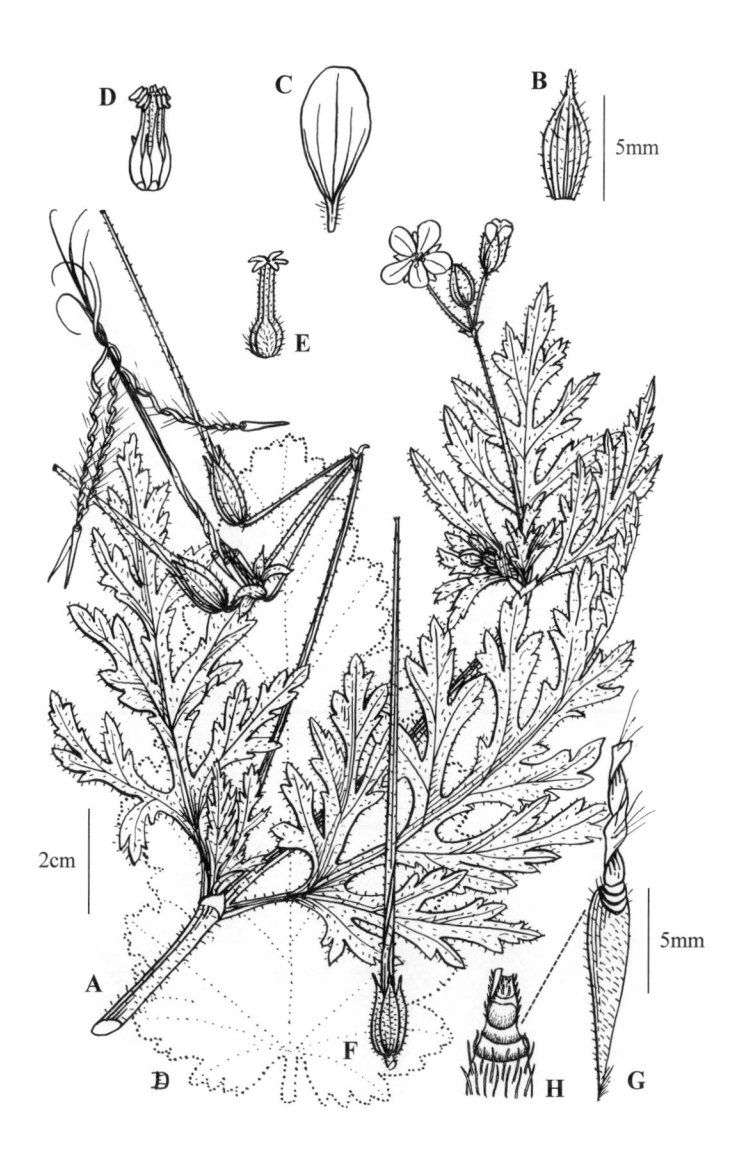

Erodium botrys (Cav.) Bertol. (Mediterranean Stork's-bill). A habit and basal leaf; B sepal; C petal; D flower minus petals and sepals; E gynoecium; F fruit; G mericarp and base of spiral beak; H apex of mericarp. (Blackmoor, N Hants). STACE 573/485.

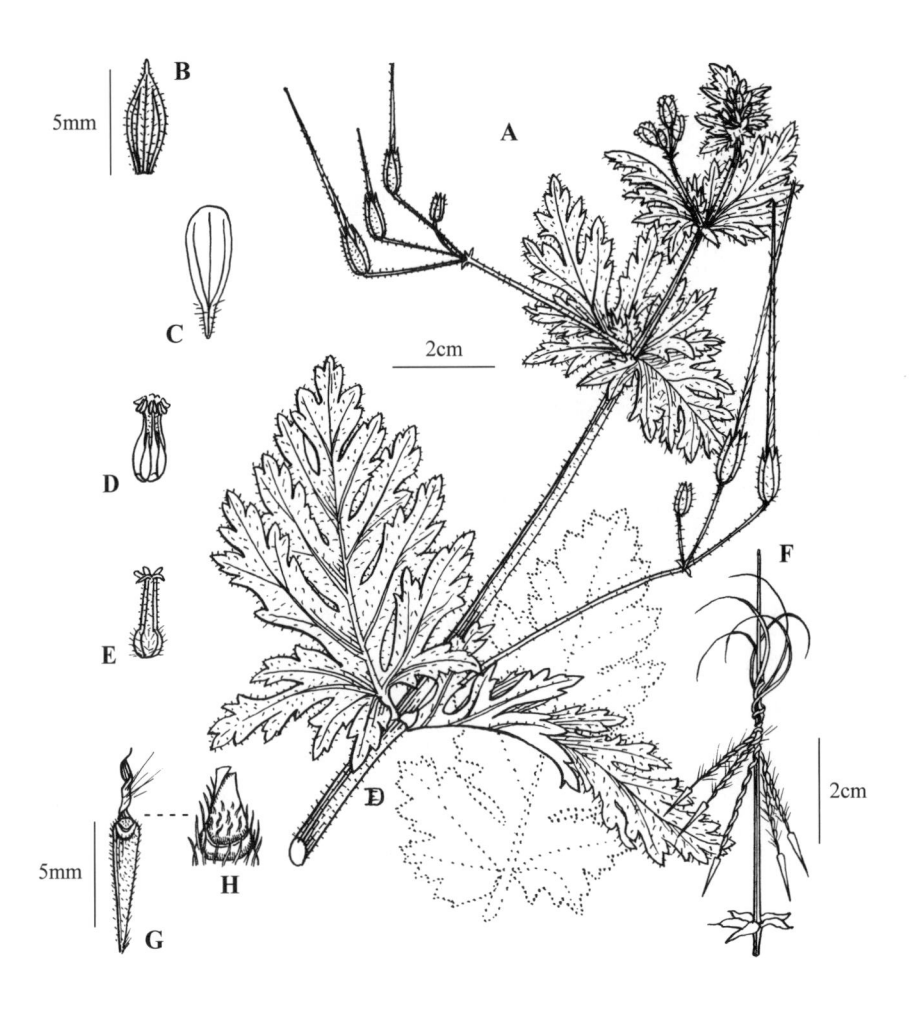

Erodium brachycarpum (Godr.) Thell. (Hairy-pitted Stork's-bill). A habit and basal leaf; B sepal; C petal; D flower minus petals and sepals; E gynoecium; F fruit dehiscing; G mericarp and base of spiral beak; H apex of mericarp. (Blackmoor, N Hants). STACE 575/485.

106. GERANIACEAE

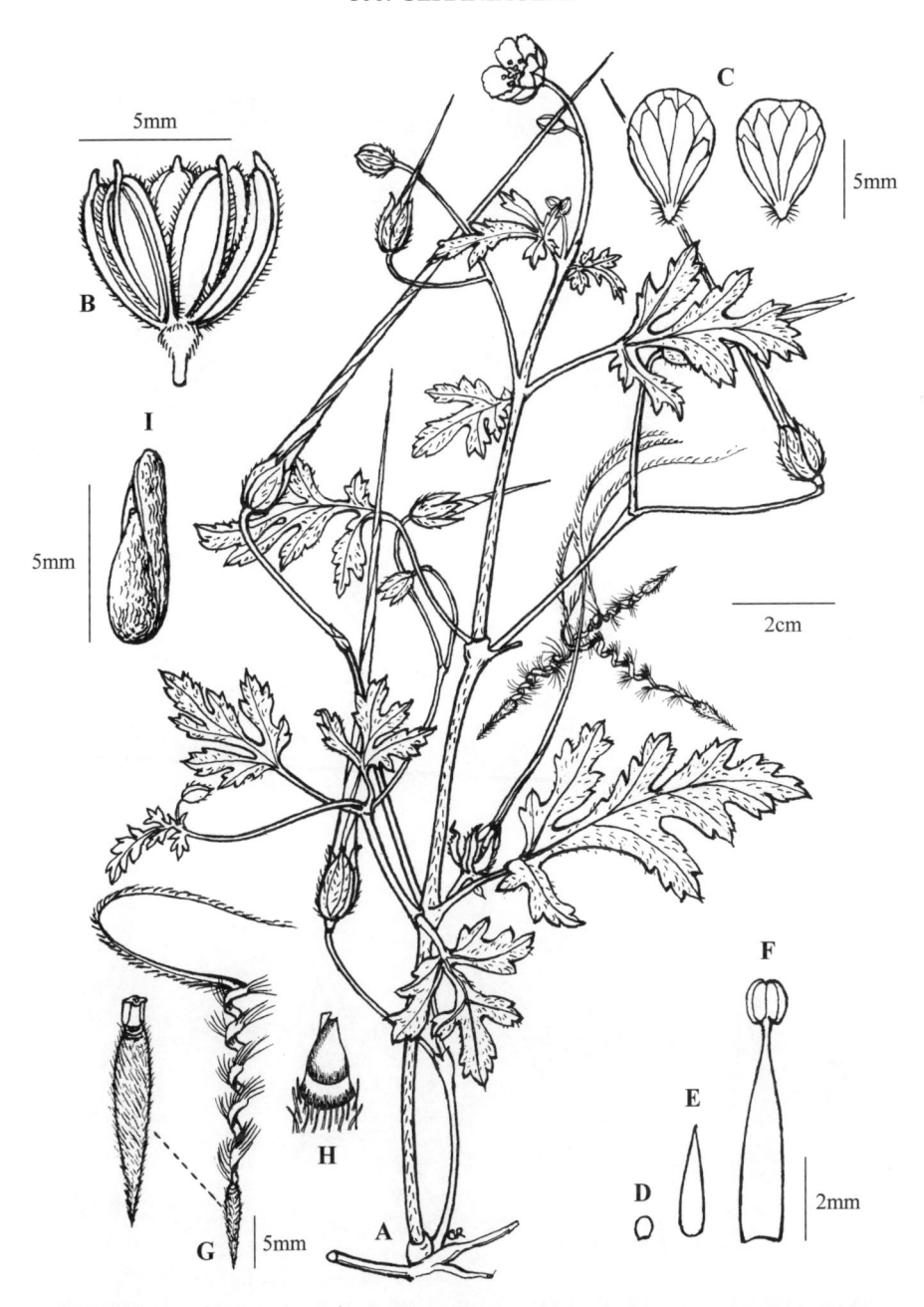

Erodium crinitum Carolin (Eastern Stork's-bill). A habit; B calyx; C petals; D nectary; E staminode; F stamen; G mericarp and spiral awn; H apex of mericarp; I seed. (Source unknown). STACE 575/485.

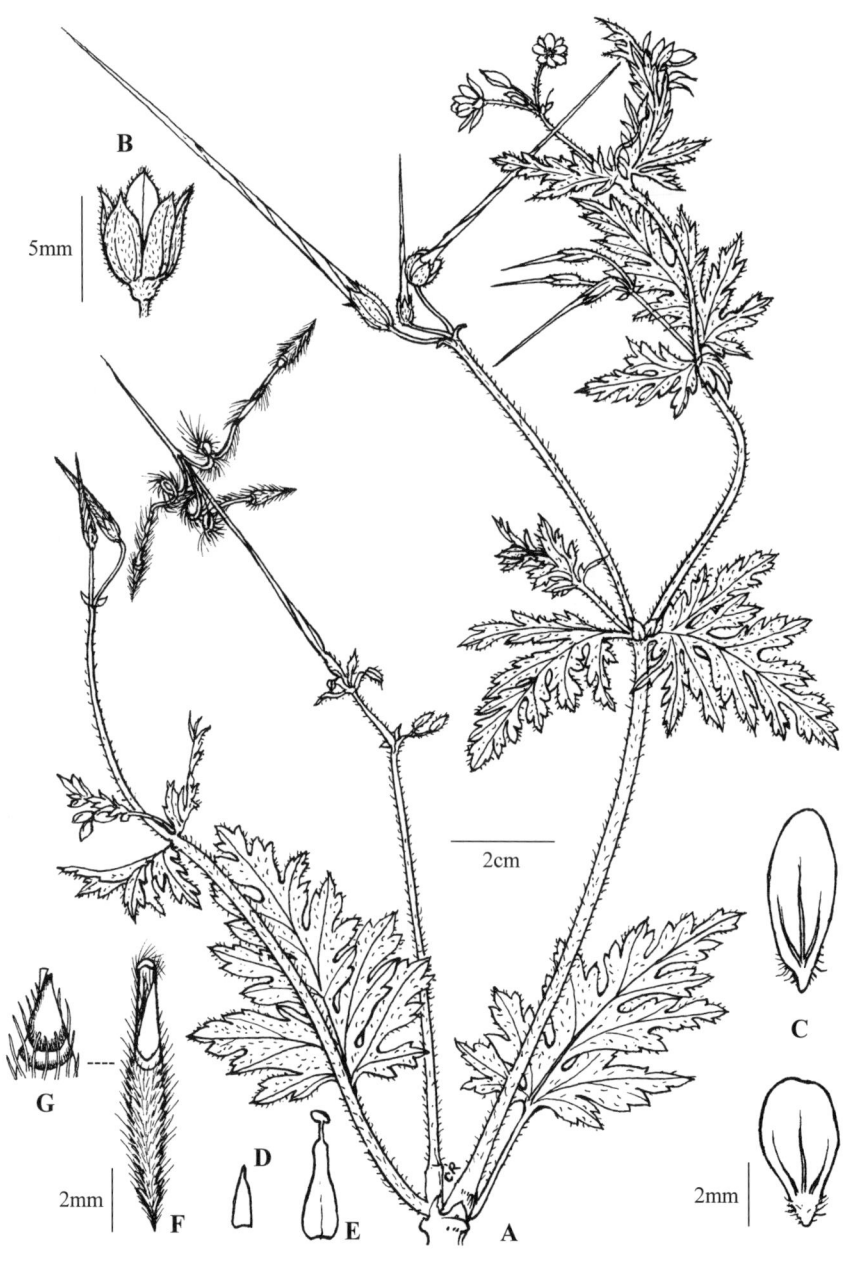

Erodium cygnorum Nees (Western Stork's-bill). A habit; B calyx; C petals; D staminode; E stamen; F mericarp and base of spiral awn; G apex of mericarp. (Blackmoor, N Hants). STACE 575/485.

110. ARALIACEAE

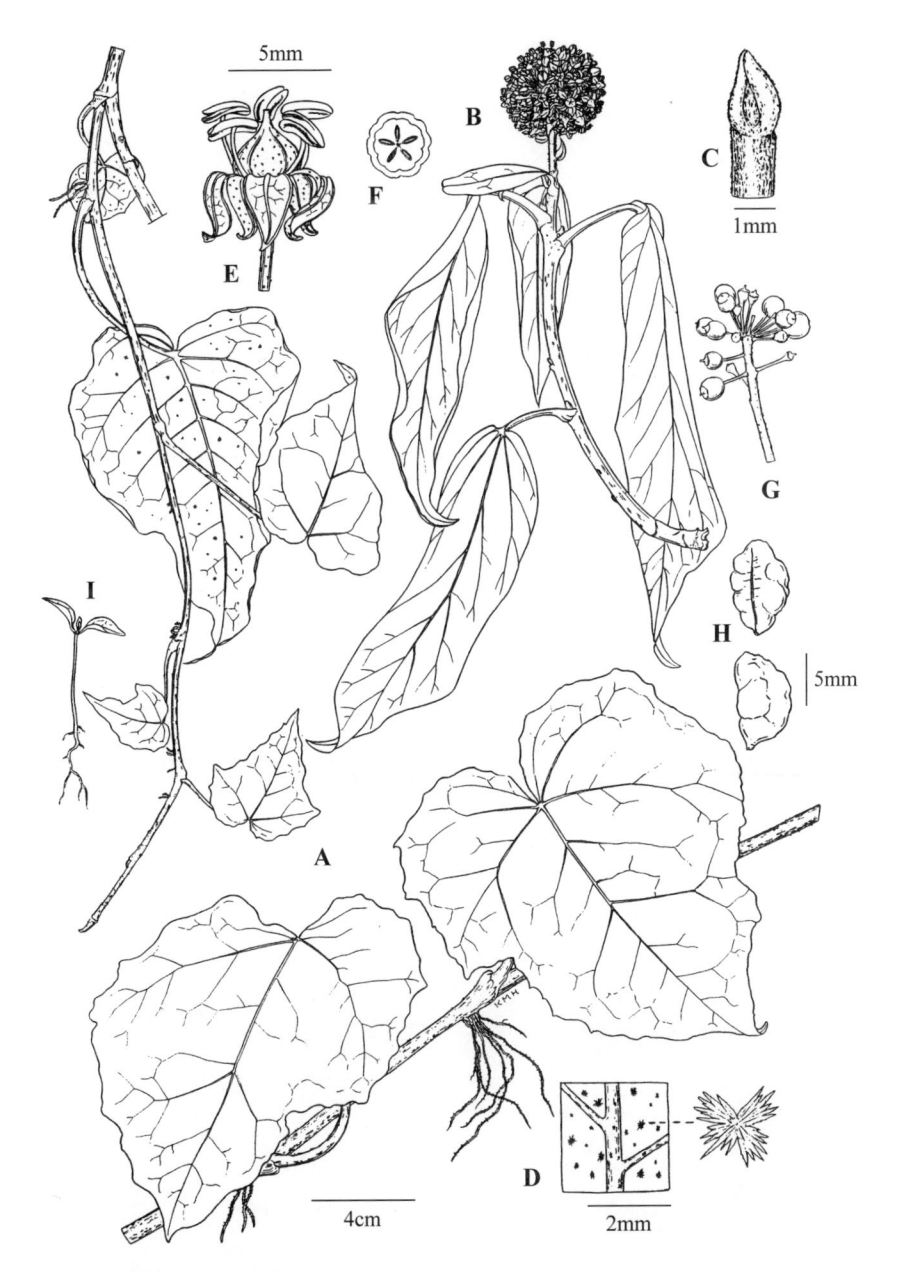

Hedera colchica (K. Koch) K. Koch (Persian Ivy). A barren branches; B flowering branch; C tip of stem; D underside of young leaf with semi-peltate hairs; E flower; F TS of ovary; G fruiting head; H seed; I seedling. (Ardrishaig, Kintyre). STACE 579/490.

111. APIACEAE

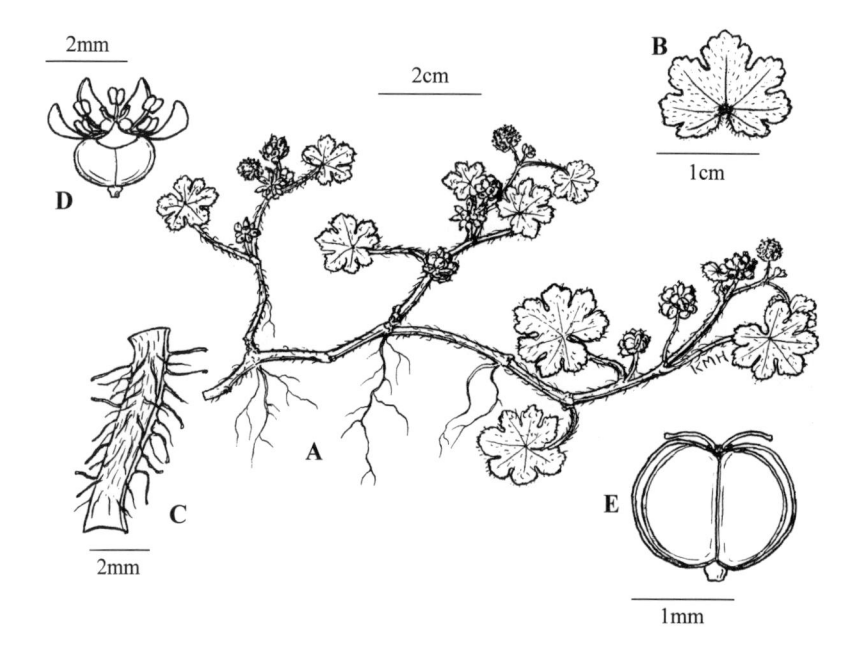

Hydrocotyle moschata G. Forst. (Hairy Pennywort). A habit; B leaf blade; C petiole; D flower; E fruit. (Valencia Island, S Kerry). STACE 588/497.

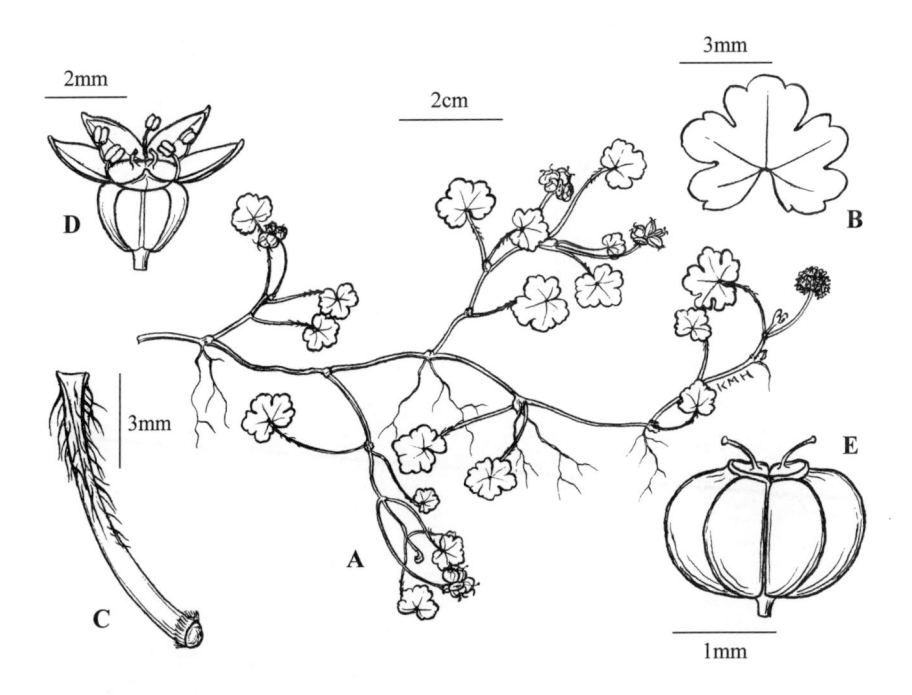

111. APIACEAE

Hydrocotyle novae-zeelandiae DC. (New Zealand Pennywort). A habit; B leaf blade; C petiole; D flower; E fruit. (Source unknown). STACE 588/497.

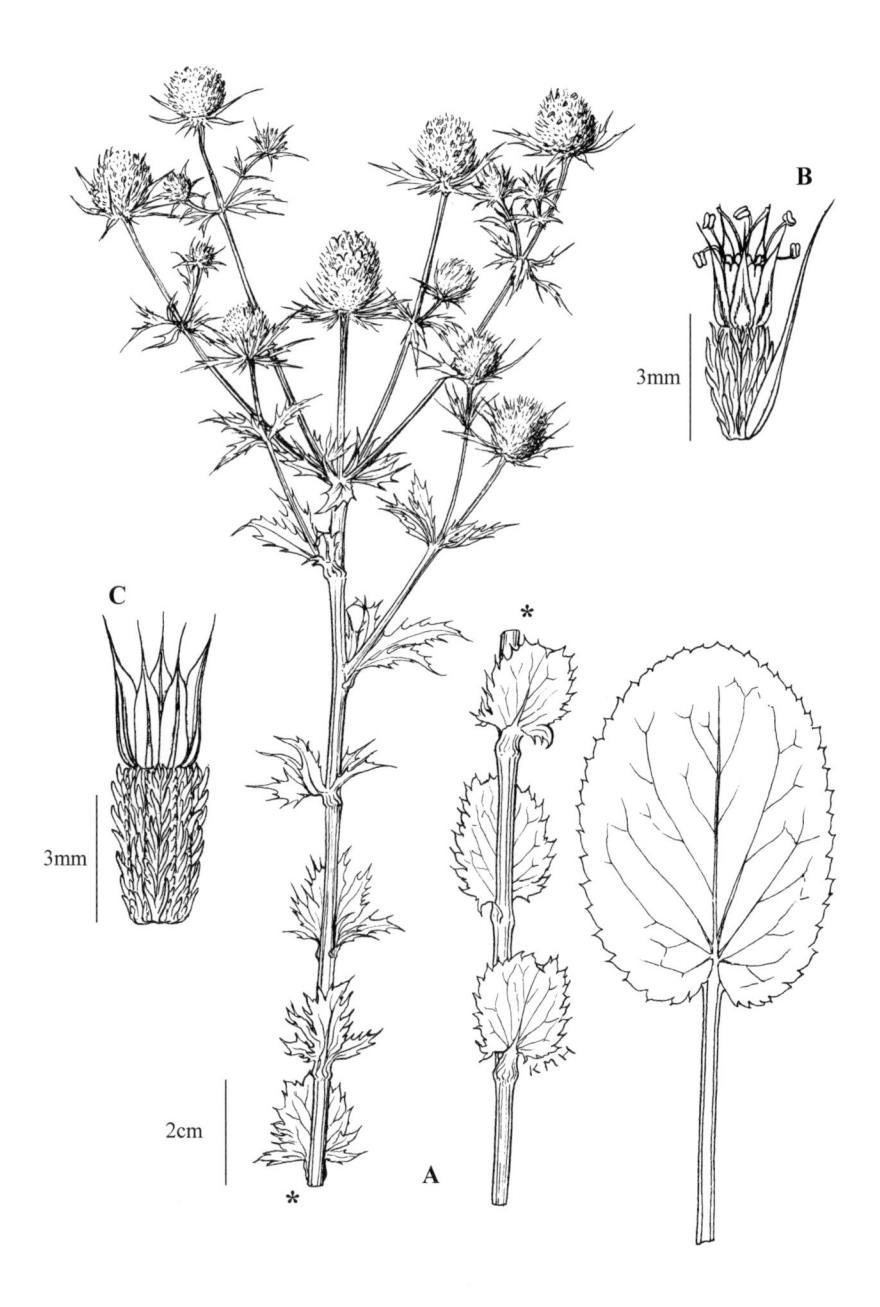

Eryngium planum L. (Blue Eryngo). A flowering stem and basal leaf; B flower; C fruit. (Littlestone-on-Sea, E Kent). STACE 591/500.

111. APIACEAE

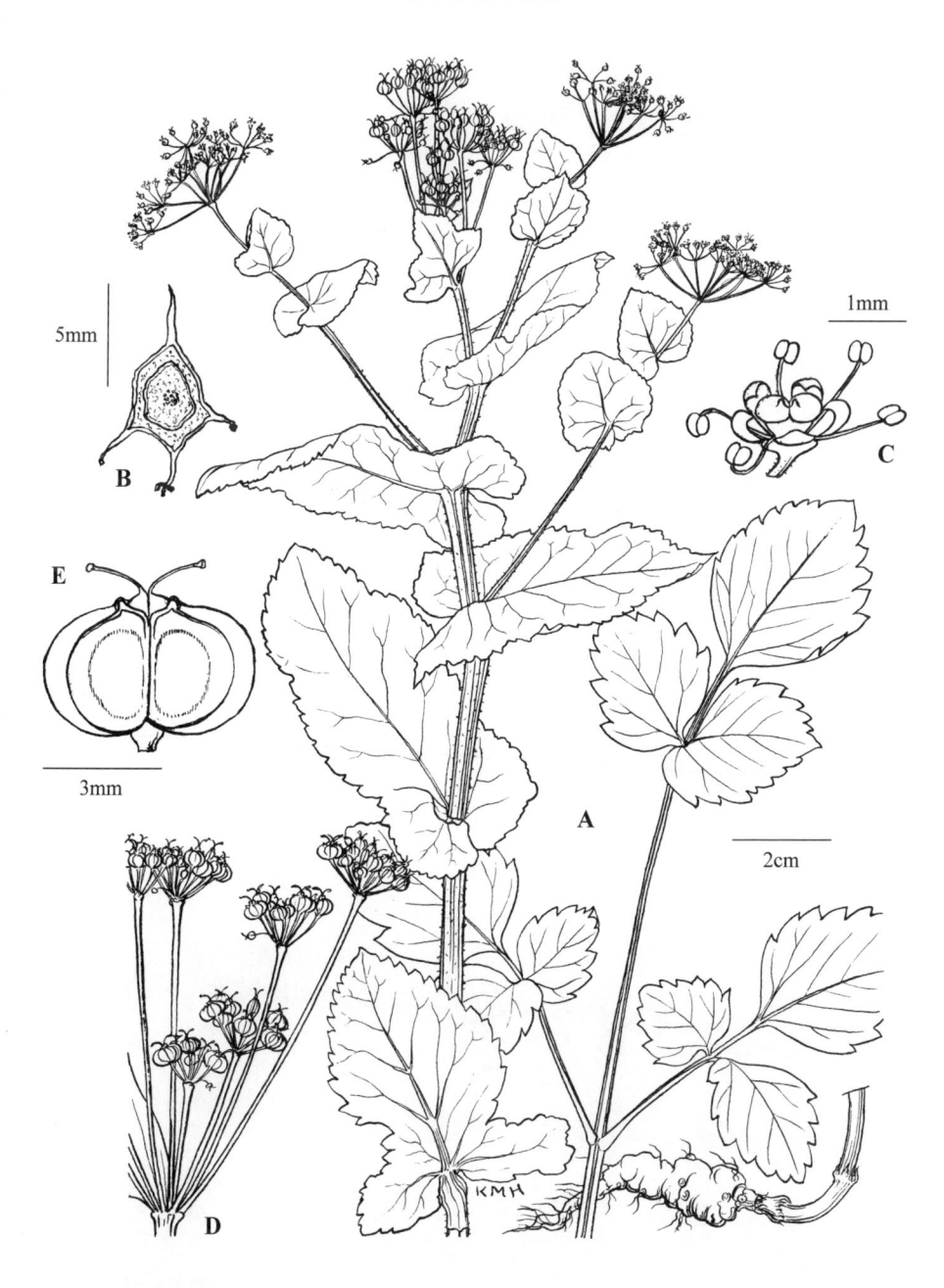

Smyrnium perfoliatum L. (Perfoliate Alexanders). A flowering stem and basal leaf; B TS of stem; C flower; D umbel in fruit; E fruit. (Chelsea, Middlesex). STACE 594/502.

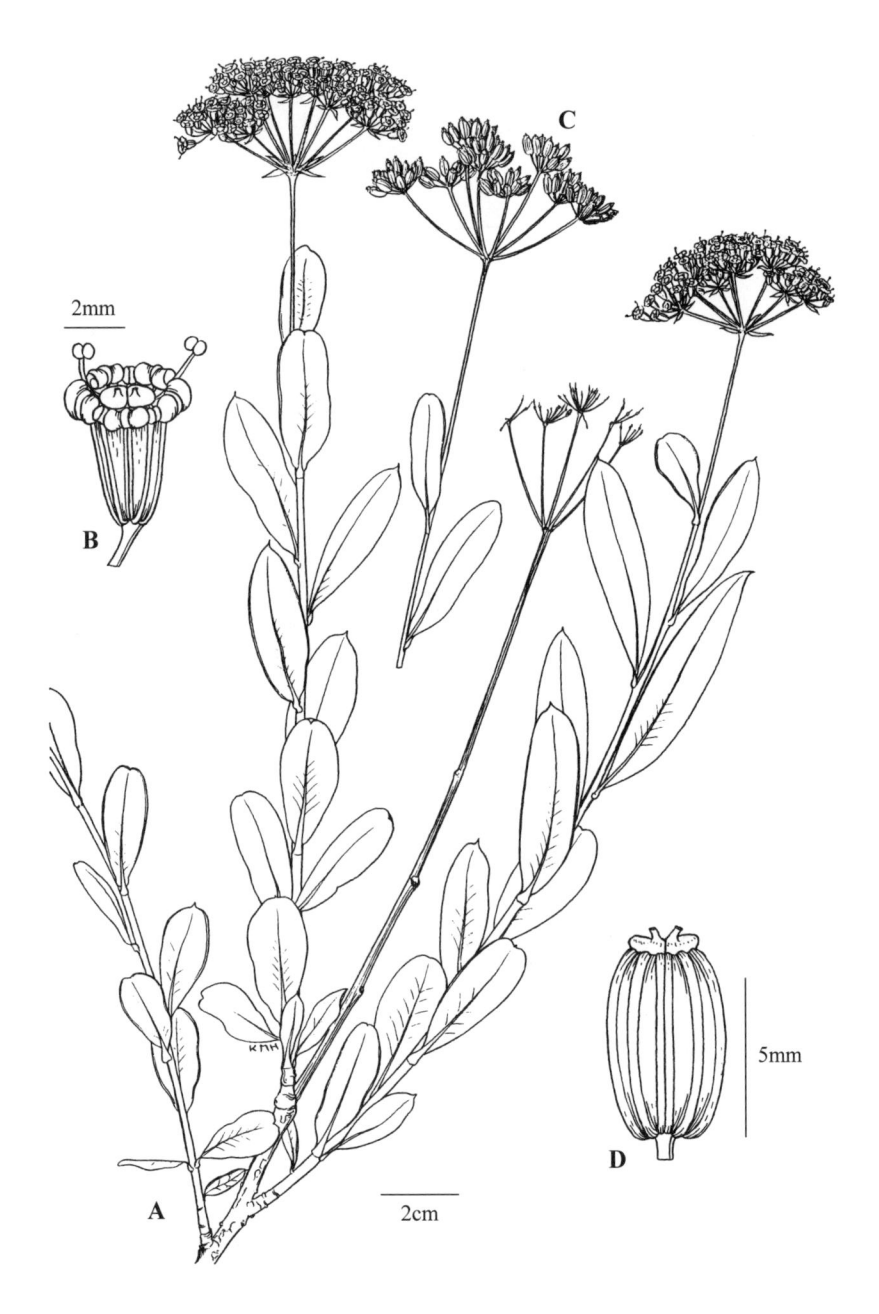

Bupleurum fruticosum L. (Shrubby Hare's-ear). A flowering branch; B flower; C fruiting stem; D fruit. (South Darenth, W Kent). STACE 601/508.

111. APIACEAE

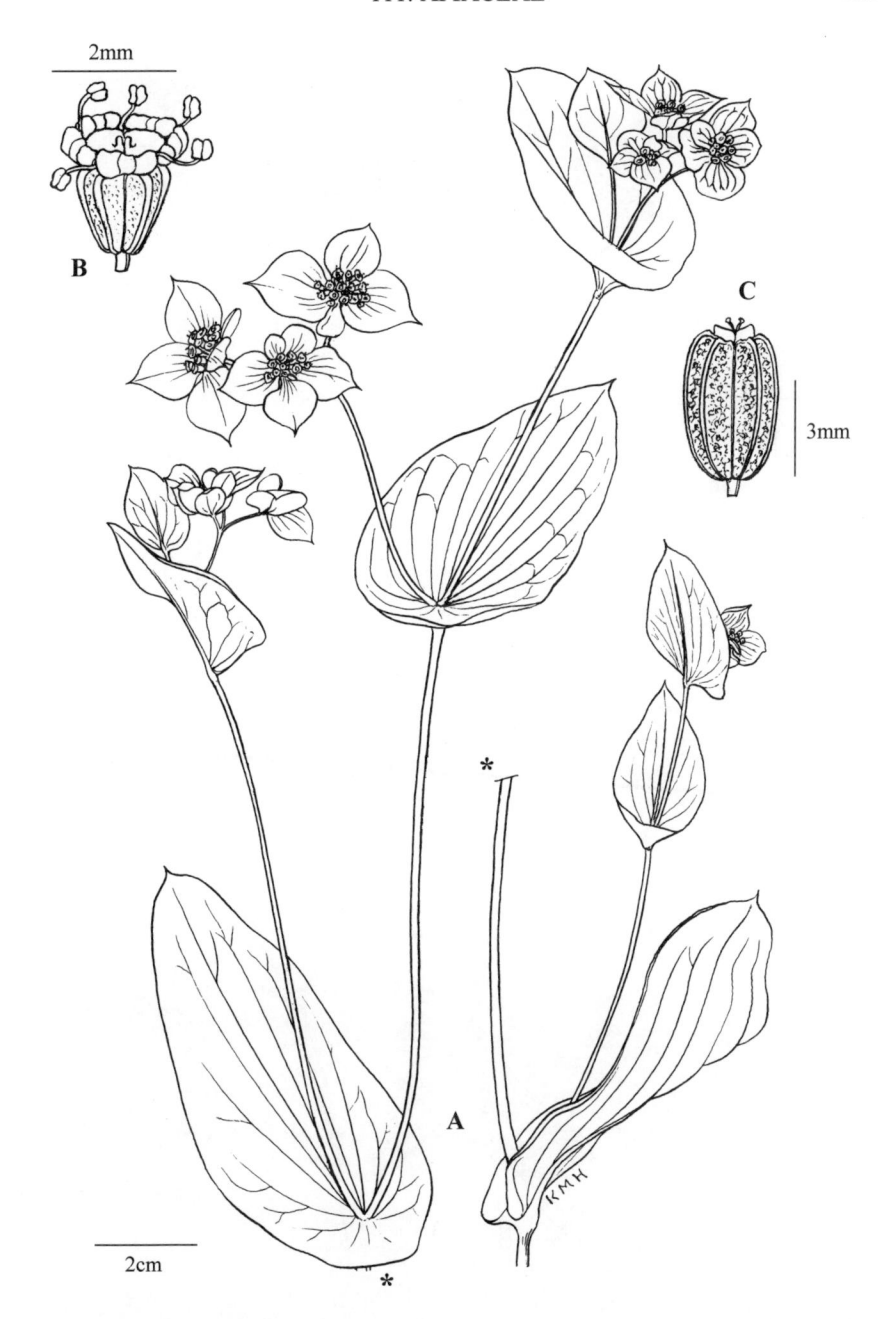

Bupleurum subovatum Link ex Spreng. (False Thorow-wax). A habit; B flower; C fruit. (Strood, W Kent). STACE 602/510.

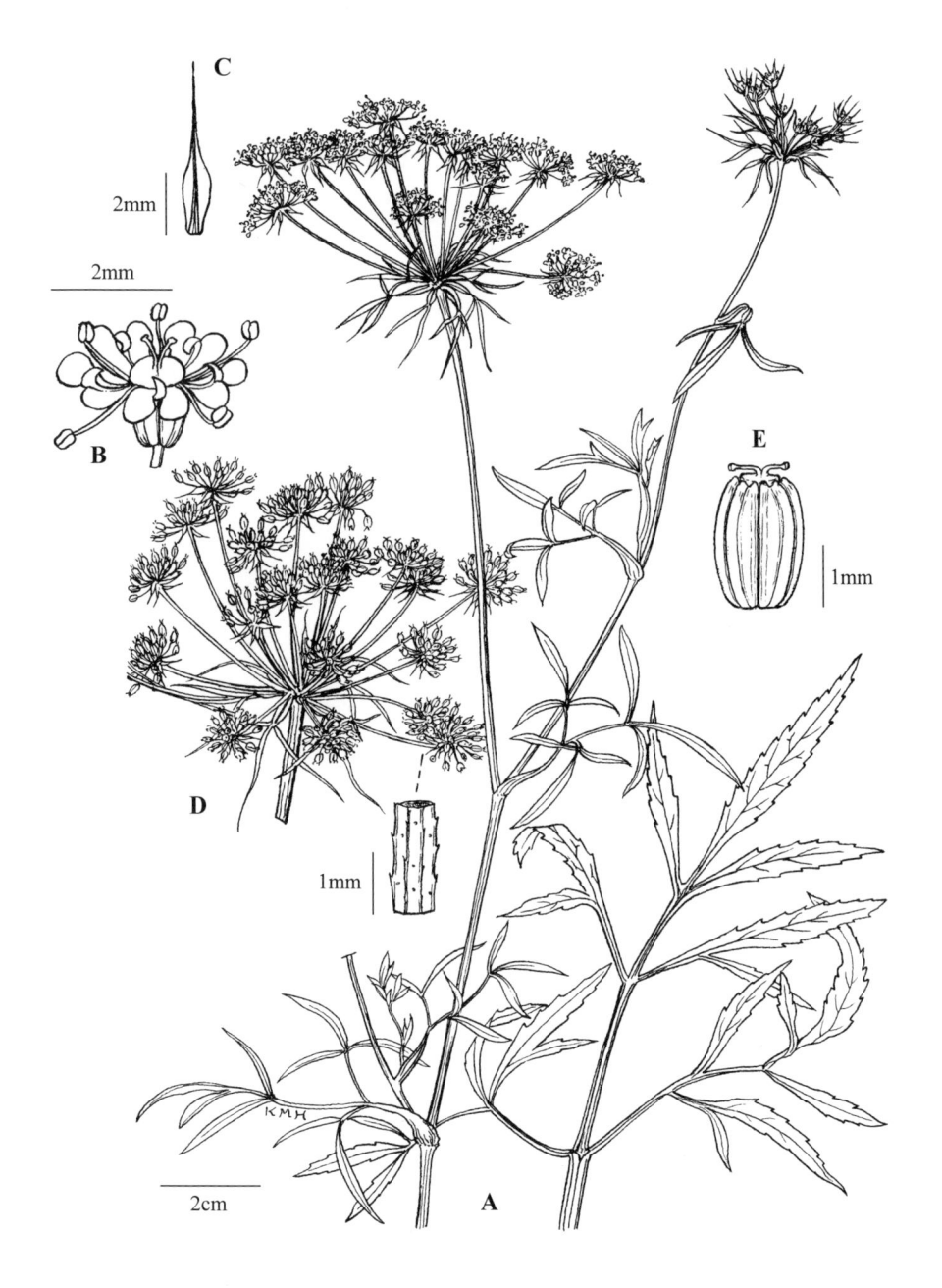

Ammi majus L. (Bullwort). A habit; B flower; C bracteole; D umbel in fruit; E fruit. (Source unknown). STACE 606/512.

111. APIACEAE 205

Ammi visnaga (L.) Lam. (Toothpick-plant). A habit; B flower; C umbel in fruit; D fruit. (Source unknown). STACE 607/512.

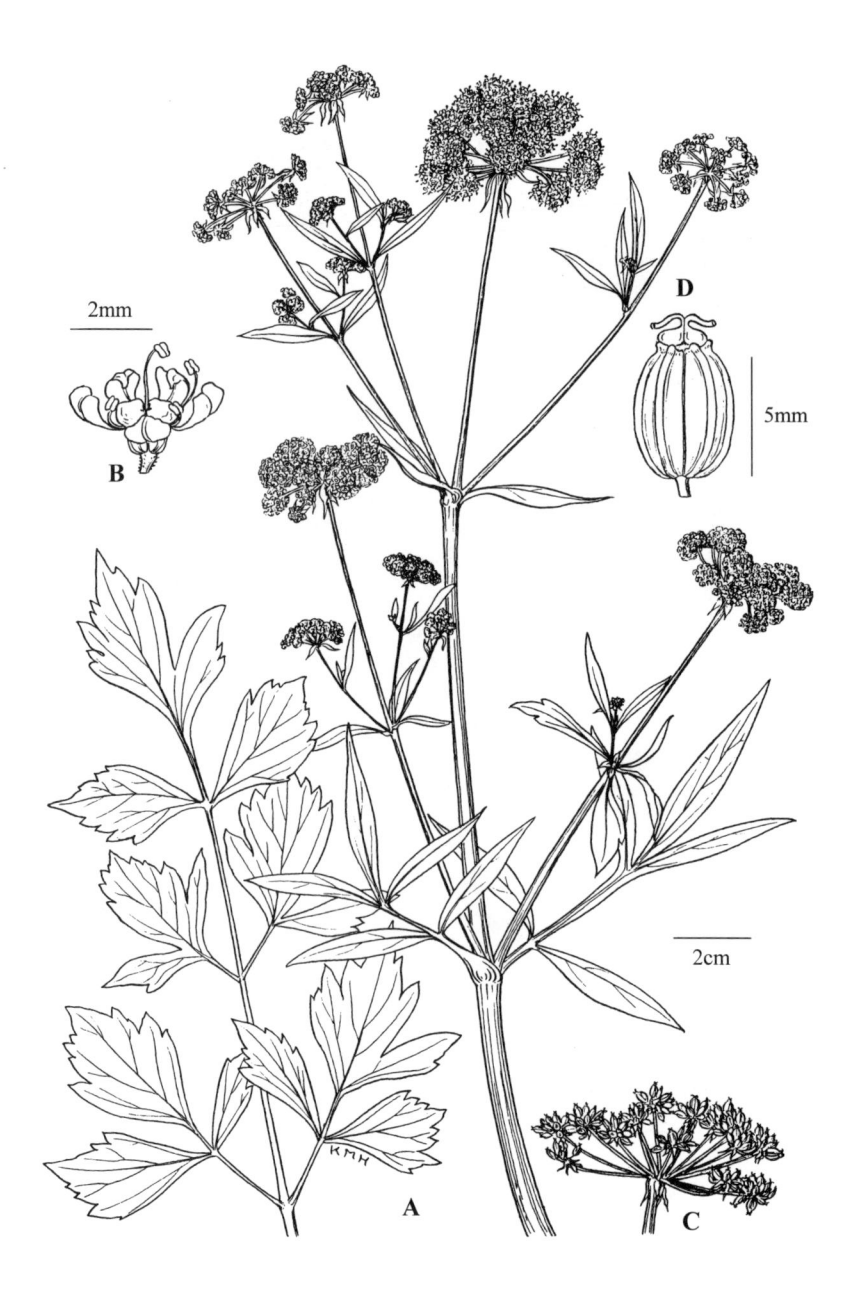

Levisticum officinale W.D.J. Koch (Lovage). A flowering stem and basal leaf; B flower; C umbel in fruit; D fruit. (Near Leeds, SW Yorks). STACE 608/514.

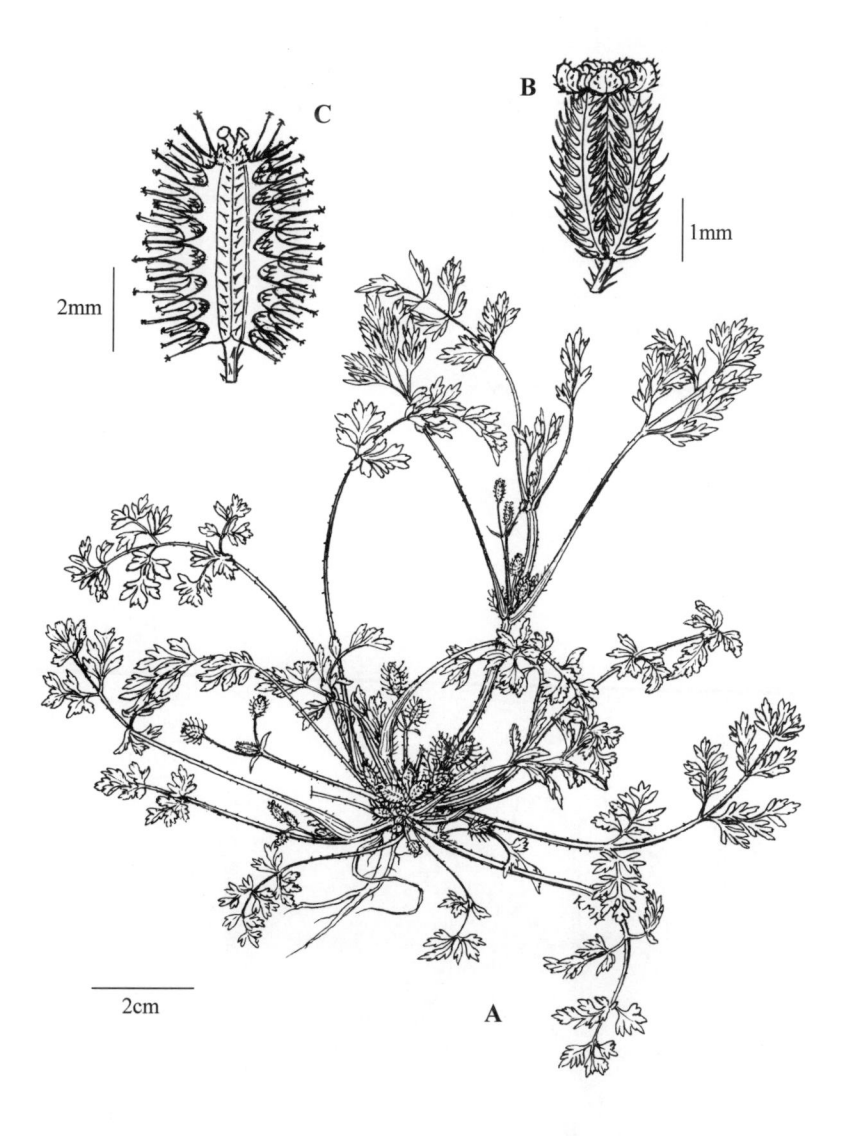

Daucus glochidiatus (Labill.) Fisch., C.A. Mey. & Avé-Lall. (Australian Carrot). A habit; B flower; C fruit. (Blackmoor, N Hants). STACE 613/518.

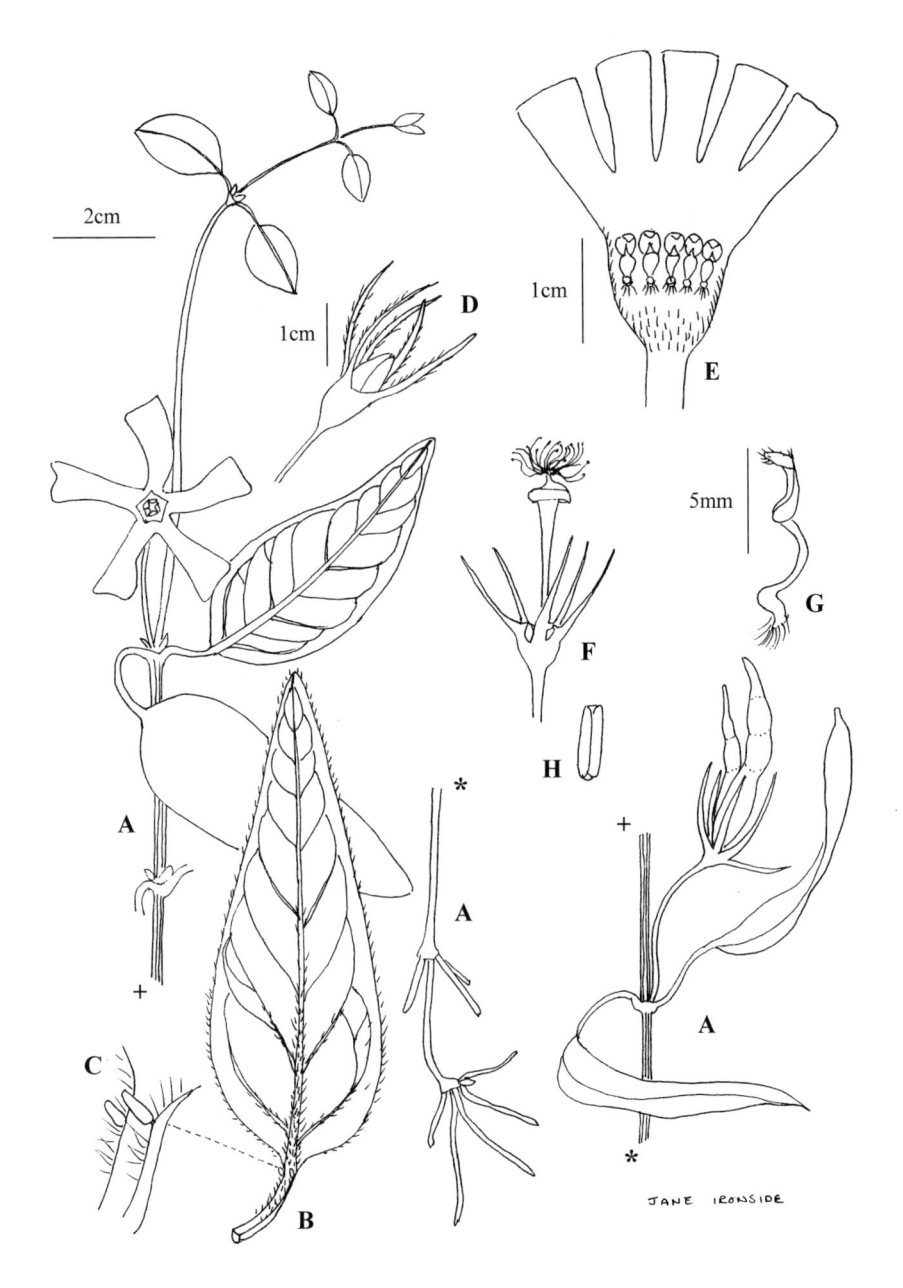

Vinca major L. var. **oxyloba** Stearn (Greater Periwinkle variety). A habit showing roots, flower and fruits; B leaf; C glands at base of leaf blade; D sepals; E flower opened out; F gynoecium; G stamen; H seed. (Stanhoe, W Norfolk). STACE 622/525.

114. SOLANACEAE

Nicandra physalodes (L.) Gaertn. (Apple-of-Peru). A habit; B stamen; C style; D calyx opened out showing fruit. (Maulden, Beds). STACE 623/526.

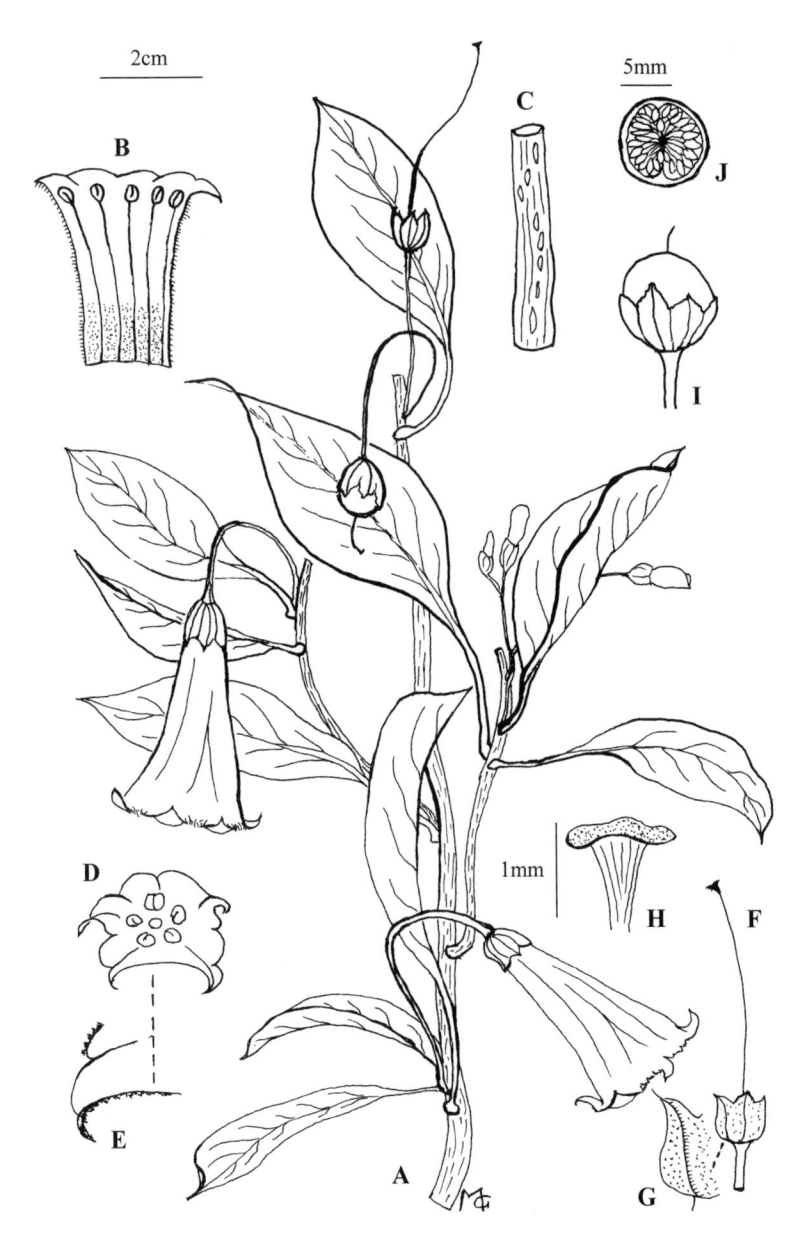

Iochroma australe Griseb. (Argentine-pear). A habit; B flower opened out; C part of woody stem showing lenticels; D corolla; E corolla margin; F flower minus corolla; G part of calyx; H stigma; I fruit; J TS of fruit. (*Hort.* Bexley, W Kent). STACE —/527.

114. SOLANACEAE

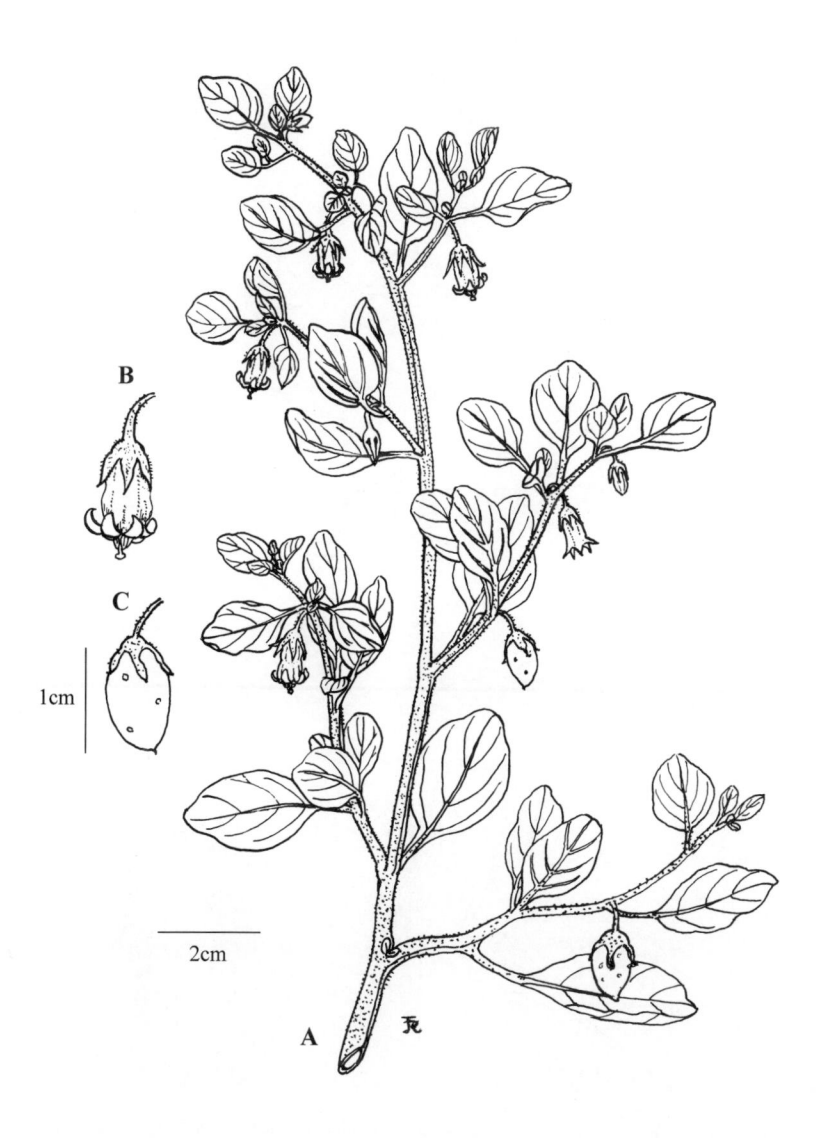

Salpichroa origanifolia (Lam.) Thell. (Cock's-eggs). A habit; B flower; C fruit. (Rosse's Tower, Guernsey). STACE 625/527.

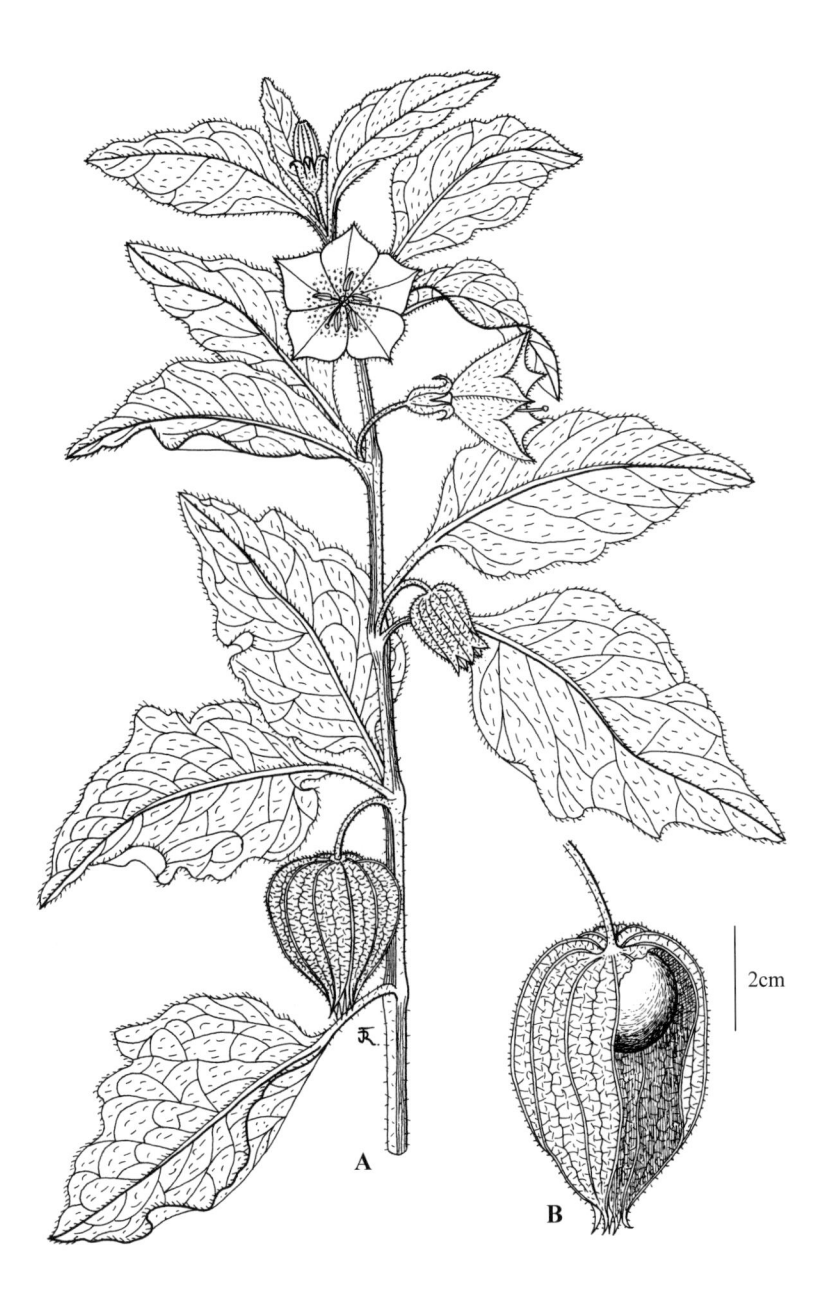

Physalis alkekengi L. (Japanese-lantern). A flowering stem; B part of calyx removed showing fruit. (*Hort*. Duton Hill, N Essex). STACE 625/528.

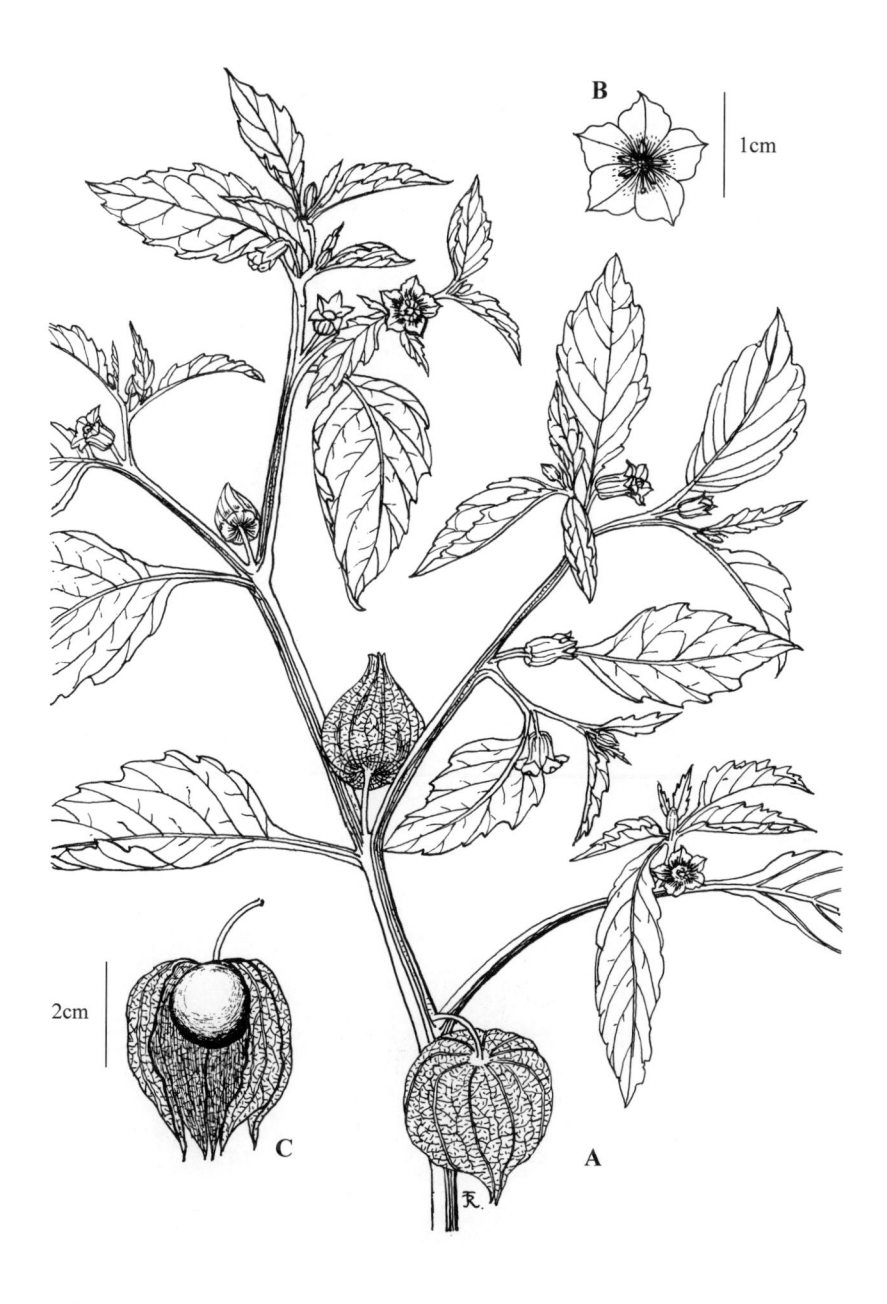

114. SOLANACEAE

Physalis ixocarpa Brot. ex Hornem. (Tomatillo). A habit; B flower; C part of calyx removed showing fruit. (Maulden, Beds). STACE 625/528.

114. SOLANACEAE

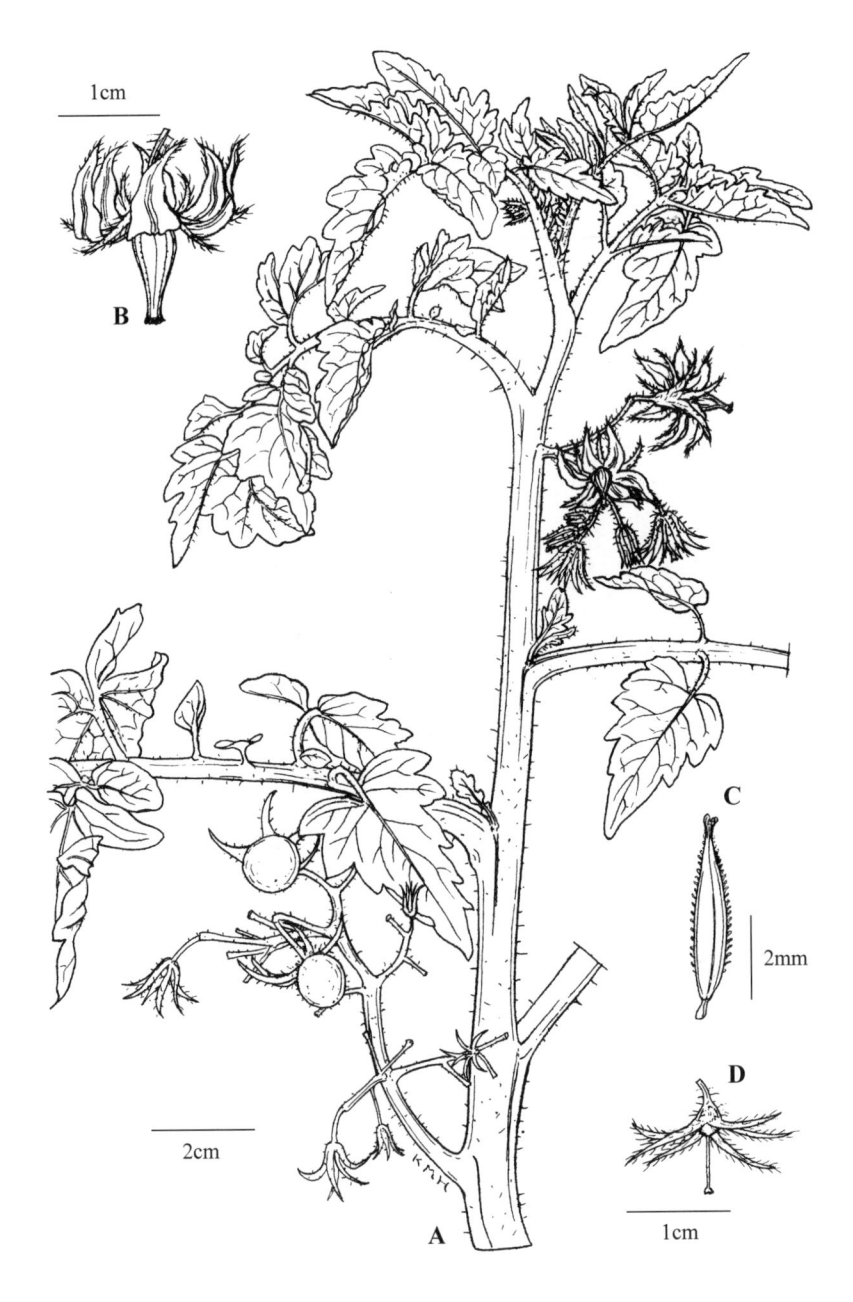

Lycopersicon esculentum Mill. (Tomato). A habit; B flower; C stamen; D flower minus corolla. (Ashbourne, Derbys). STACE 626/528.

114. SOLANACEAE

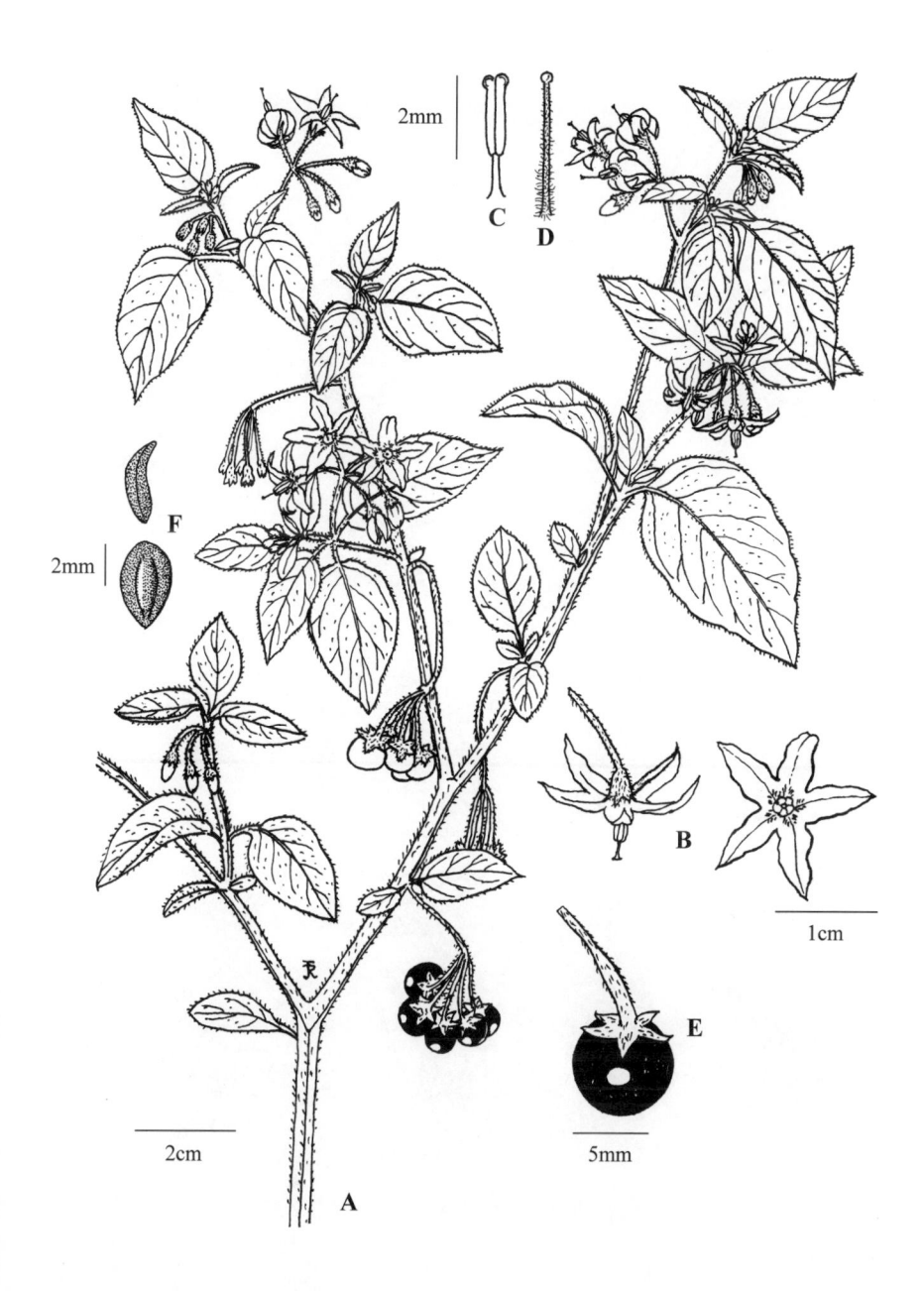

Solanum chenopodioides Lam. (Tall Nightshade). A habit; B flower; C stamen; D style; E fruit; F seed. (Guernsey). STACE 629/531.

114. SOLANACEAE

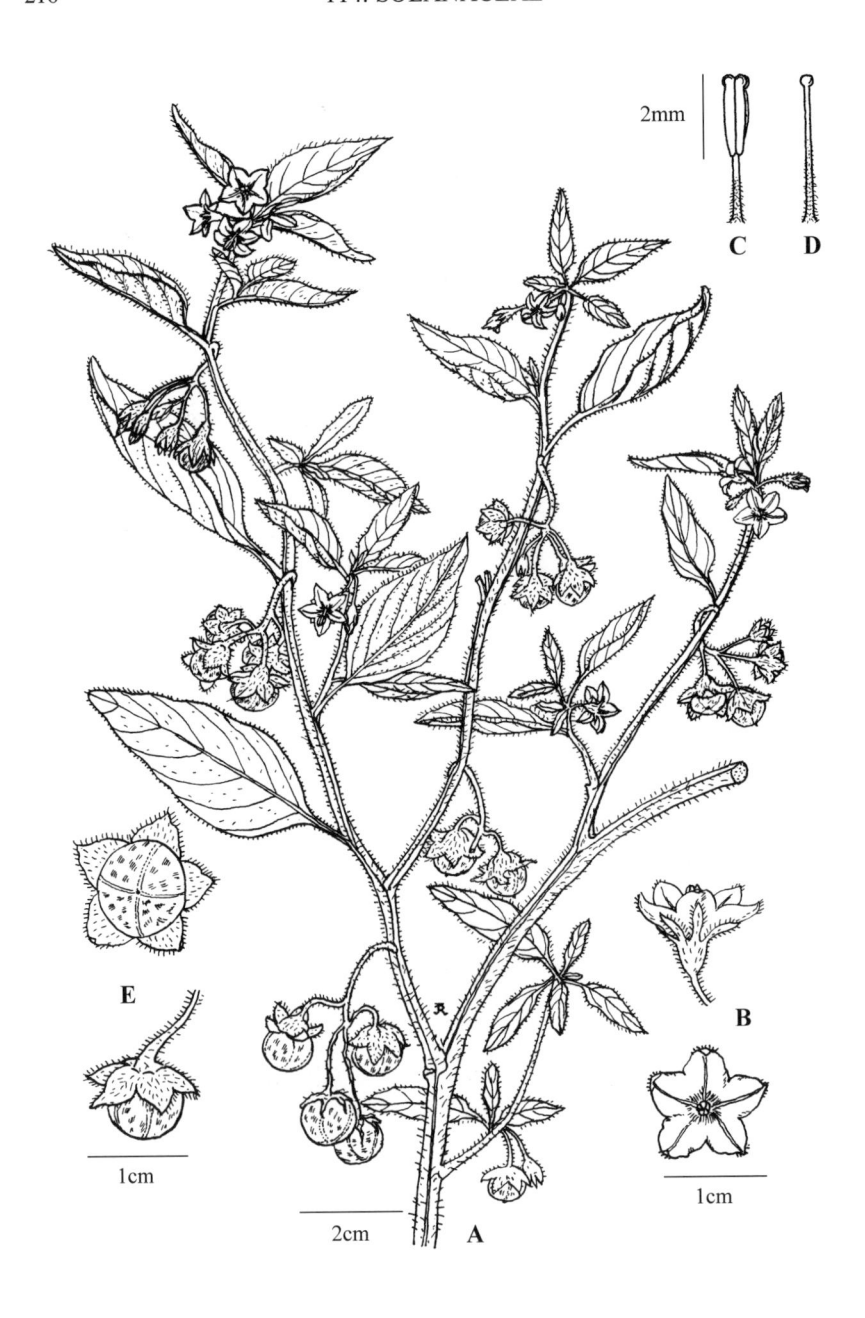

Solanum physalifolium Rusby var. **nitidibaccatum** (Bitter) Edmonds (Green Nightshade). A habit; B flower; C stamen; D style; E fruit. (Flitton, Beds). STACE 629/531.

Solanum sarachoides Sendtn. (Leafy-fruited Nightshade). A habit; B calyx opened out; C flower; D stamen; E style; F fruit. (Portsmouth, S Hants). STACE 629/531.

Solanum triflorum Nutt. (Small Nightshade). A flowering stem; B fruiting stem; C flower; D stamen; E style; F fruit. (W Norfolk). STACE 629/531.

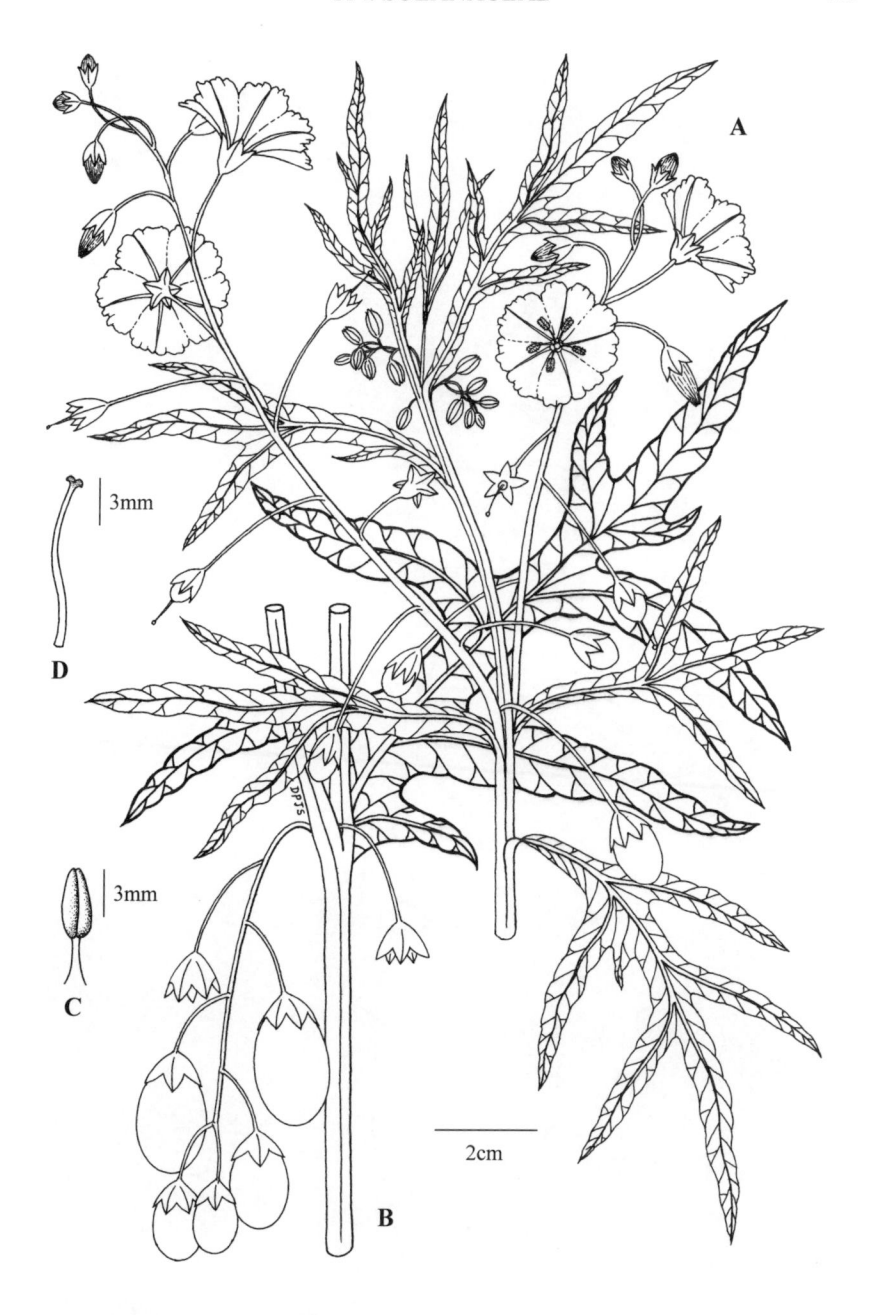

Solanum laciniatum Aiton (Kangaroo-apple). A flowering stem; B fruiting stem; C stamen; D style. (SW England). STACE 630/532.

Solanum sisymbriifolium Lam. (Red Buffalo-bur). A flowering stem; B fruit. (Avonmouth, W Gloucs). STACE 630/532.

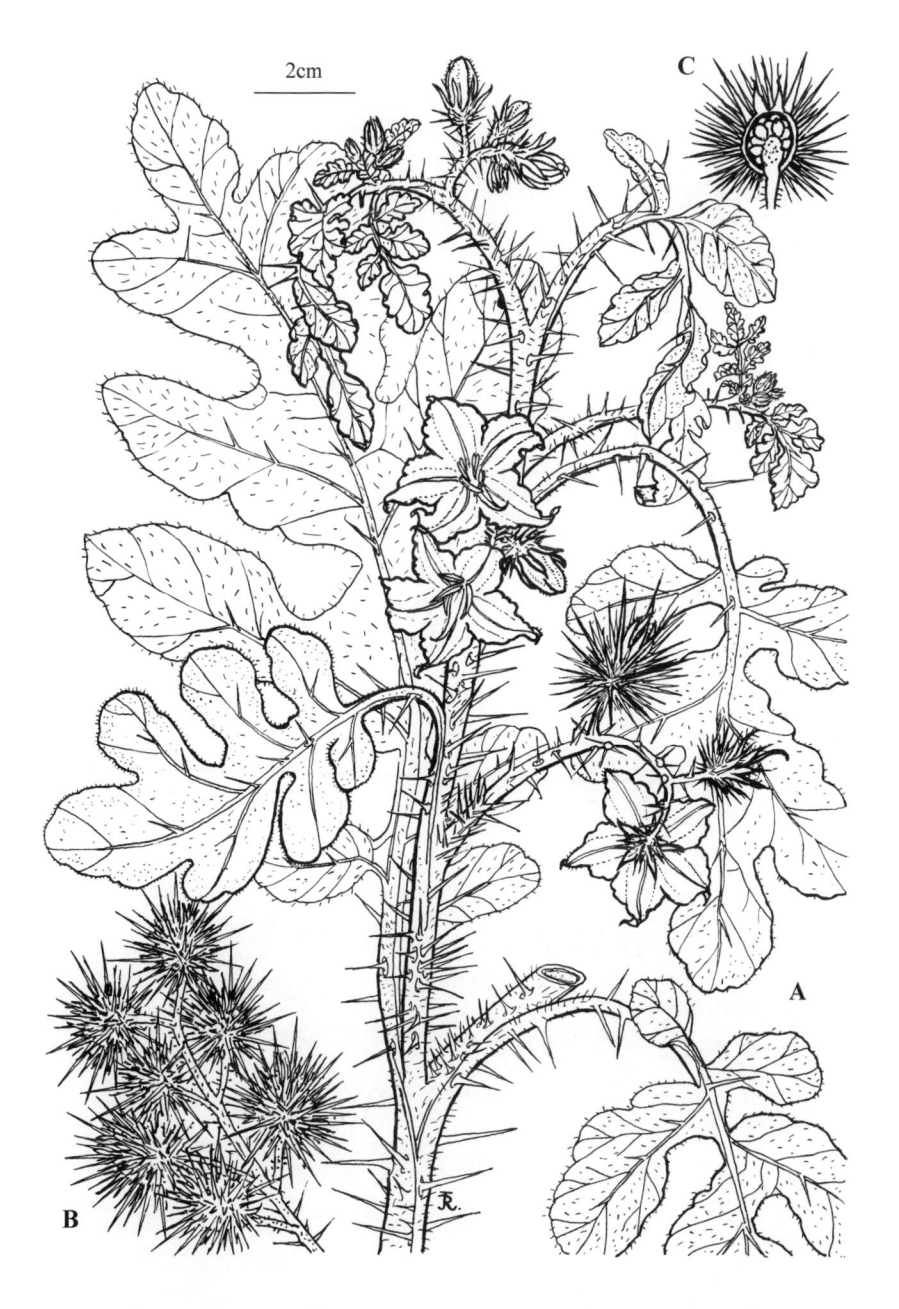

Solanum rostratum Dunal (Buffalo-bur). A habit; B cluster of fruits; C LS of fruit. (Poulton-le-Fylde, W Lancs). STACE 630/532.

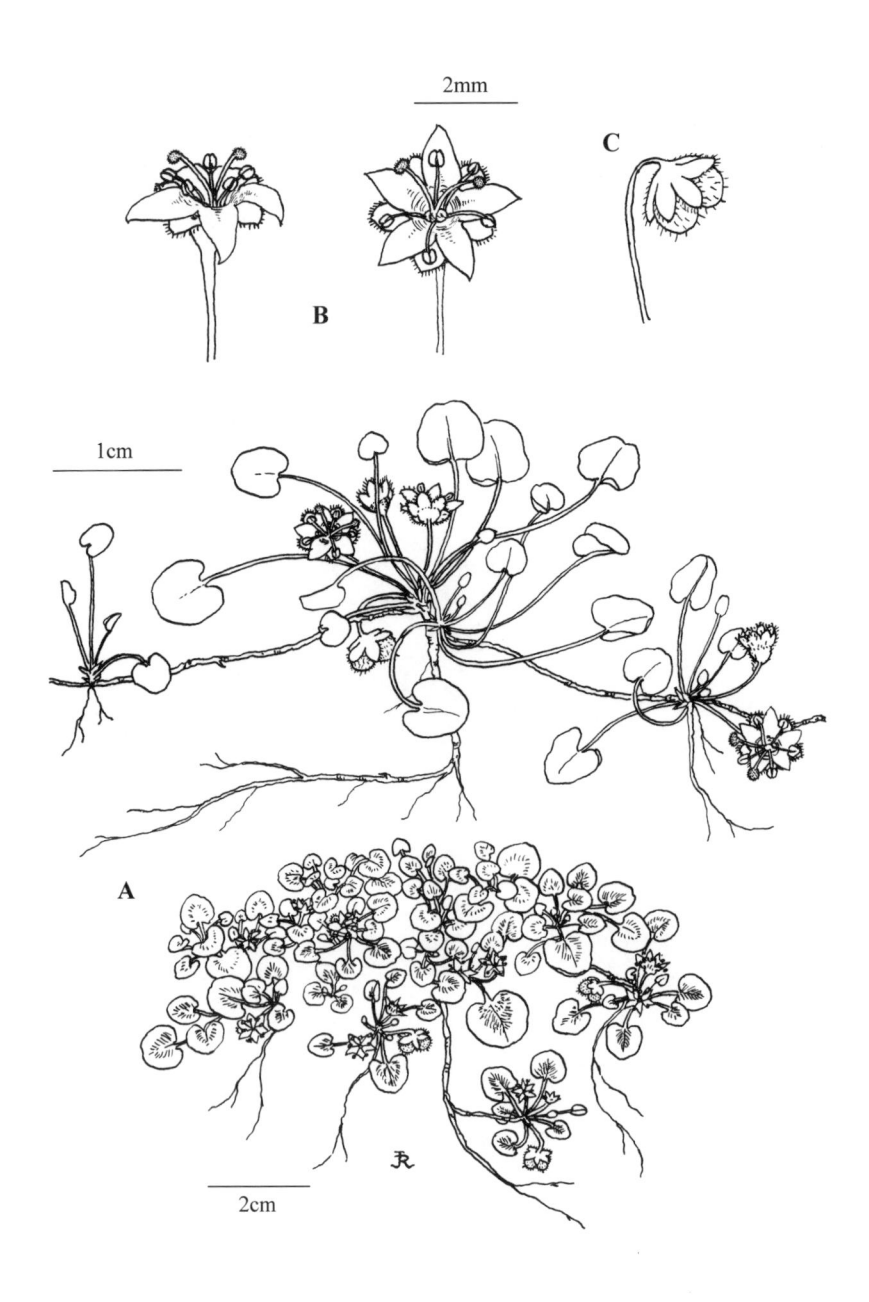

Dichondra micrantha Urb. (Kidneyweed). A habit; B flower; C fruit. (*Hort.* ex Hayle, W Cornwall). STACE 633/534.

115. CONVOLVULACEAE

Calystegia pulchra Brummitt & Heywood (Hairy Bindweed). A habit; B flower opened out. (Peel, Man). STACE 634/535.

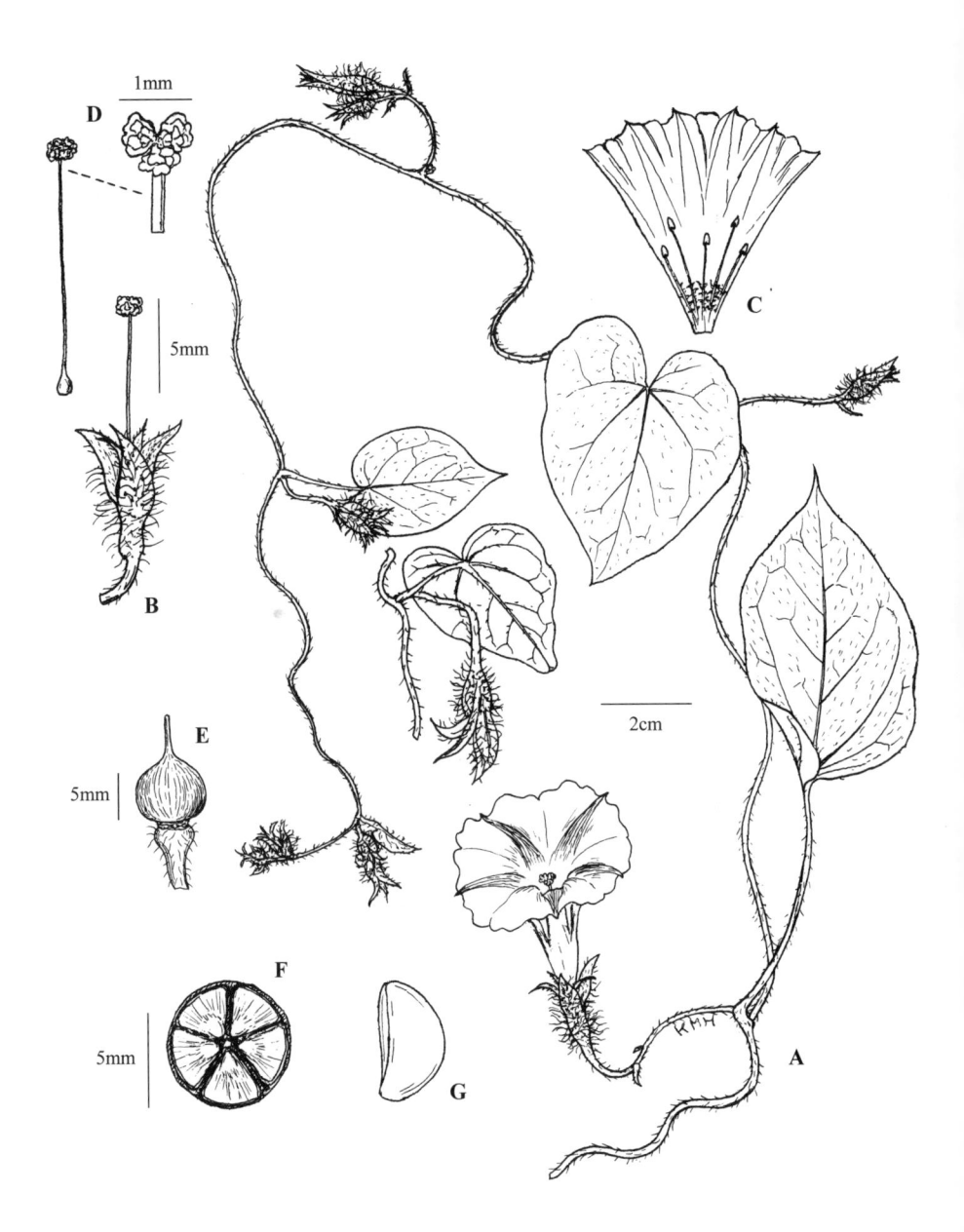

Ipomoea purpurea Roth (Common Morning-glory). A habit; B flower minus corolla; C corolla opened out; D gynoecium; E ovary; F TS of ovary; G seed. (*Hort*. ex Stone, W Kent). STACE 636/537.

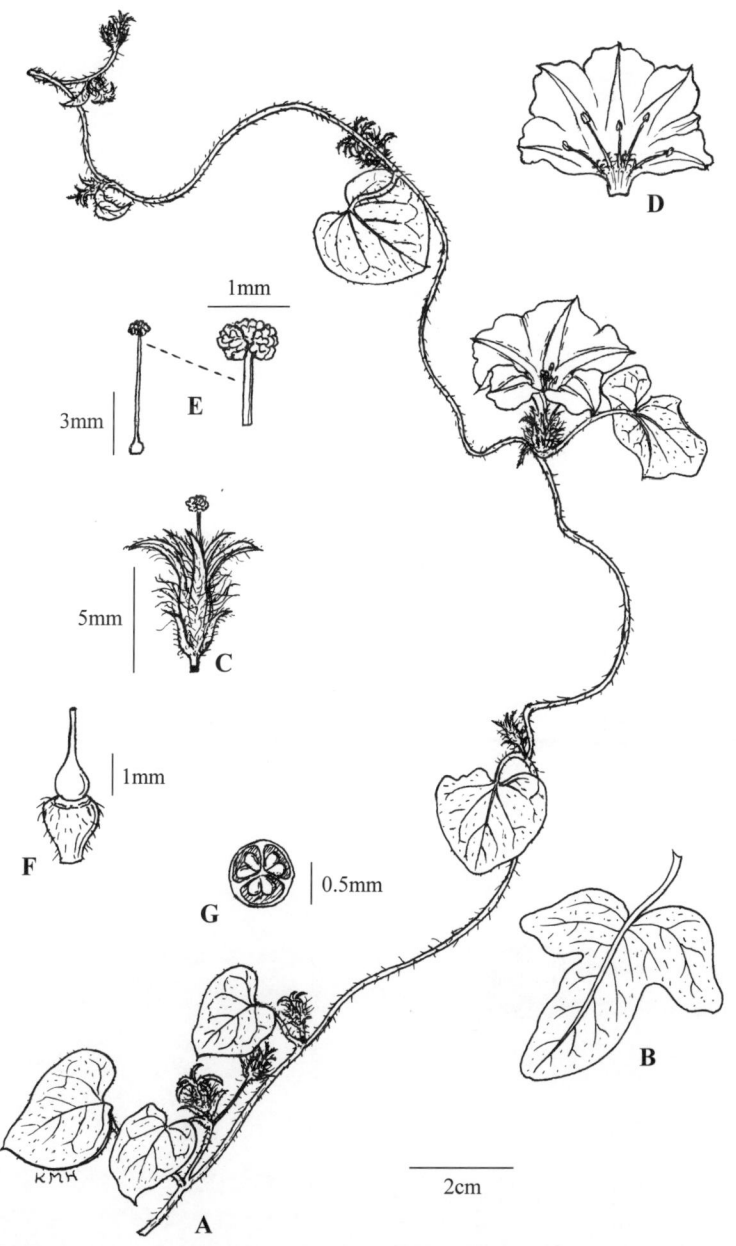

115. CONVOLVULACEAE 225

Ipomoea hederacea Jacq. var. **integriuscula** A. Gray (Ivy-leaved Morning-glory). A habit; B leaf of var. **hederacea**; C flower minus corolla; D corolla opened out; E gynoecium; F ovary; G TS of ovary. (*Hort.* ex Stone, W Kent). STACE 636/537.

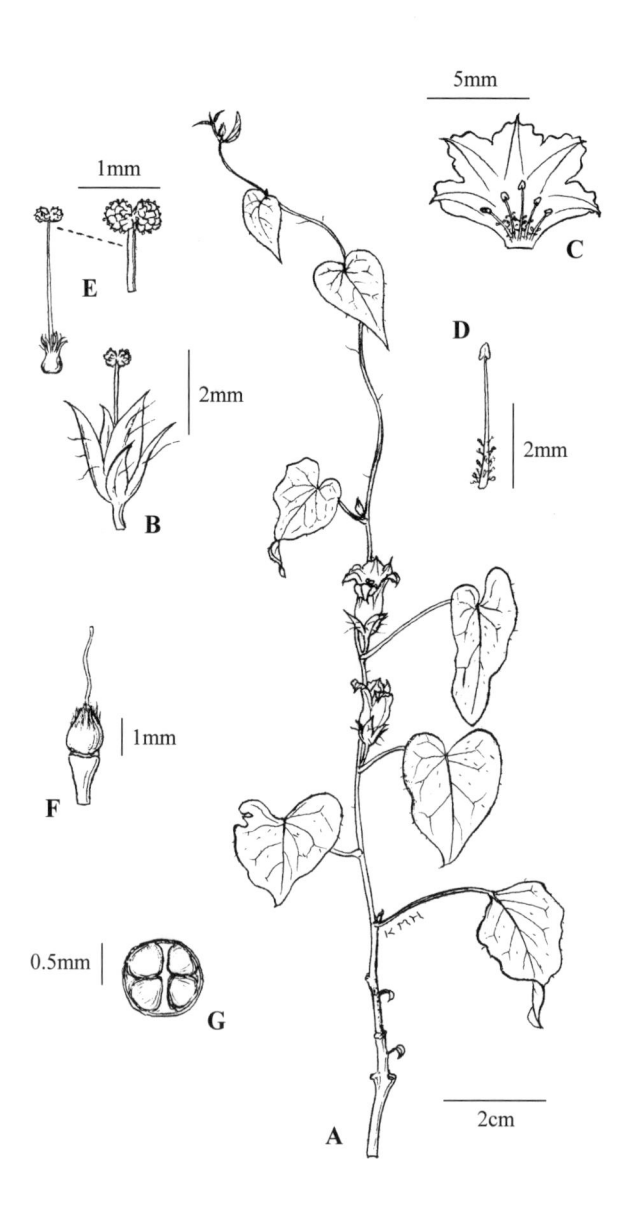

Ipomoea lacunosa L. (White Morning-glory). A habit; B flower minus corolla; C corolla opened out; D stamen; E gynoecium; F ovary; G TS of ovary. (*Hort.* ex Stone, W Kent). STACE 636/537.

116. CUSCUTACEAE

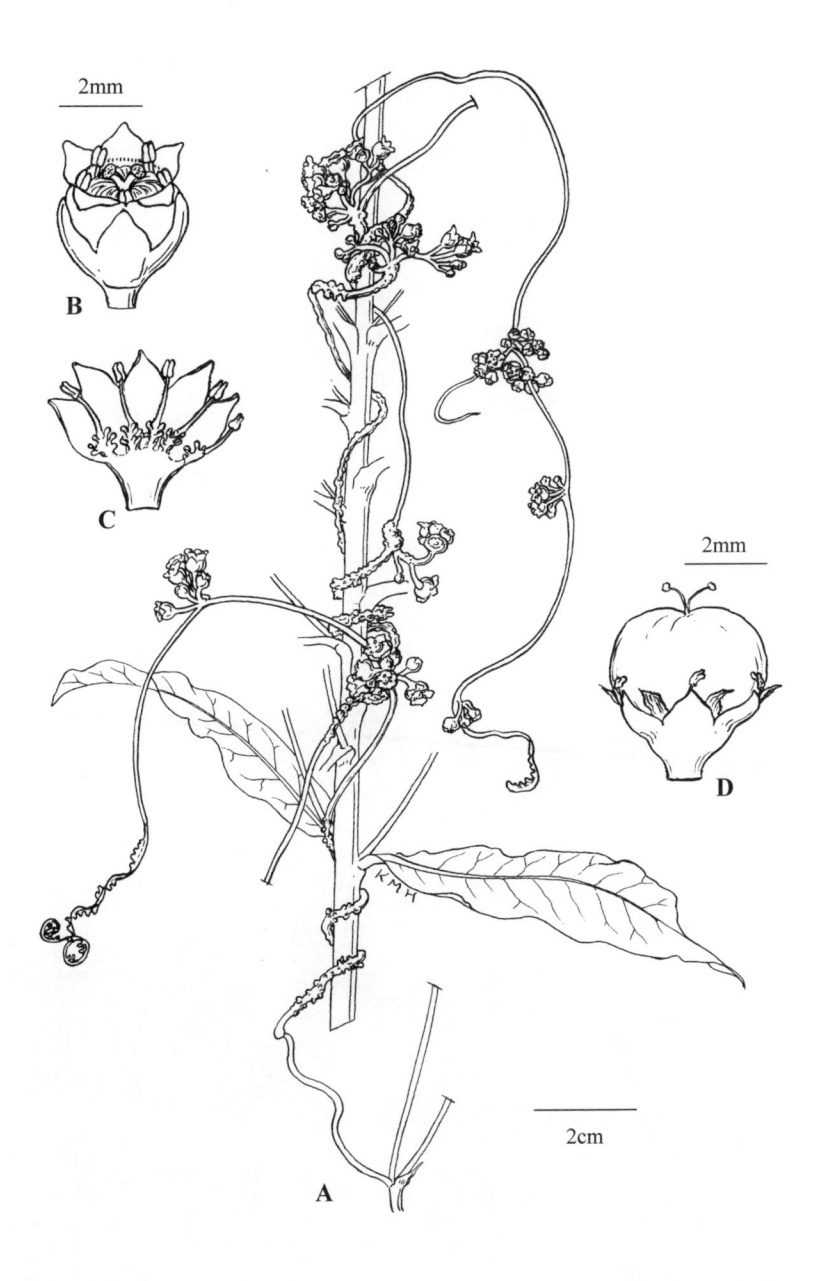

Cuscuta campestris Yunck. (Yellow Dodder). A habit; B flower; C corolla opened out; D fruit. (Esholt, NW Yorks). STACE 637/538.

Echium rosulatum Lange (Lax Viper's-bugloss). A habit; B flower minus corolla; C corolla opened out; D fruit within calyx. (Barry Docks, Glam). STACE 643/542.

120. BORAGINACEAE

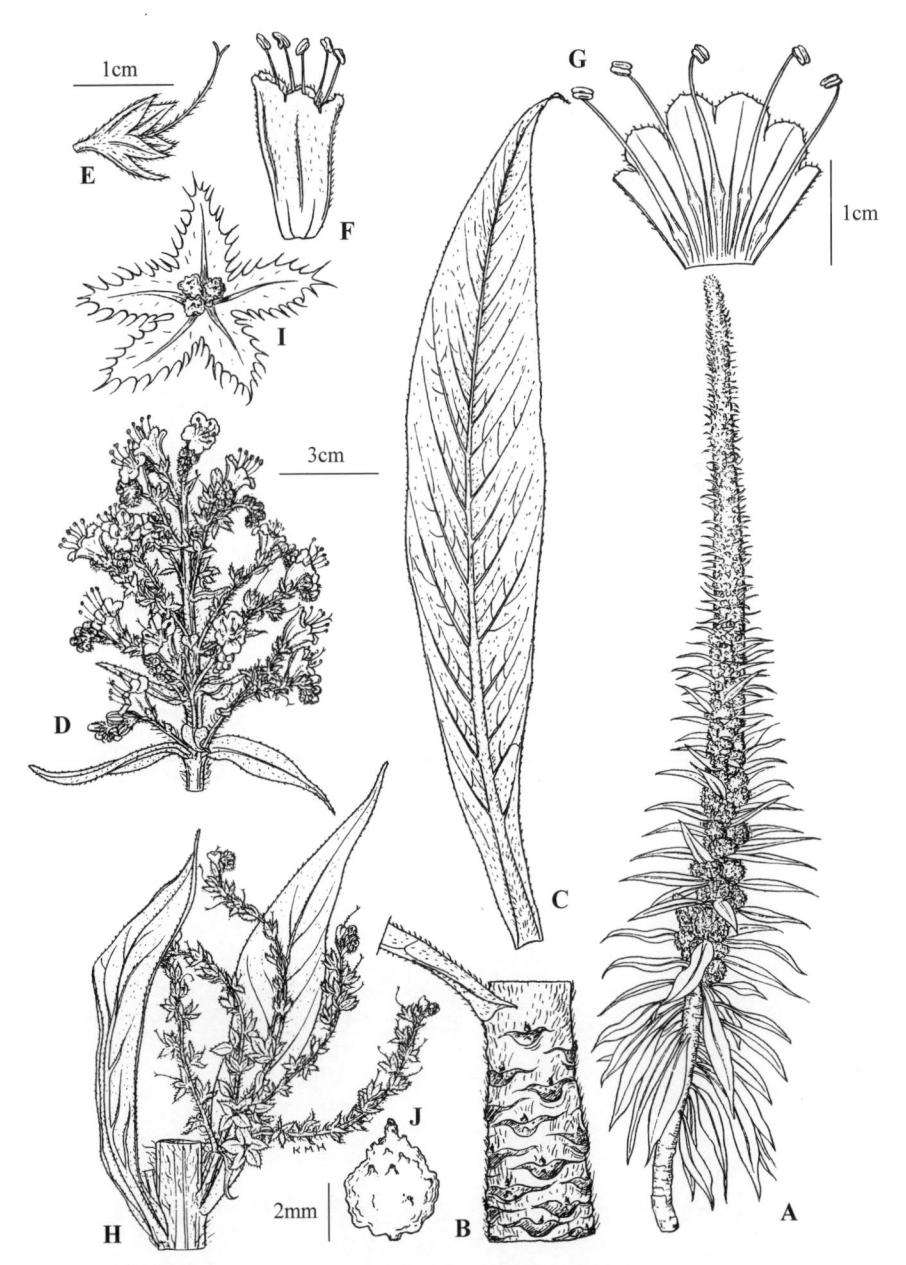

Echium pininana Webb & Berthel. (Giant Viper's-bugloss). A habit; B stem with leaf traces; C stem leaf; D partial inflorescence; E calyx and style; F corolla and stamens; G corolla opened out; H part of fruiting stem; I fruit within calyx; J nutlet. (Guernsey). STACE 643/542.

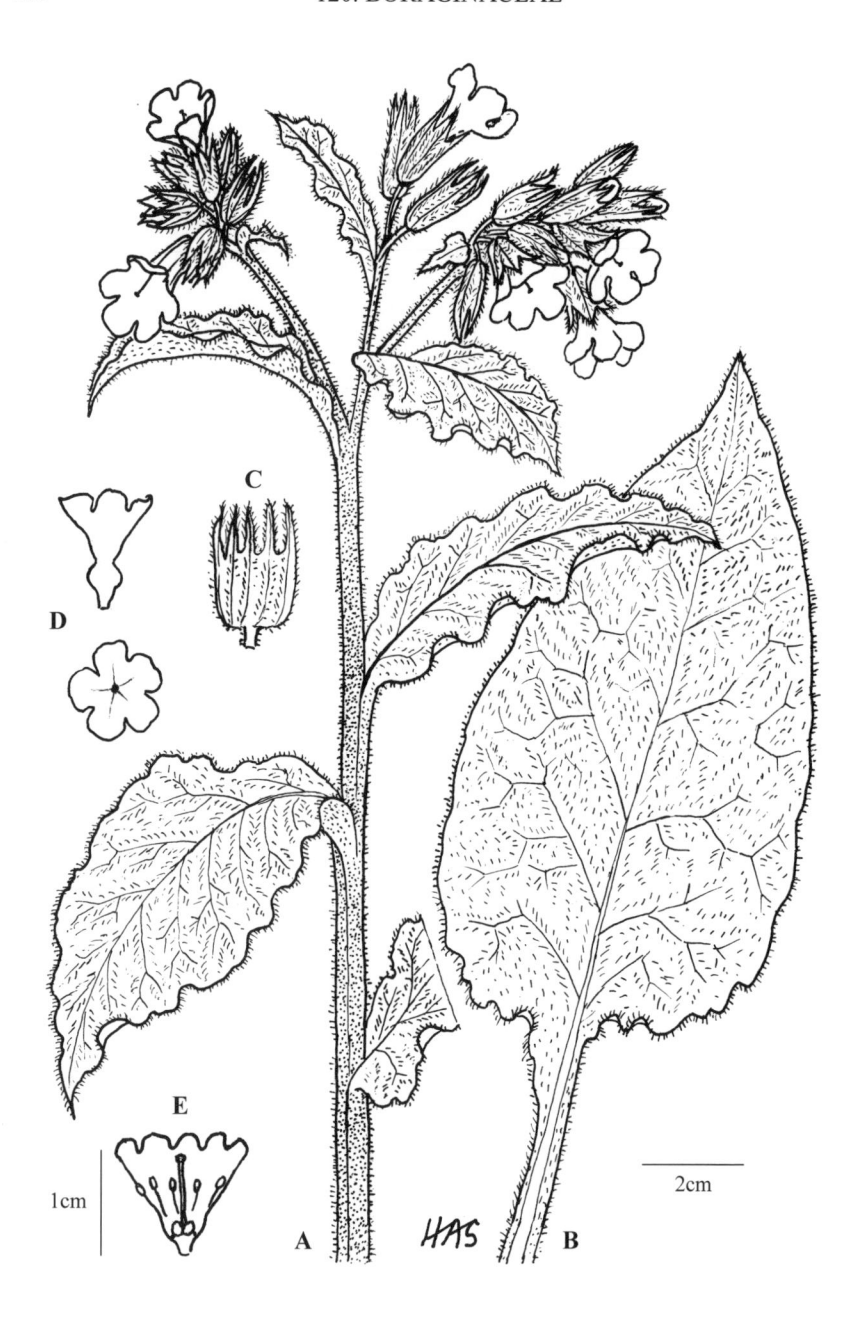

Pulmonaria rubra Schott (Red Lungwort). A flowering stem; B basal leaf; C calyx; D side and plan views of flower; E flower opened out. (*Hort.* Aberdeen, S Aberdeen). STACE 644/543.

120. BORAGINACEAE

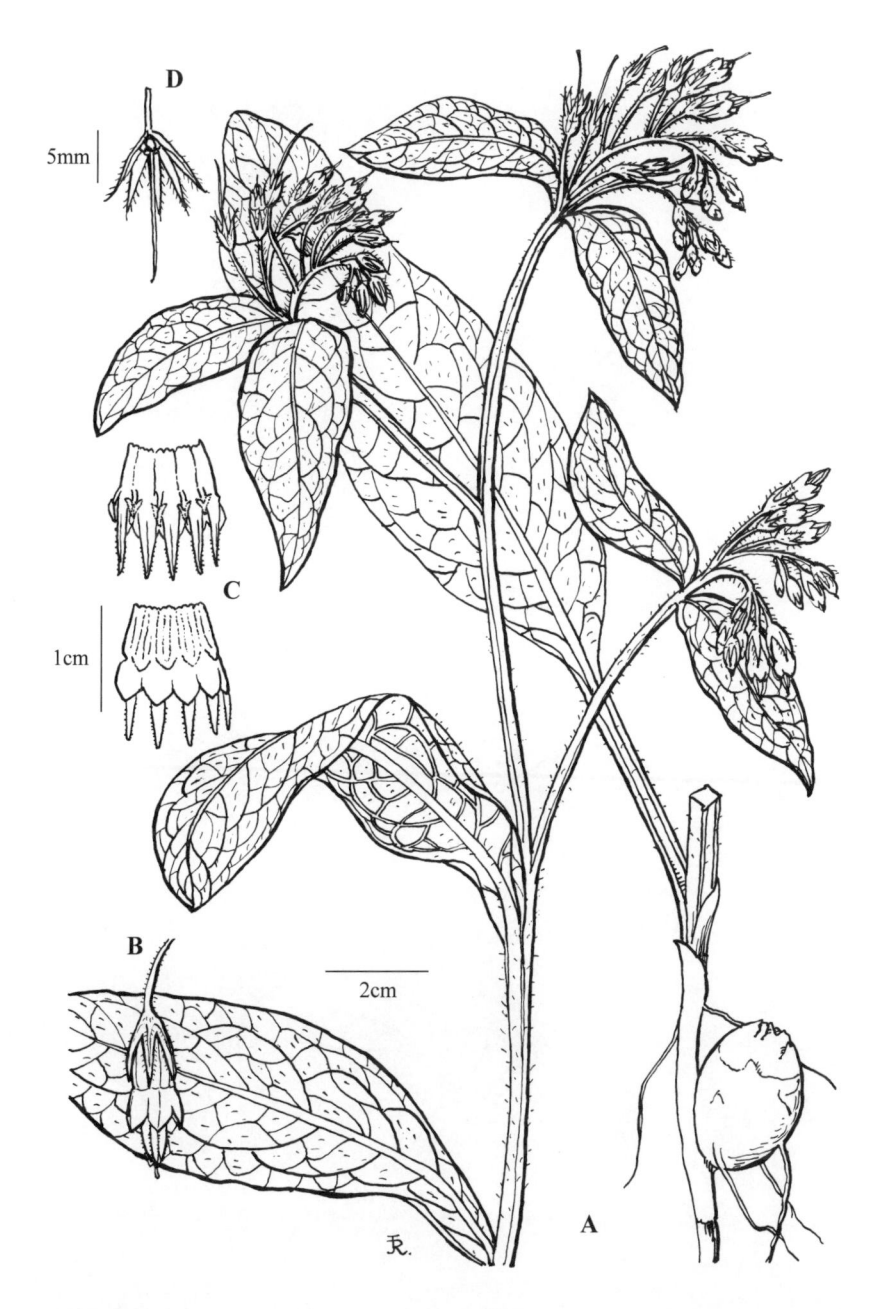

Symphytum bulbosum K.F. Schimper (Bulbous Comfrey). A habit including tuber; B flower; C corolla opened out; D fruit with calyx opened out. (Abbotsbury, Dorset). STACE 648/546.

Anchusa officinalis L. (Alkanet). A habit; B flower; C corolla; D corolla opened out; E corolla top view; F calyx cut away to show fruit. (Upton Towans, W Cornwall). STACE 649/547.

120. BORAGINACEAE

Anchusa ochroleuca M. Bieb. (Yellow Alkanet). A habit; B flower; C corolla; D corolla opened out; E corolla top view; F calyx cut away to show fruit. (Upton Towans, W Cornwall). STACE 649/547.

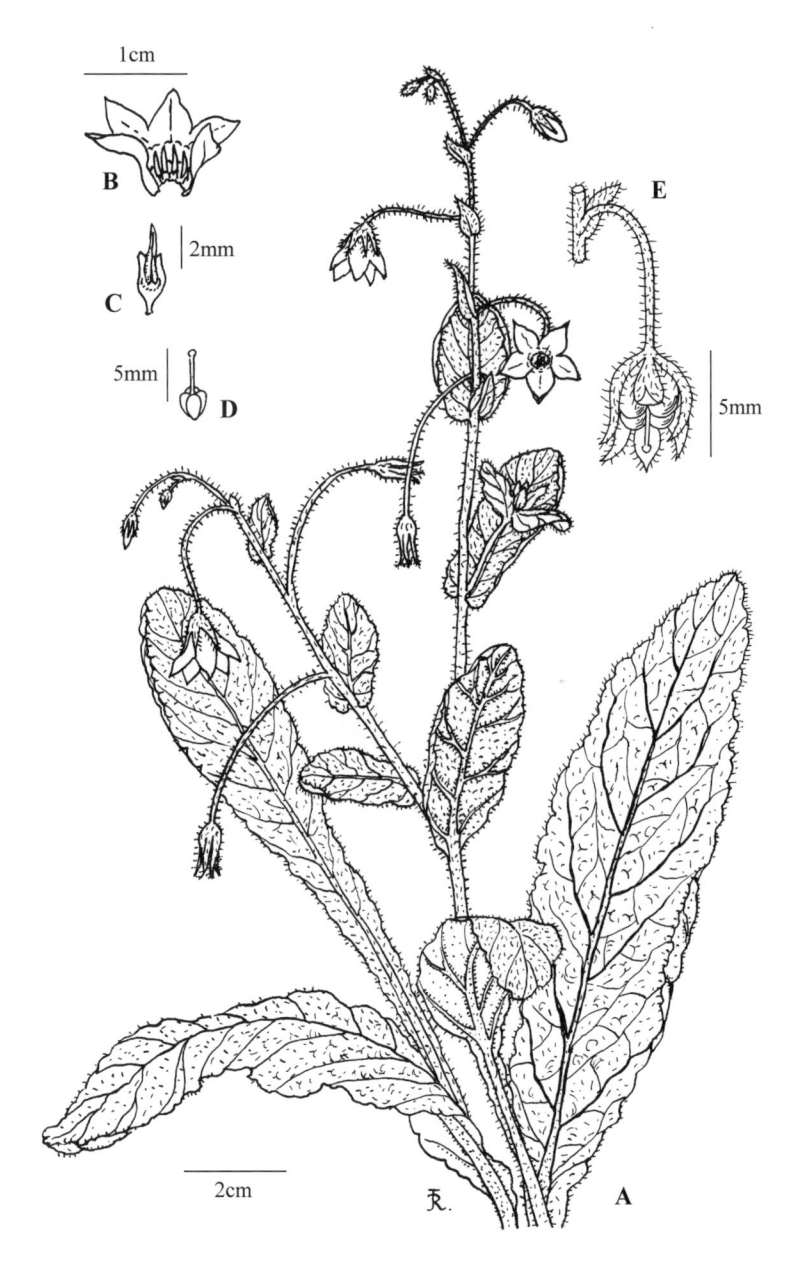

Borago pygmaea (DC.) Chater & Greuter (Slender Borage). A habit; B corolla opened out to show stamens; C stamen; D gynoecium; E fruiting calyx enclosing nutlets. (Jethou, Channel Islands). STACE 650/548.

120. BORAGINACEAE

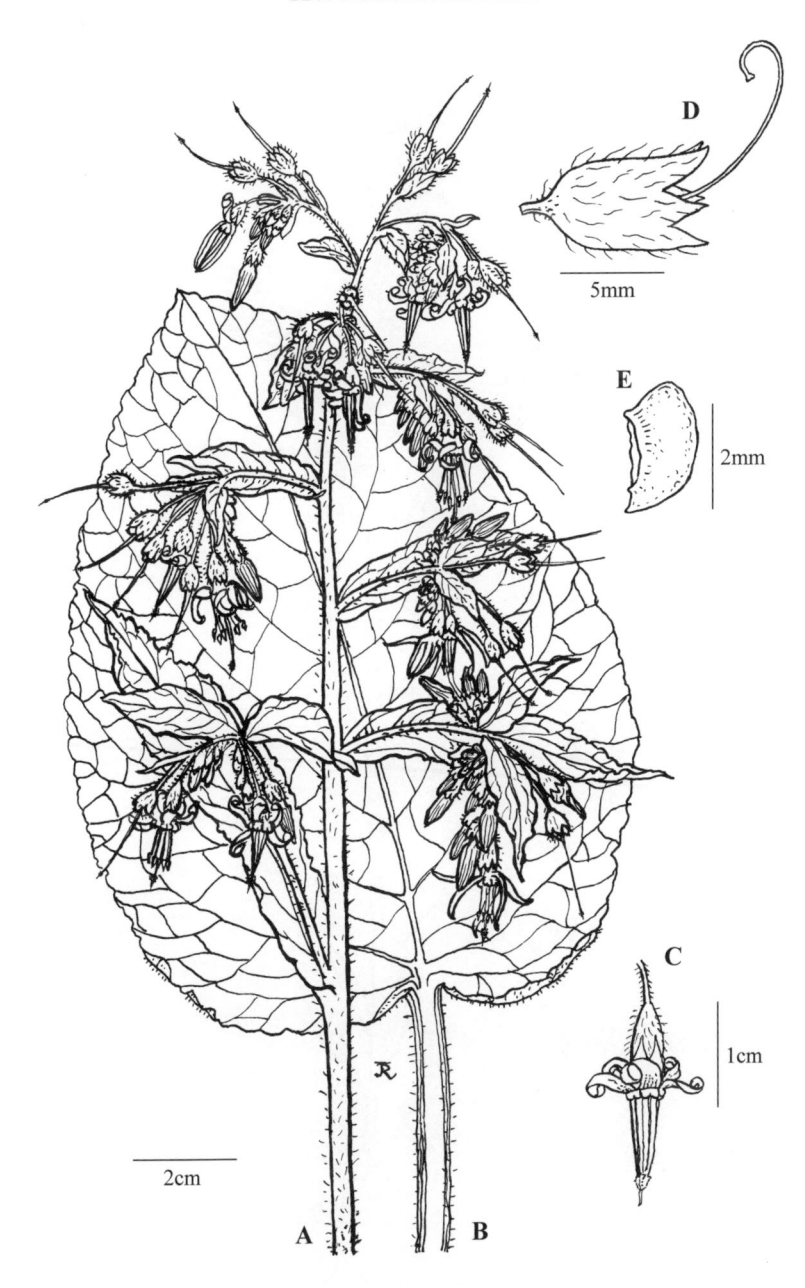

Trachystemon orientalis (L.) G. Don (Abraham-Isaac-Jacob). A habit; B basal leaf; C flower; D fruiting calyx and style; E nutlet. (*Hort*. Dunmow, N Essex). STACE 651/548.

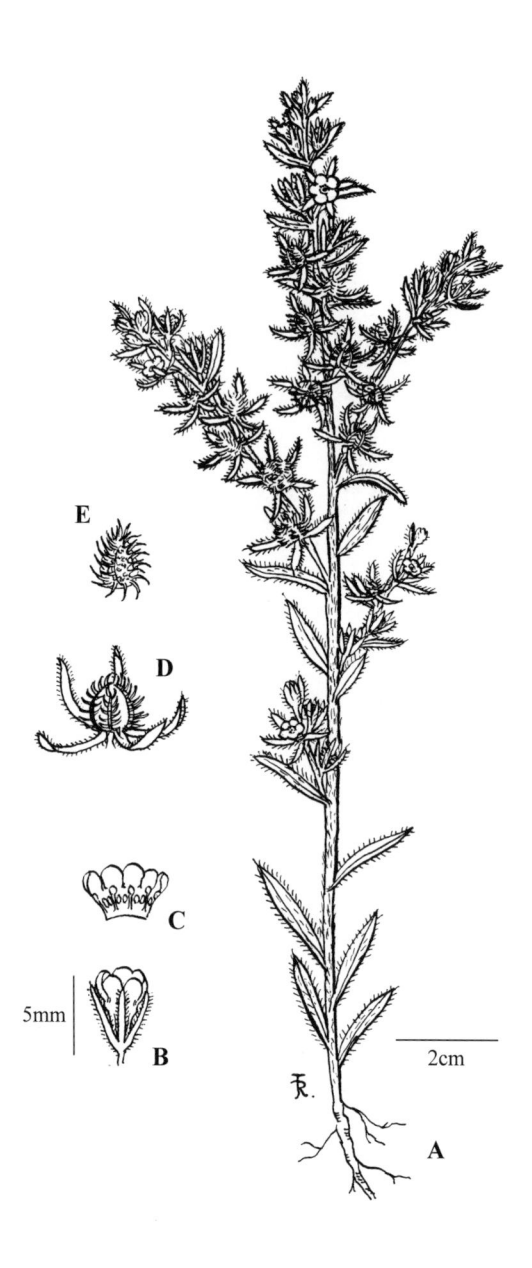

Lappula squarrosa (Retz.) Dumort. (Bur Forget-me-not). A habit; B flower; C corolla opened out; D fruit within calyx; E nutlet. (Source unknown). STACE 656/553.

122. LAMIACEAE 237

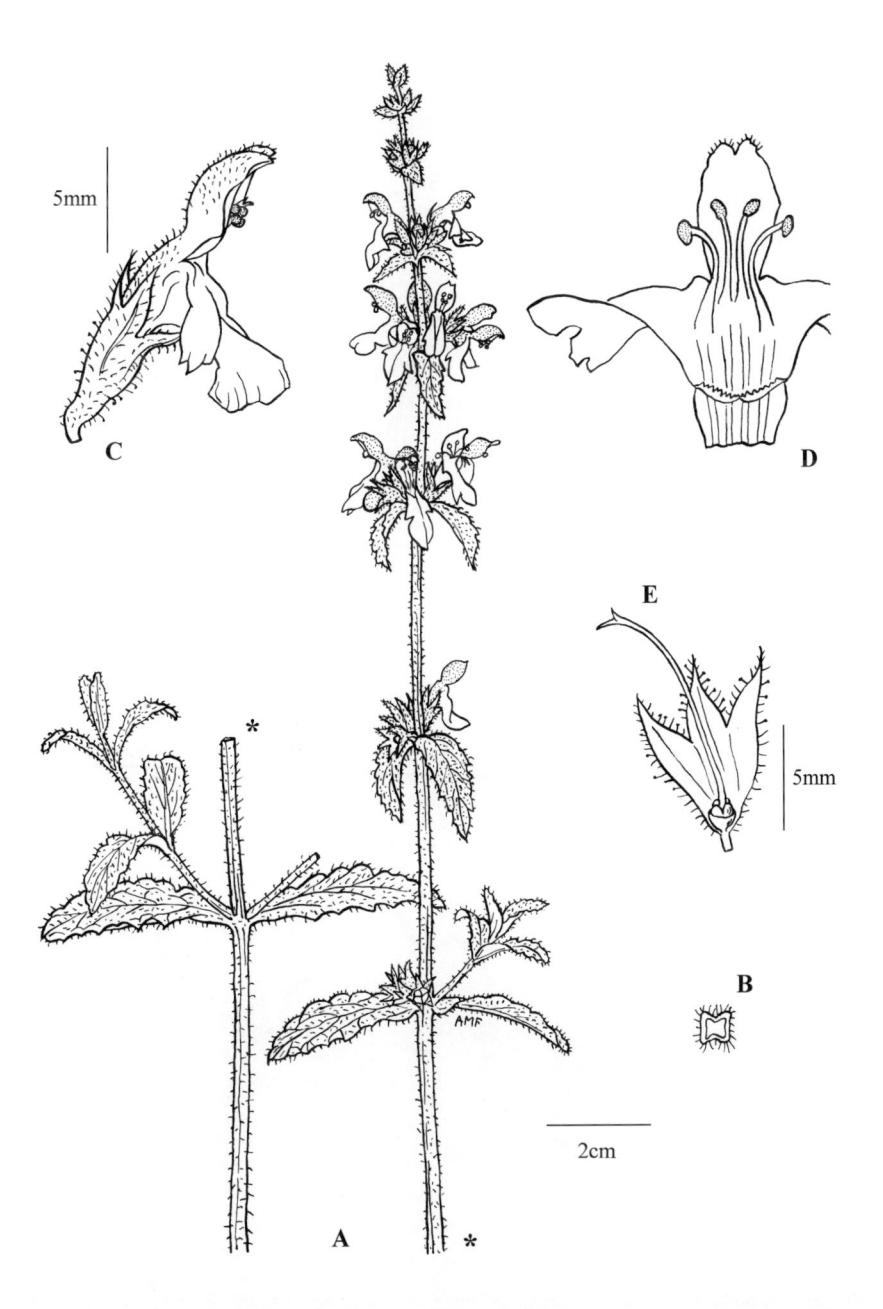

Stachys recta L. ssp. **labiosa** (Bertol.) Briq. (Perennial Yellow-woundwort). A habit; B TS of stem; C flower; D corolla opened out; E gynoecium within calyx. (Barry Docks, Glam). STACE 664/559.

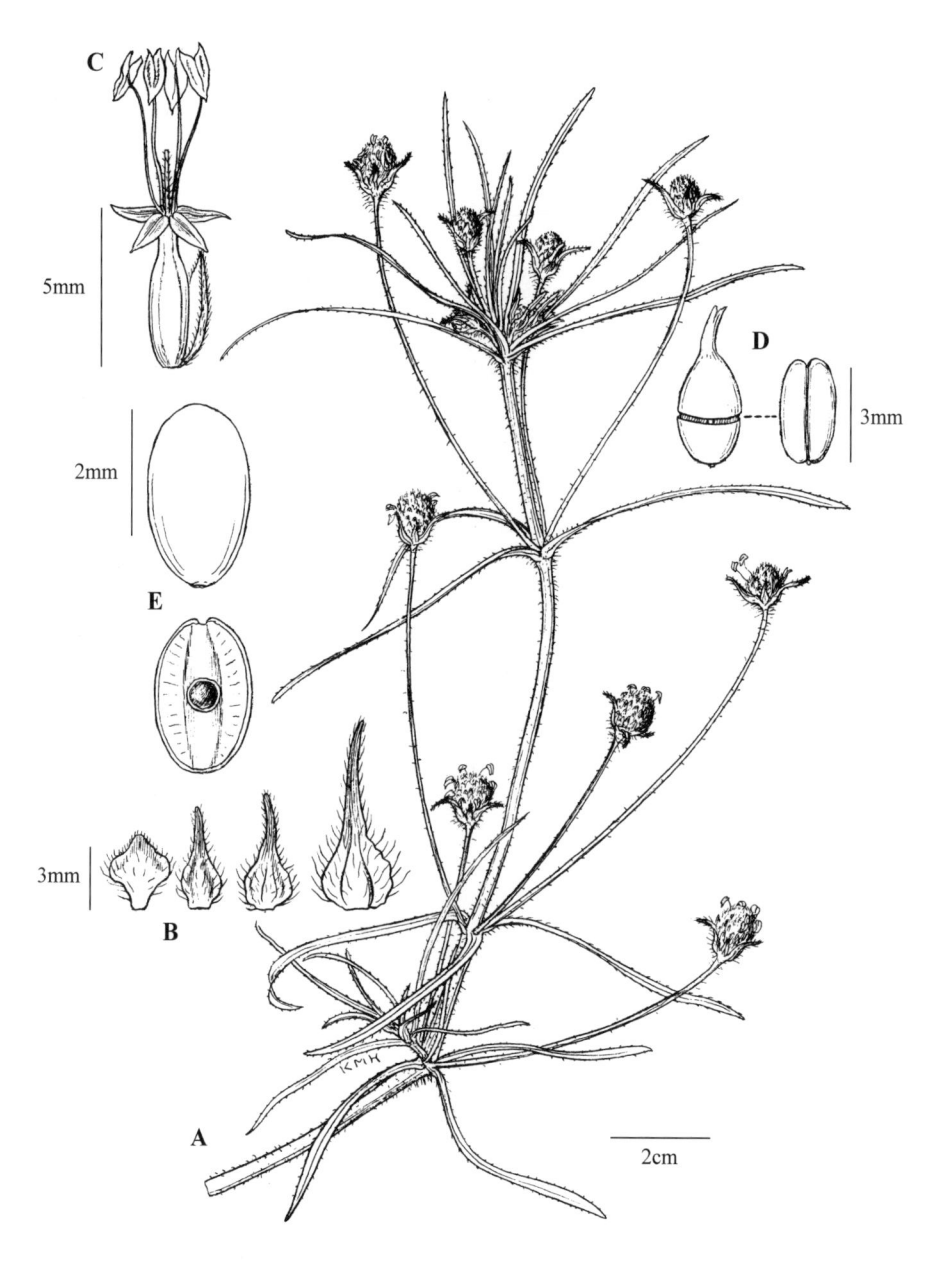

Plantago arenaria Waldst. & Kit. (Branched Plantain). A habit; B bracts from base of flower spike; C flower and bracteole; D capsule dehiscing to expose two seeds; E seed. (Breckland). STACE 697/584.

126. BUDDLEJACEAE

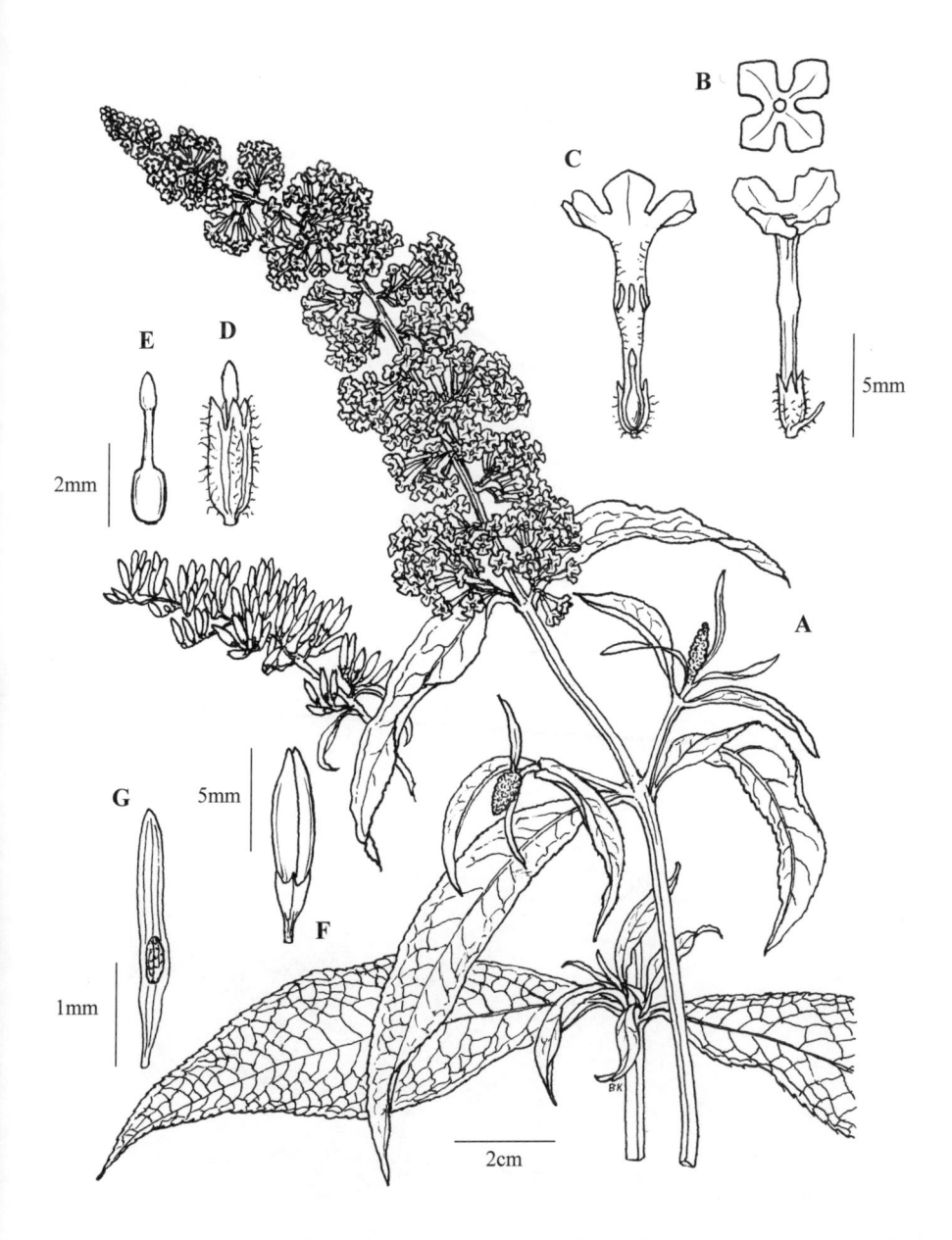

Buddleja davidii Franchet (Butterfly-bush). A flowering and fruiting branches; B flower and corolla lobes; C flower with part of corolla cut away; D calyx and stigma; E gynoecium; F fruit; G seed. (Source unknown). STACE 698/585.

127. OLEACEAE

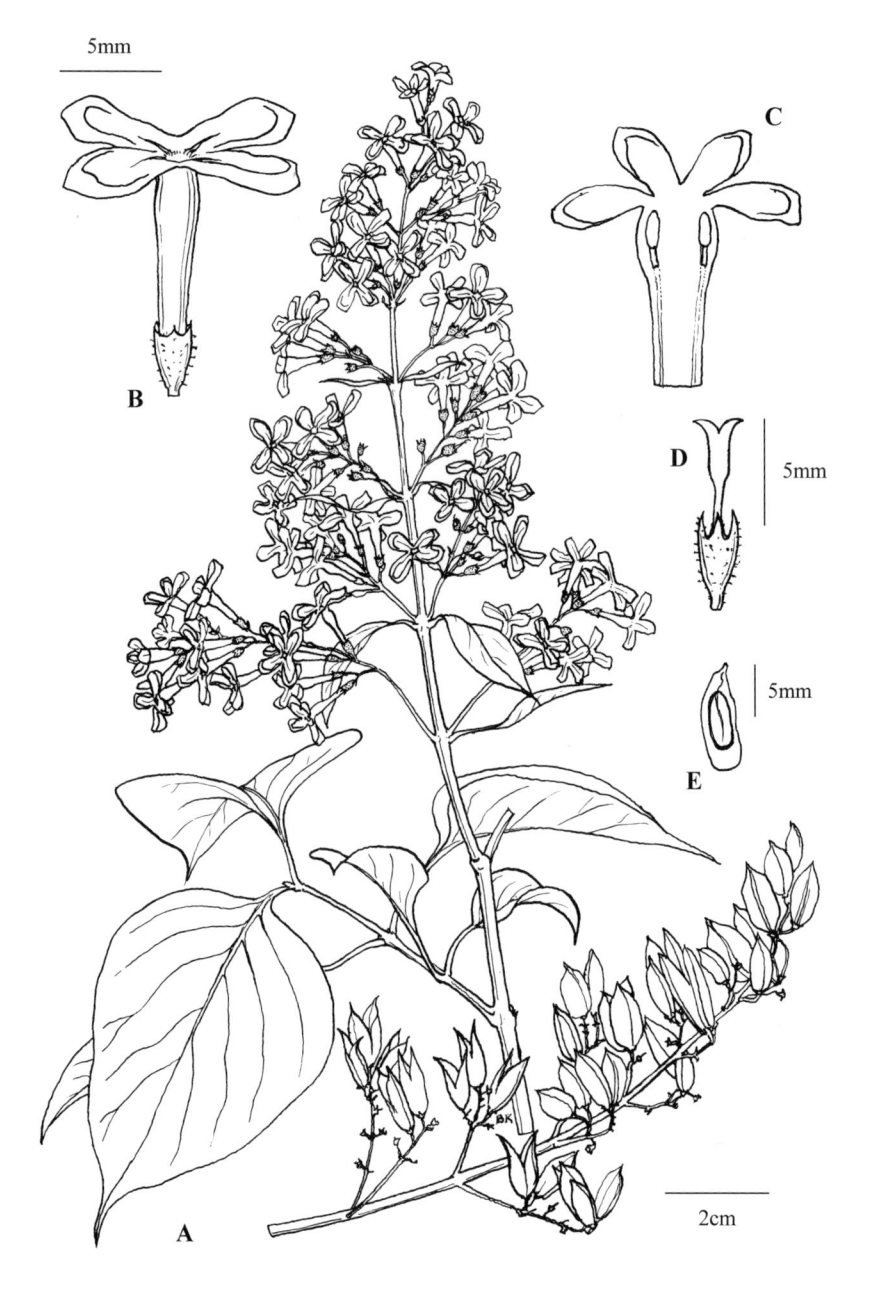

Syringa vulgaris L. (Lilac). A flowering and fruiting branches; B flower; C corolla opened out; D gynoecium within calyx; E seed. (Source unknown). STACE 700/587.

127. OLEACEAE   241

Ligustrum ovalifolium Hassk. (Garden Privet). A flowering branch; B fruiting branch; C flower; D corolla opened out; E gynoecium within calyx; F seed. (Source unknown). STACE 700/587.

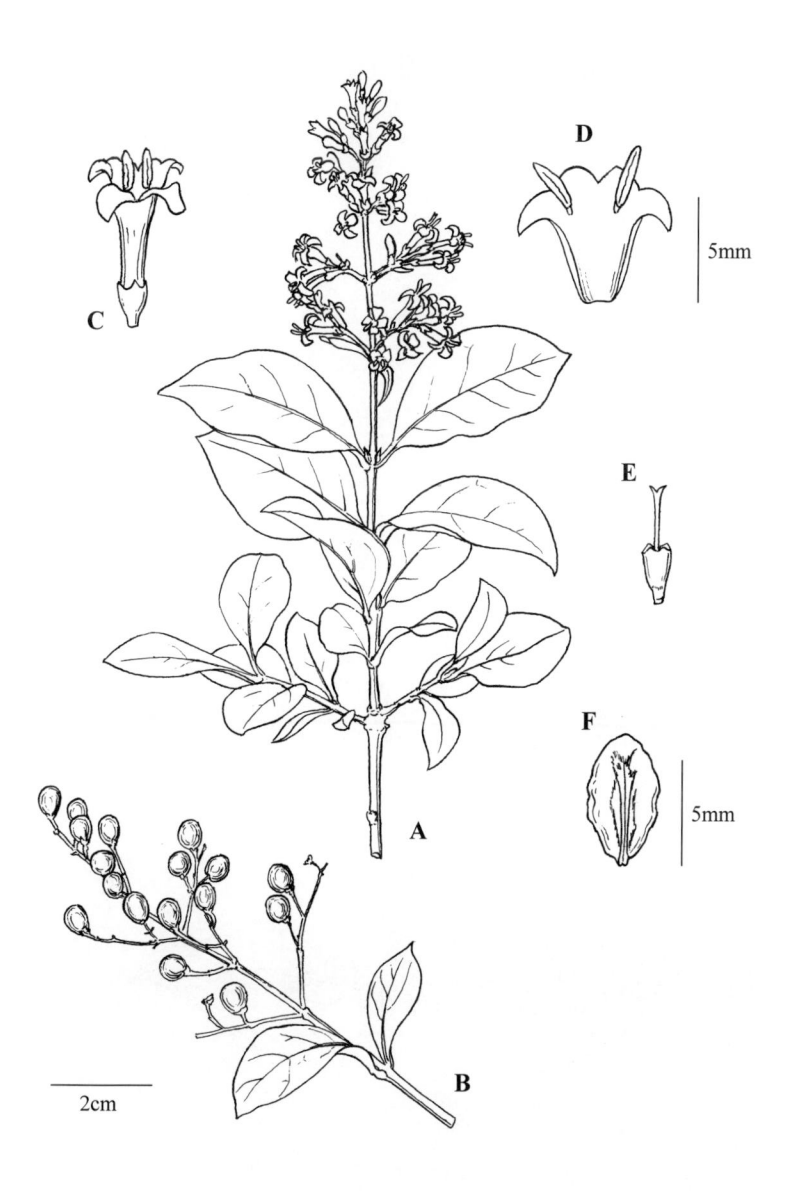

128. SCROPHULARIACEAE

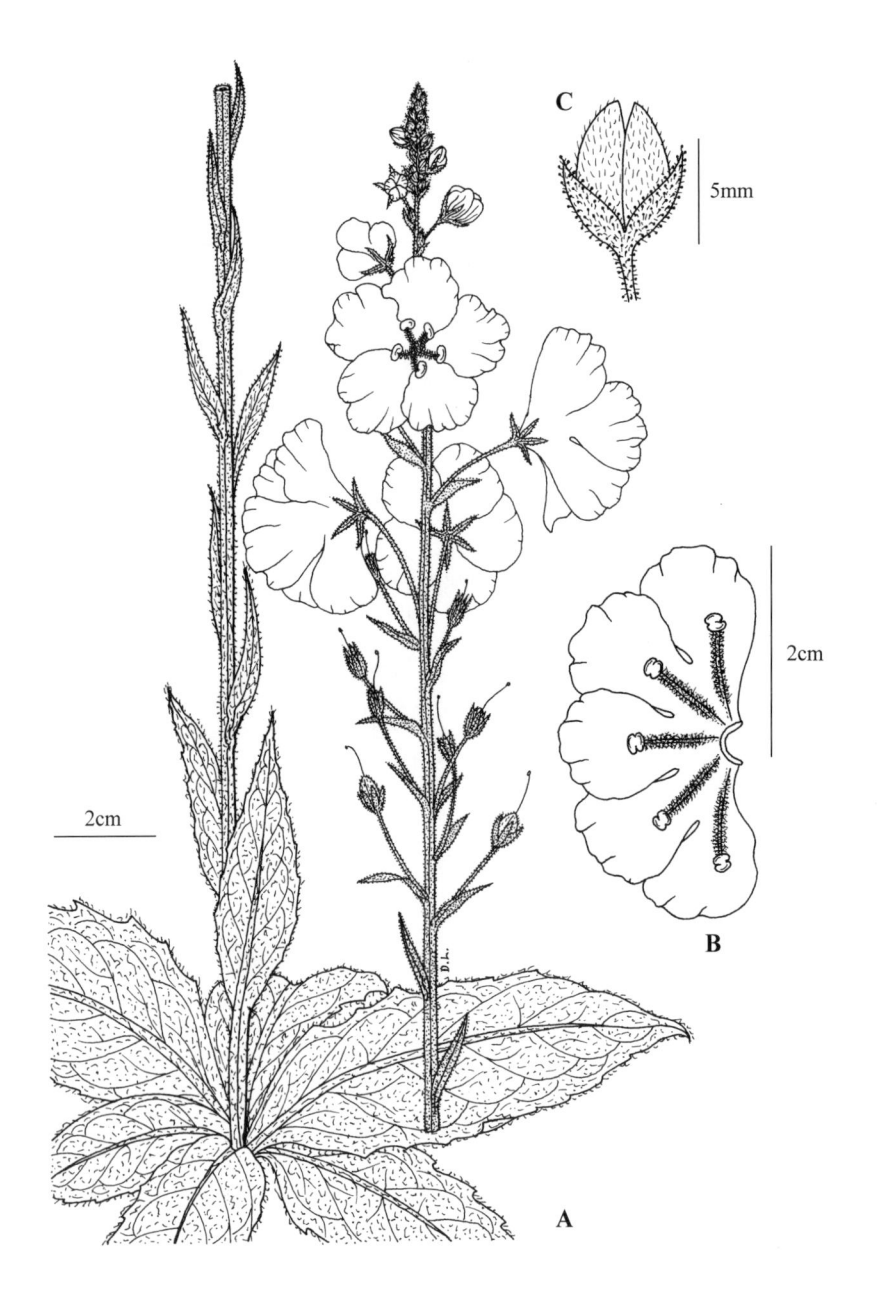

Verbascum phoeniceum L. (Purple Mullein). A habit; B corolla opened out; C fruit. (Oxford, Oxon). STACE 705/590.

128. SCROPHULARIACEAE

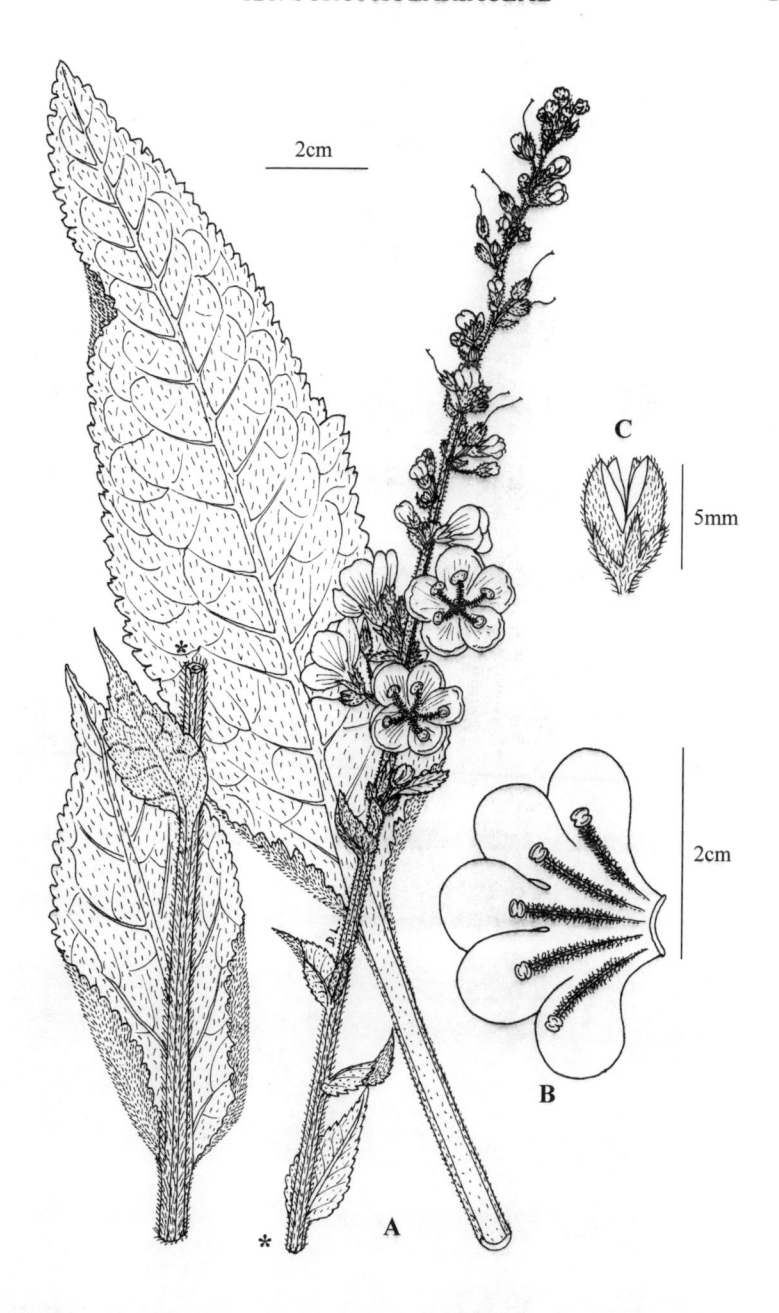

Verbascum chaixii Vill. (Nettle-leaved Mullein). A habit and basal leaf; B corolla opened out; C fruit. (N Lincs). STACE 706/592.

128. SCROPHULARIACEAE

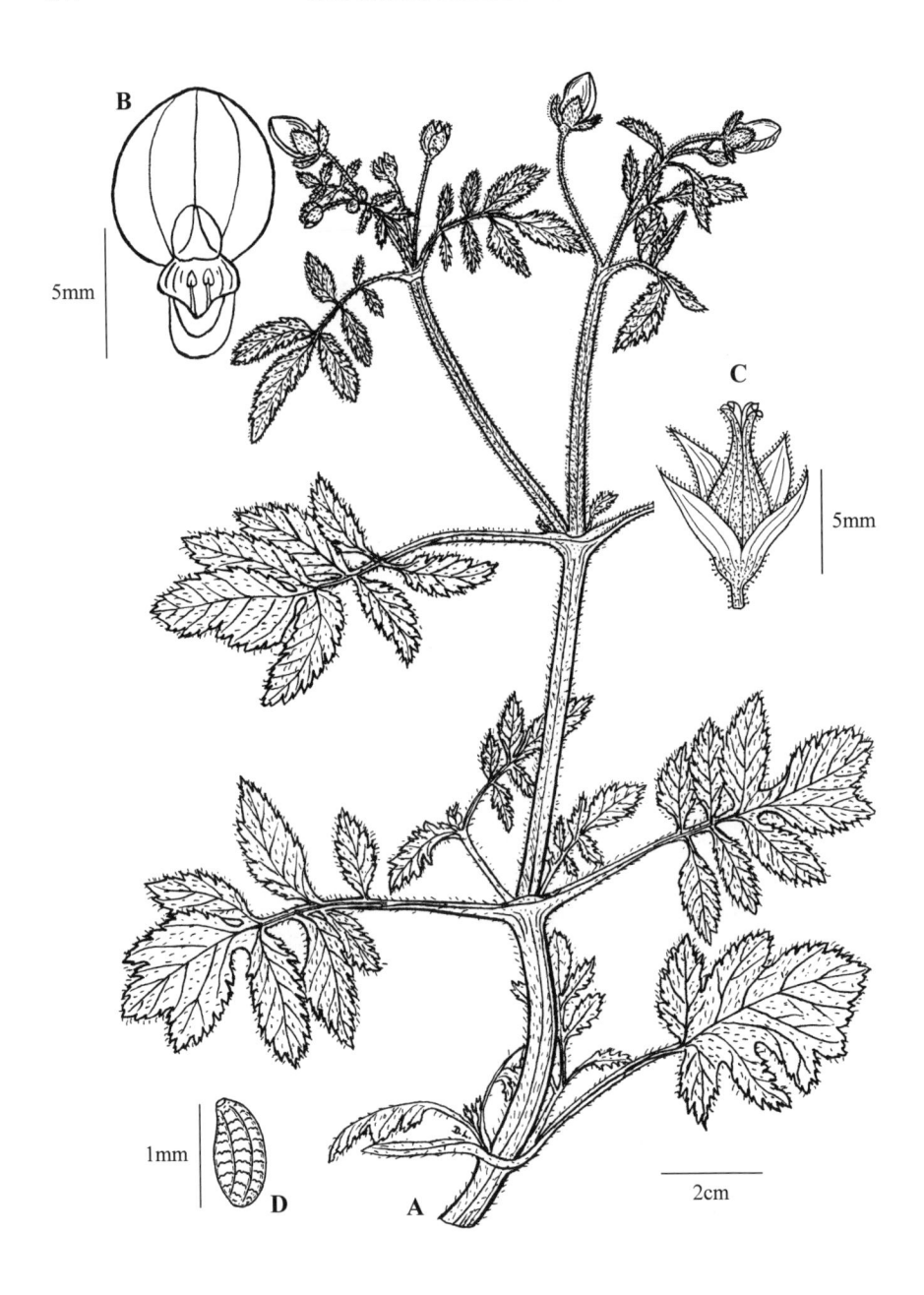

Calceolaria chelidonioides Kunth (Slipperwort). A habit; B corolla with lips separated; C capsule; D seed. (Battersea Park, Surrey). STACE 712/596.

128. SCROPHULARIACEAE

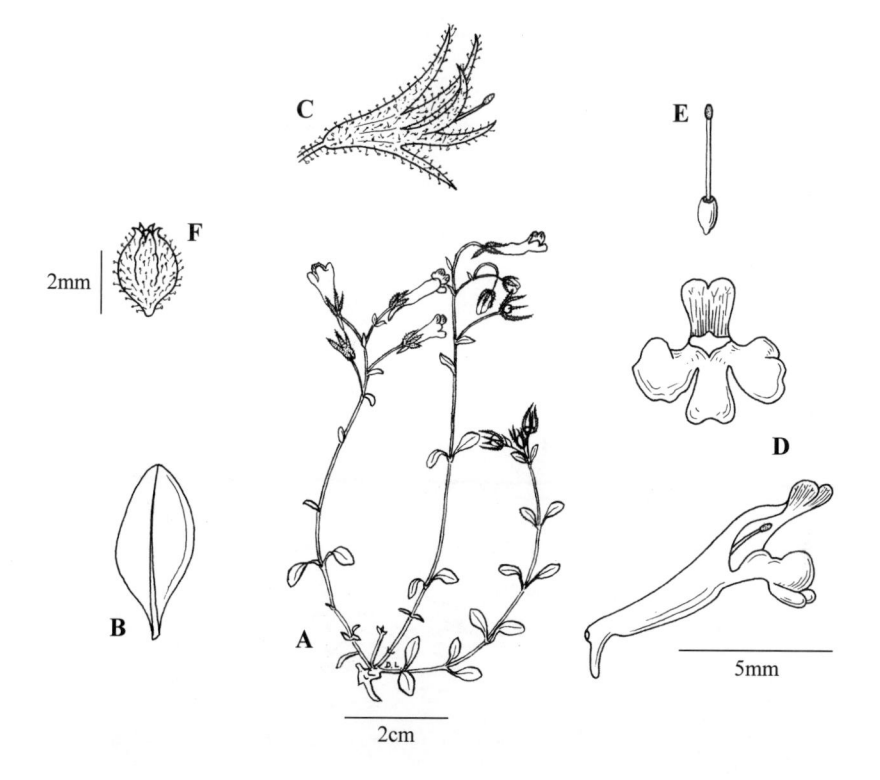

Chaenorhinum origanifolium (L.) Kostel. ssp. **crassifolium** (Cav.) Rivas Goday & Borja (Malling Toadflax). A habit; B leaf; C calyx; D side and front views of corolla; E gynoecium; F fruit. (West Malling, W Kent). STACE 712/597.

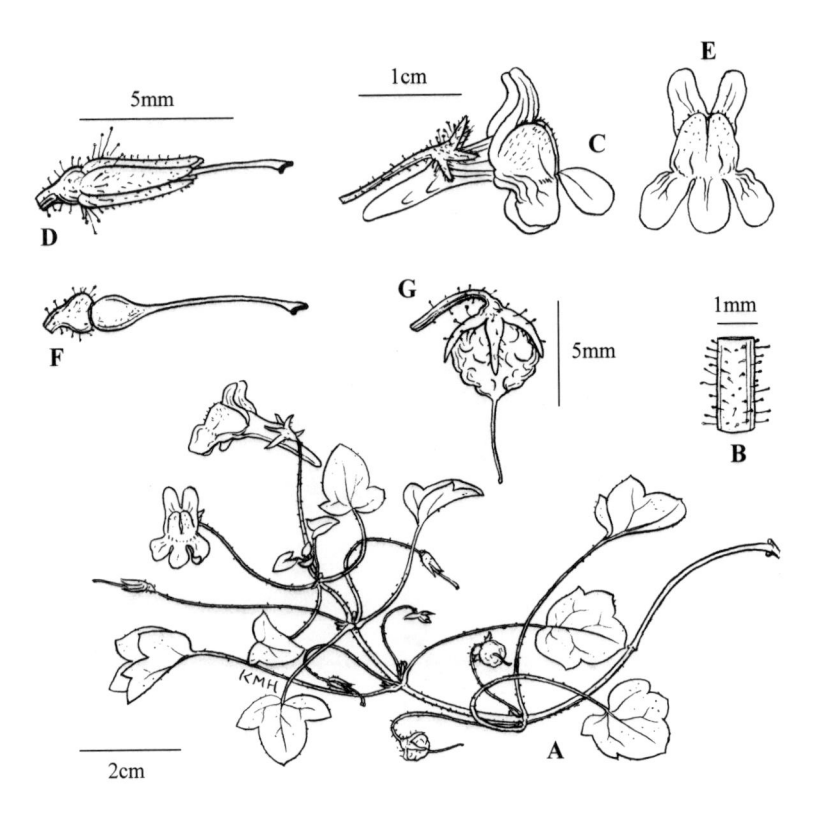

Cymbalaria pallida (Ten.) Wettst. (Italian Toadflax). A habit; B part of stem; C flower; D calyx; E corolla; F gynoecium; G fruit. (Preston, W Lancs). STACE 714/598.

128. SCROPHULARIACEAE

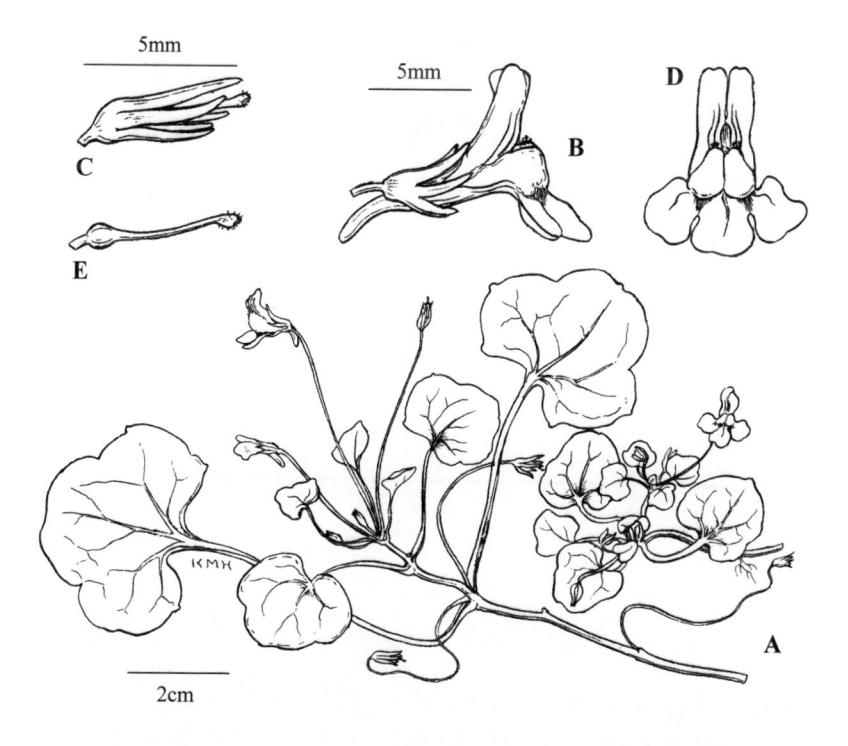

Cymbalaria hepaticifolia (Poir.) Wettst. (Corsican Toadflax). A habit; B flower; C calyx; D corolla; E gynoecium. (Edinburgh Botanical Garden, Midlothian). STACE 714/598.

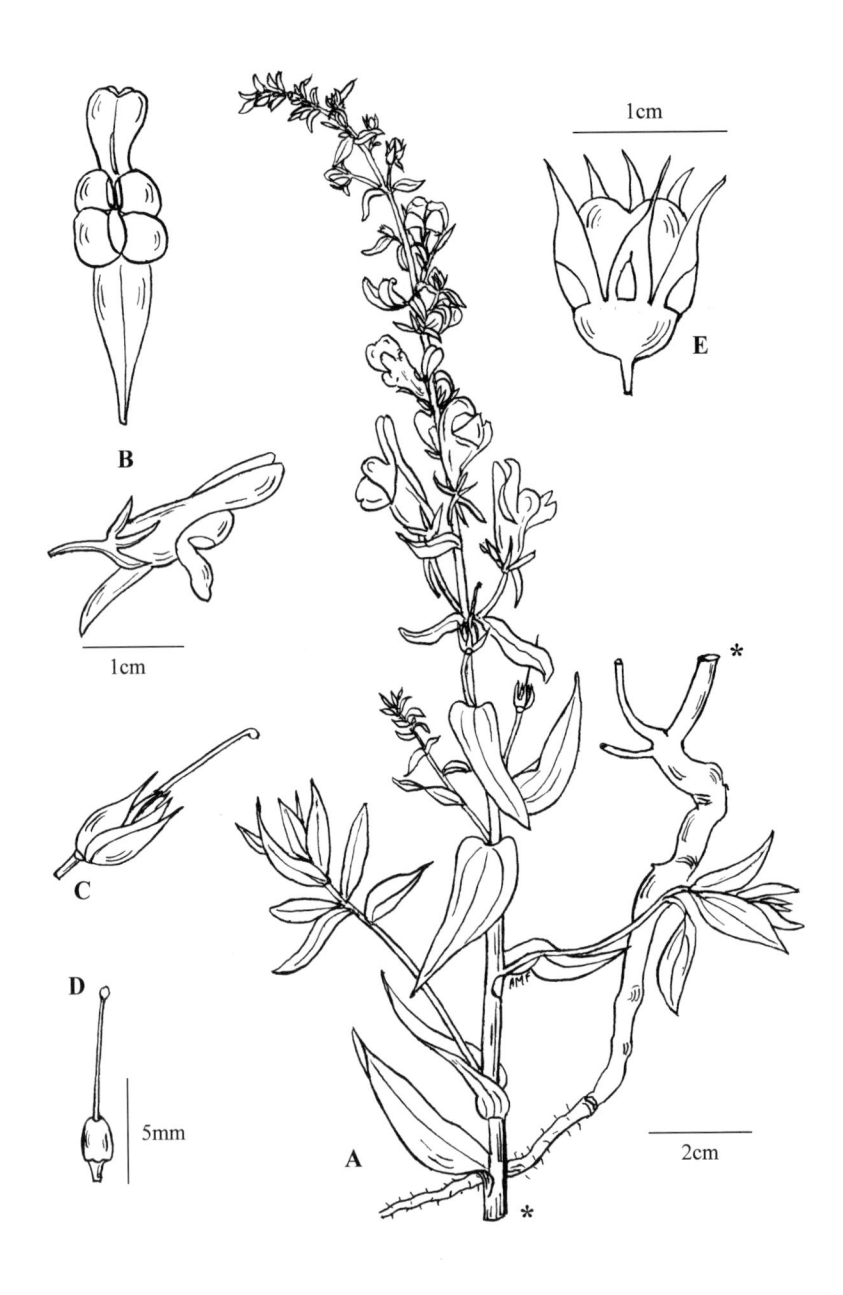

Linaria dalmatica (L.) Mill. (Balkan Toadflax). A habit; B flower; C flower minus corolla; D gynoecium; E fruit. (S Somerset). STACE 715/599.

128. SCROPHULARIACEAE

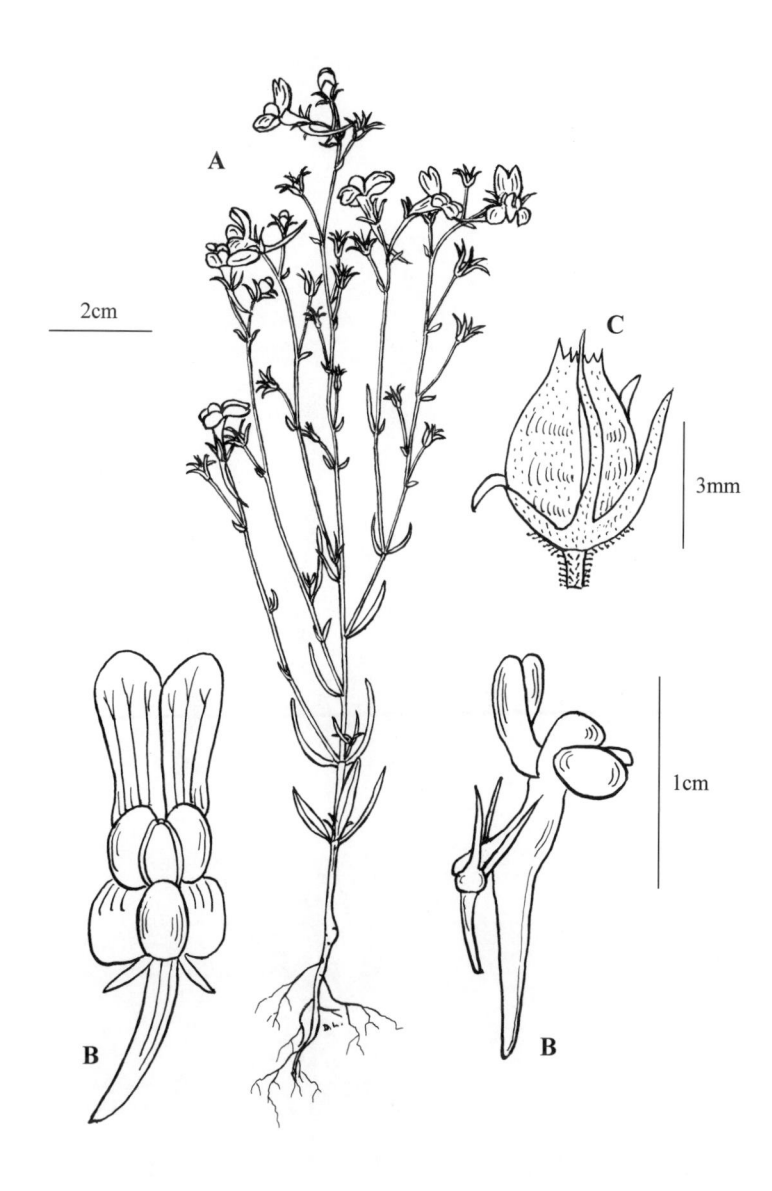

Linaria maroccana Hook. f. (Annual Toadflax). A habit; B front and side view of flower; C dehisced capsule. (*Hort*. Canterbury, E Kent). STACE 716/600.

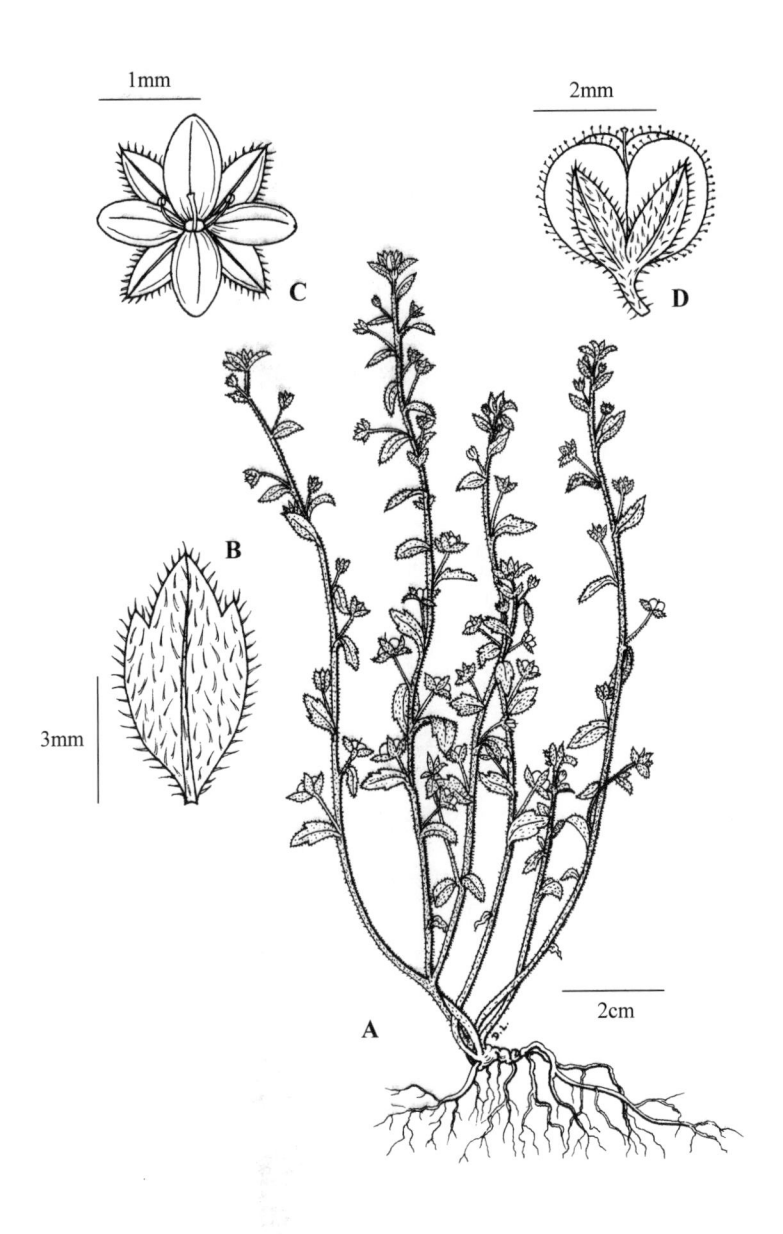

Veronica acinifolia L. (French Speedwell). A habit; B lower surface of leaf; C flower; D fruit with calyx. (Dorset). STACE 722/604.

128. SCROPHULARIACEAE

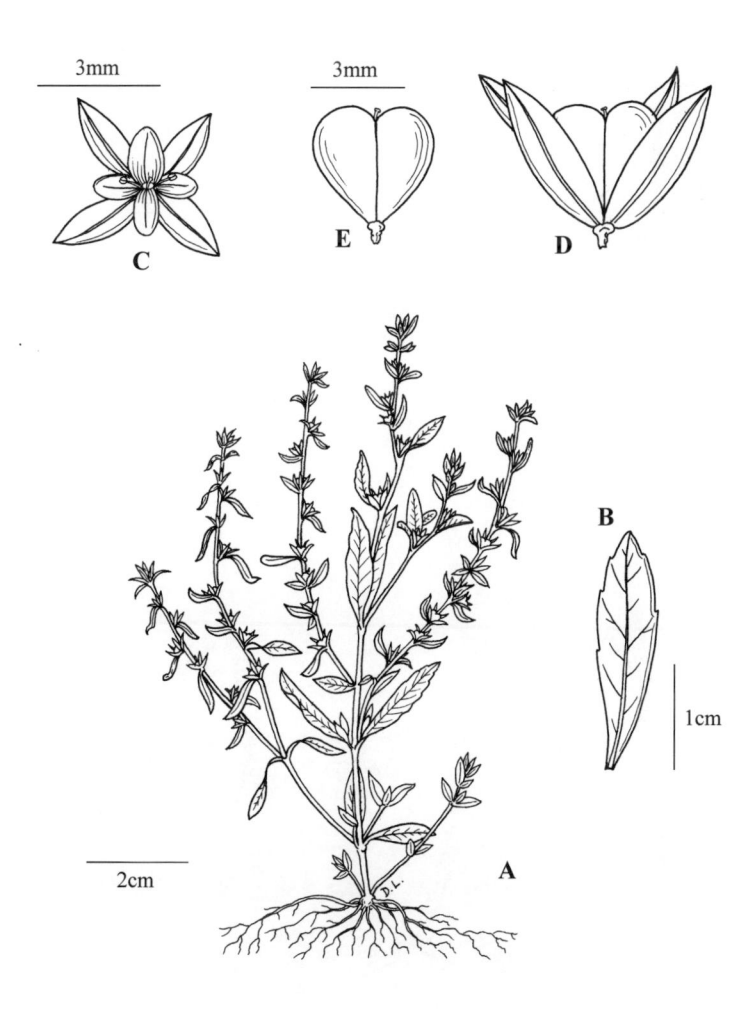

Veronica peregrina L. (American Speedwell). A habit; B leaf; C flower; D fruit with calyx; E capsule. (Withyham, E Sussex). STACE 723/604.

128. SCROPHULARIACEAE

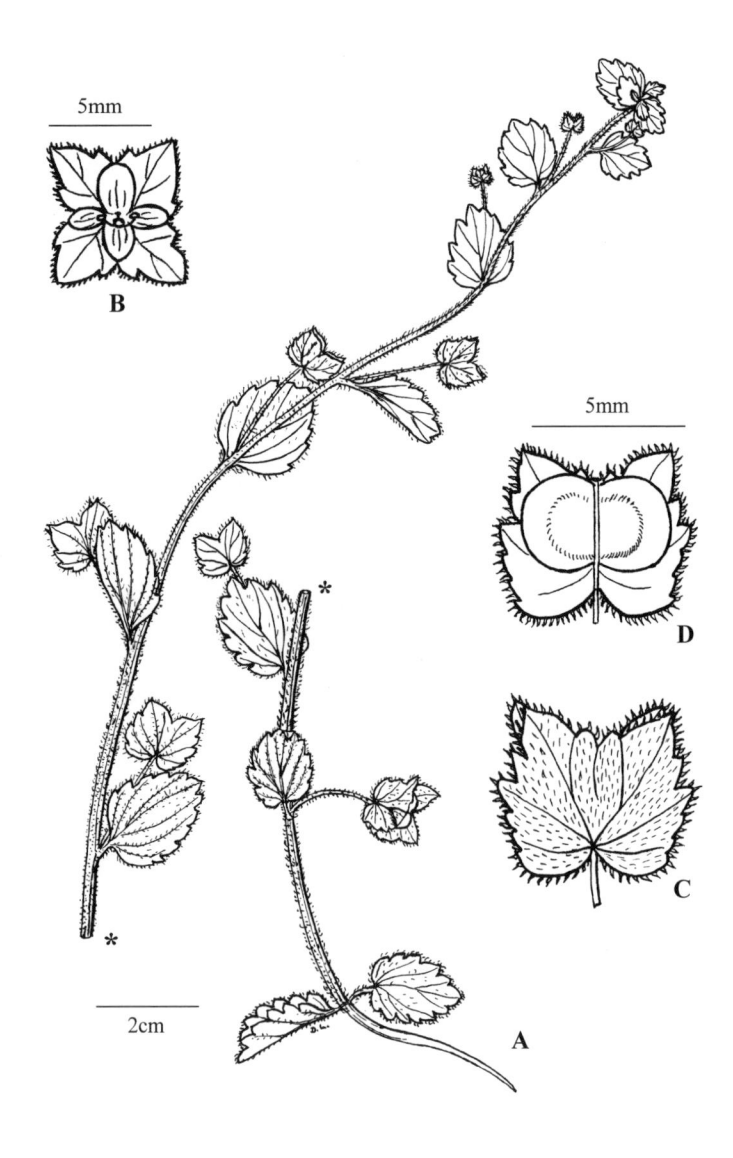

Veronica crista-galli Steven (Crested Field-speedwell). A habit; B flower; C fruit with calyx; D fruit with calyx partly removed. (Henfield, W Sussex). STACE 723/605.

128. SCROPHULARIACEAE

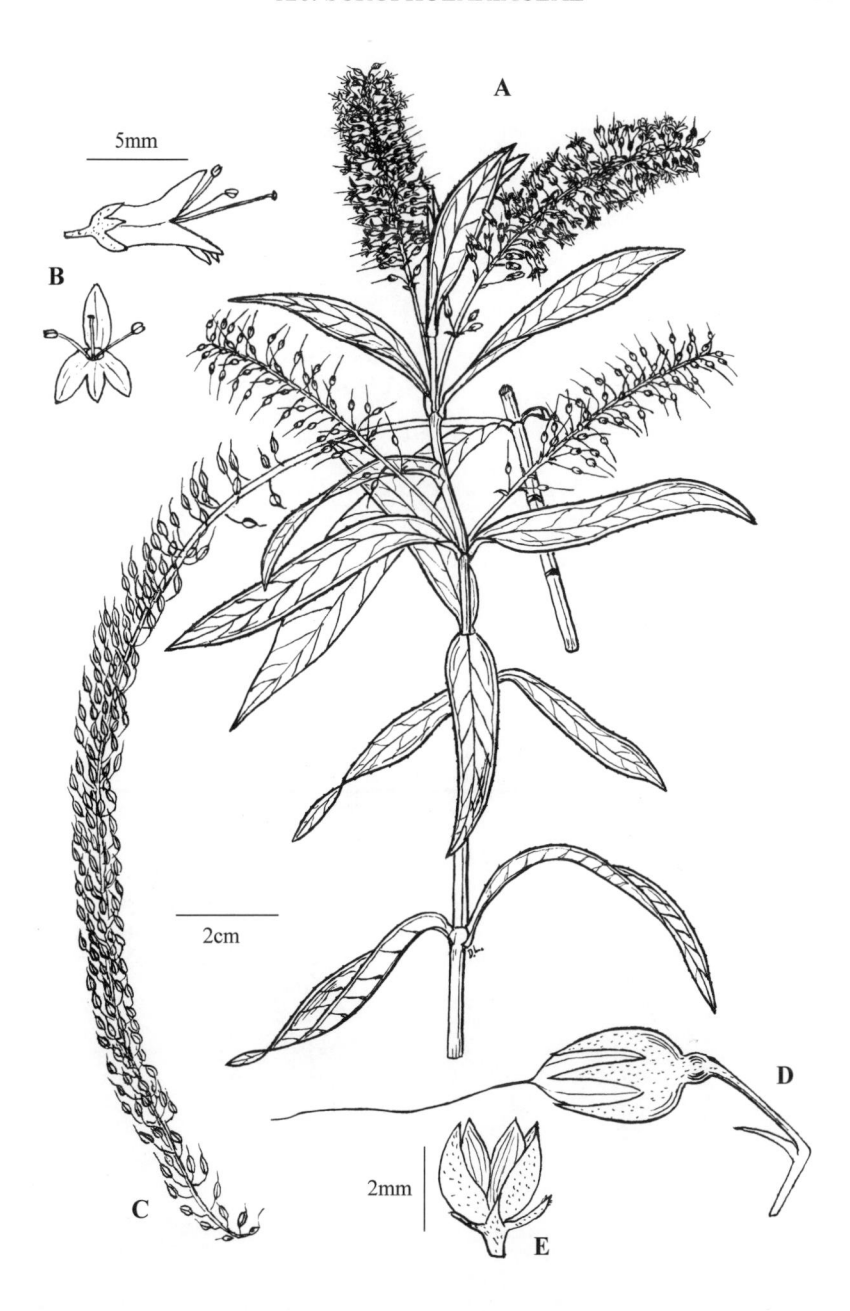

Hebe salicifolia (G. Forst.) Pennell (Koromiko). A flowering branch; B front and side views of flower; C fruiting raceme; D capsule; E dehisced capsule. (Laxey, Isle of Man). STACE 727/606.

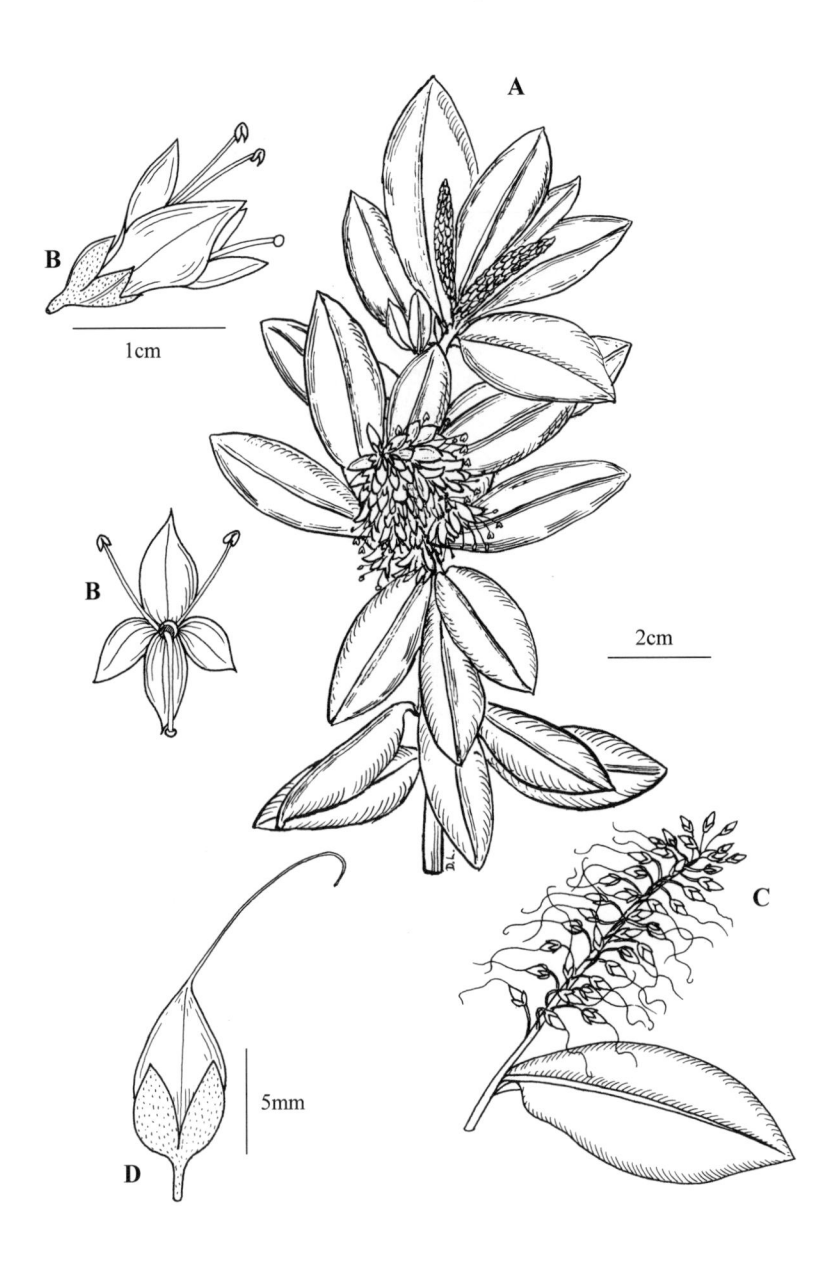

Hebe x franciscana (Eastw.) Souster (Hedge Veronica). A flowering branch; B front and side views of flower; C fruiting raceme; D fruit. (Cornwall). STACE 727/608.

128. SCROPHULARIACEAE

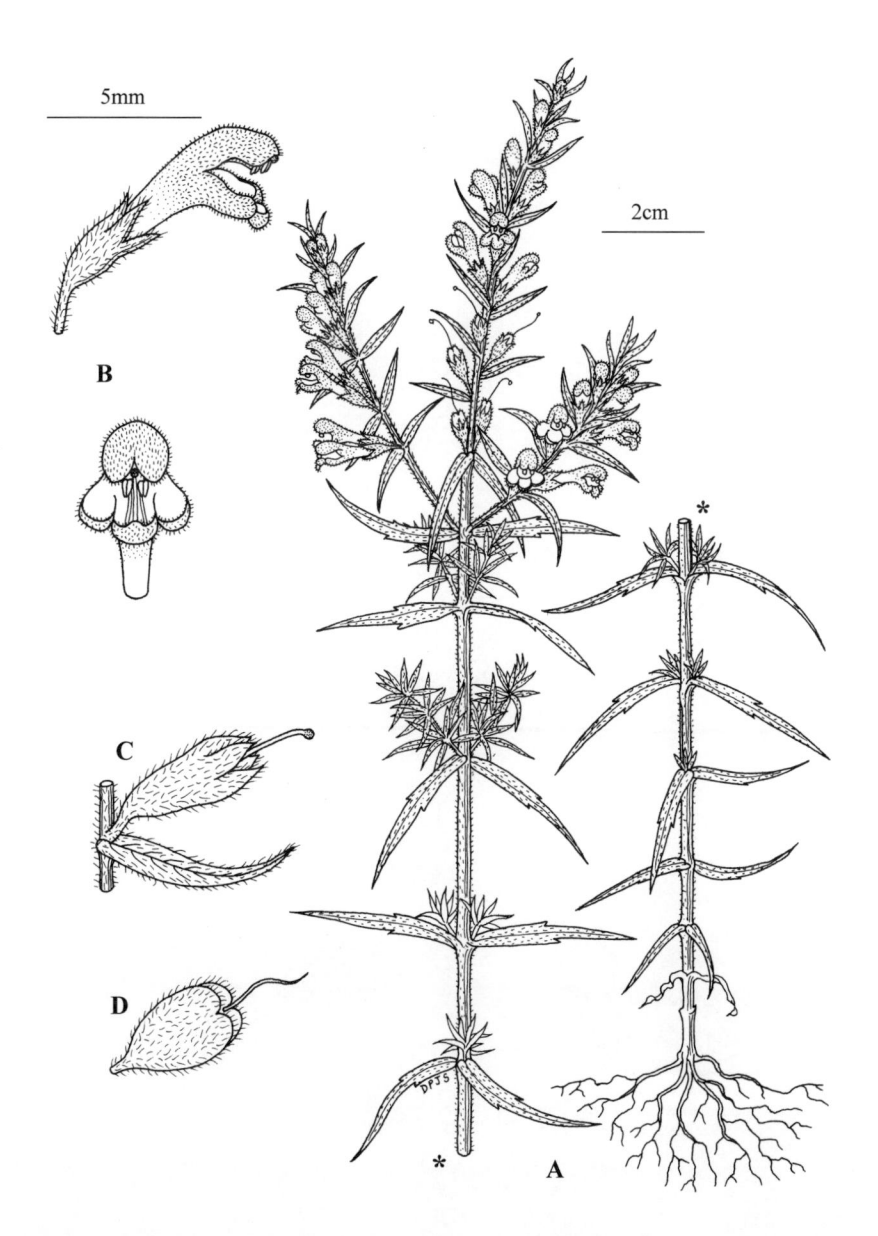

Odontites jaubertianus (Boreau) D. Dietr. ex Walp. ssp. **chrysanthus** (Boreau) P. Fourn. (French Bartsia). A habit; B front and side views of flower; C flower minus corolla; D capsule. (Aldermaston, Berks). STACE 739/622.

129. OROBANCHACEAE

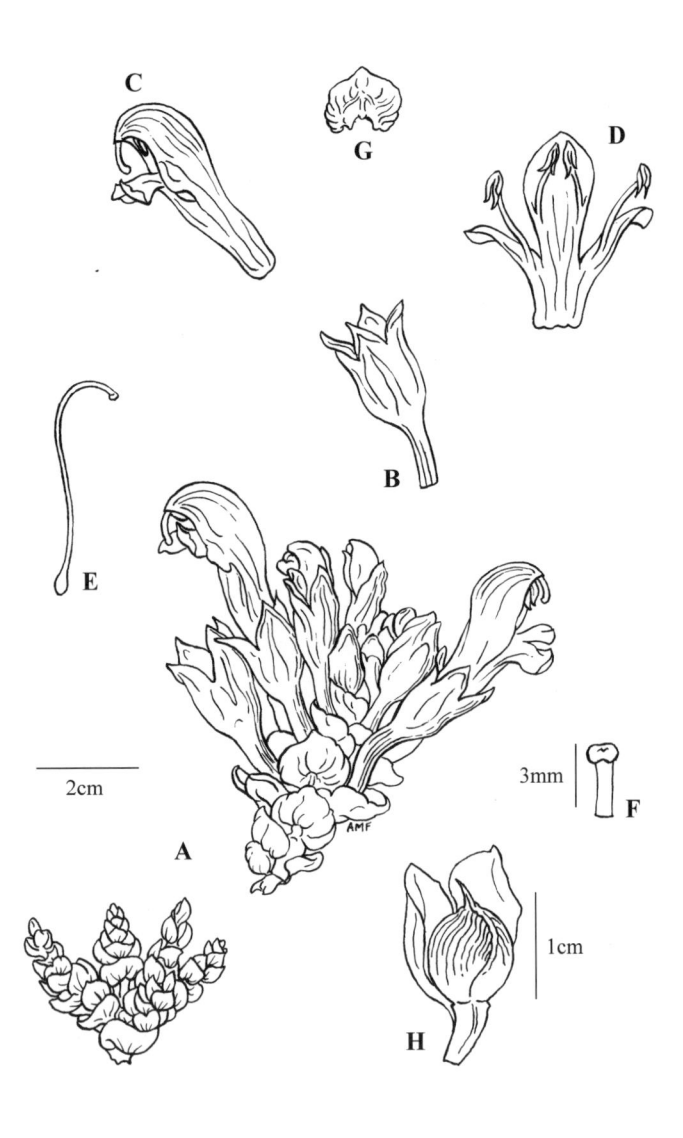

Lathraea clandestina L. (Purple Toothwort). A habit of young plant and flowering stem; B calyx; C corolla; D corolla opened out; E gynoecium; F stigma; G bract; H fruit with part of calyx removed. (Source unknown). STACE 746/627.

131. ACANTHACEAE

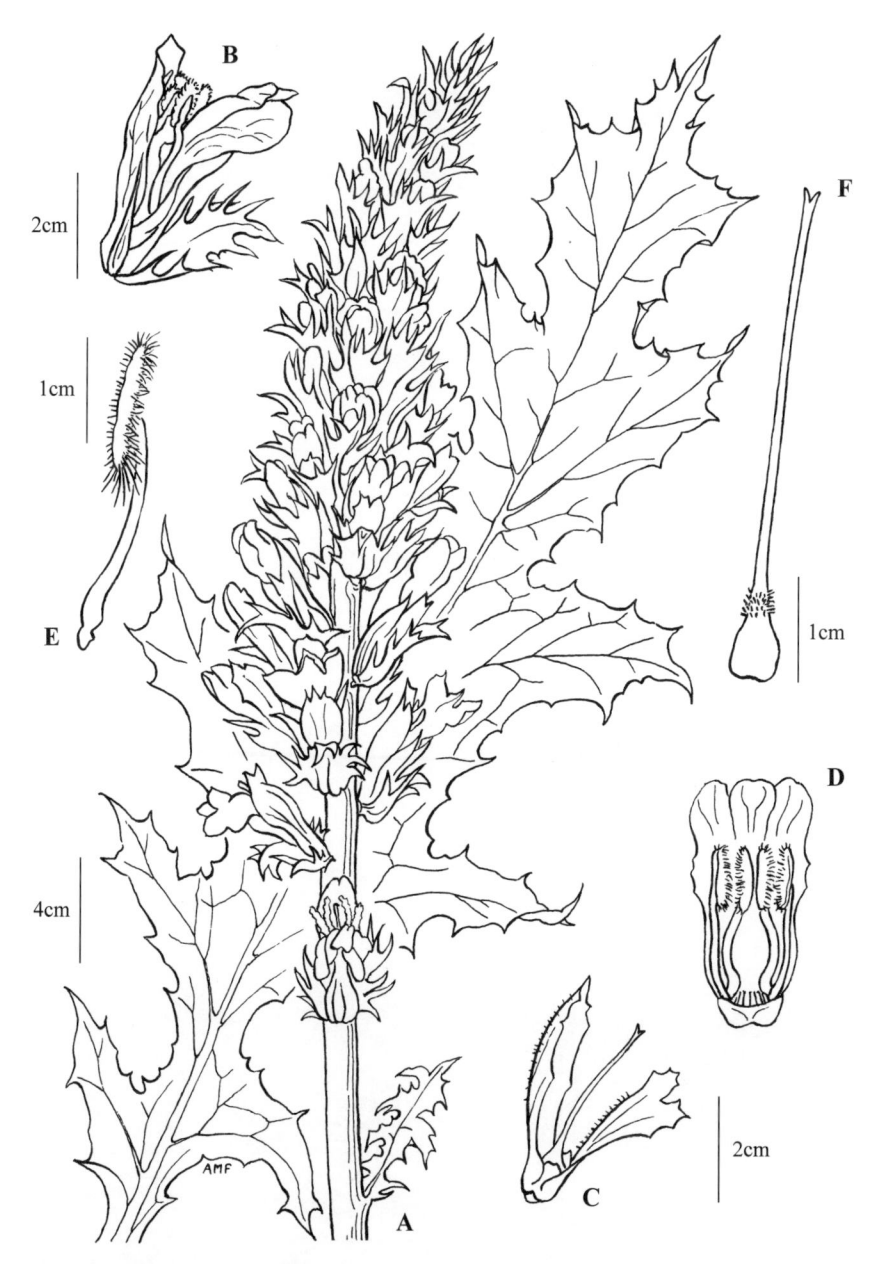

Acanthus mollis L. (Bear's-breech). A flower spike and basal leaf; B flower, bracteoles and bract; C flower with corolla removed; D corolla with stamens; E stamen; F gynoecium. (Source unknown). STACE 751/631.

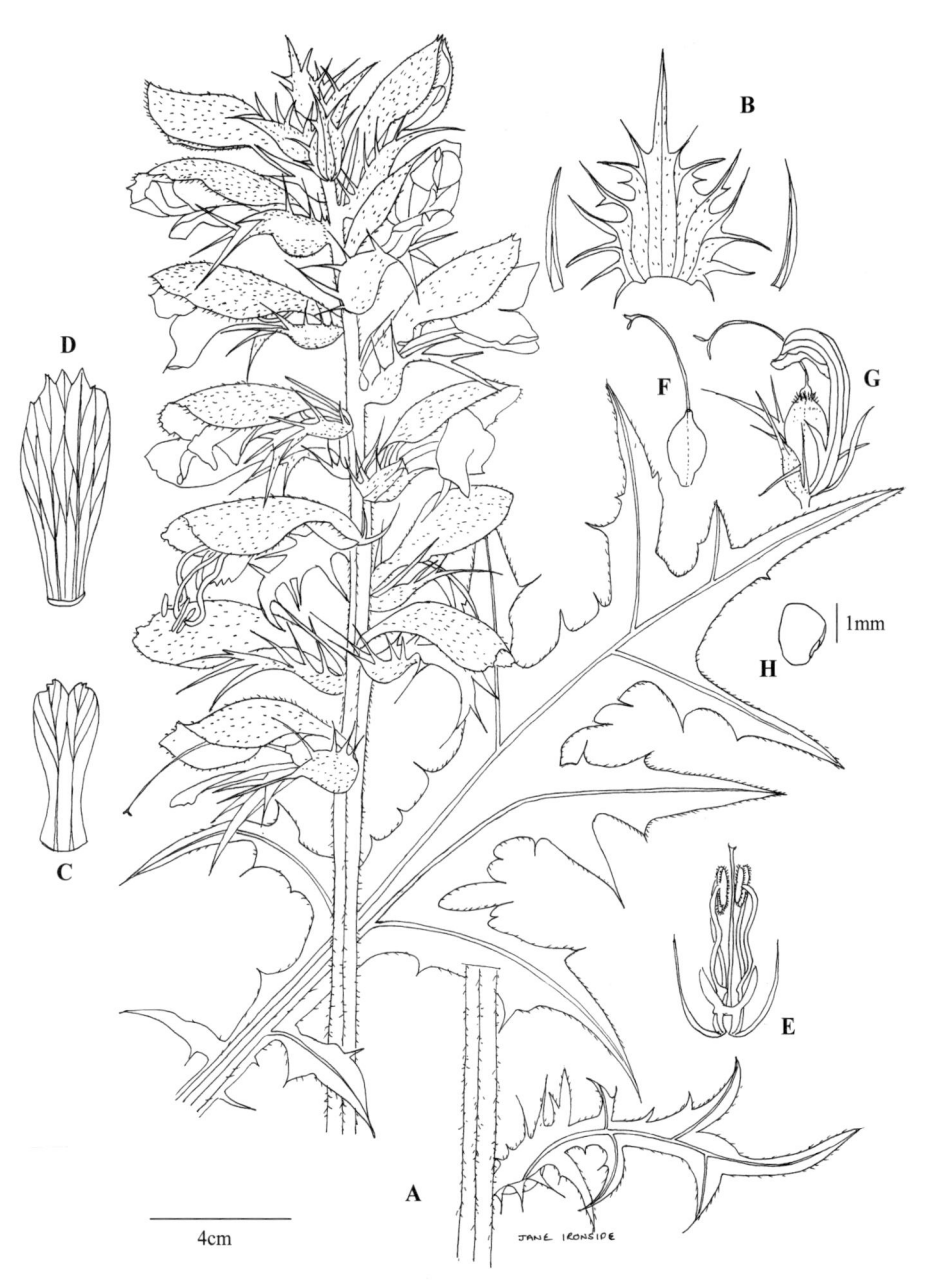

Acanthus spinosus L. (Spiny Bear's-breech). A flower spike and basal leaf; B bracteoles and bract; C lower calyx lobe; D upper calyx lobe; E flower with corolla removed; F gynoecium; G young fruit within enveloping bract; H seed. (*Hort*. Kettlestone, W Norfolk). STACE 751/631.

133. CAMPANULACEAE

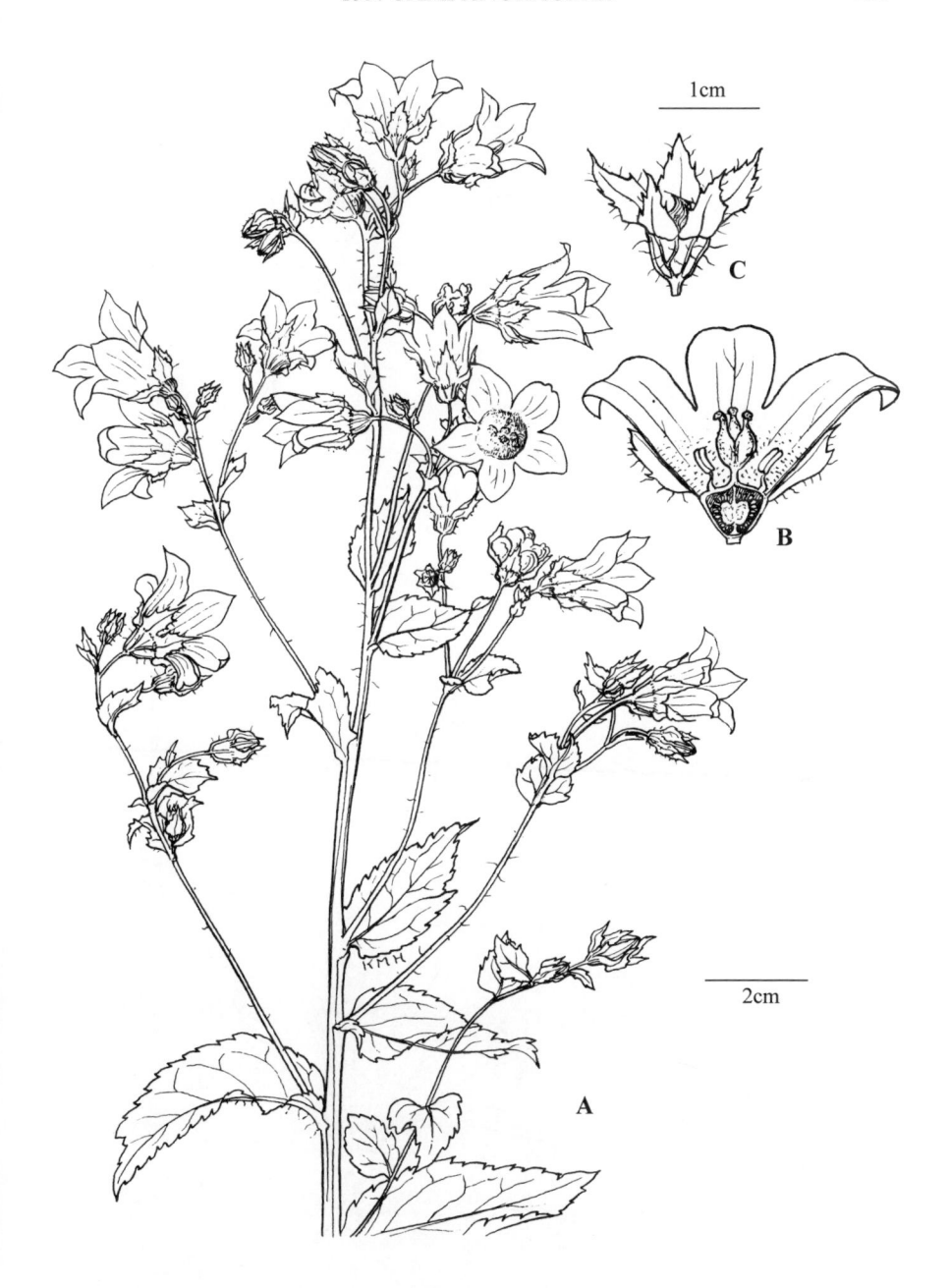

Campanula lactiflora M. Bieb. (Milky Bellflower). A habit; B LS of flower; C fruit. (Ballater, S Aberdeen). STACE 759/639.

260   133. CAMPANULACEAE

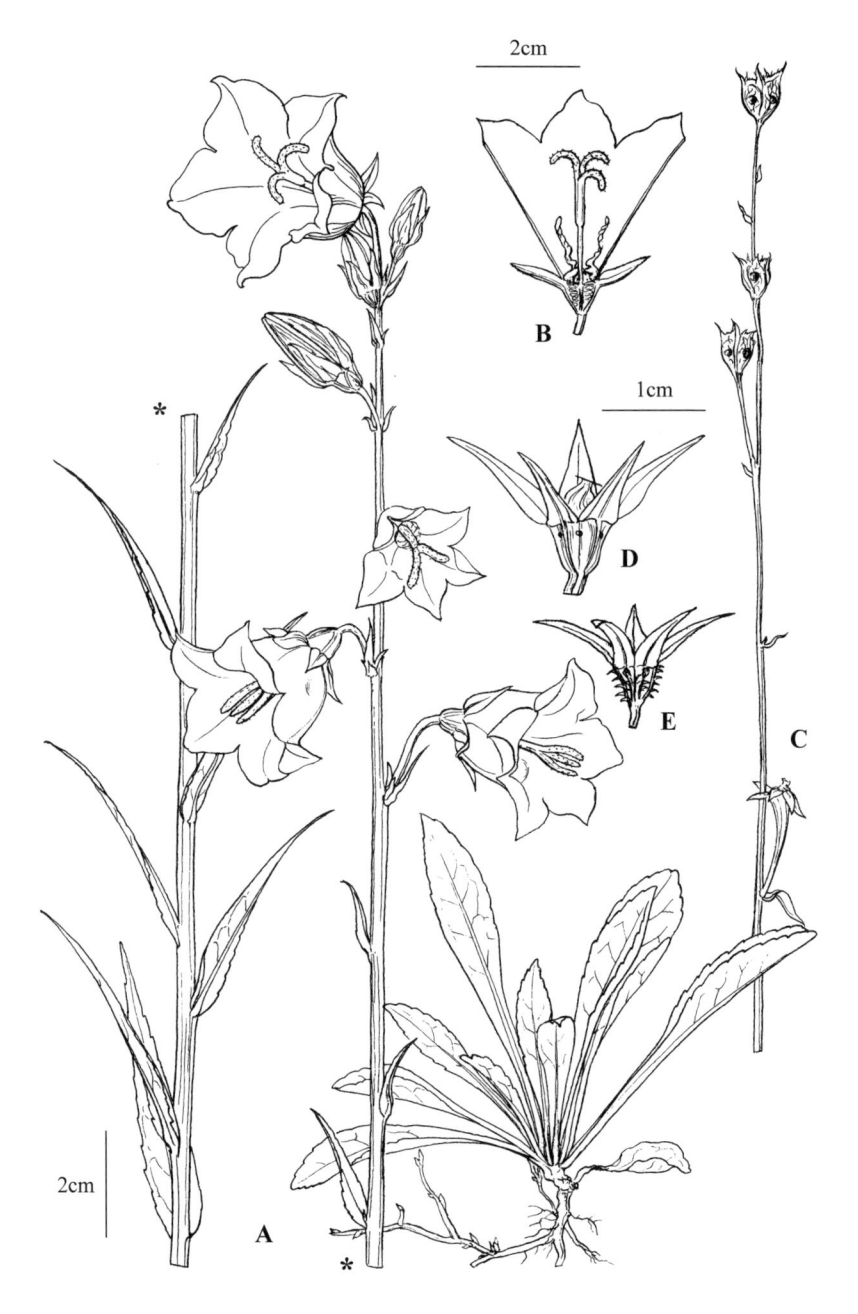

Campanula persicifolia L. (Peach-leaved Bellflower). A habit; B LS of flower; C fruiting stem; D fruit; E fruit of variant. (*Hort*. Ashbourne, Derbys). STACE 759/639.

133. CAMPANULACEAE

Campanula medium L. (Canterbury-bells). A habit; B LS of flower; C fruit. (Shoreham, W Kent). STACE 761/639.

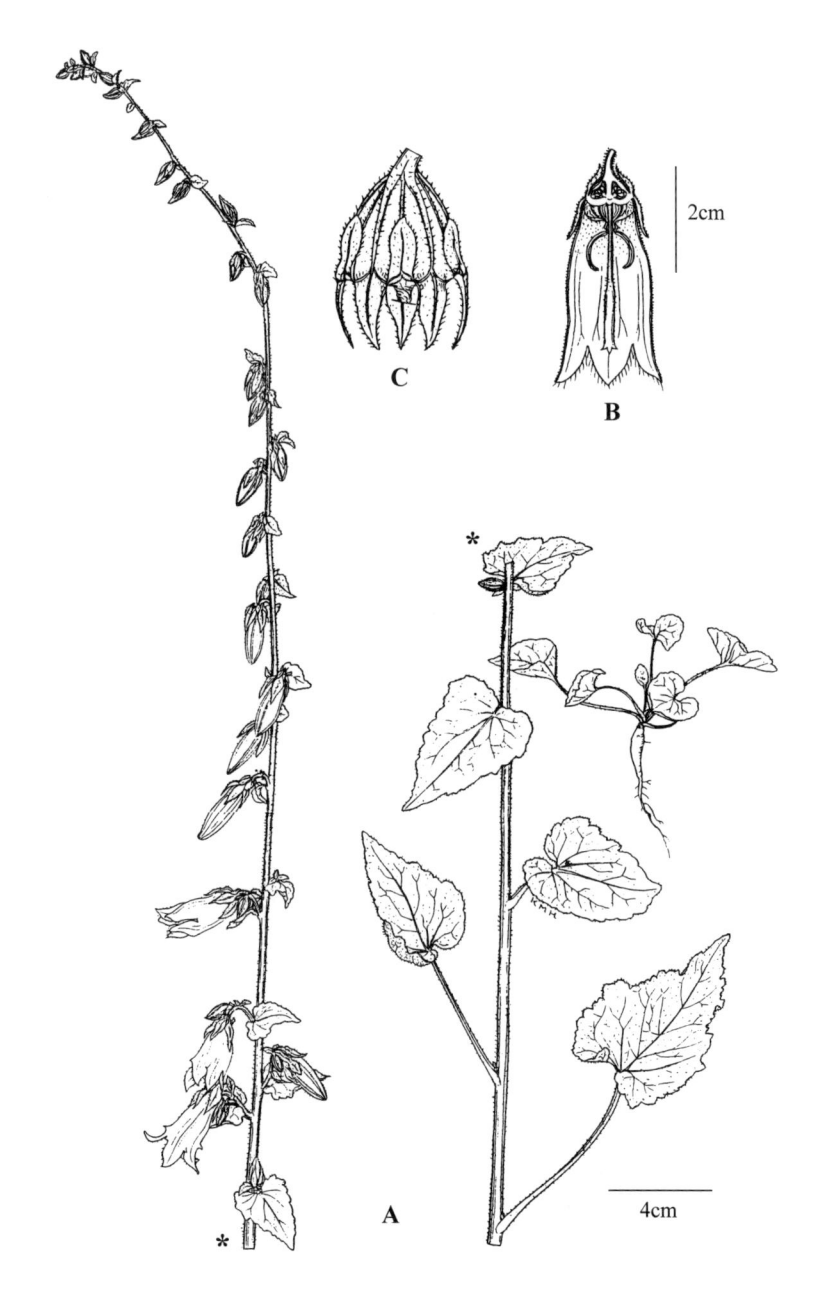

Campanula alliariifolia Willd. (Cornish Bellflower). A flowering stem and young plant; B LS of flower; C fruit. (Source unknown). STACE 761/639.

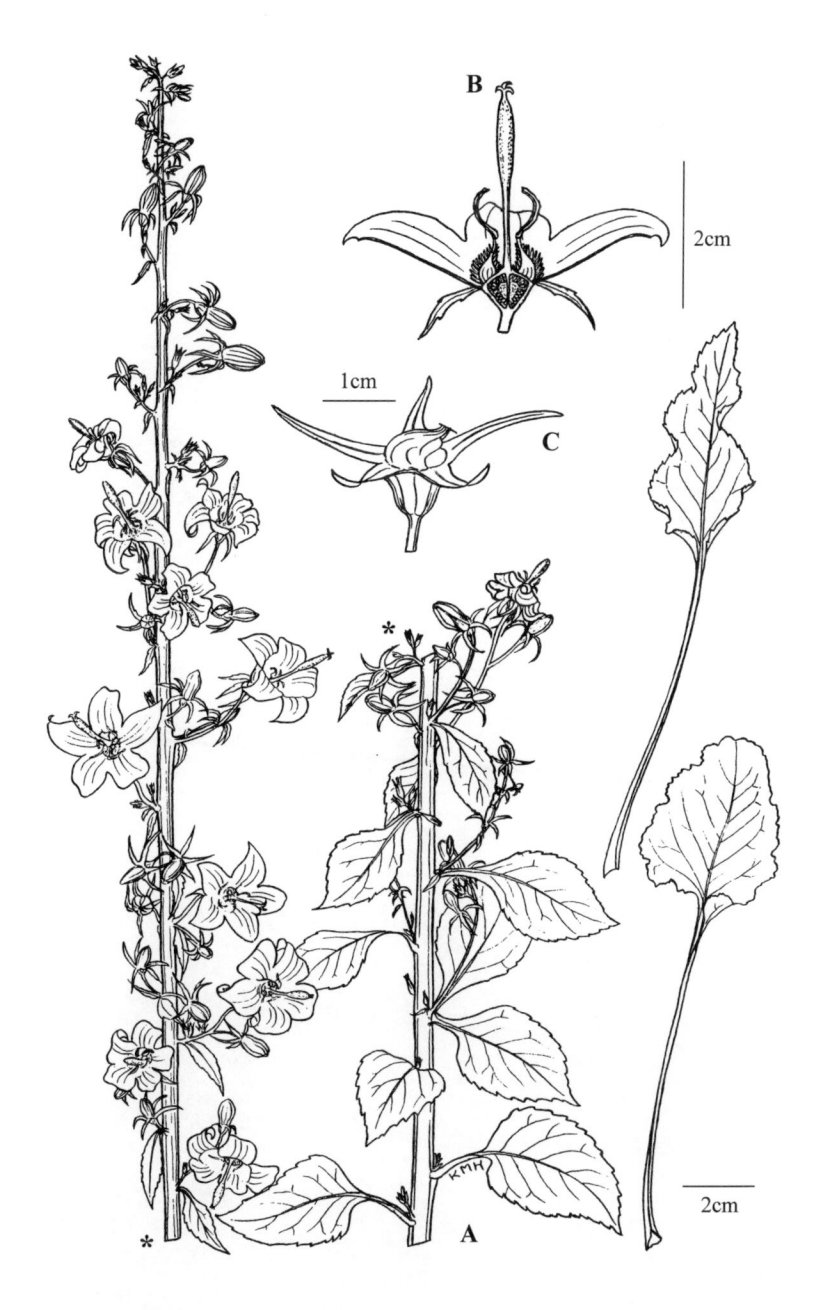

133. CAMPANULACEAE

Campanula pyramidalis L. (Chimney Bellflower). A flowering stem and basal leaves; B flower; C fruit. (Guernsey). STACE 761/639.

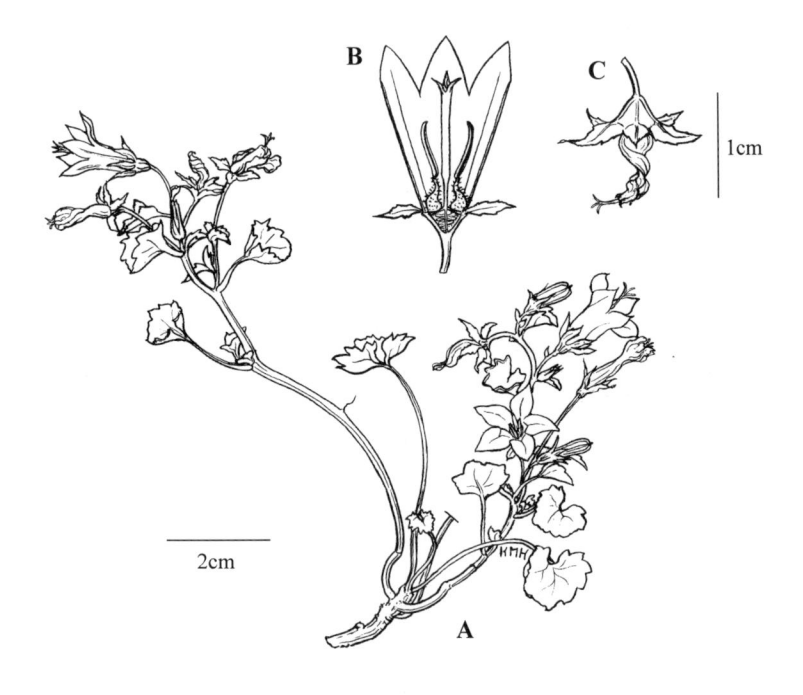

Campanula portenschlagiana Schult. (Adria Bellflower). A habit; B LS of flower; C fruit. (St Ouen, Jersey). STACE 761/639.

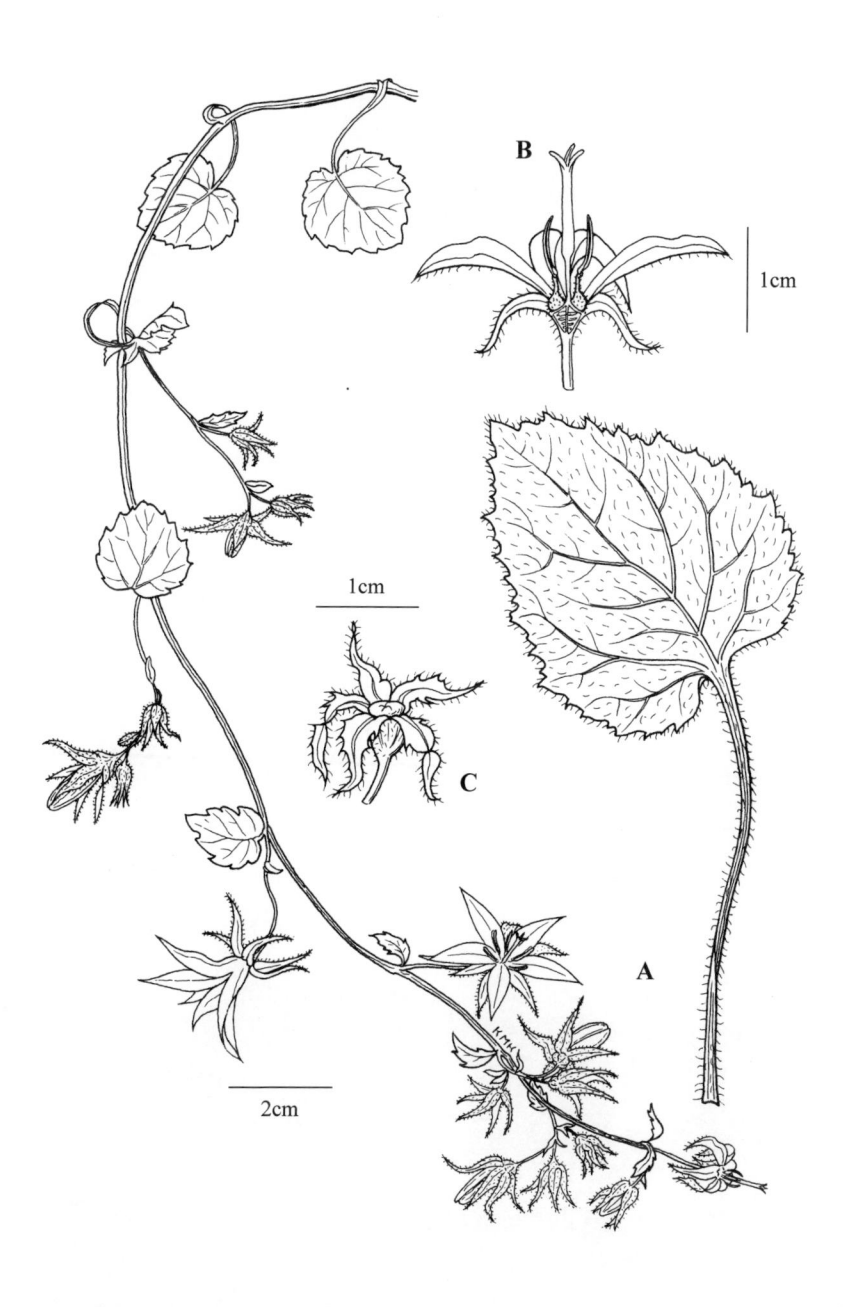

Campanula poscharskyana Degen (Trailing Bellflower). A habit and basal leaf; B LS of flower; C fruit. (Godstone, Surrey). STACE 761/639.

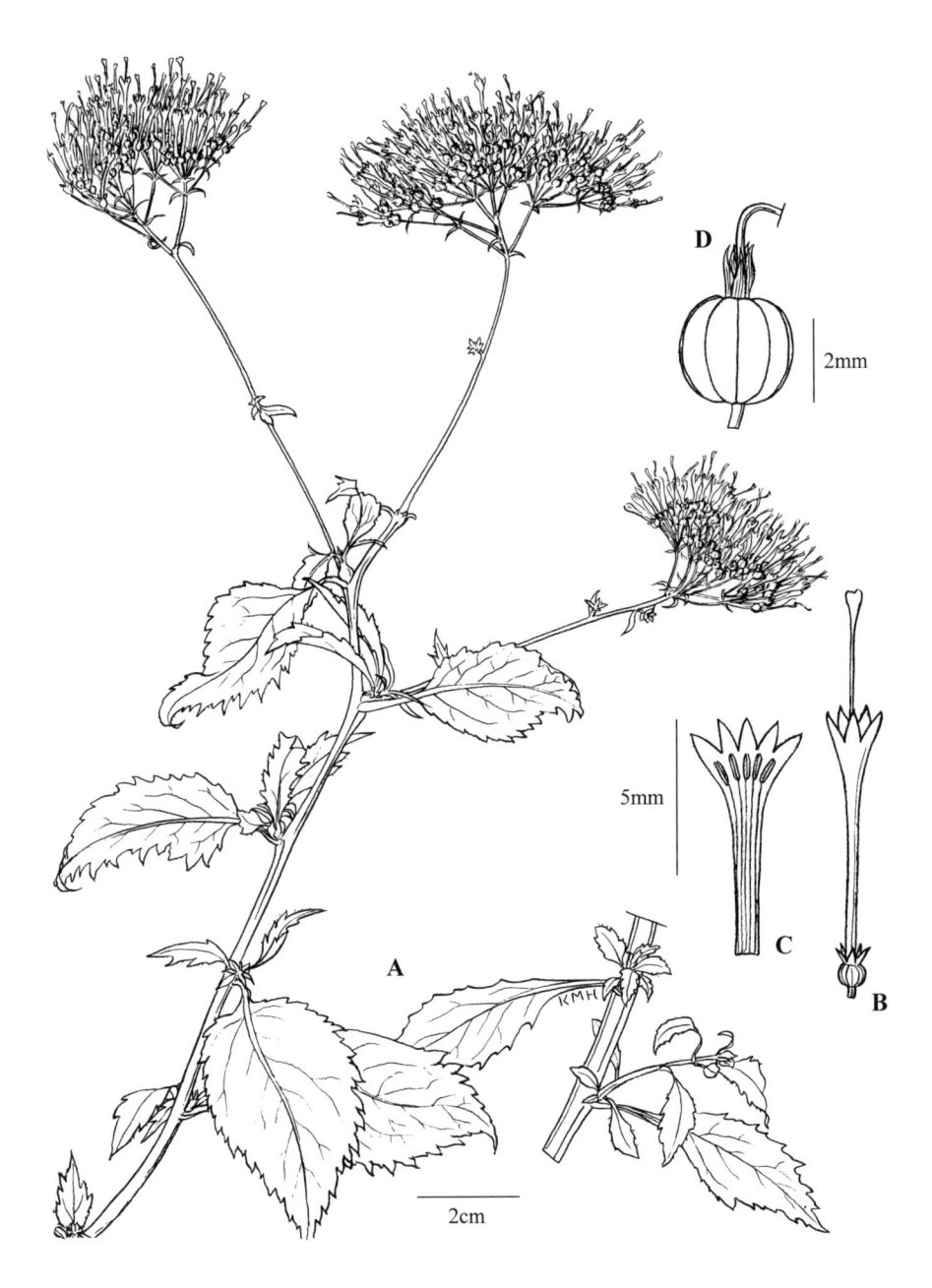

Trachelium caeruleum L. (Throatwort). A habit; B flower; C corolla opened out to show stamens; D fruit. (Guernsey). STACE 763/641.

133. CAMPANULACEAE

Phyteuma scheuchzeri All. (Oxford Rampion). A habit; B flower; C fruit. (Oxford, Oxon). STACE 764/641.

133. CAMPANULACEAE

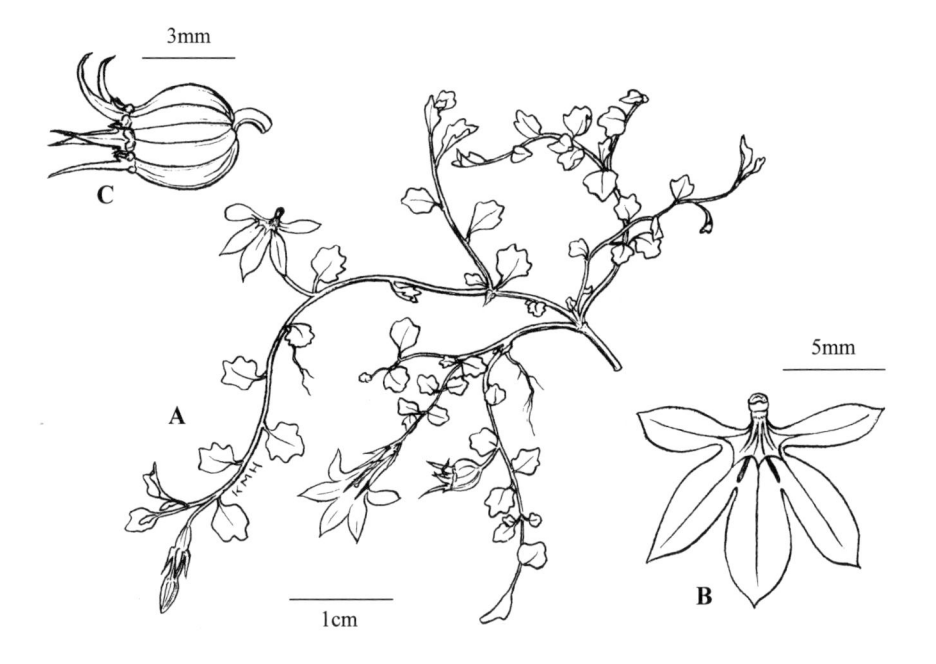

Pratia angulata (G. Forst.) Hook. f. (Lawn Lobelia). A habit; B flower; C fruit. (Hever Castle, W Kent). STACE 765/642.

134. RUBIACEAE

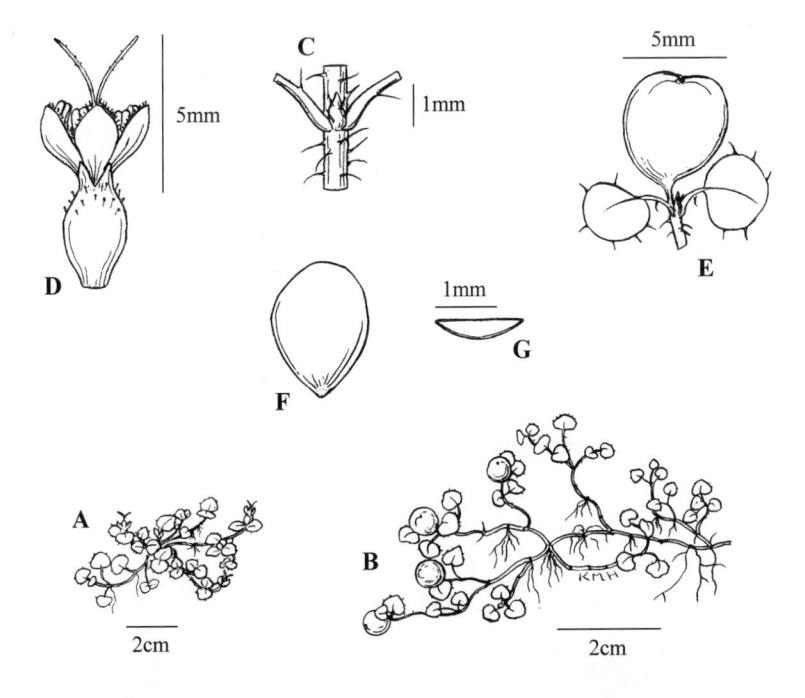

Nertera granadensis (Mutis ex L.f.) Druce (Beadplant). A flowering habit; B fruiting habit; C stem node with stipules; D flower; E fruiting shoot; F seed; G TS of seed. (Helensburgh, Dunbarton). STACE 767/644.

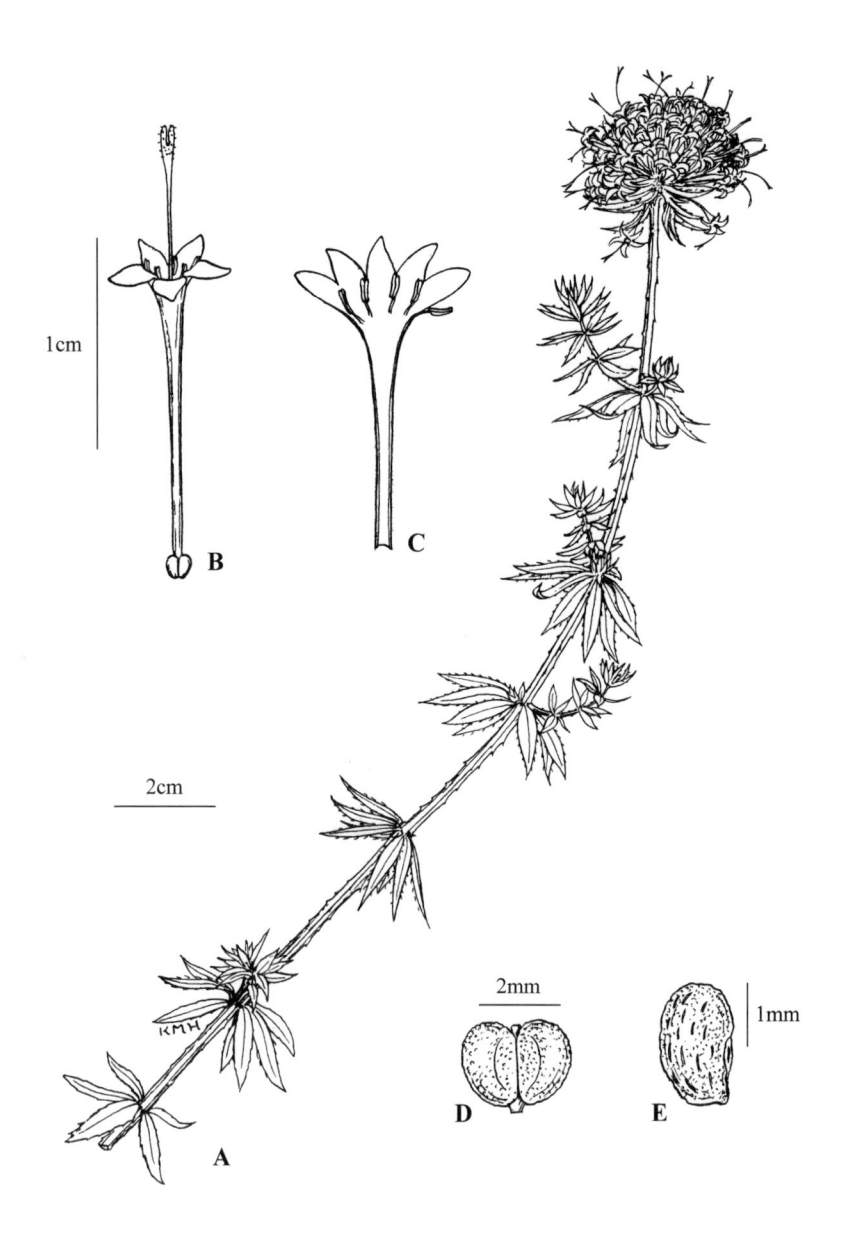

Phuopsis stylosa (Trin.) Benth. & Hook. f. ex B.D. Jacks. (Caucasian Crosswort). A habit; B flower; C corolla opened out; D fruit; E seed. (Harrietsham, E Kent). STACE 767/644.

134. RUBIACEAE

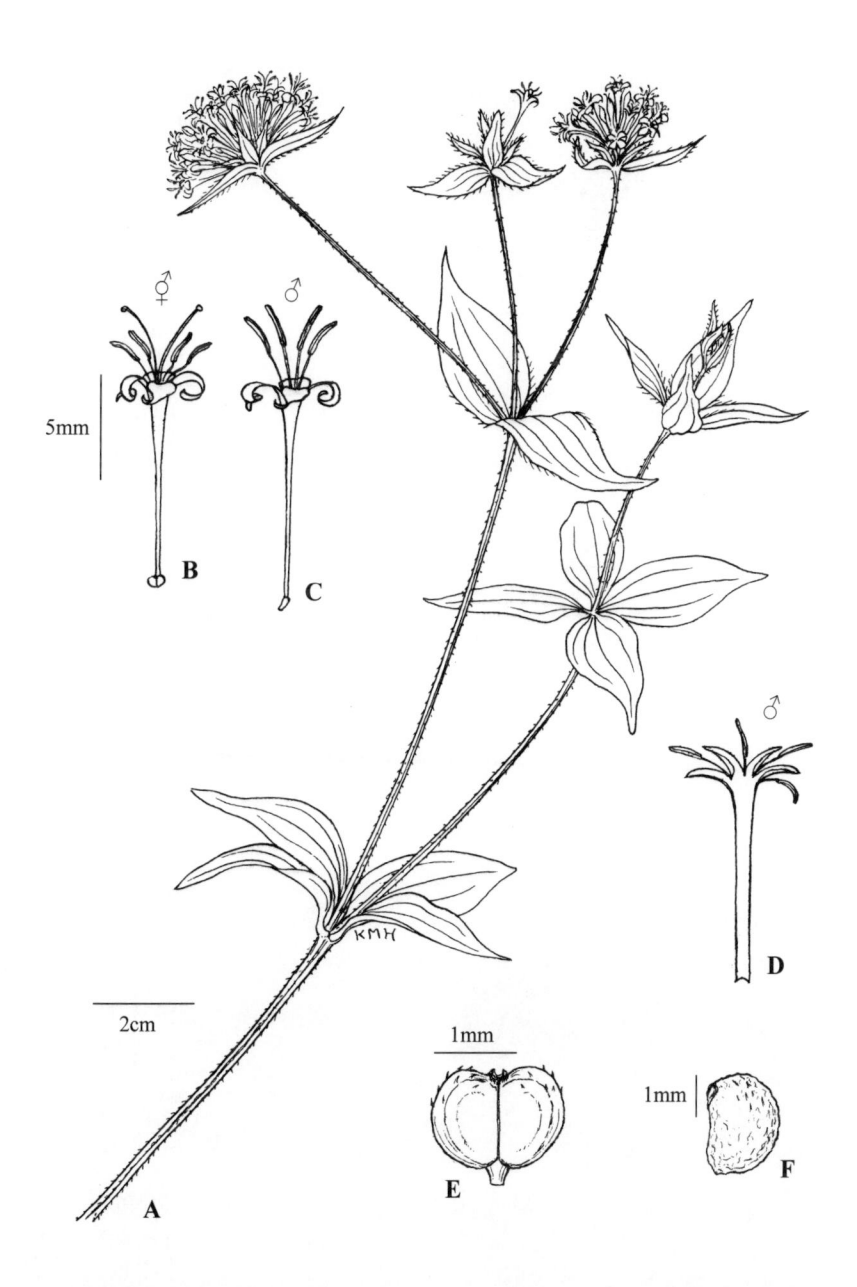

Asperula taurina L. (Pink Woodruff). A habit; B bisexual flower; C male flower; D male flower corolla opened out; E fruit; F seed. (Hopetown, SW Yorks). STACE 768/645.

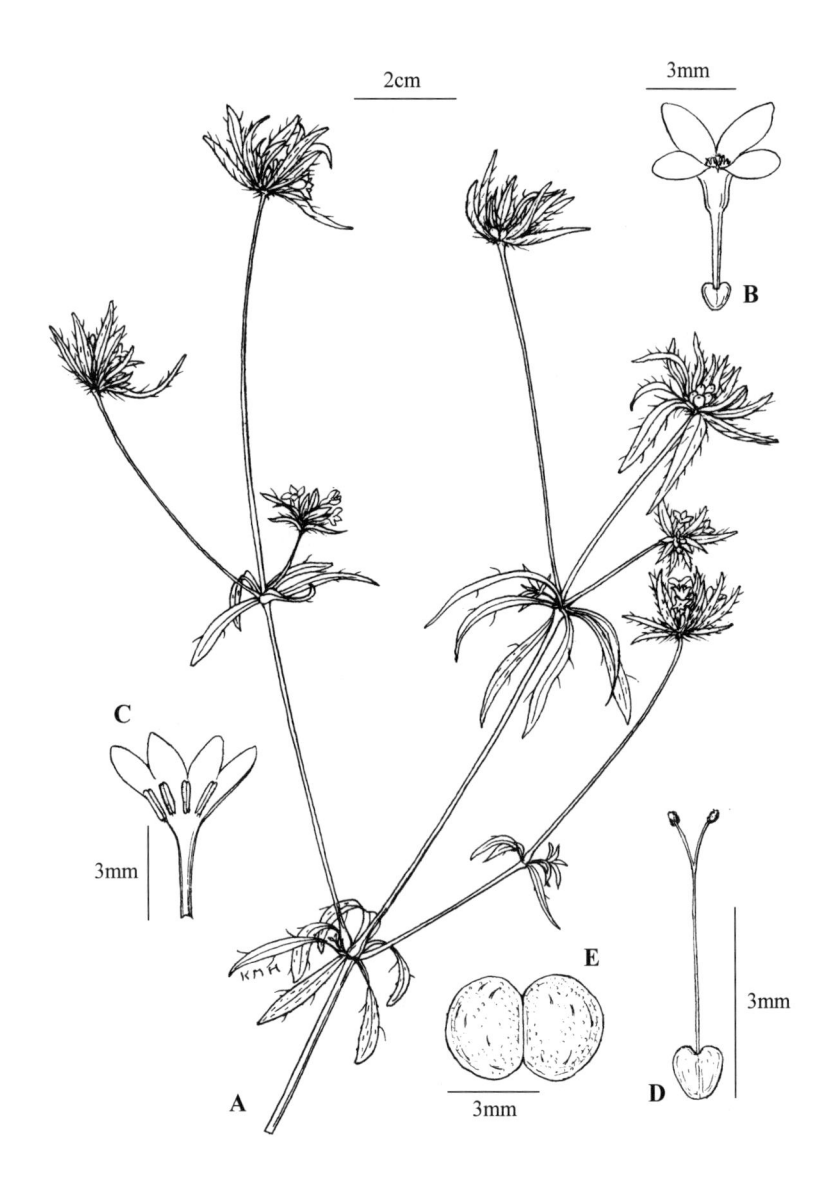

Asperula arvensis L. (Blue Woodruff). A habit; B flower; C corolla opened out; D gynoecium; E fruit. (Hants). STACE 768/645.

135. CAPRIFOLIACEAE 273

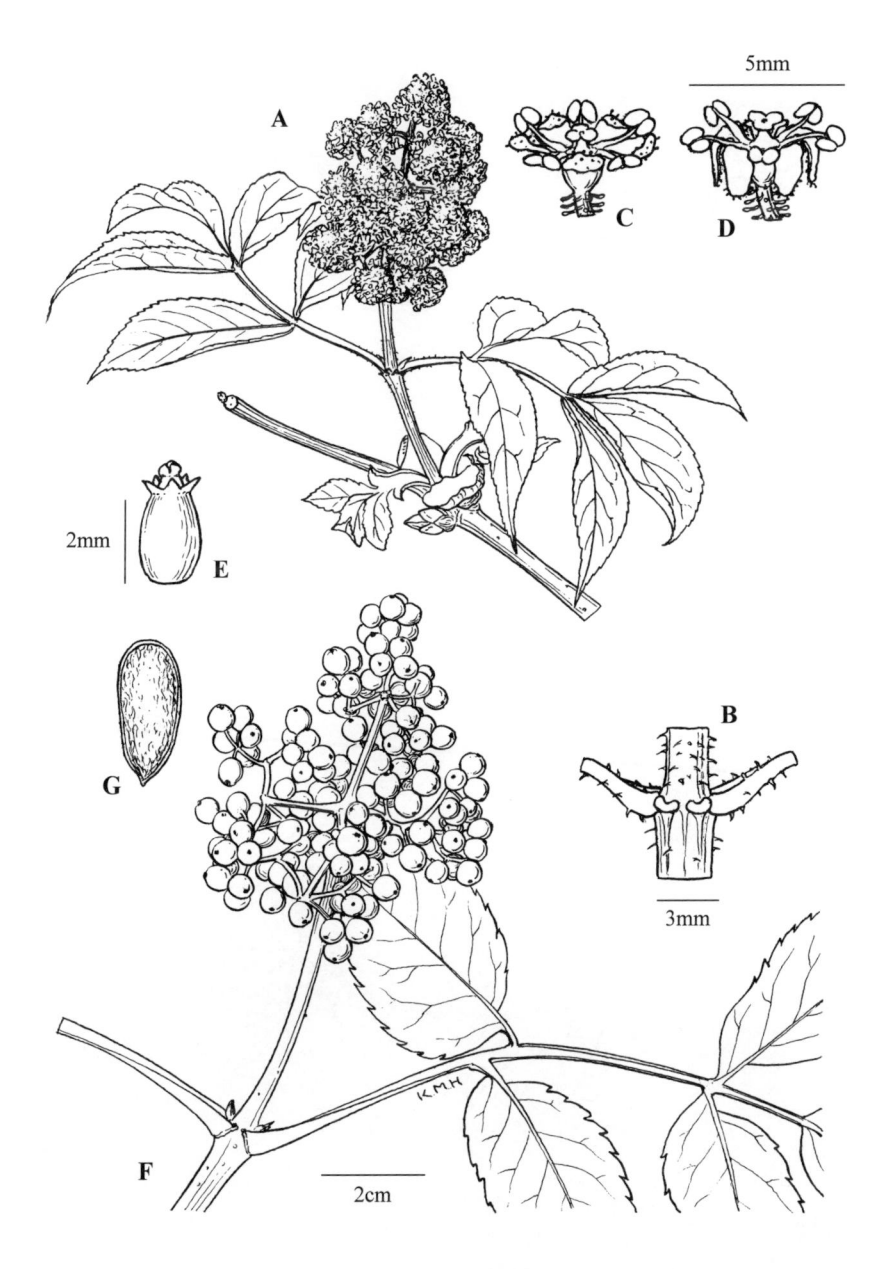

Sambucus racemosa L. (Red-berried Elder). A flowering branch; B stem node with stipules; C flower at anthesis; D flower after anthesis; E young fruit; F fruiting branch; G seed. (Wray, Westmorland). STACE 776/651.

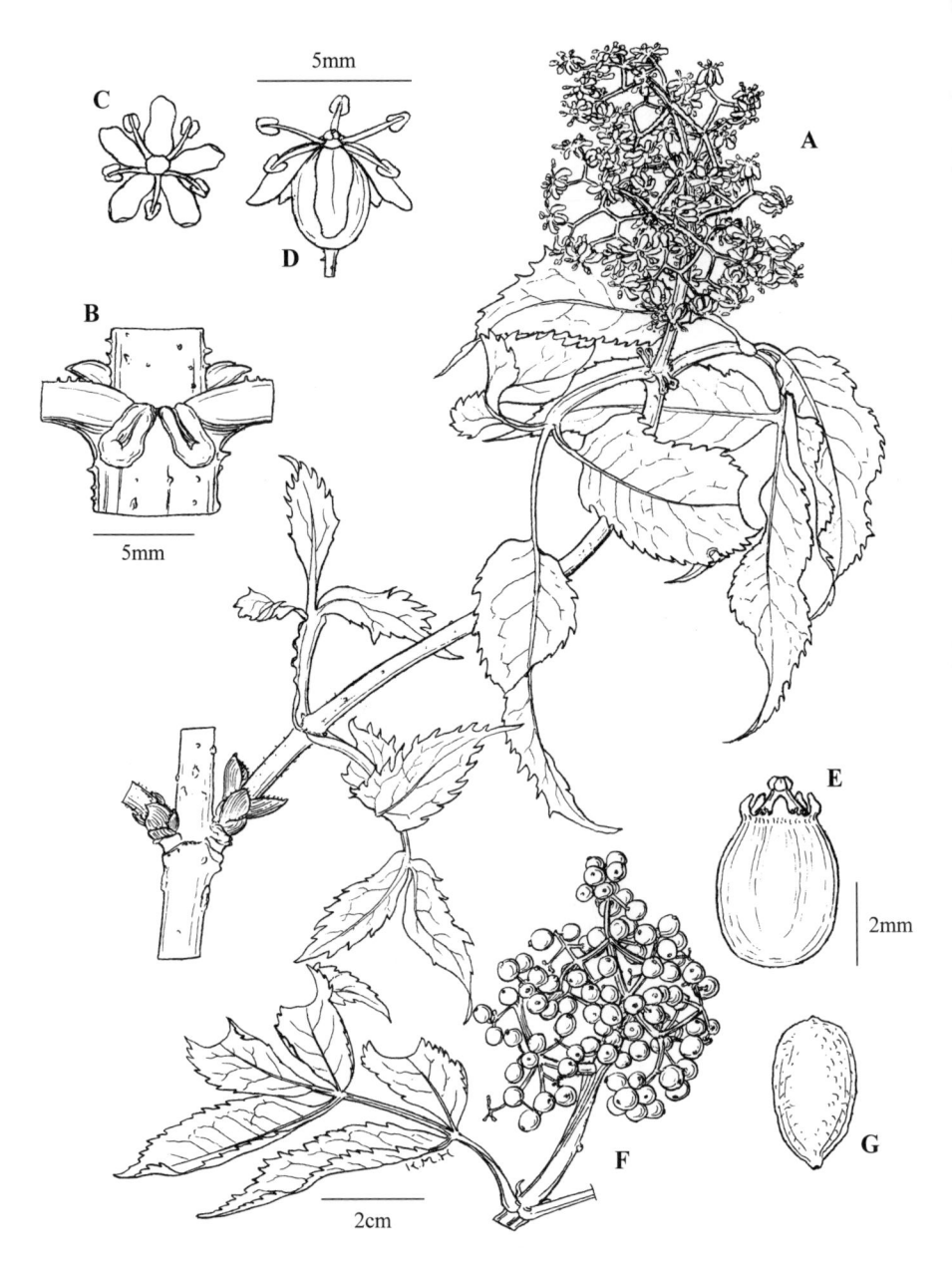

274   135. CAPRIFOLIACEAE

Sambucus racemosa L. var. **pubens** (Michx.) Koehne 'Dissecta' (Red-berried Elder cultivar). A flowering branch; B stem node with stipules; C flower at anthesis; D flower after anthesis; E young fruit; F fruiting branch; G seed. (Dolphinston, Roxburghs). STACE 776/651.

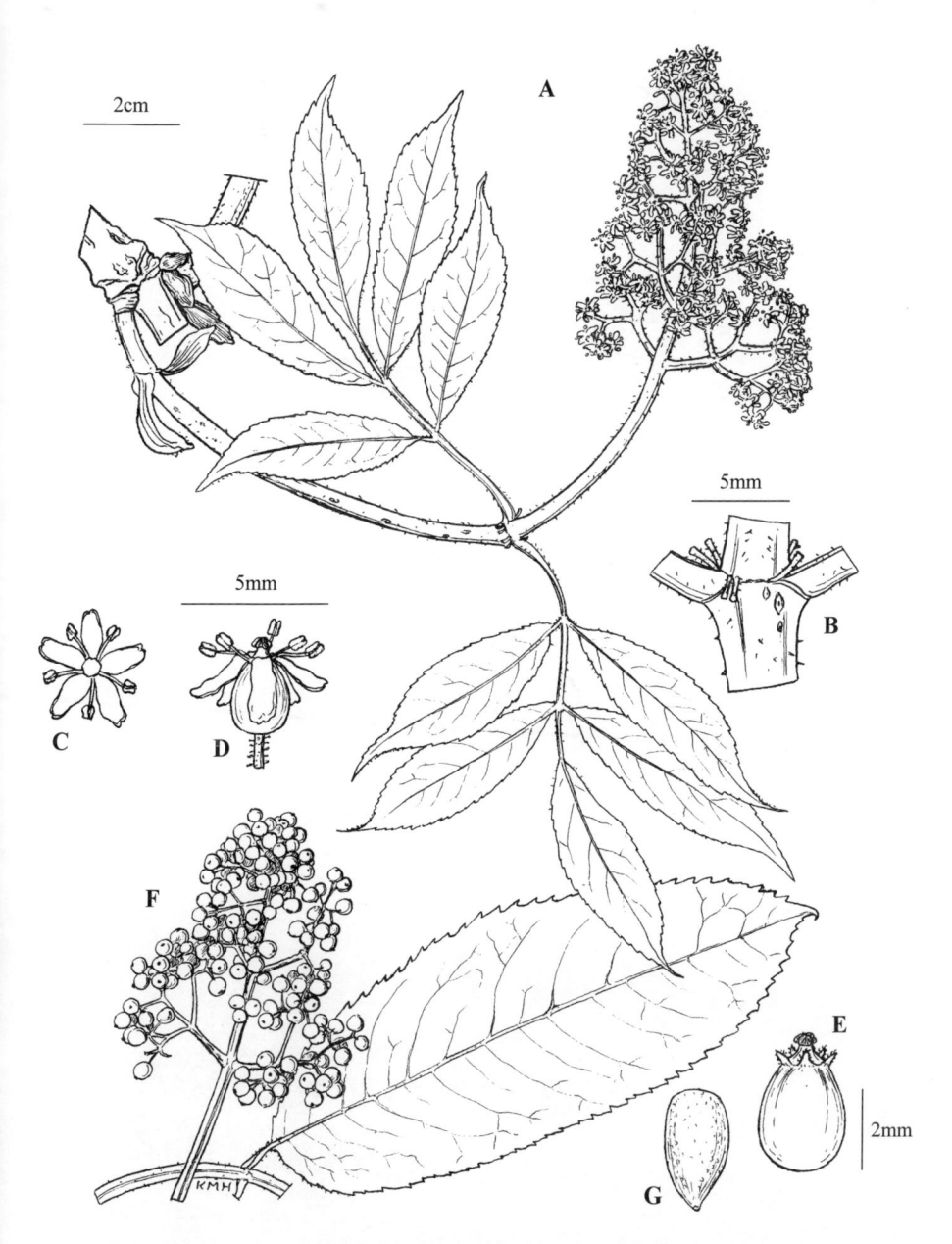

135. CAPRIFOLIACEAE

Sambucus racemosa L. var. **sieboldiana** Miq. (Red-berried Elder variety). A flowering branch; B stem node with stipules; C flower at anthesis; D flower after anthesis; E young fruit; F fruiting branch; G seed. (Ludham, E Norfolk). STACE 776/651.

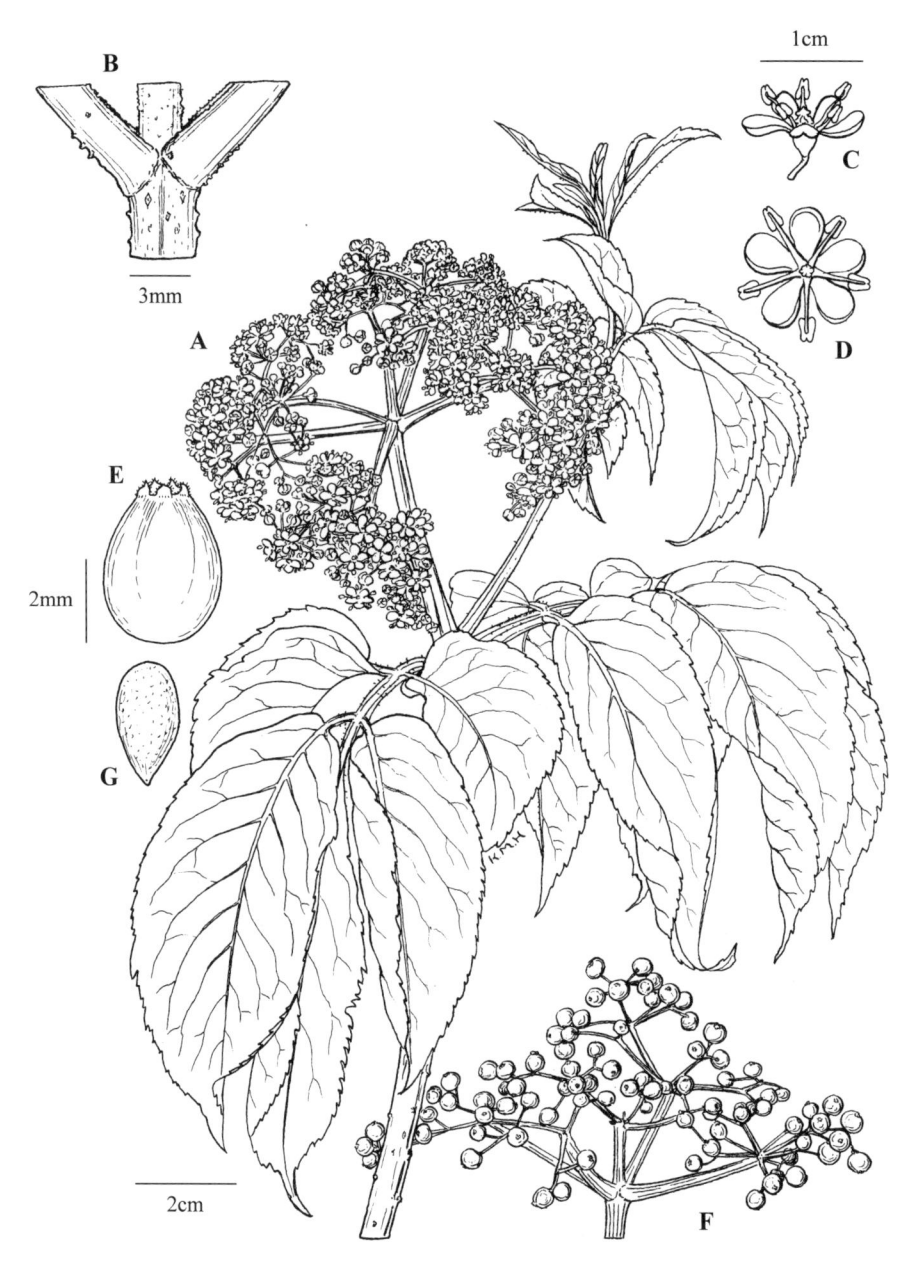

Sambucus canadensis L. (American Elder). A flowering branch; B stem node with stipules; C flower at anthesis; D flower after anthesis; E young fruit; F fruiting branch; G seed. (Alsop, Derbys). STACE 776/651.

135. CAPRIFOLIACEAE 277

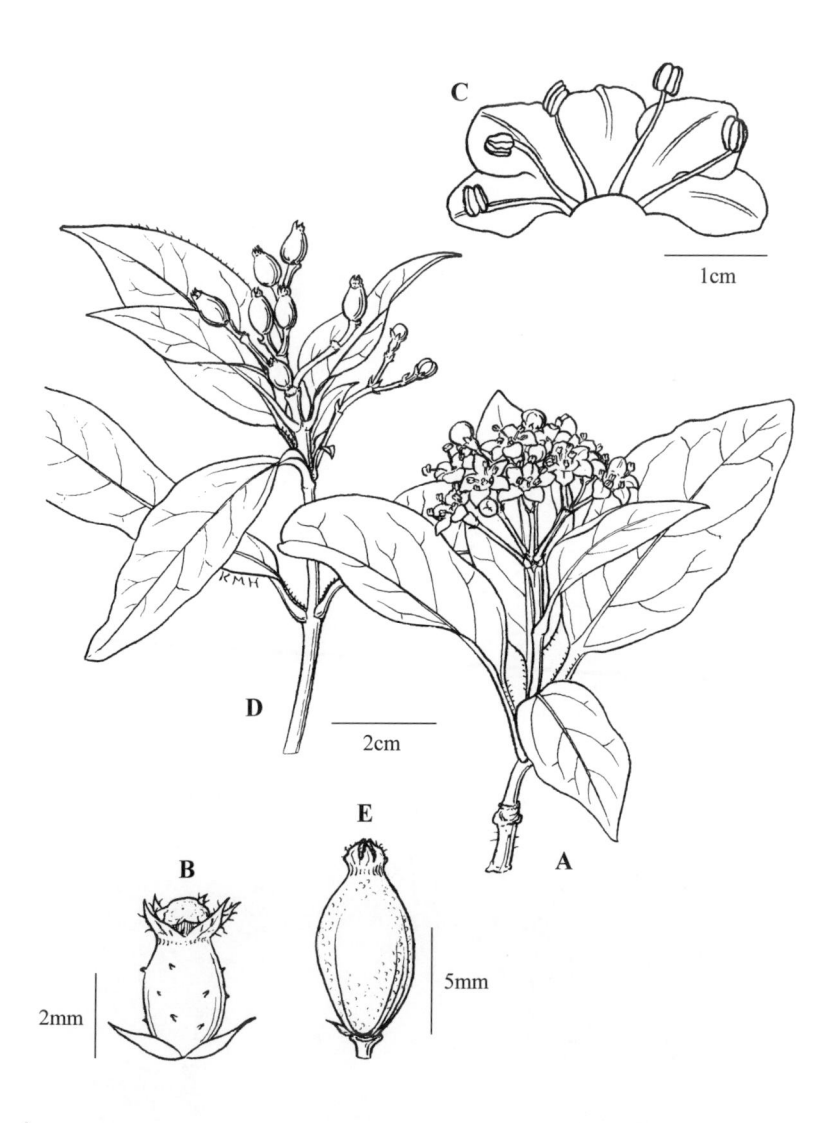

Viburnum tinus L. (Laurustinus). A flowering branch; B flower bud with bracteoles; C corolla opened out to show stamens; D fruiting branch; E young fruit. (Torquay, S Devon). STACE 777/652.

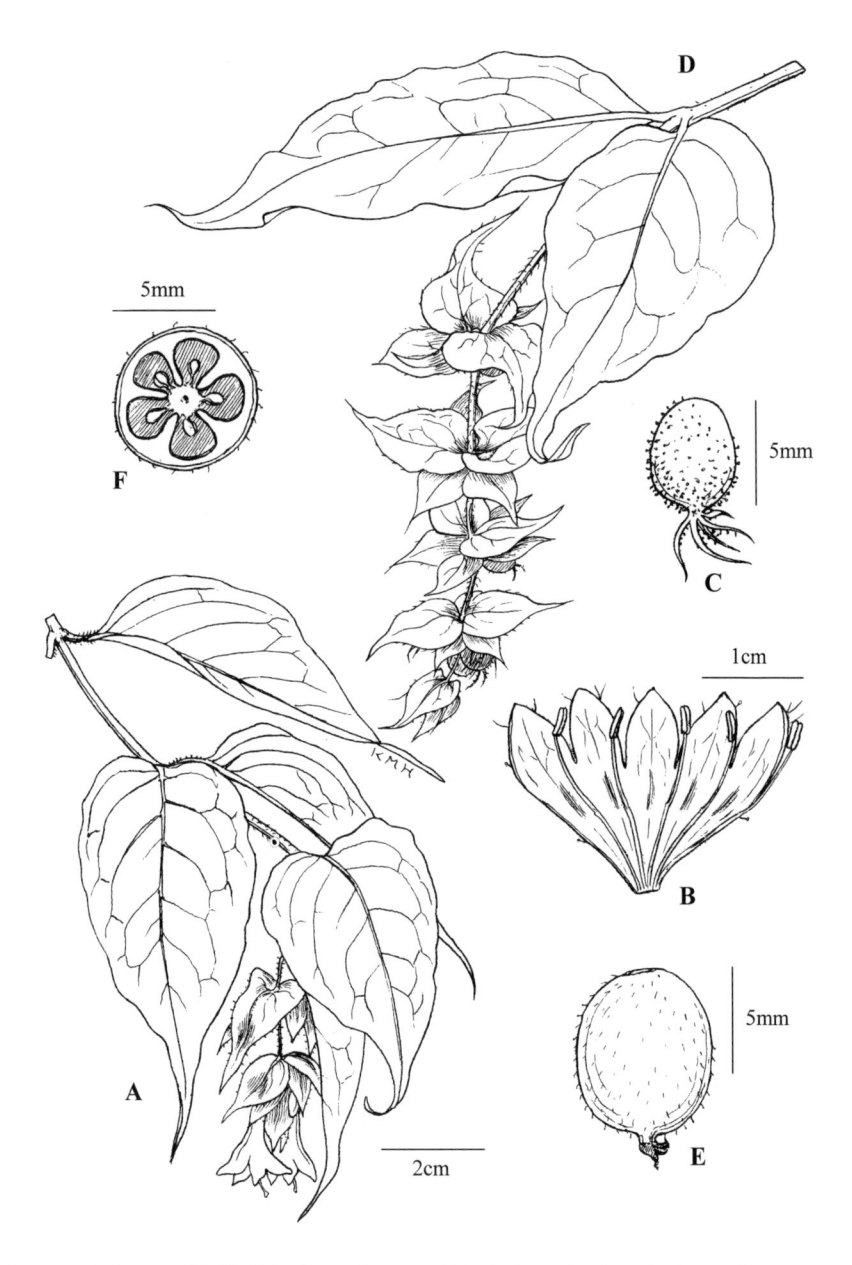

Leycesteria formosa Wall. (Himalayan Honeysuckle). A flowering branch; B corolla opened out to show stamens; C young fruit; D fruiting branch; E fruit; F TS of fruit. (Hatchet Pond, S Hants). STACE 779/653.

135. CAPRIFOLIACEAE

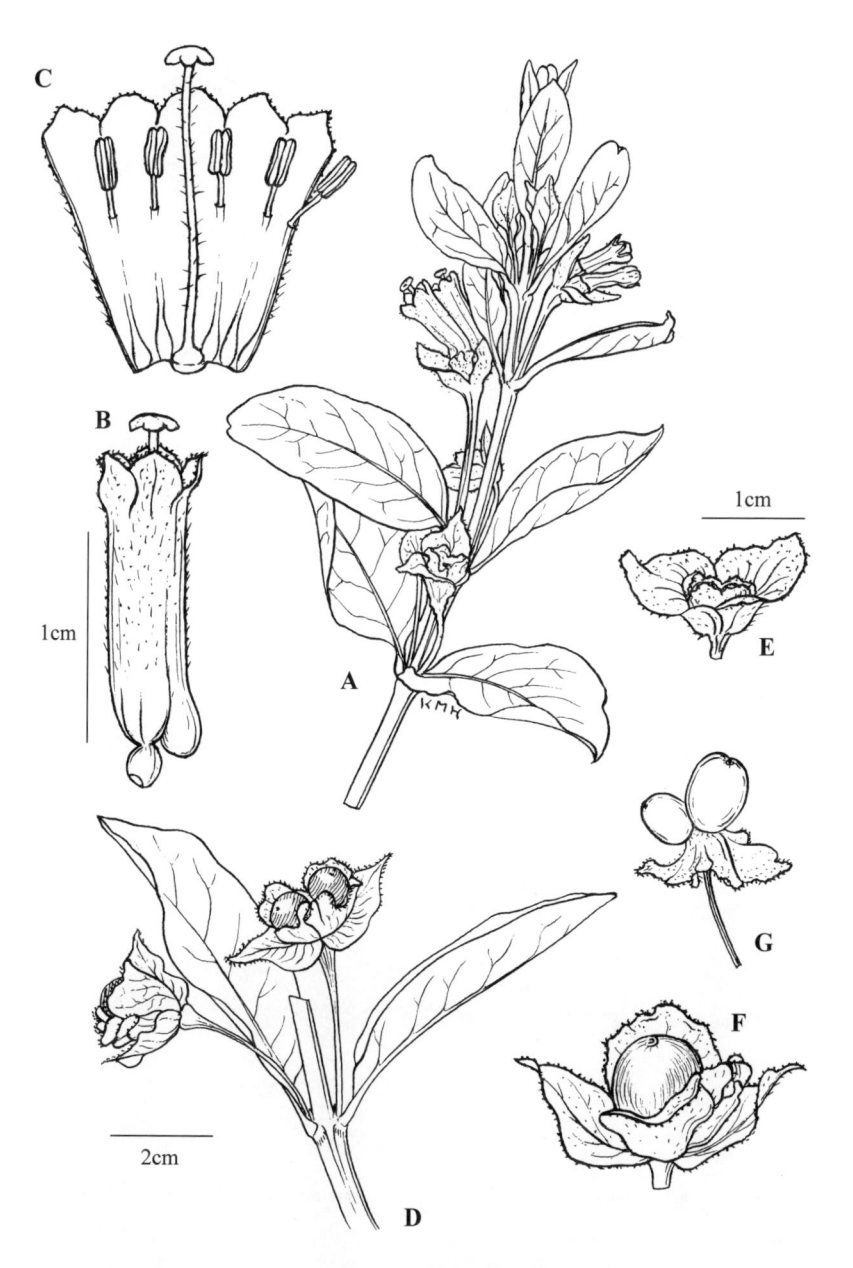

Lonicera involucrata (Richardson) Banks ex Spreng. (Californian Honeysuckle). A flowering branch; B flower; C flower opened out; D fruiting stem; E bracts and bracteoles around young fruit; F fruit; G older fruits. (*Hort*. Cubley, Derbys). STACE 780/654.

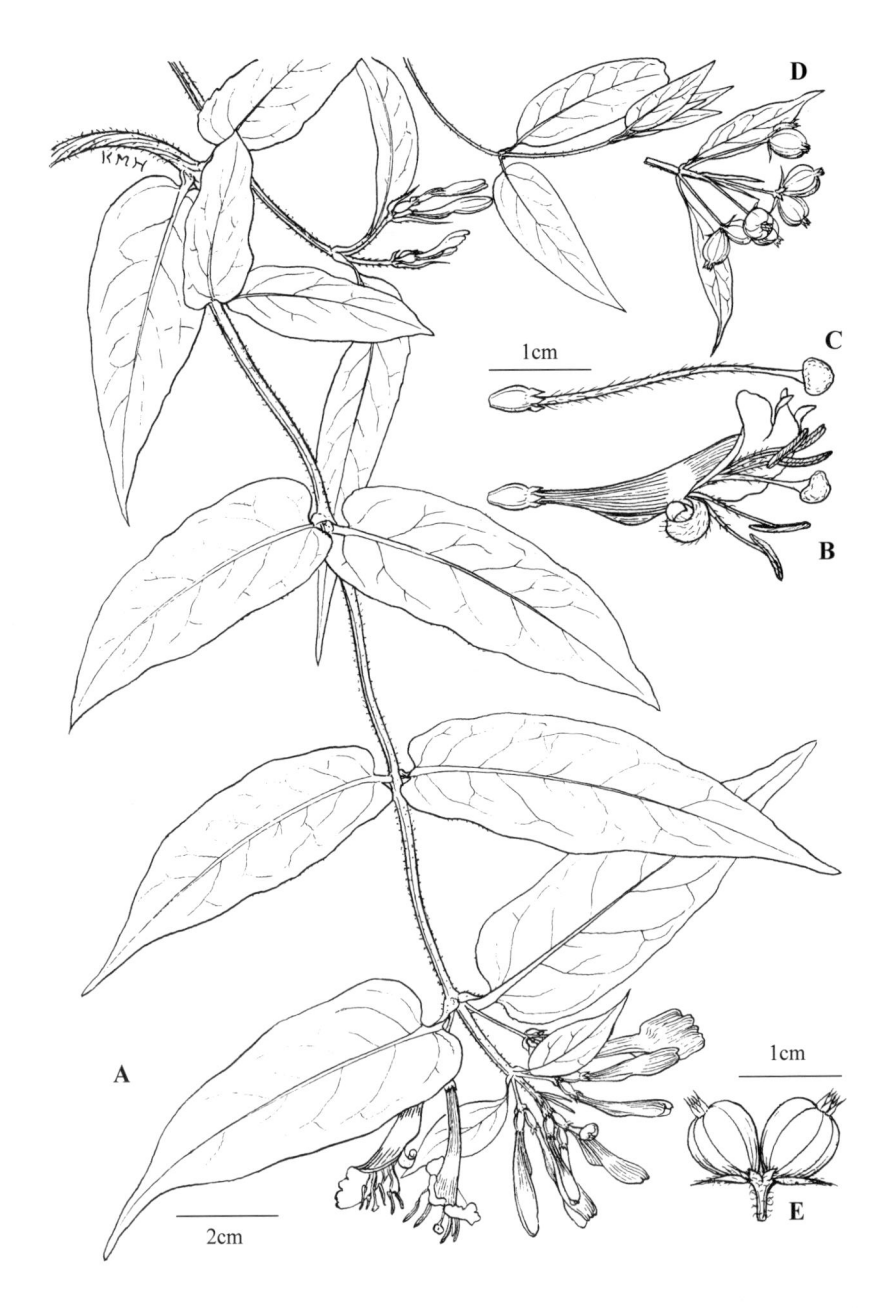

Lonicera henryi Hemsl. (Henry's Honeysuckle). A habit; B flower; C gynoecium; D fruiting stem; E pair of fruits. (Holmwood, Surrey). STACE 780/654.

135. CAPRIFOLIACEAE

Lonicera japonica Thunb. ex Murray (Japanese Honeysuckle). A barren and flowering branches; B flower; C gynoecium; D four fruits at a node; E pair of fruits; F TS of fruit. (Bere Ferrers, S Devon). STACE 780/656.

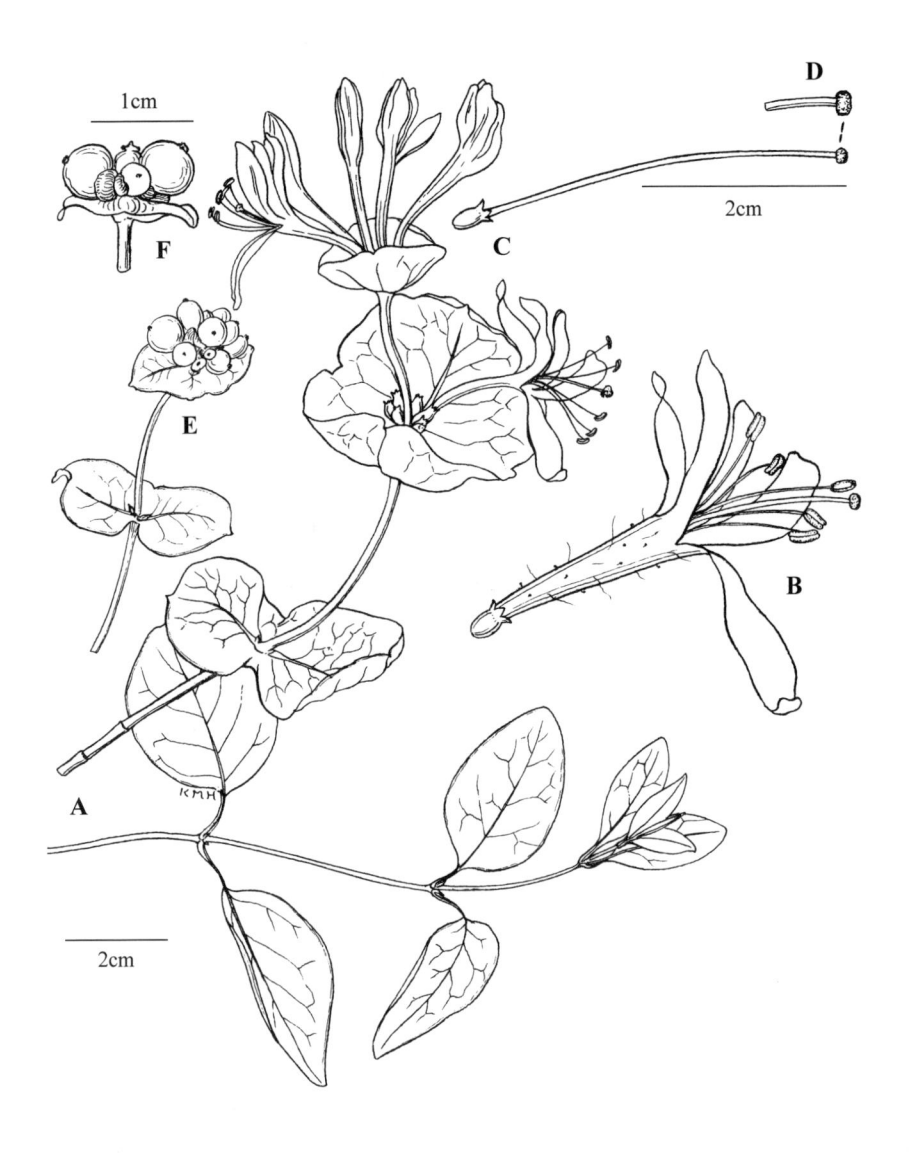

Lonicera caprifolium L. (Perfoliate Honeysuckle). A barren and flowering branches; B flower; C gynoecium; D stigma; E fruiting head; F fruits. (Source unknown). STACE 782/656.

138. DIPSACACEAE

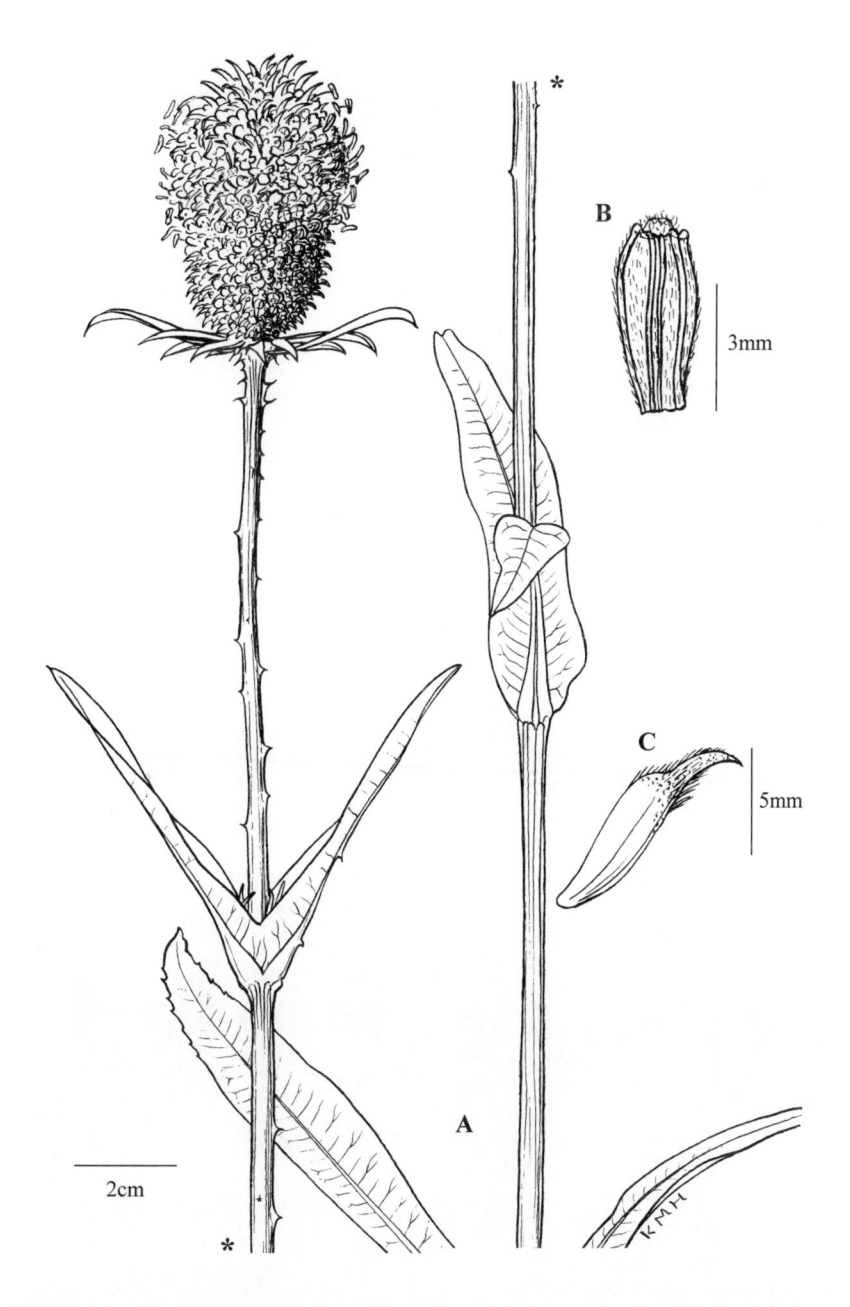

Dipsacus sativus (L.) Honck. (Fuller's Teasel). A habit; B achene; C receptacular bract. (*Hort*. Ashbourne, Derbys). STACE 788/661.

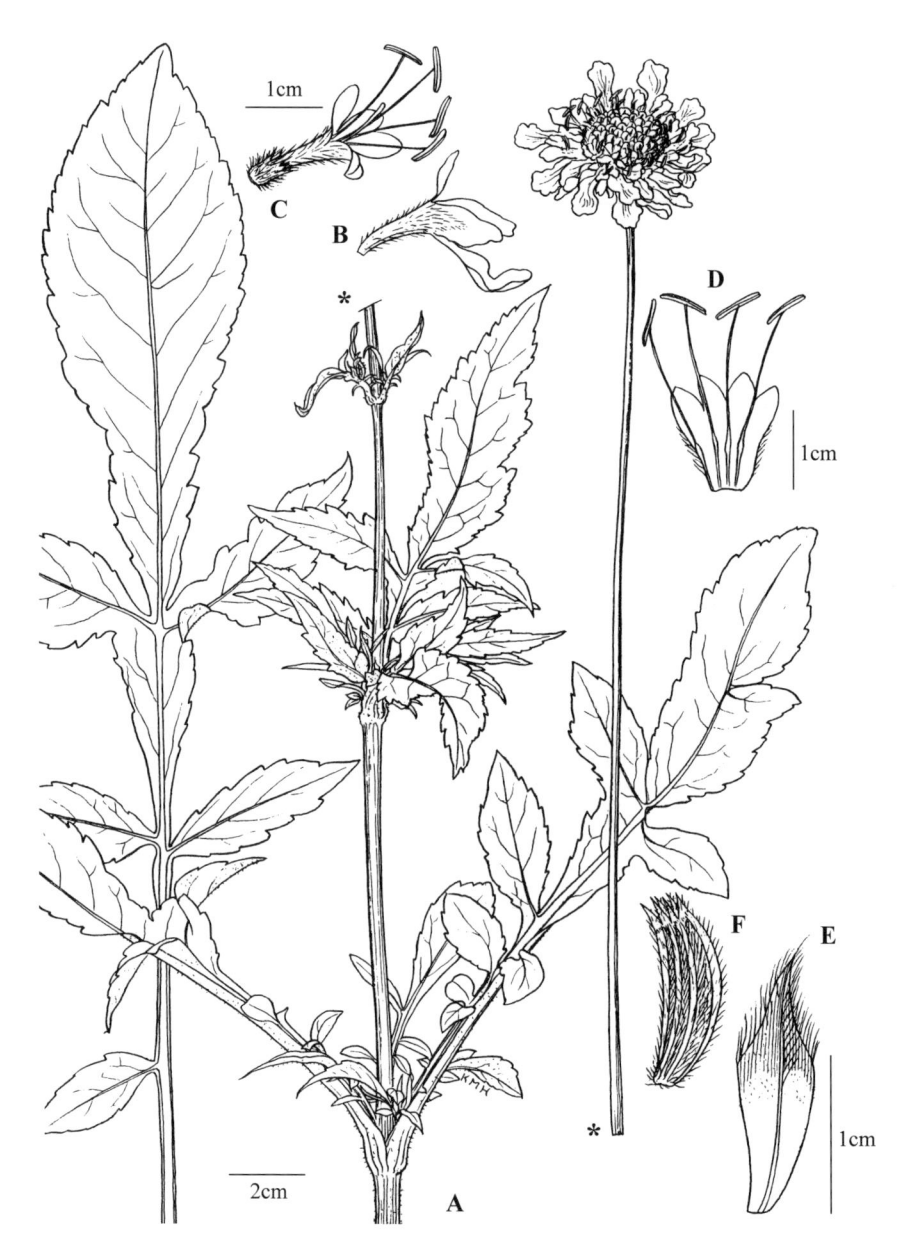

Cephalaria gigantea (Ledeb.) Bobrov (Giant Scabious). A habit and lower leaf; B outer flower; C inner flower; D inner flower opened out; E receptacular bract; F achene. (Source unknown). STACE 789/661.

138. DIPSACACEAE 285

Scabiosa atropurpurea L. (Sweet Scabious). A habit; B outer flower; C inner flower; D corolla opened out; E fruiting capitulum; F fruit; G receptacular bract. (Folkestone, E Kent). STACE 790/662.

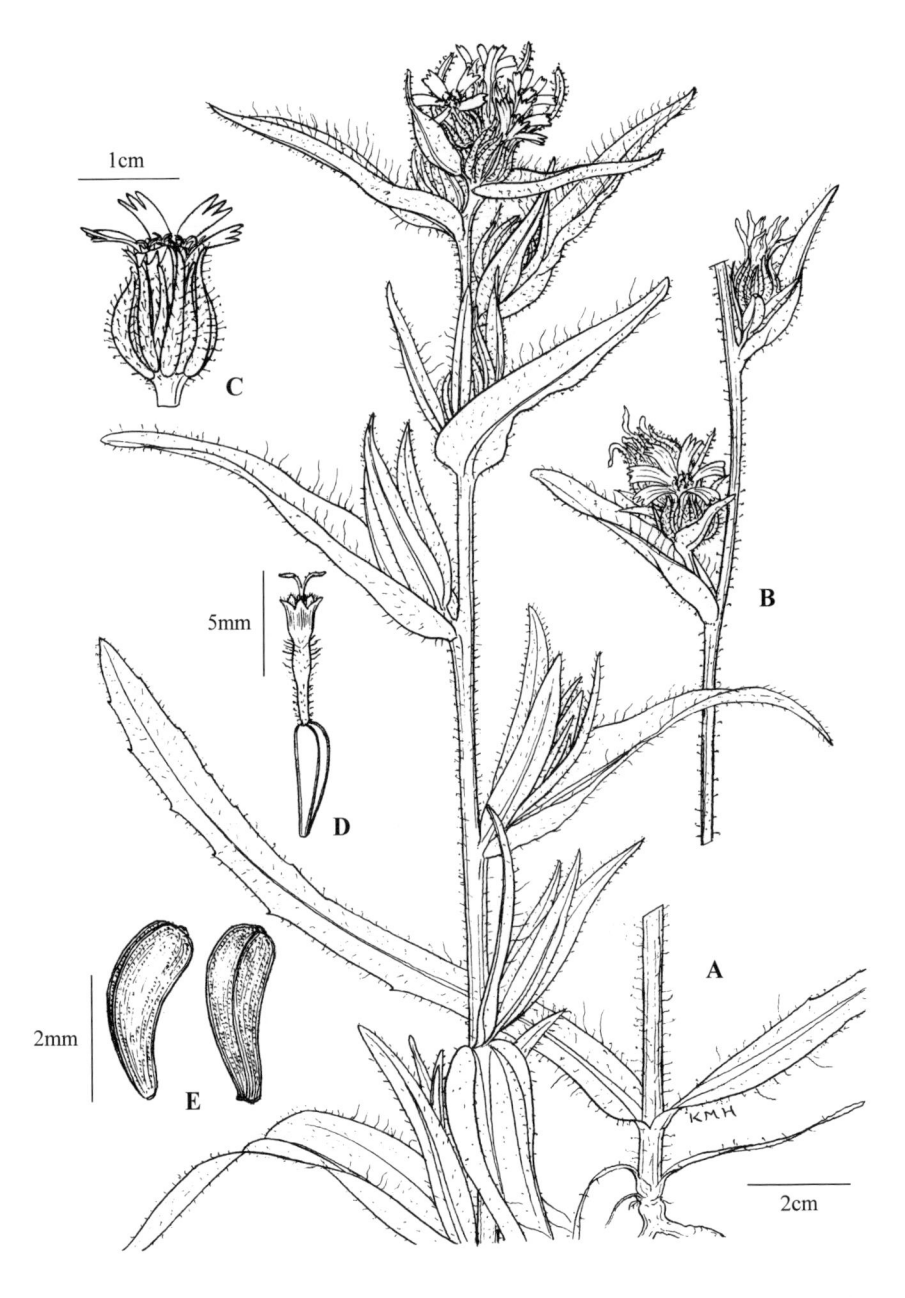

Madia sativa Molina (Coast Tarweed). A habit; B axillary capitula; C capitulum; D disc floret; E inner and outer achenes. (Brittany, France). STACE —/—.

139. ASTERACEAE

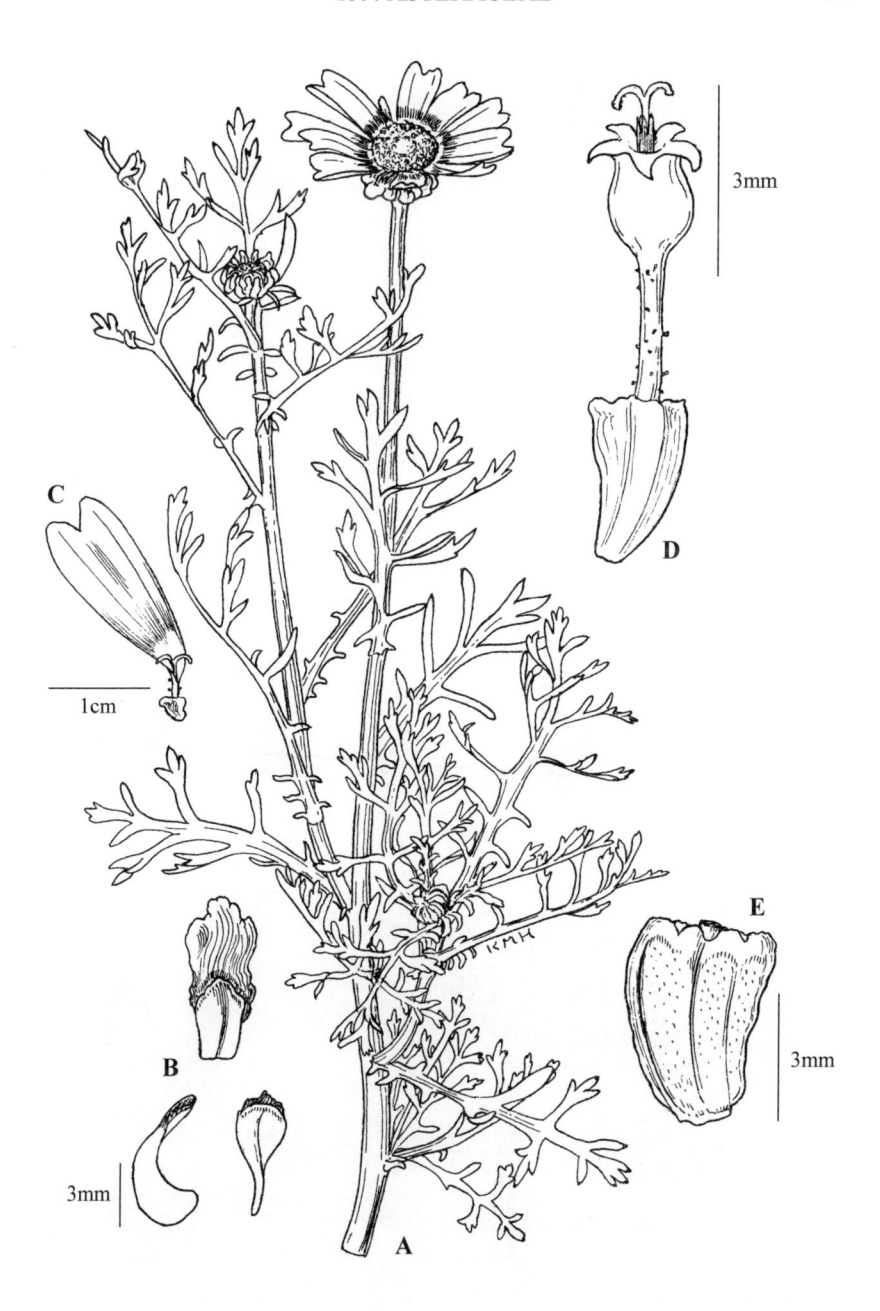

Ismelia carinata (Schousb.) Sch. Bip. (Tricolor Chrysanthemum). A habit; B phyllaries; C ray floret; D disc floret; E achene. (*Hort.* Ashbourne, Derbys). STACE 802/672.

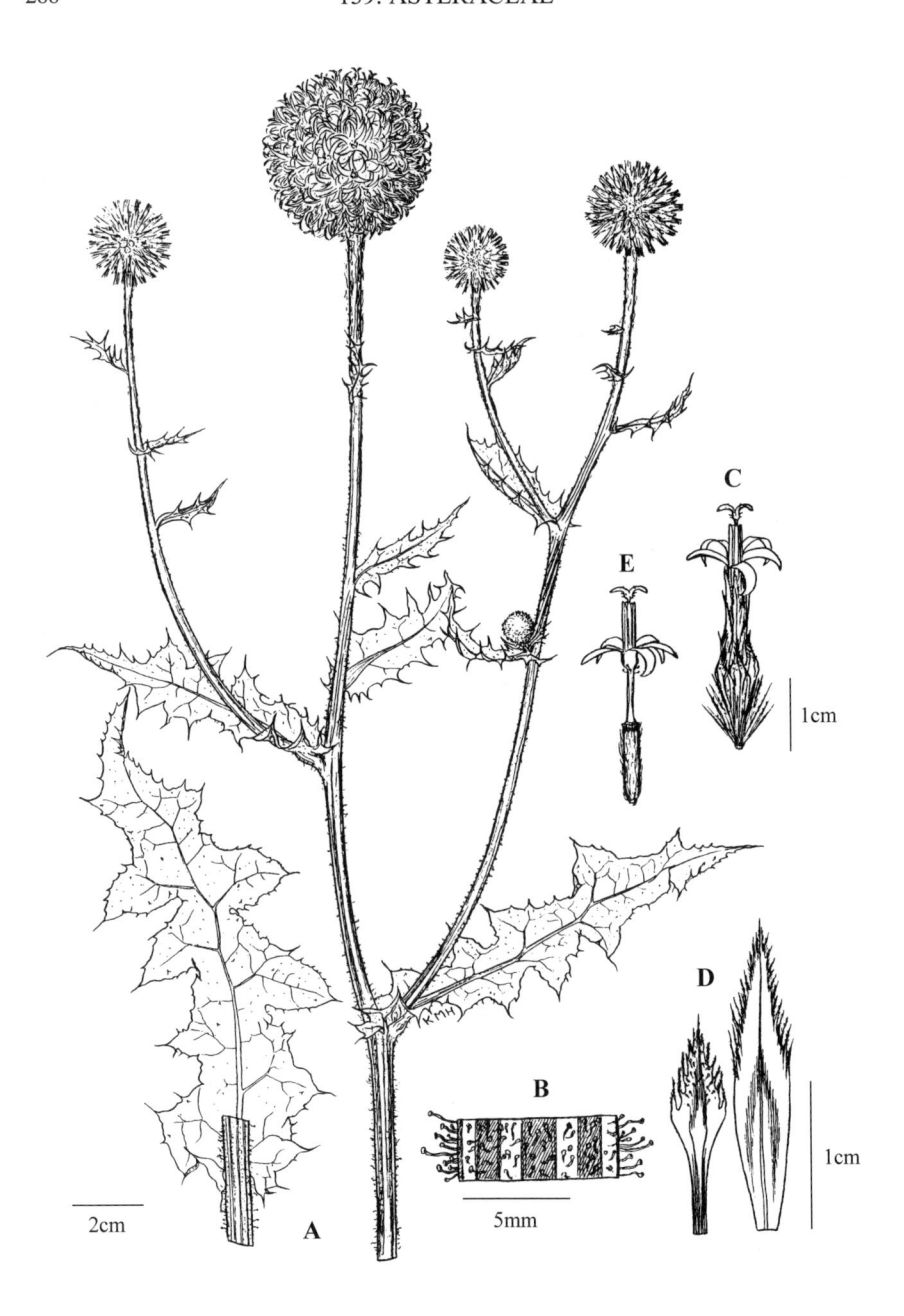

Echinops sphaerocephalus L. (Glandular Globe-thistle). A habit; B part of stem; C capitulum; D phyllaries; E floret. (New Hythe Way, W Kent). STACE 803/672.

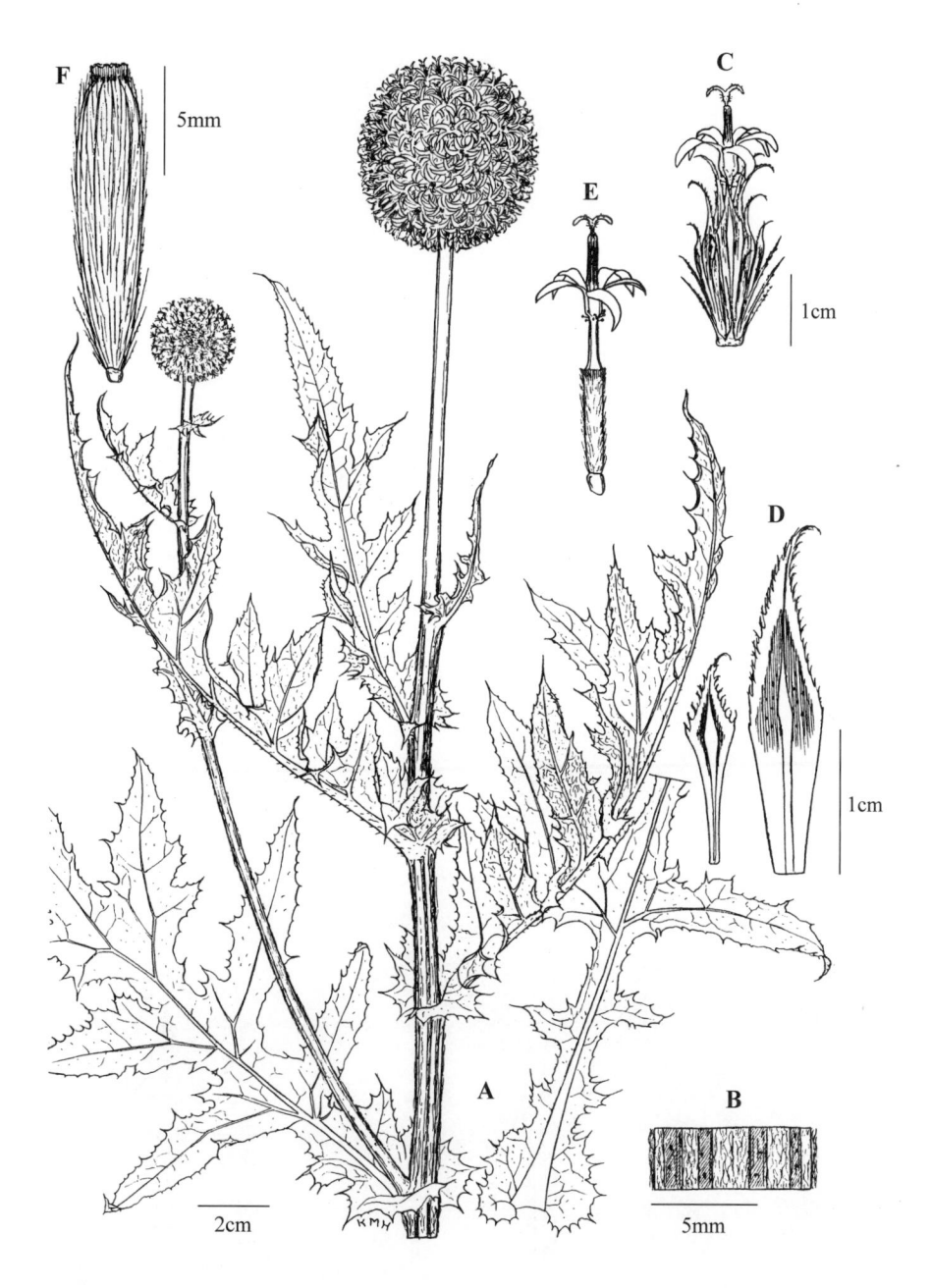

Echinops exaltatus Schrad. (Globe-thistle). A habit; B part of stem; C capitulum; D phyllaries; E floret; F achene. (*Hort*. Ashbourne, Derbys). STACE 803/673.

Echinops bannaticus Rochel ex Schrad. (Blue Globe-thistle). A habit; B part of stem; C capitulum; D phyllaries; E floret. (*Hort*. Newark, Notts). STACE 803/673.

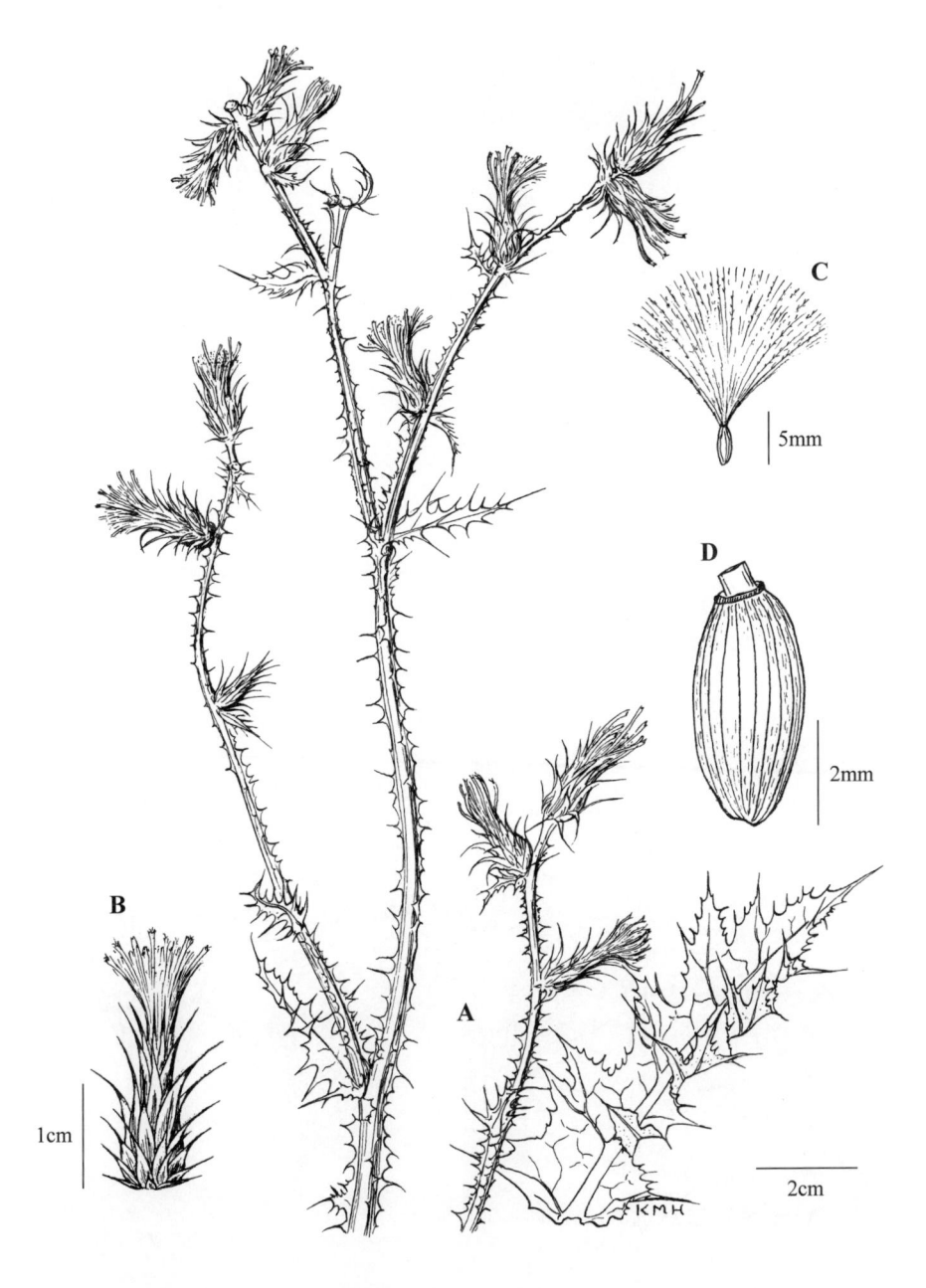

Carduus pycnocephalus L. (Plymouth Thistle). A habit; B capitulum; C fruit; D achene. (Plymouth Hoe, S Devon). STACE 806/675.

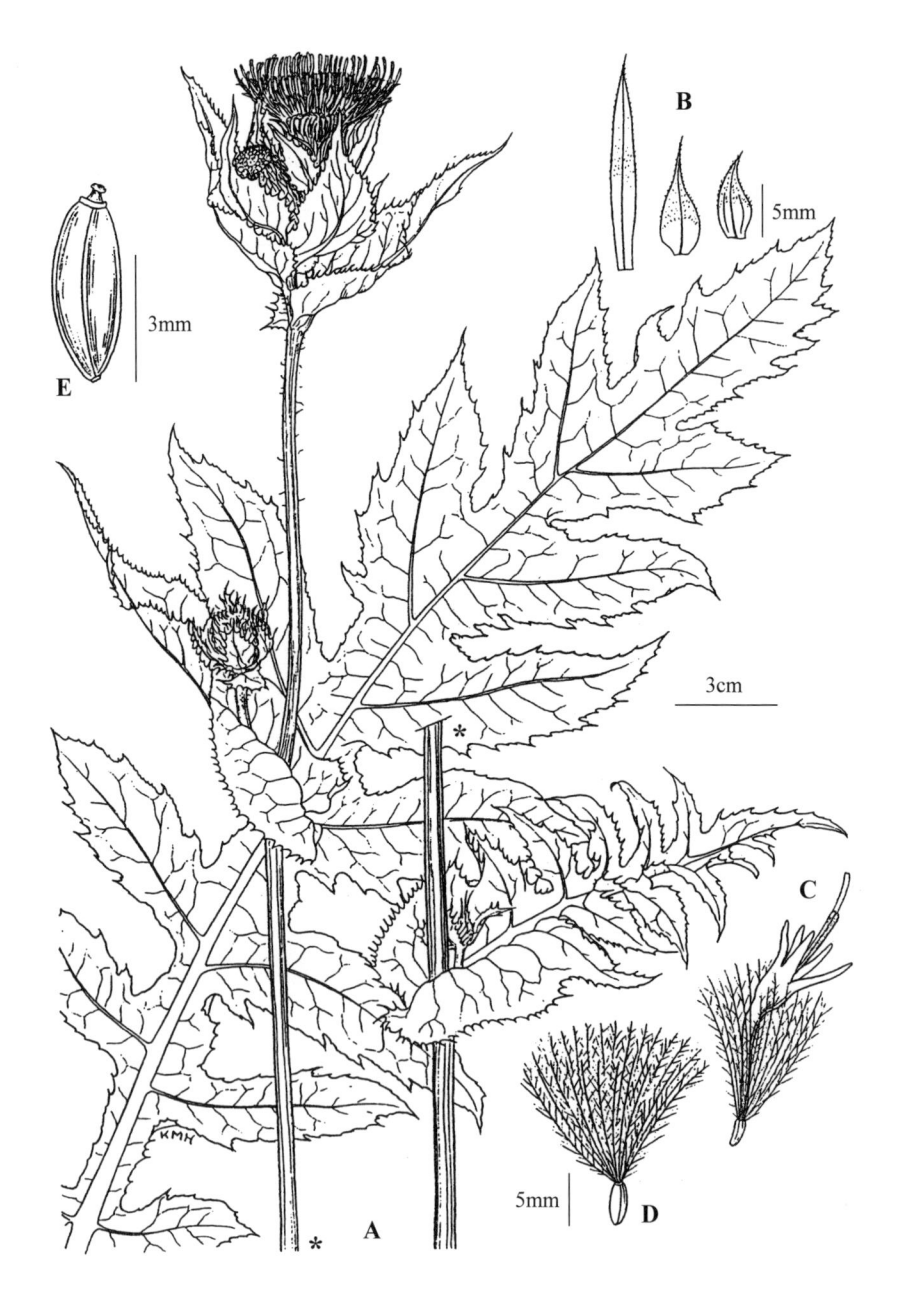

Cirsium oleraceum (L.) Scop. (Cabbage Thistle). A upper part of stem and basal leaf; B phyllaries; C floret; D fruit; E achene. (Tay Marshes, E Perth). STACE 809/678.

139. ASTERACEAE

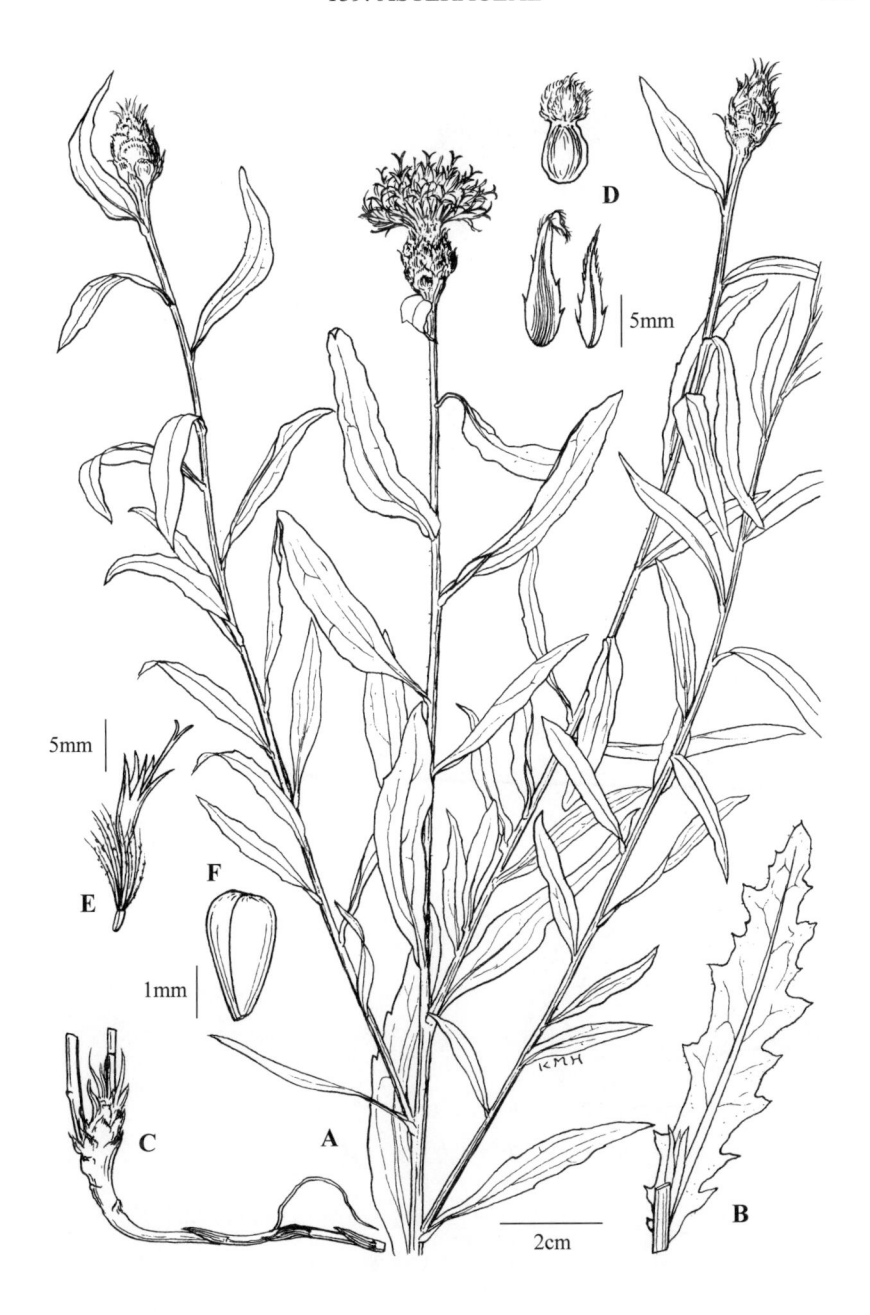

Acroptilon repens (L.) DC. (Russian Knapweed). A habit; B lower stem leaf; C rhizome; D phyllaries; E floret; F achene. (Hereford, Herefs). STACE 811/680.

294 139. ASTERACEAE

Centaurea montana L. (Perennial Cornflower). A habit; B phyllary; C ray floret; D disc floret; E achene. (Oban, Argyll). STACE 814/681.

139. ASTERACEAE

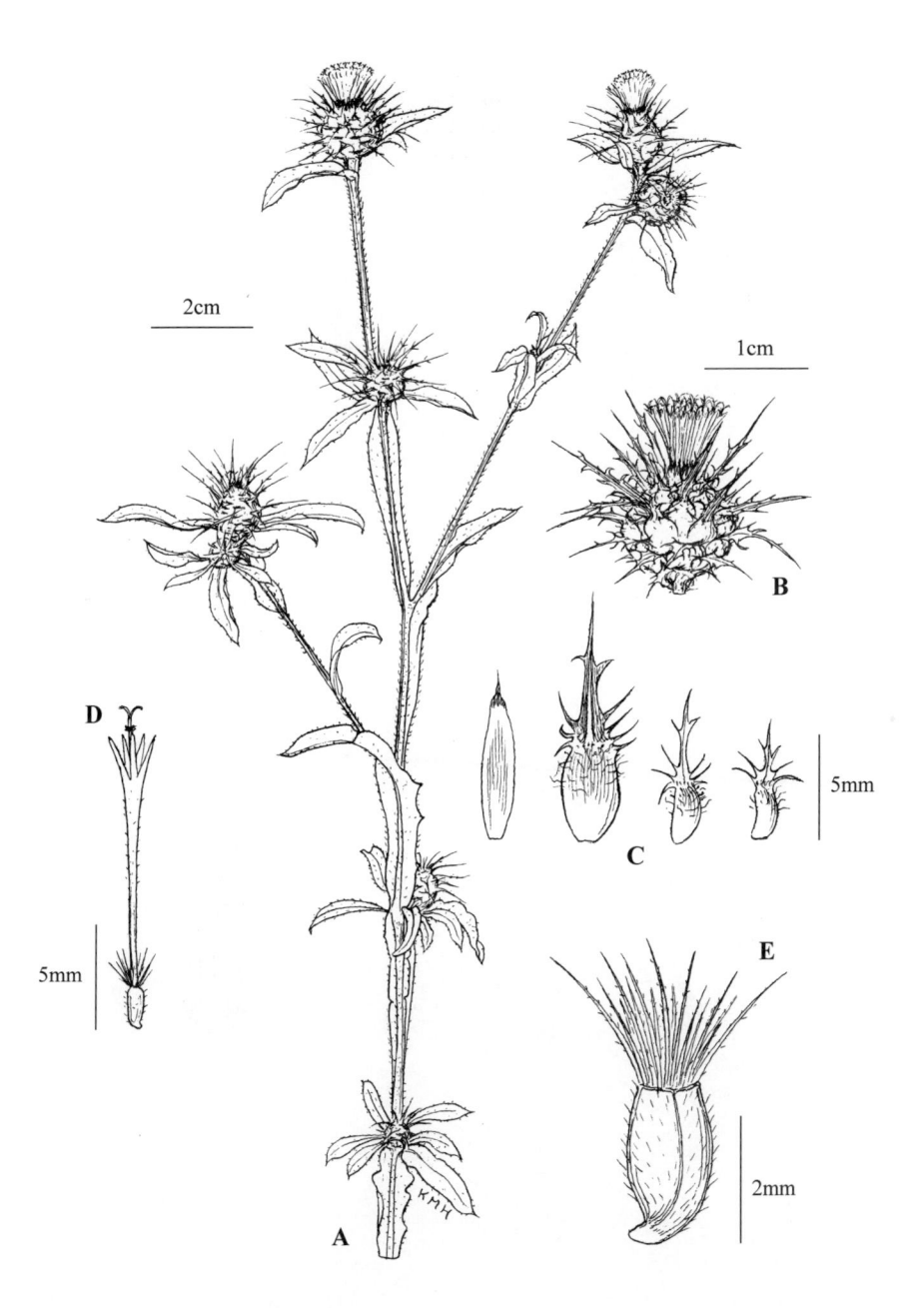

Centaurea melitensis L. (Maltese Star-thistle). A habit; B capitulum; C phyllaries; D floret; E achene. (*Hort.* Ware, Herts). STACE 814/681.

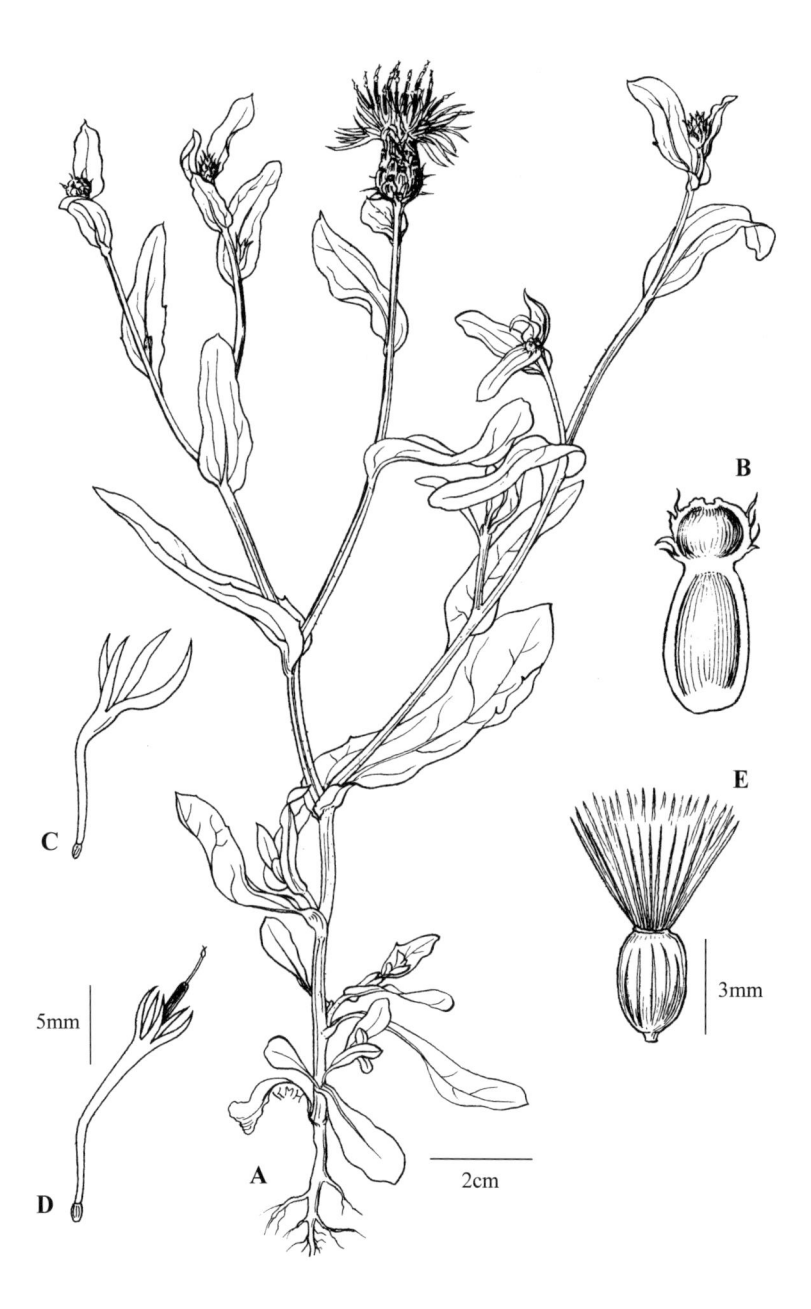

Centaurea diluta Aiton (Lesser Star-thistle). A habit; B phyllary; C ray floret; D disc floret; E achene. (Greenhithe, E Kent). STACE 814/681.

139. ASTERACEAE

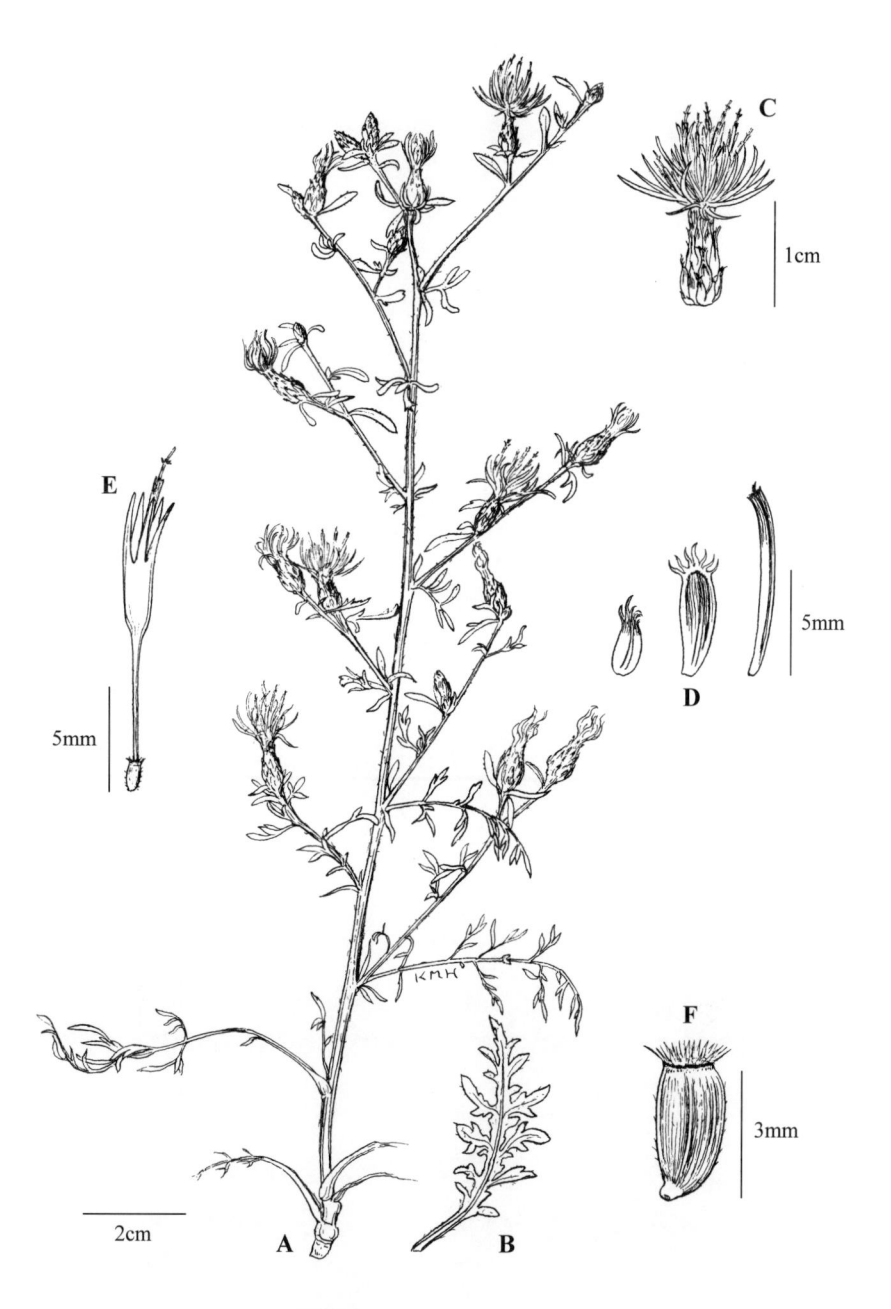

Centaurea paniculata L. (Jersey Knapweed). A habit; B rosette leaf; C capitulum; D phyllaries; E disc floret; F achene. (St Ouen, Jersey). STACE 812/681.

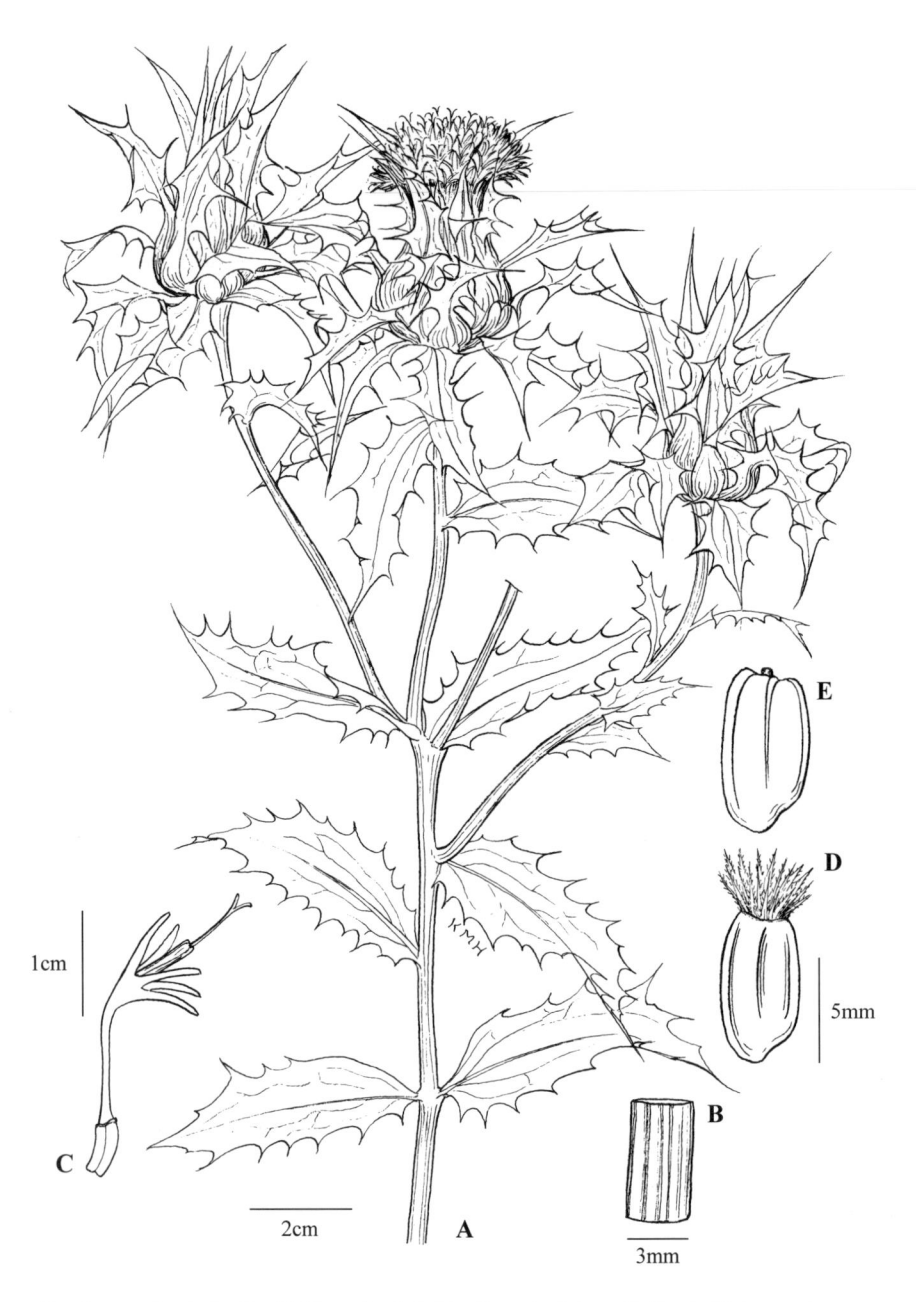

Carthamus tinctorius L. (Safflower). A habit; B part of stem; C floret; D inner achene; E outer achene. (*Hort.* Sevenoaks, W Kent). STACE 815/683.

139. ASTERACEAE

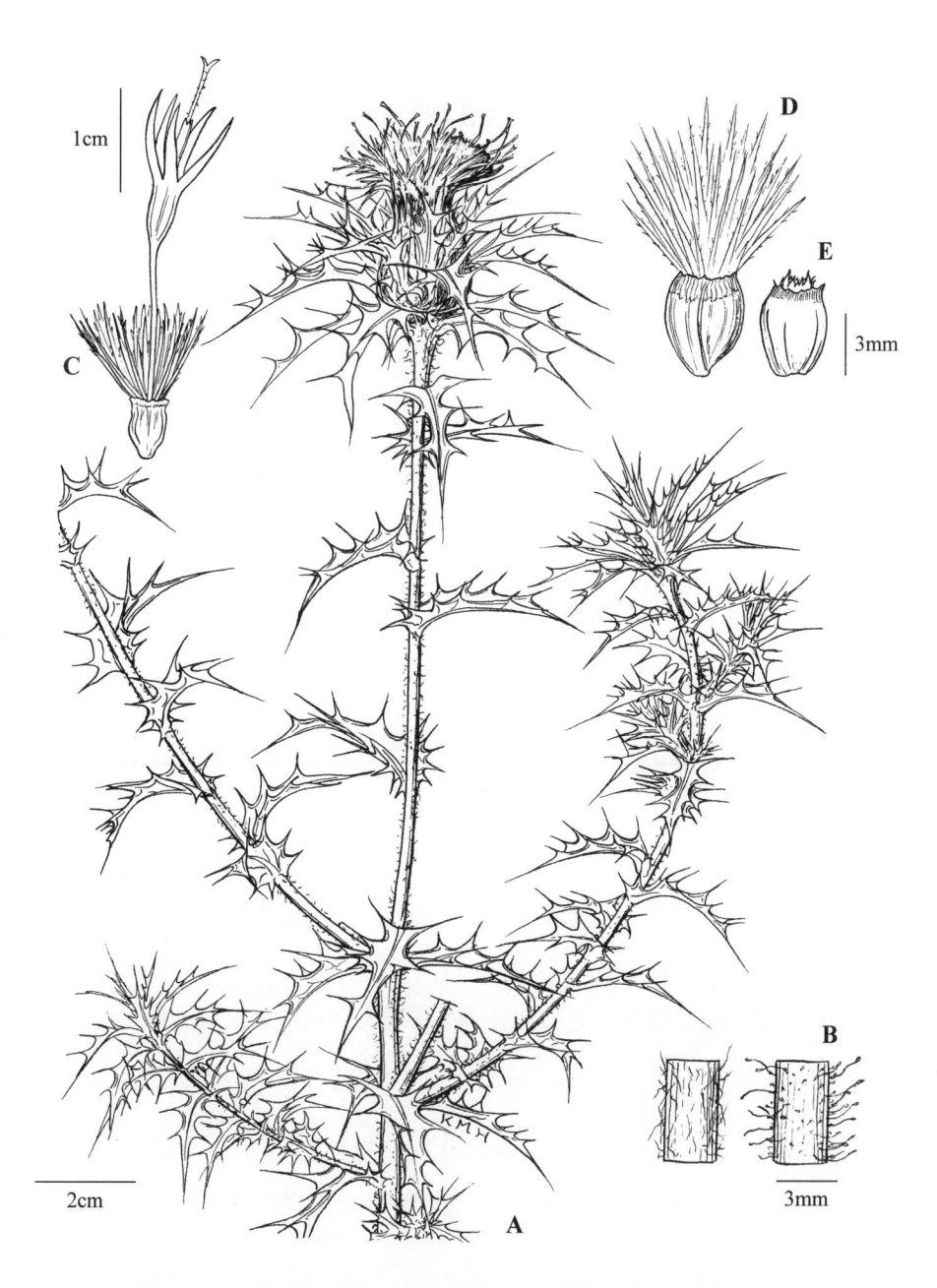

Carthamus lanatus L. (Downy Safflower). A habit; B parts of stem; C floret; D inner achene; E outer achene. (A-D, Ash, E Kent; E, Maulden, Beds). STACE 815/683.

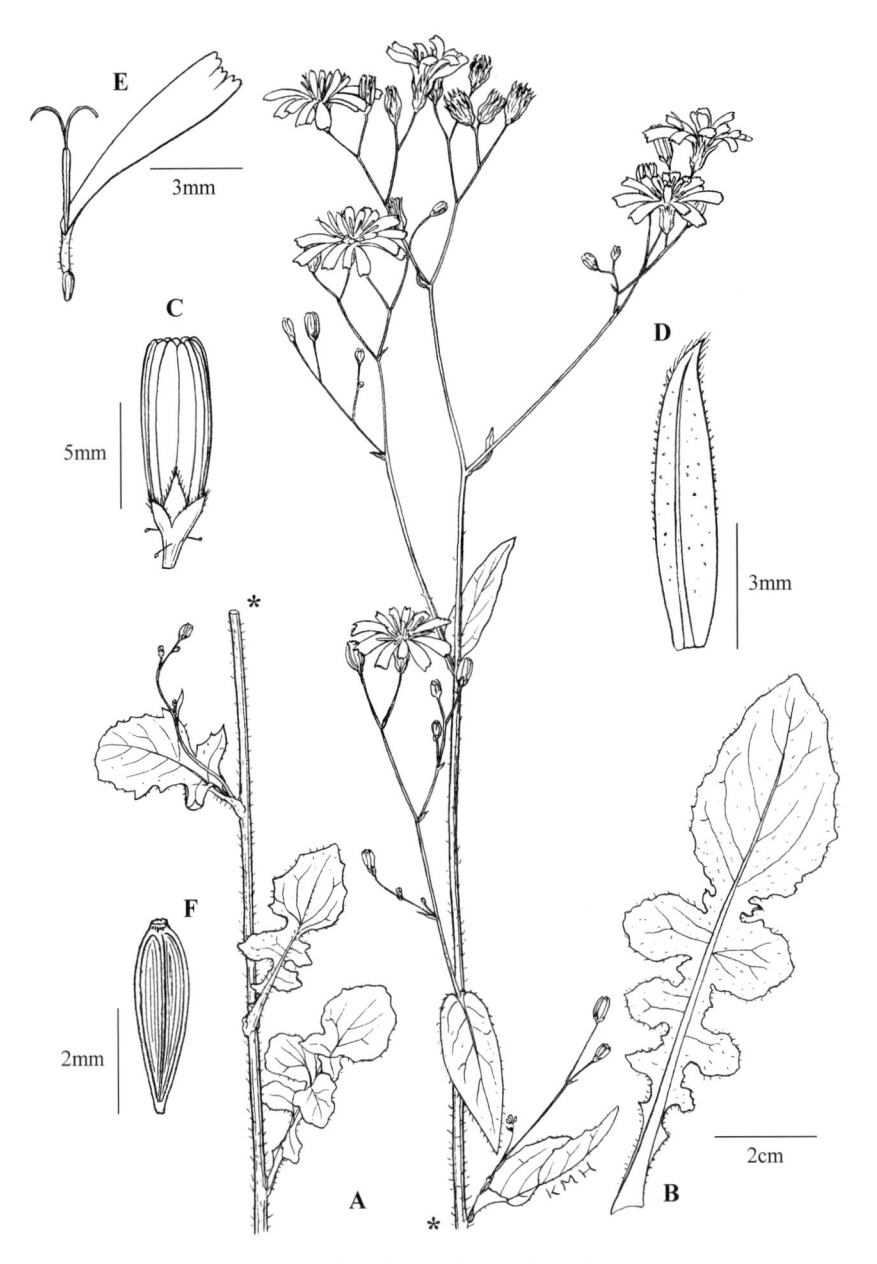

Lapsana communis L. ssp. **intermedia** (M. Bieb.) Hayek (Large Nipplewort). A habit; B rosette leaf; C capitulum in bud; D inner phyllary; E floret; F achene. (*Hort.* ex Totternhoe, Beds). STACE 817/686.

Tragopogon hybridus L. (Slender Salsify). A habit; B phyllary; C floret; D capitulum in fruit; E achene. (Tunbridge Wells, W Kent). STACE 822/689.

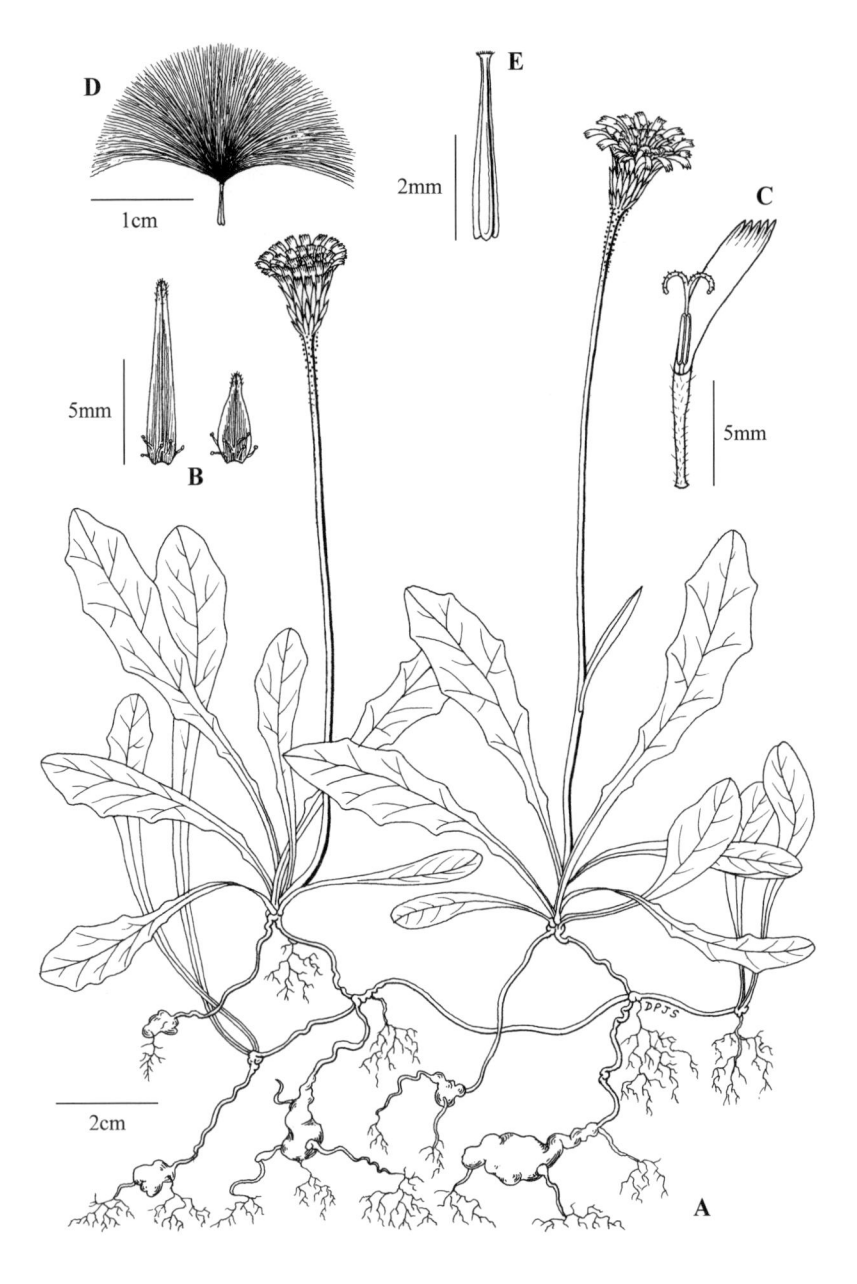

Aetheorhiza bulbosa (L.) Cass. (Tuberous Hawk's-beard). A habit; B phyllaries; C floret; D fruit; E achene. (A-C, Tyrone Abbey, Tyrone; D-E, S Spain). STACE 822/689.

139. ASTERACEAE

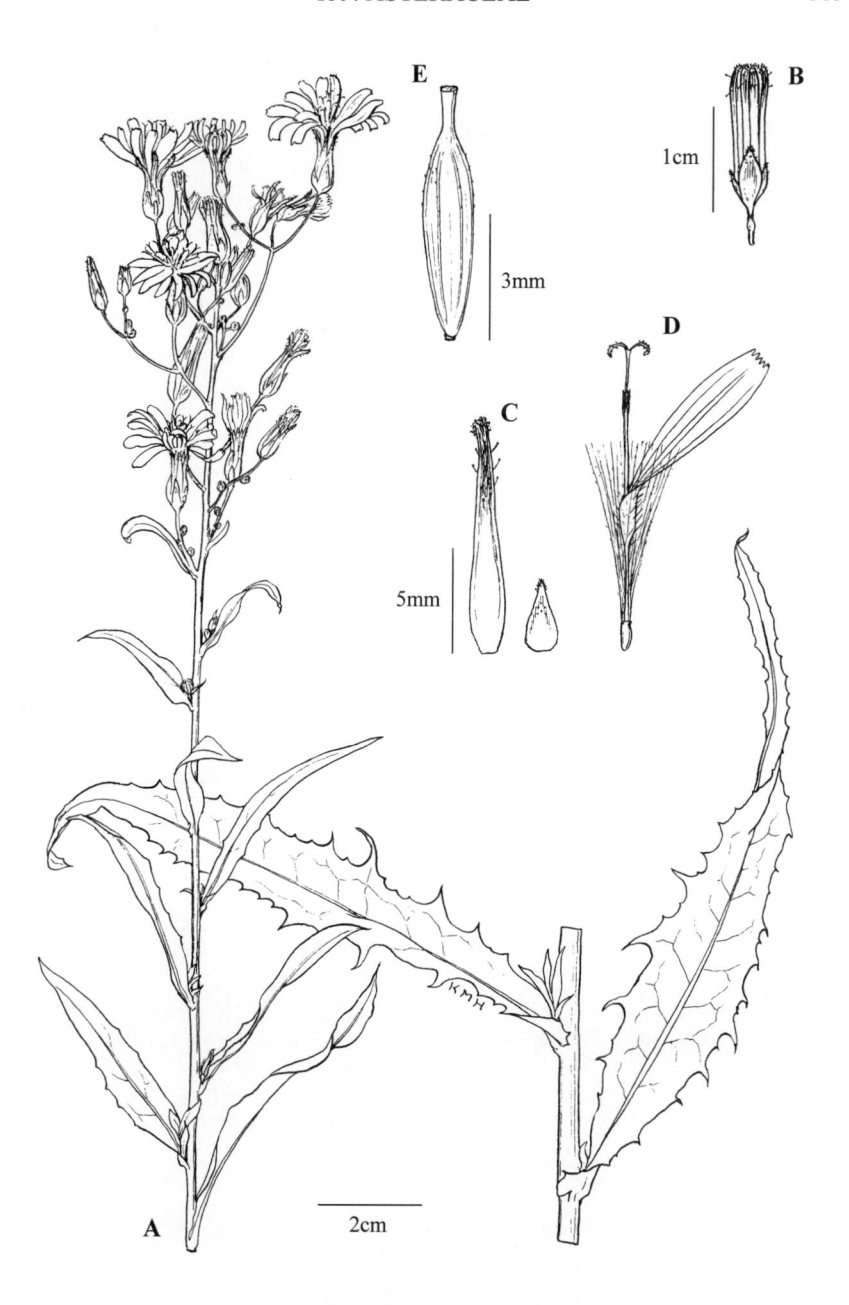

Lactuca tatarica (L.) C.A. Mey. (Blue Lettuce). A habit; B capitulum in bud; C phyllaries; D floret; E achene. (*Hort.* Aldershot, N Hants). STACE 826/692.

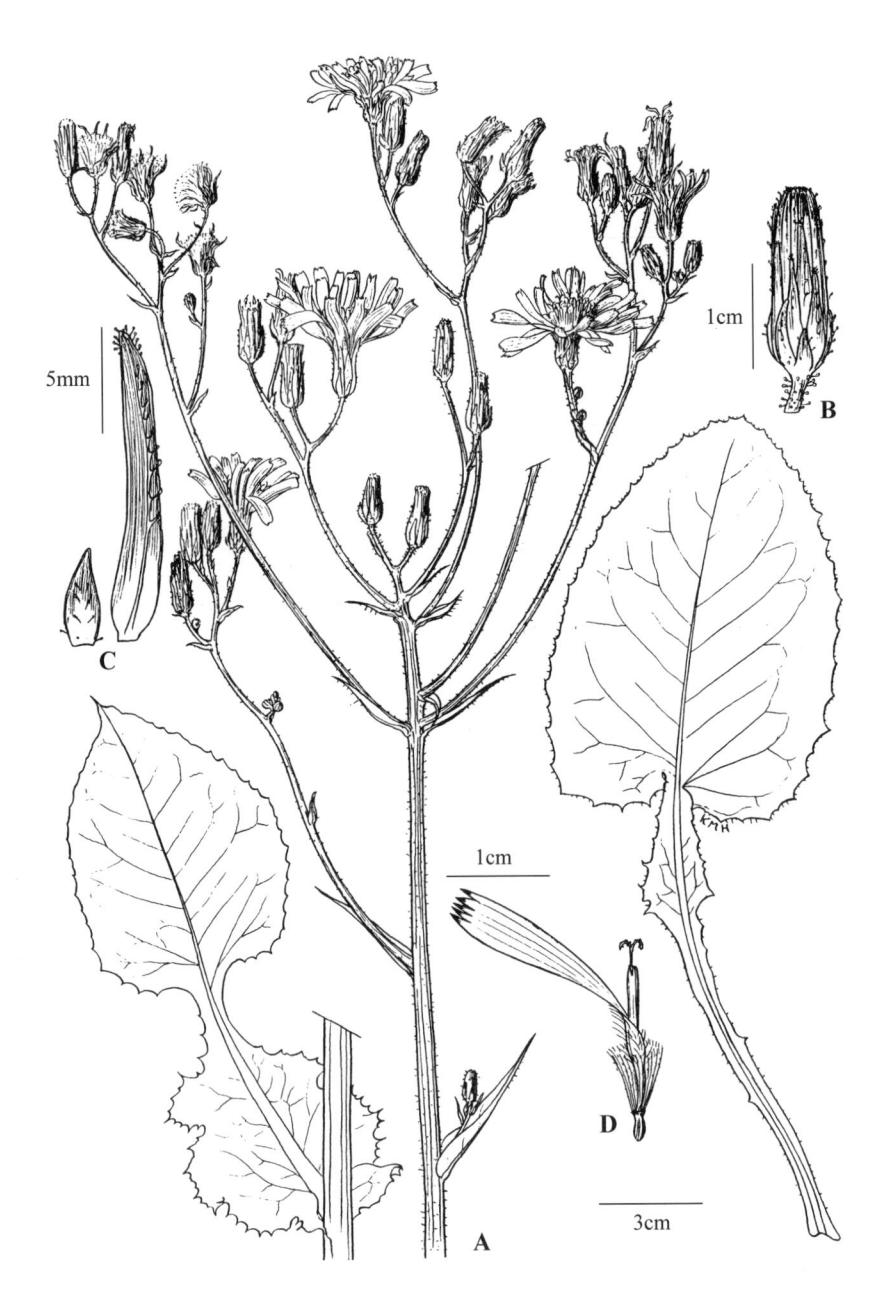

Cicerbita macrophylla (Willd.) Wallr. ssp. **uralensis** (Rouy) P.D. Sell (Common Blue-sow-thistle). A habit; B capitulum in bud; C phyllaries; D floret. (Otley, MW Yorks). STACE 826/692.

139. ASTERACEAE

Cicerbita bourgaei (Boiss.) Beauverd (Pontic Blue-sow-thistle). A habit; B capitulum in bud; C phyllaries; D floret; E achene. (*Hort*. Reigate, Surrey). STACE 827/692.

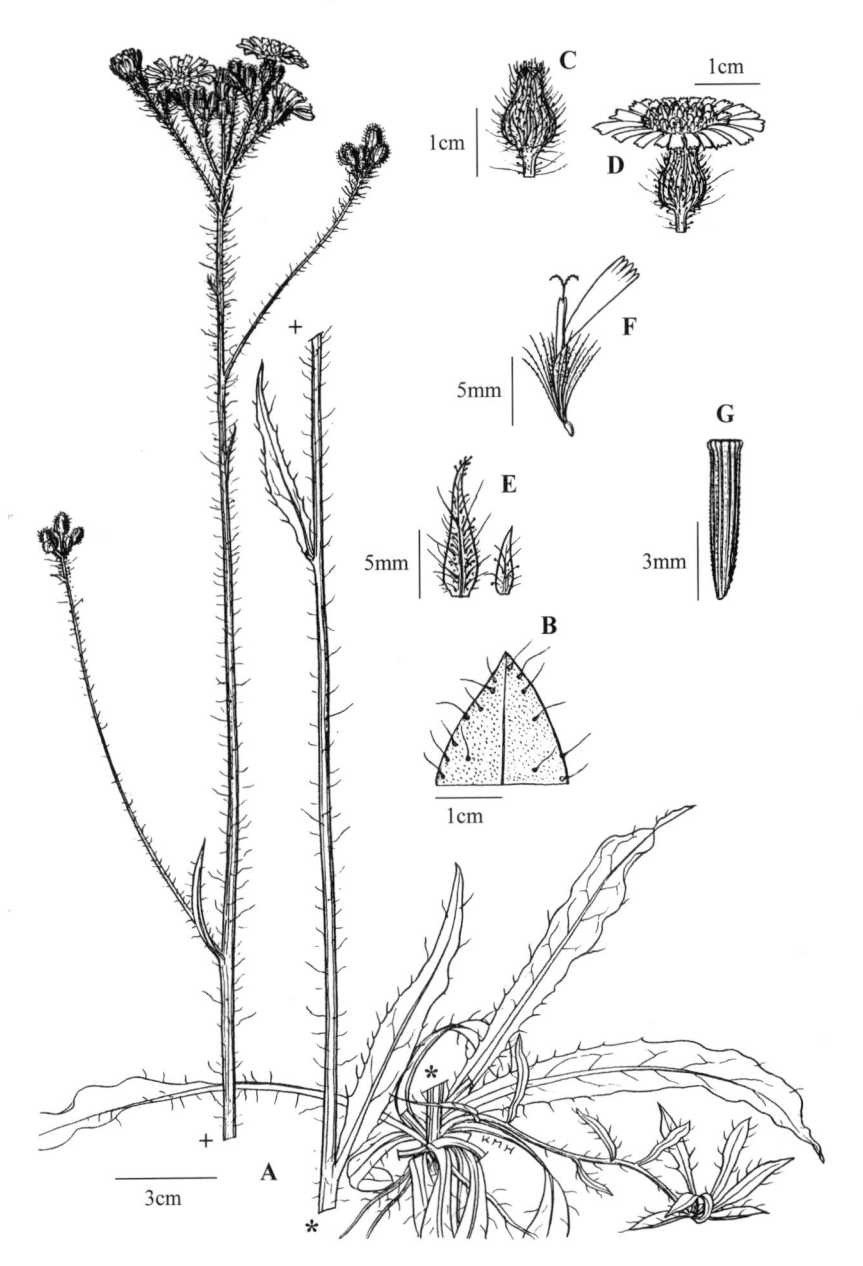

Pilosella praealta (Vill. ex Gochnat) F.W. Schultz & Sch. Bip. ssp. **praealta** (Tall Mouse-ear-hawkweed). A habit; B leaf apex; C capitulum in bud; D capitulum; E phyllaries; F floret; G achene. (Hungerford. Berks). STACE 835/703.

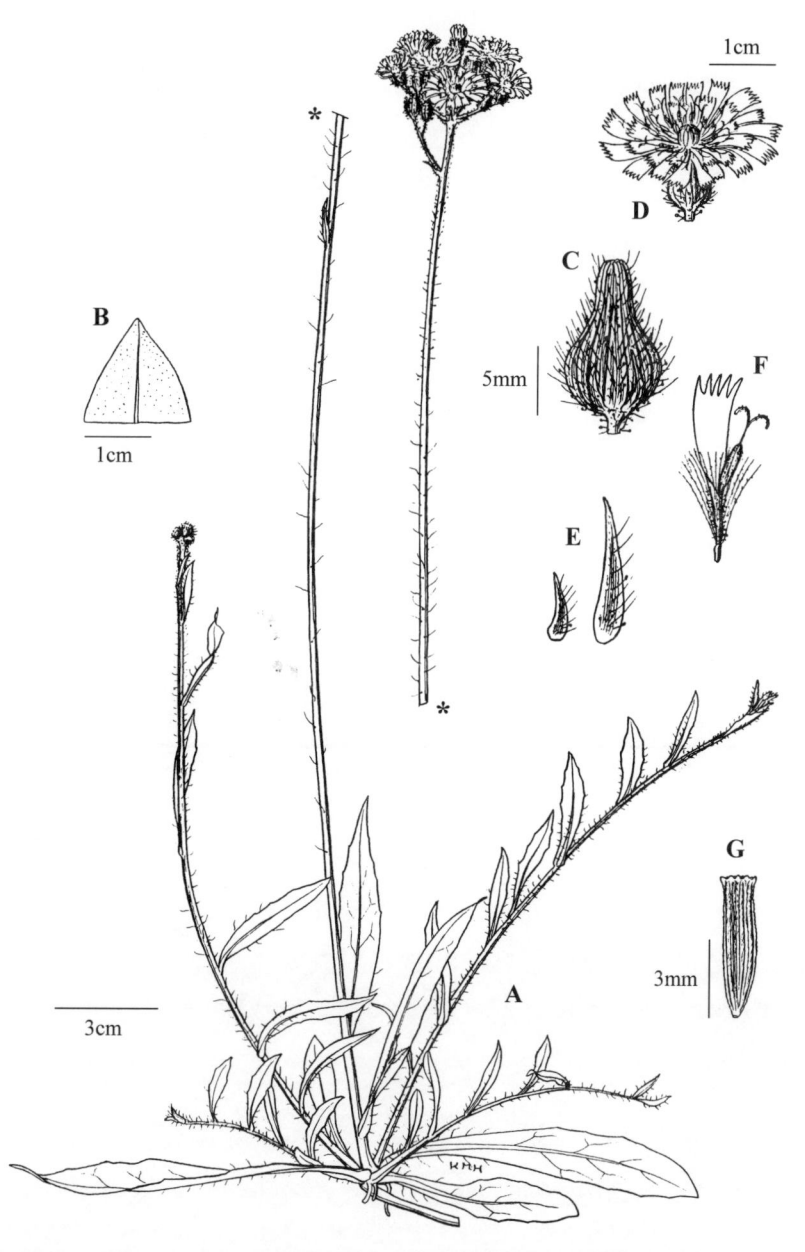

139. ASTERACEAE

Pilosella praealta (Vill. ex Gochnat) F.W. Schultz & Sch. Bip. ssp. **thaumasia** (Peter) P.D.Sell (Tall Mouse-ear-hawkweed subspecies). A habit; B leaf apex; C capitulum in bud; D capitulum; E phyllaries; F floret; G achene. (*Hort*. ex Chorley Wood, Herts). STACE 835/703.

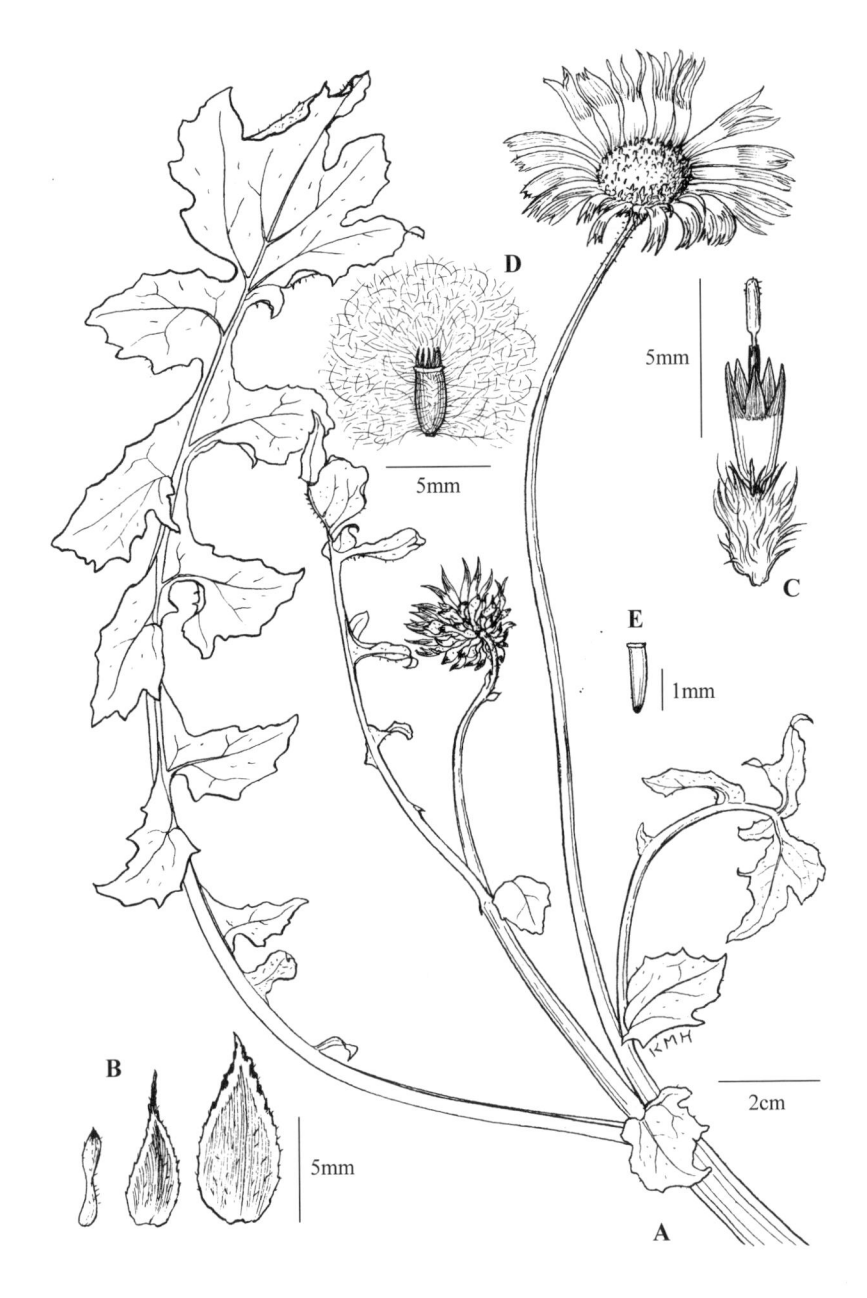

Arctotheca calendula (L.) Levyns (Plain Treasureflower). A habit; B phyllaries; C disk floret; D disc achene; E ray achene. (*Hort.* Sevenoaks, W Kent). STACE 839/712.

139. ASTERACEAE

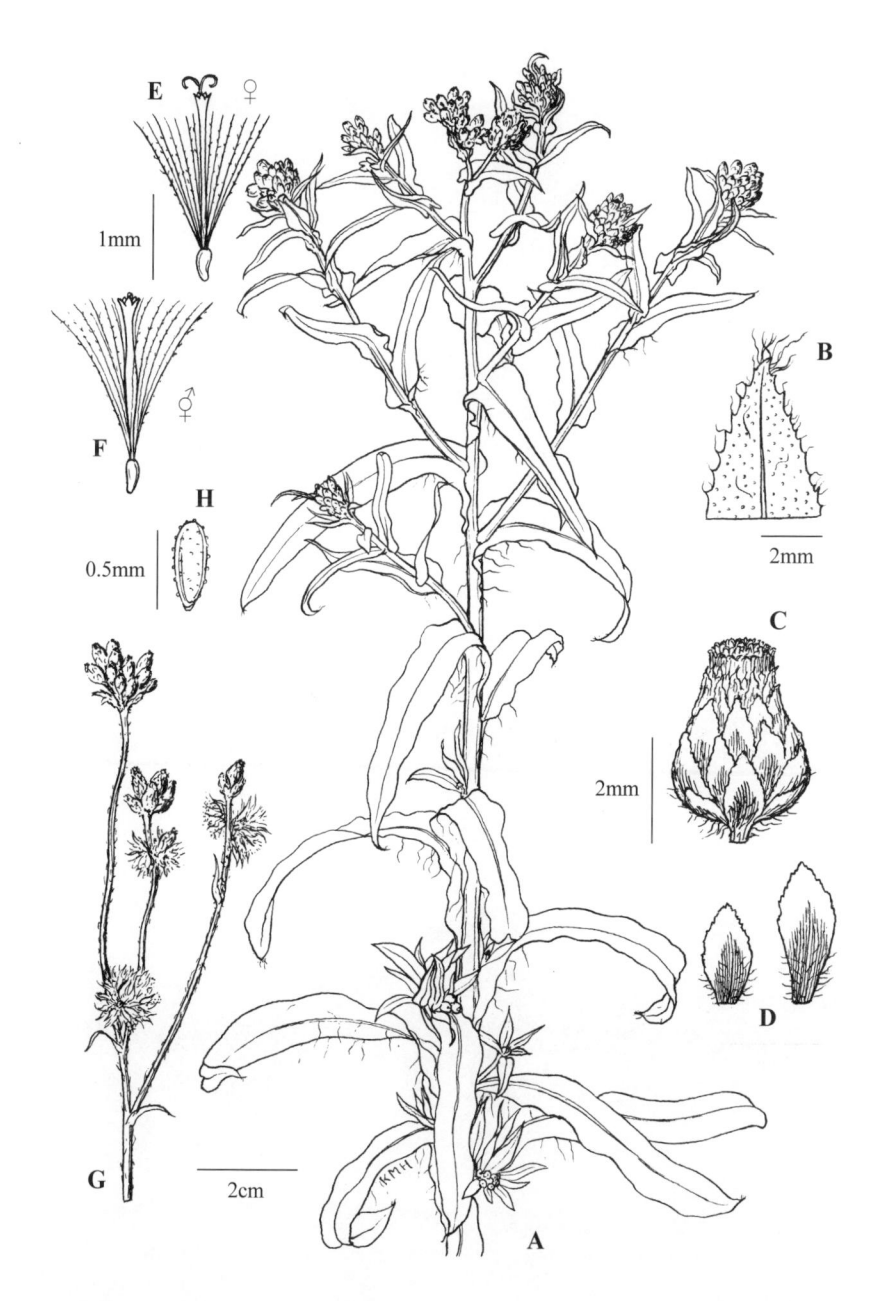

Gnaphalium undulatum L. (Cape Cudweed). A habit; B leaf apex; C capitulum; D phyllaries; E outer floret; F inner floret; G fruiting stem; H achene. (*Hort.* ex Jersey). STACE 842/715.

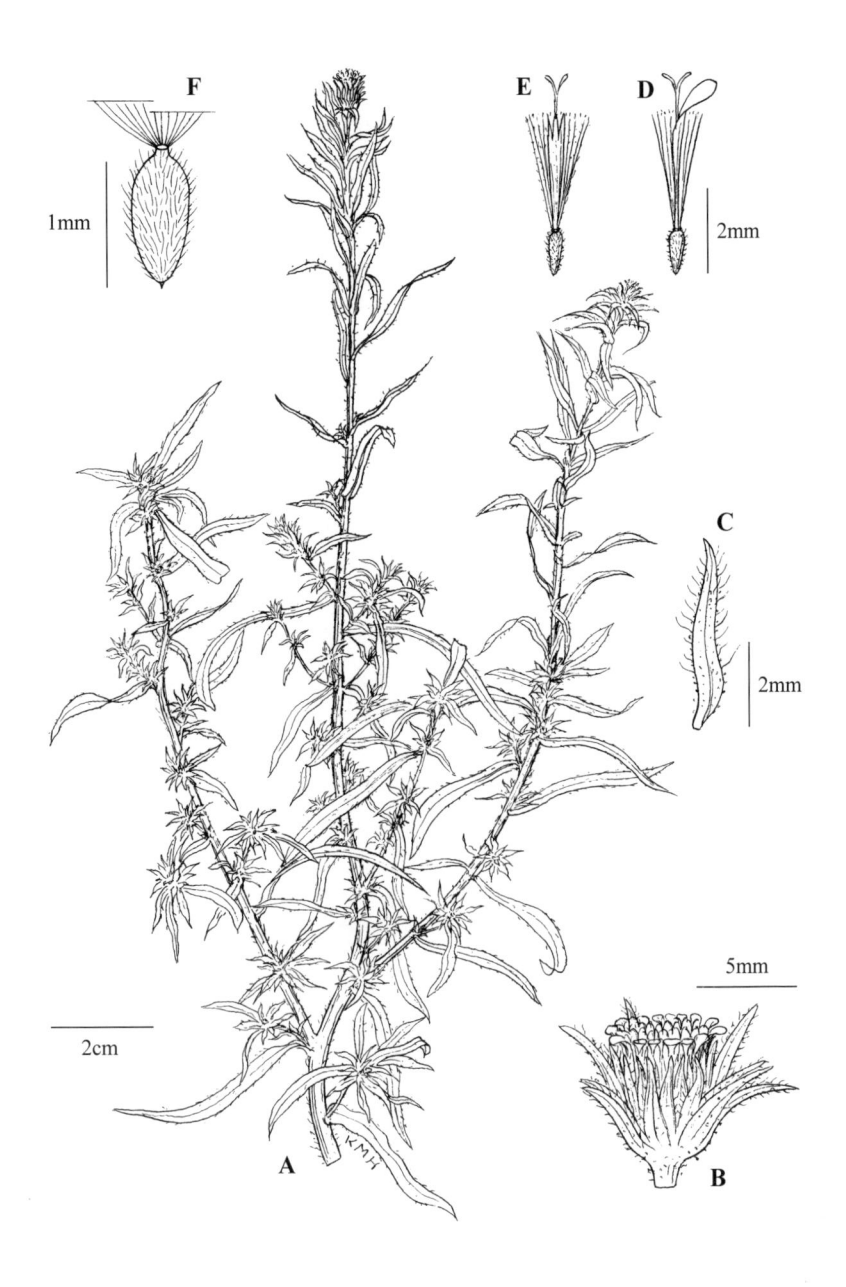

Dittrichia graveolens (L.) Greuter (Stinking Fleabane). A habit; B capitulum; C phyllary; D ray floret; E disc floret; F achene. (Sandwich, E Kent). STACE 845/717.

139. ASTERACEAE 311

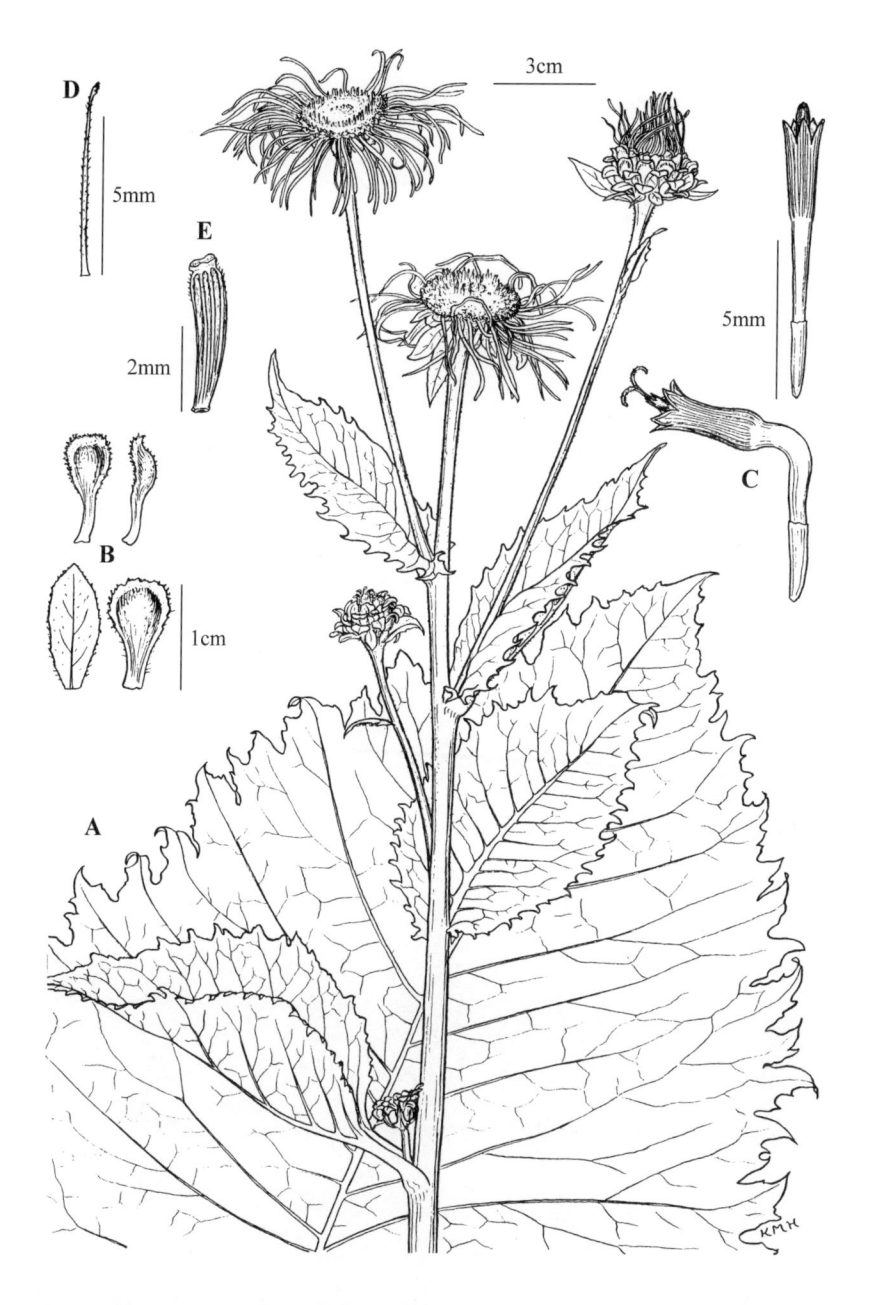

Telekia speciosa (Schreb.) Baumg. (Yellow Oxeye). A habit and basal leaf; B phyllaries; C disc florets; D receptacular scale; E achene. (Scotland). STACE 846/718.

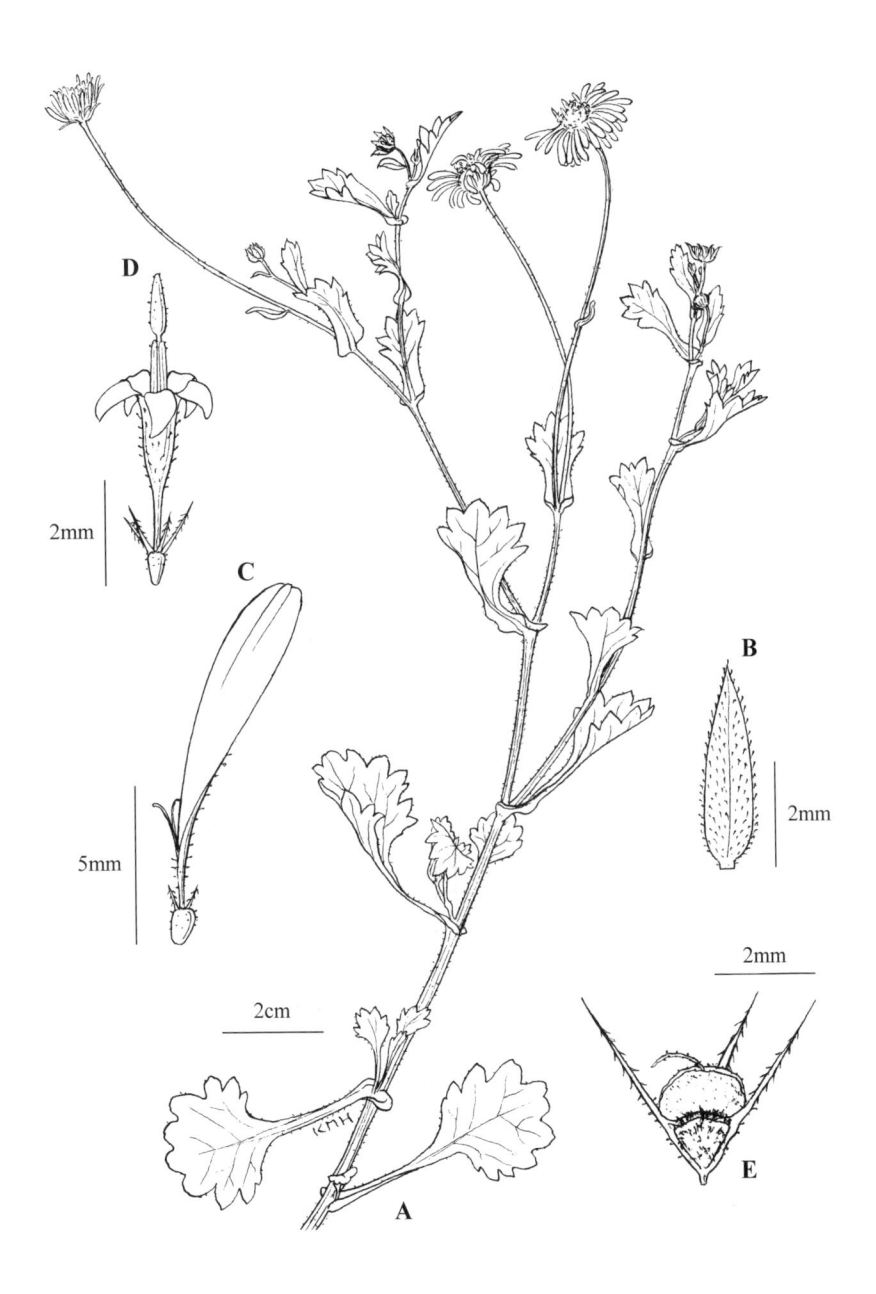

Calotis cuneifolia R. Br. (Bur Daisy). A habit; B phyllary; C ray floret; D disc floret; E achene. (E Kent). STACE 847/719.

139. ASTERACEAE

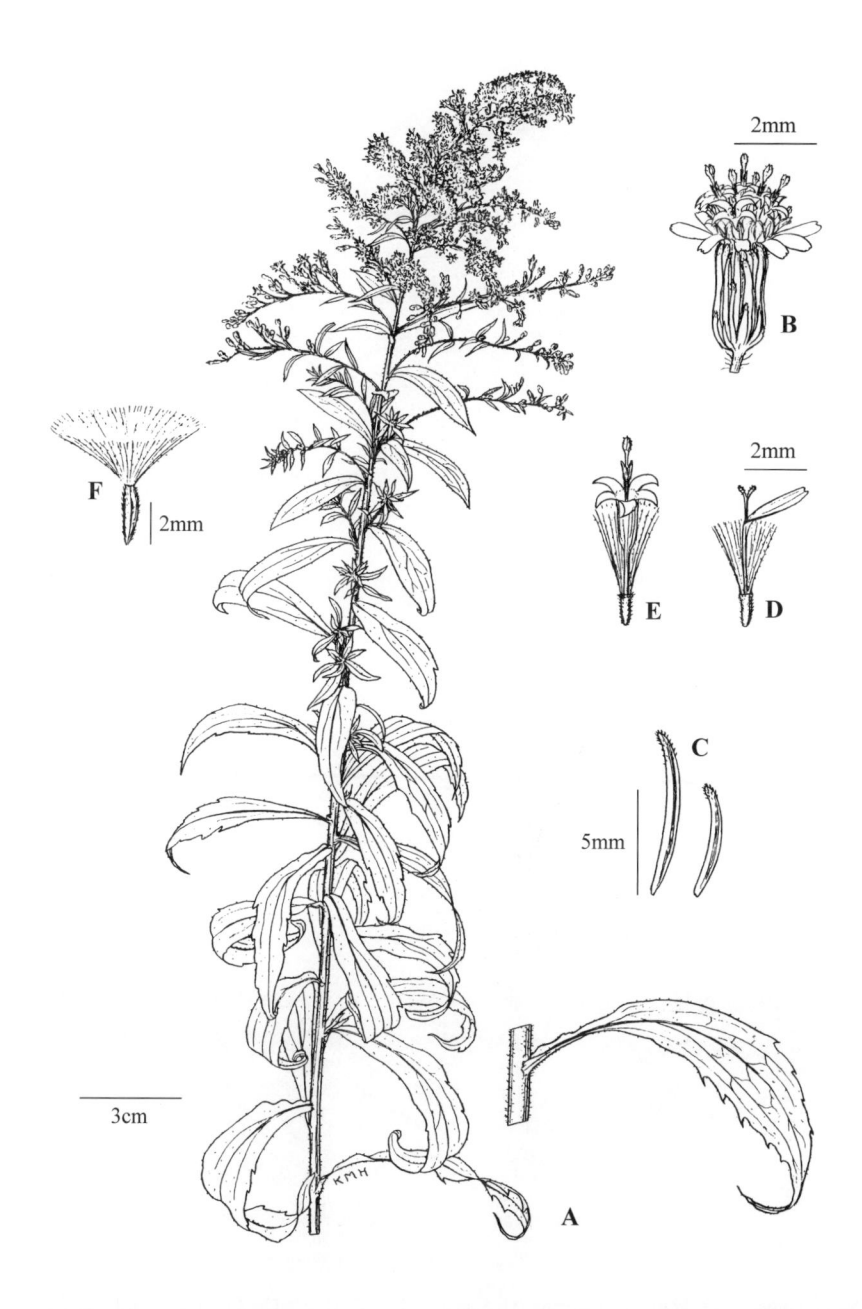

Solidago canadensis L. (Canadian Goldenrod). A habit and stem leaf; B capitulum; C phyllaries; D ray floret; E disc floret; F achene. (*Hort*. Ashbourne, Derbys). STACE 848/719.

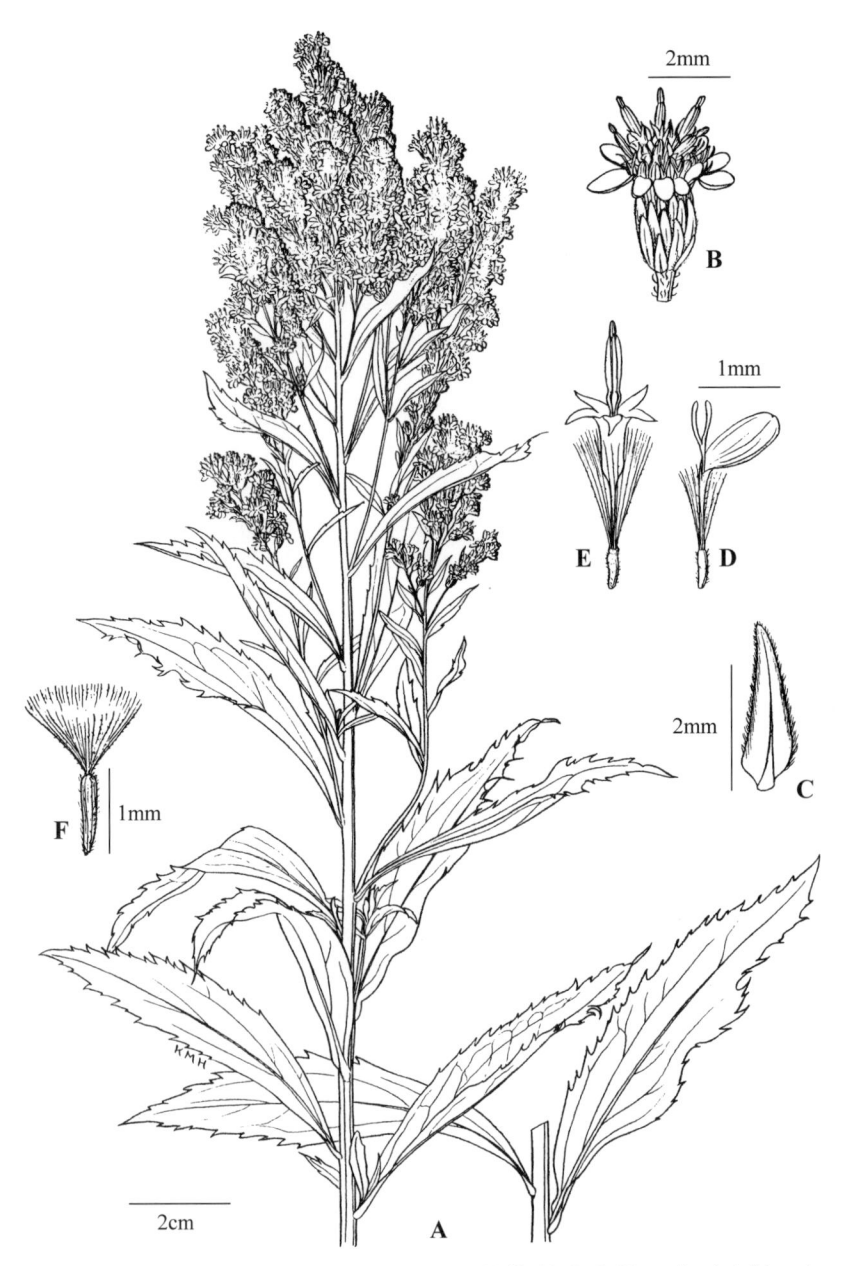

Solidago gigantea Aiton ssp. **serotina** (O. Kuntze) McNeill (Early Goldenrod). A habit and stem leaves; B capitulum; C phyllary; D ray floret; E disc floret; F achene. (Otford, W Kent). STACE 848/720.

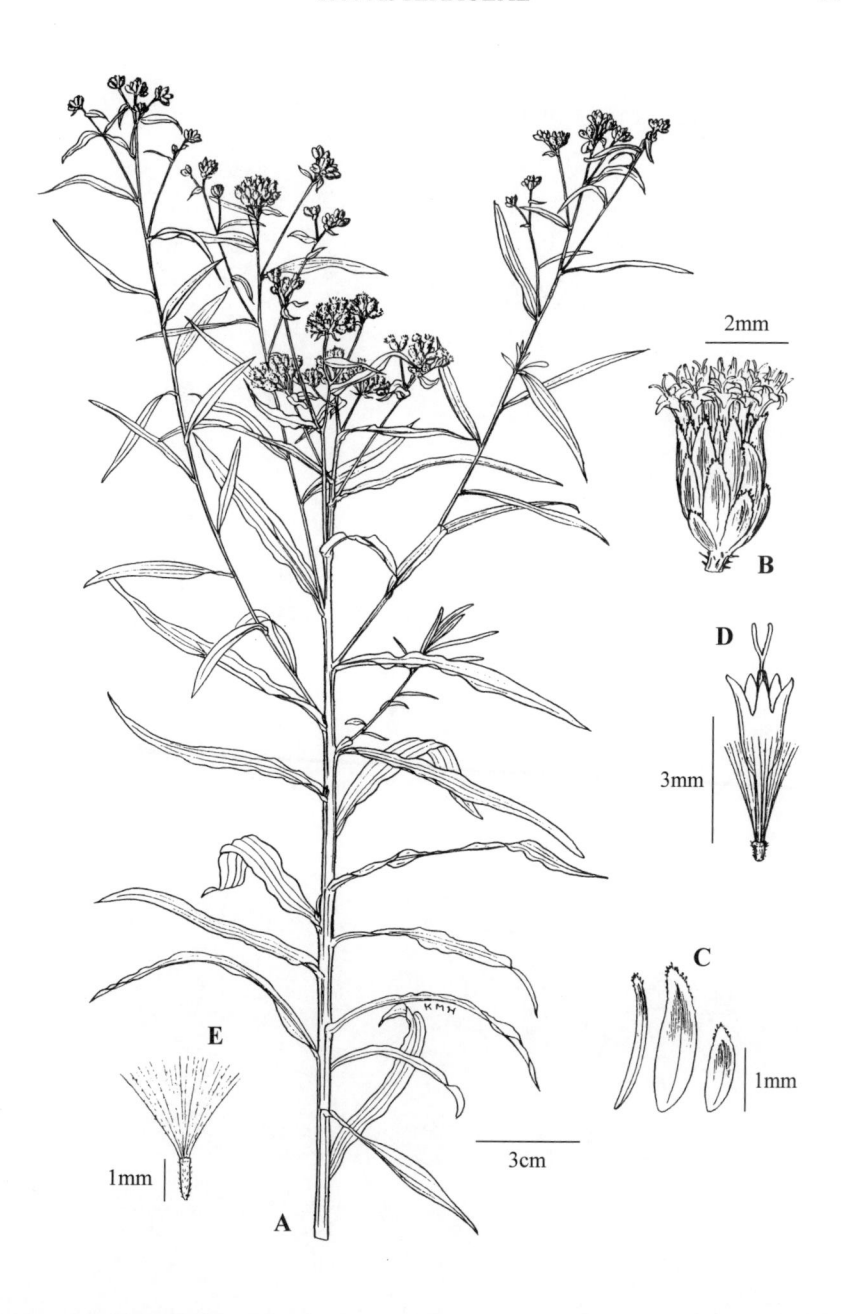

Solidago graminifolia (L.) Salisb. (Grass-leaved Goldenrod). A habit; B capitulum; C phyllaries; D floret; E achene. (*Hort.* ex Axminster, S Devon). STACE 848/720.

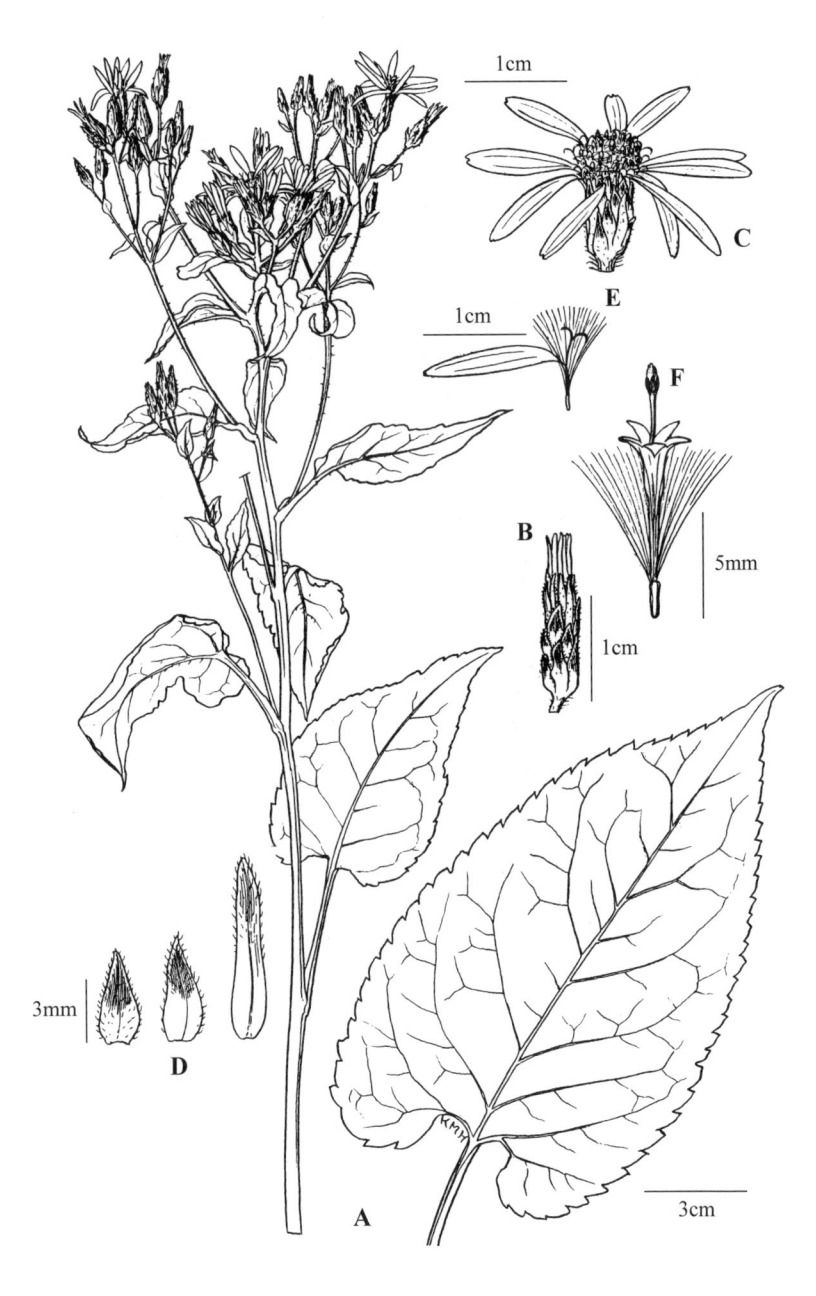

Aster schreberi Nees (Nettle-leaved Michaelmas-daisy). A habit and basal leaf; B capitulum in bud; C capitulum; D phyllaries; E ray floret; F disc floret. (Lochside, Renfrews). STACE 851/721.

139. ASTERACEAE

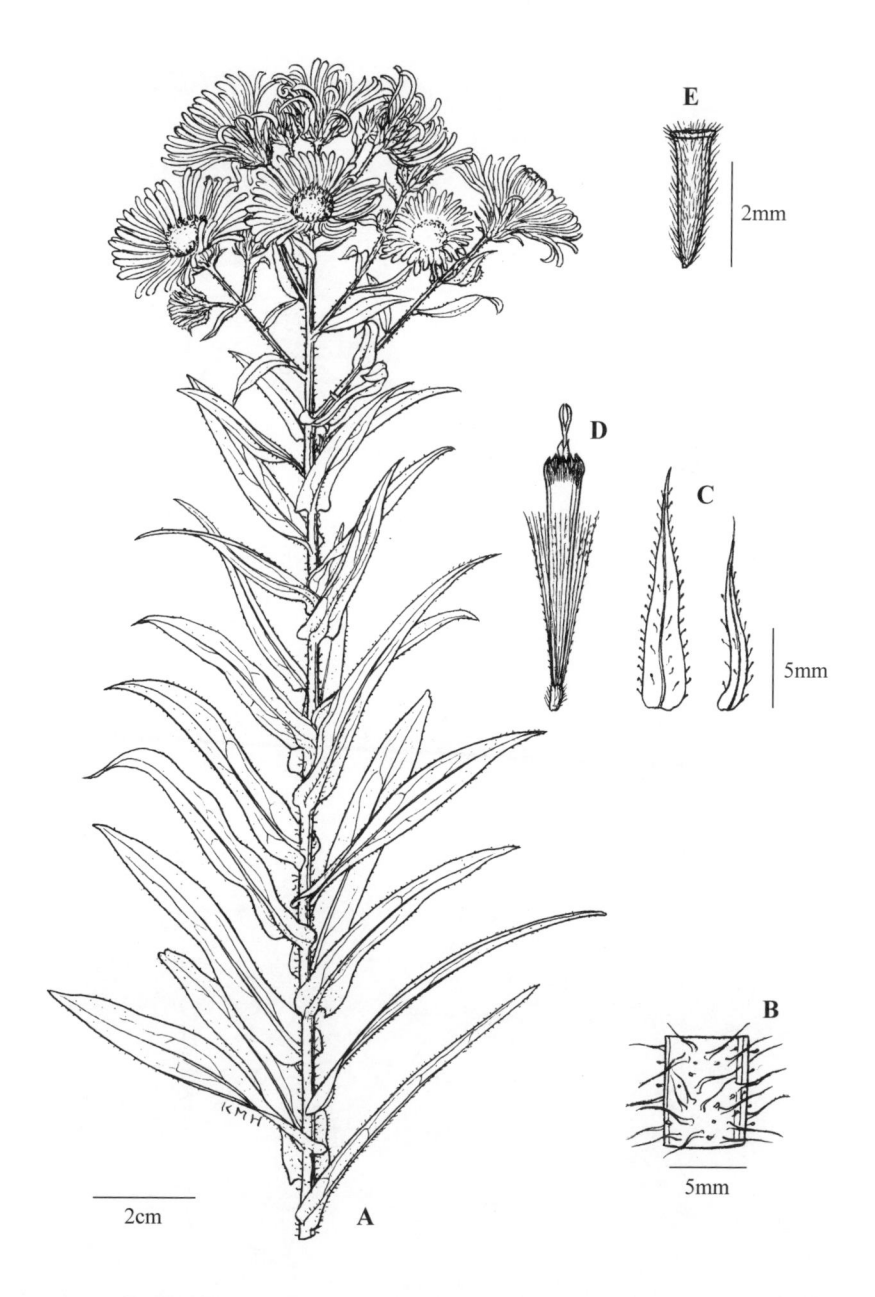

Aster novae-angliae L. (Hairy Michaelmas-daisy). A habit; B part of stem; C phyllaries; D disc floret; E achene. (*Hort.* Sevenoaks, W Kent). STACE 851/721.

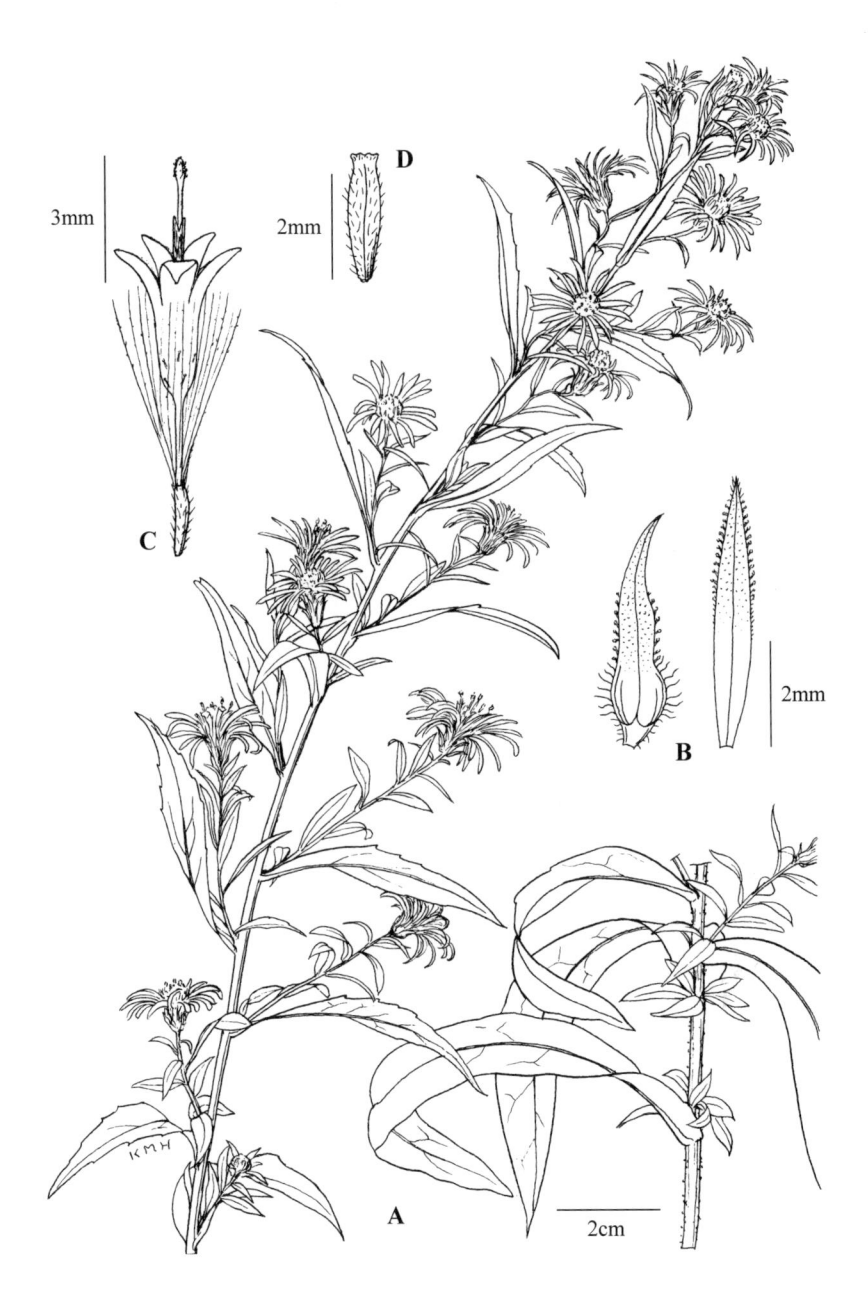

Aster x versicolor Willd. (Late Michaelmas-daisy). A habit and middle stem-leaves; B phyllaries; C disc floret; D achene. (Source unknown). STACE 851/721.

139. ASTERACEAE

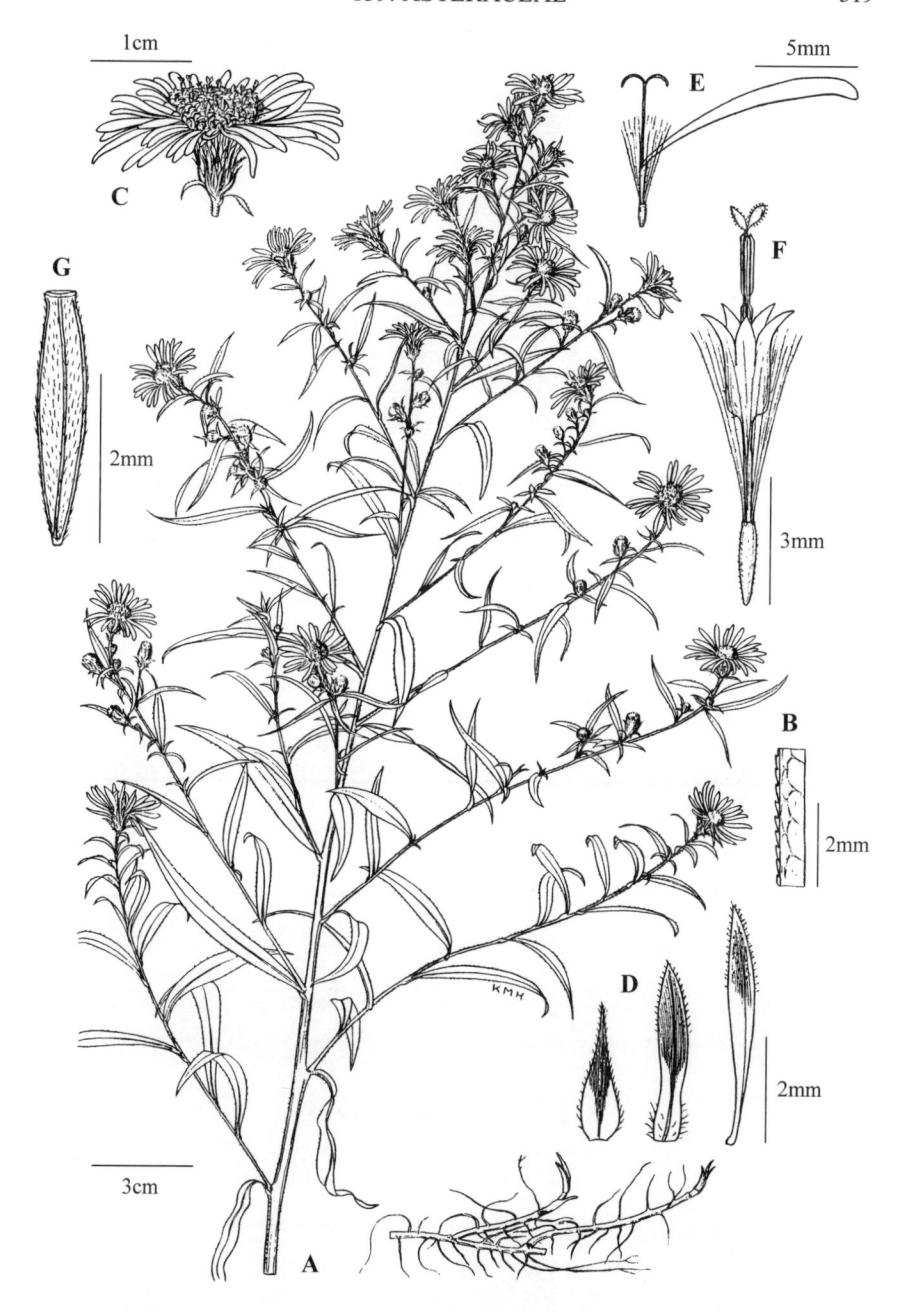

Aster lanceolatus Willd. (Narrow-leaved Michaelmas-daisy). A habit; B leaf margin; C capitulum; D phyllaries; E ray floret; F disc floret; G achene. (*Hort.* Sevenoaks, W Kent). STACE 852/721.

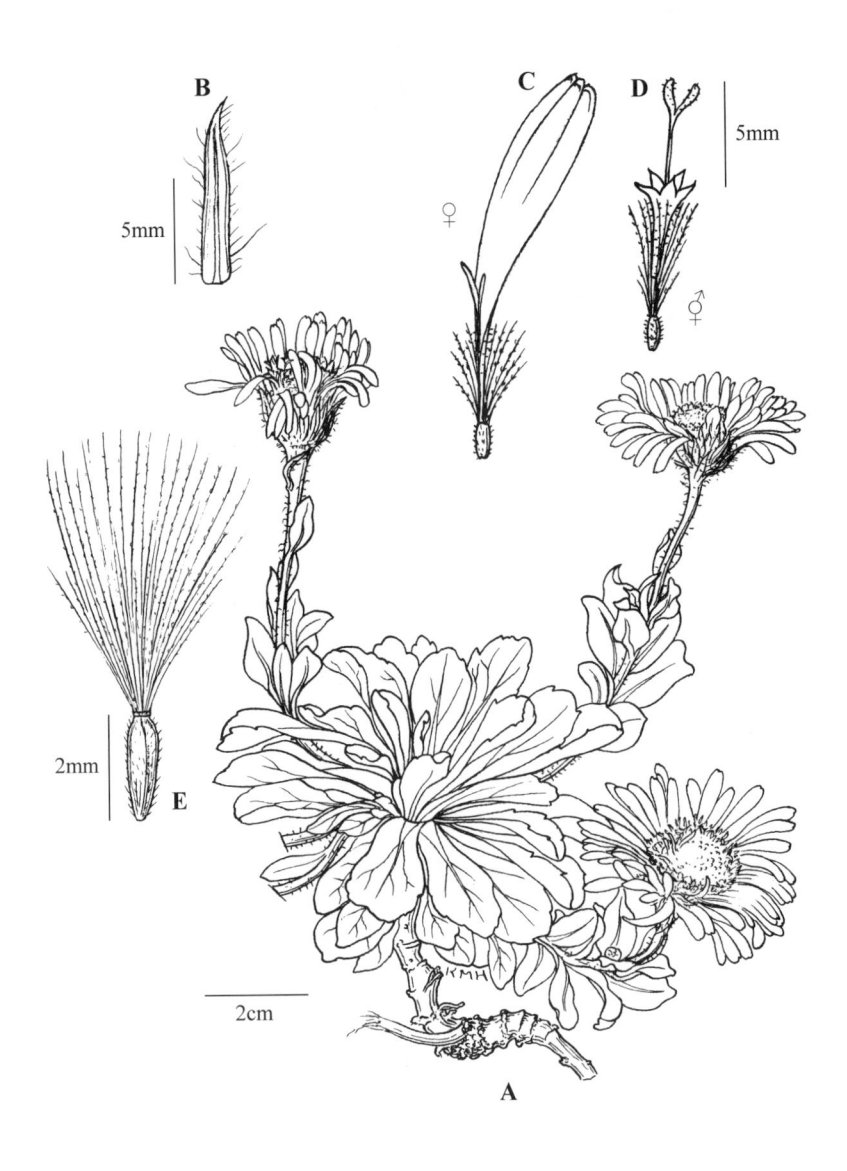

Erigeron glaucus Ker Gawl. (Seaside Daisy). A habit; B phyllary; C ray floret; D disc floret; E achene. (Croyde Bay, N Devon). STACE 853/724.

139. ASTERACEAE 321

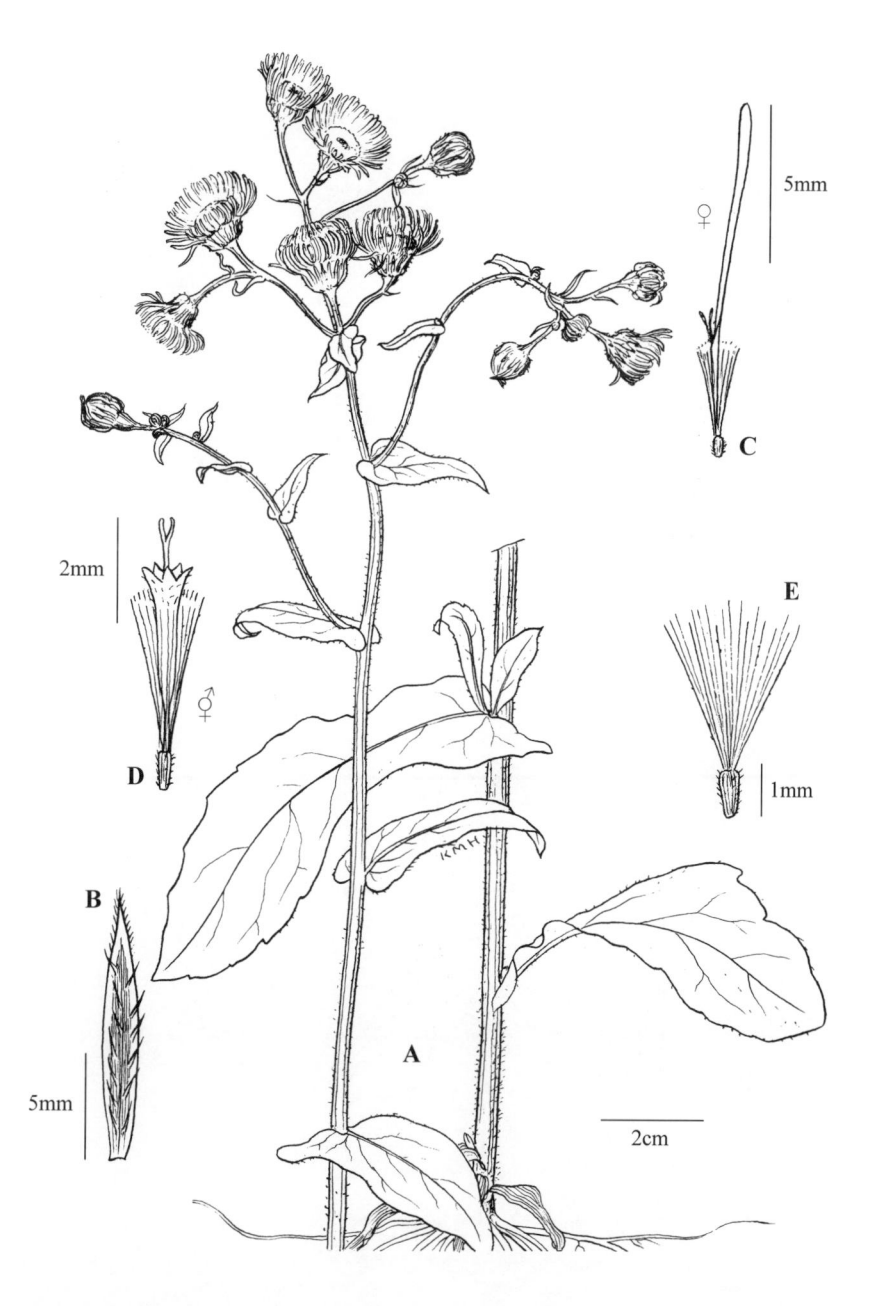

Erigeron philadelphicus L. (Robin's-plantain). A habit; B phyllary; C ray floret; D disc floret; E achene. (Poulton-le-Fylde, W Lancs). STACE 853/724.

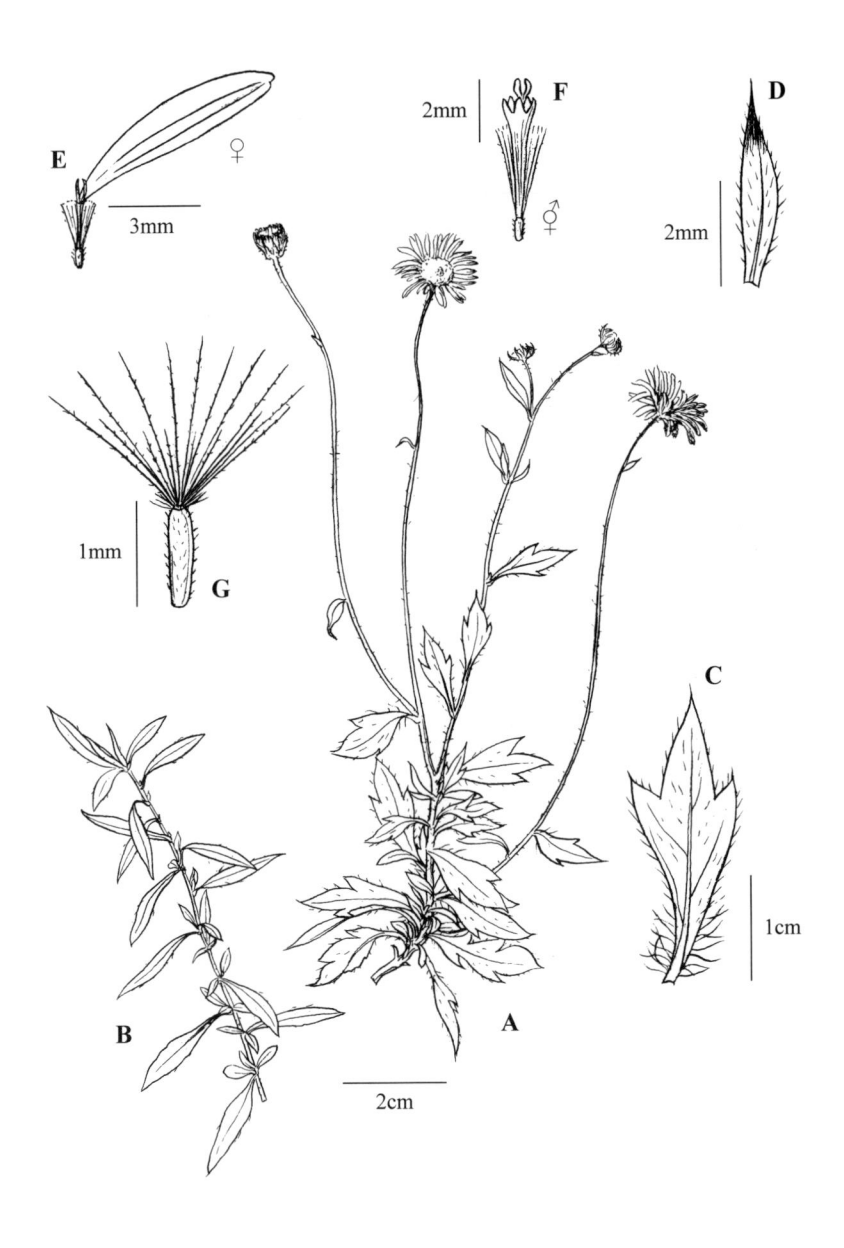

Erigeron karvinskianus DC. (Mexican Fleabane). A habit; B non-flowering shoot; C flowering stem leaf; D phyllary; E ray floret; F disc floret; G achene. (Jersey). STACE 853/724.

139. ASTERACEAE

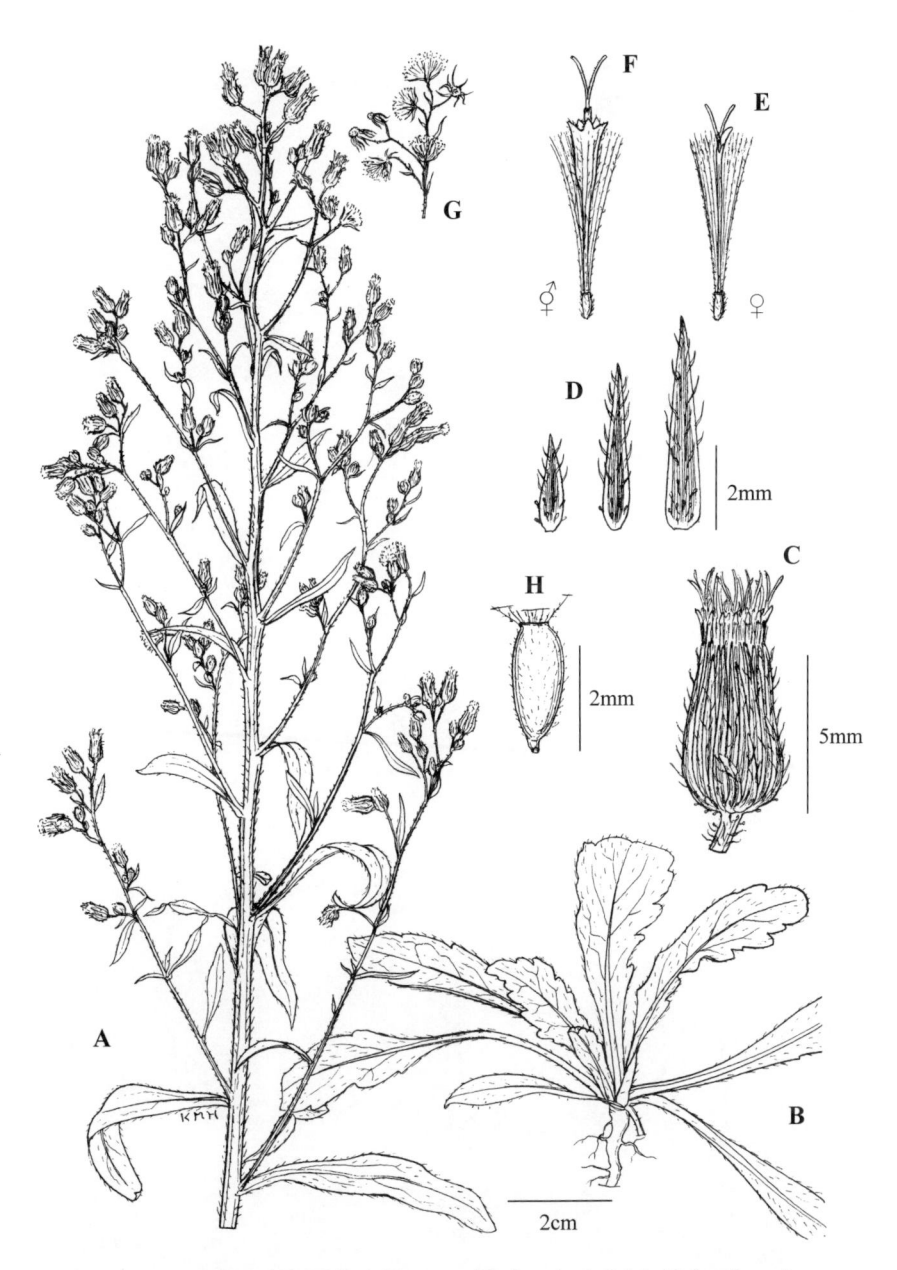

Conyza sumatrensis (Retz.) E. Walker (Guernsey Fleabane). A habit; B basal rosette; C capitulum; D phyllaries; E outer floret; F inner floret; G fruiting stem; H achene. (Guernsey). STACE 855/725.

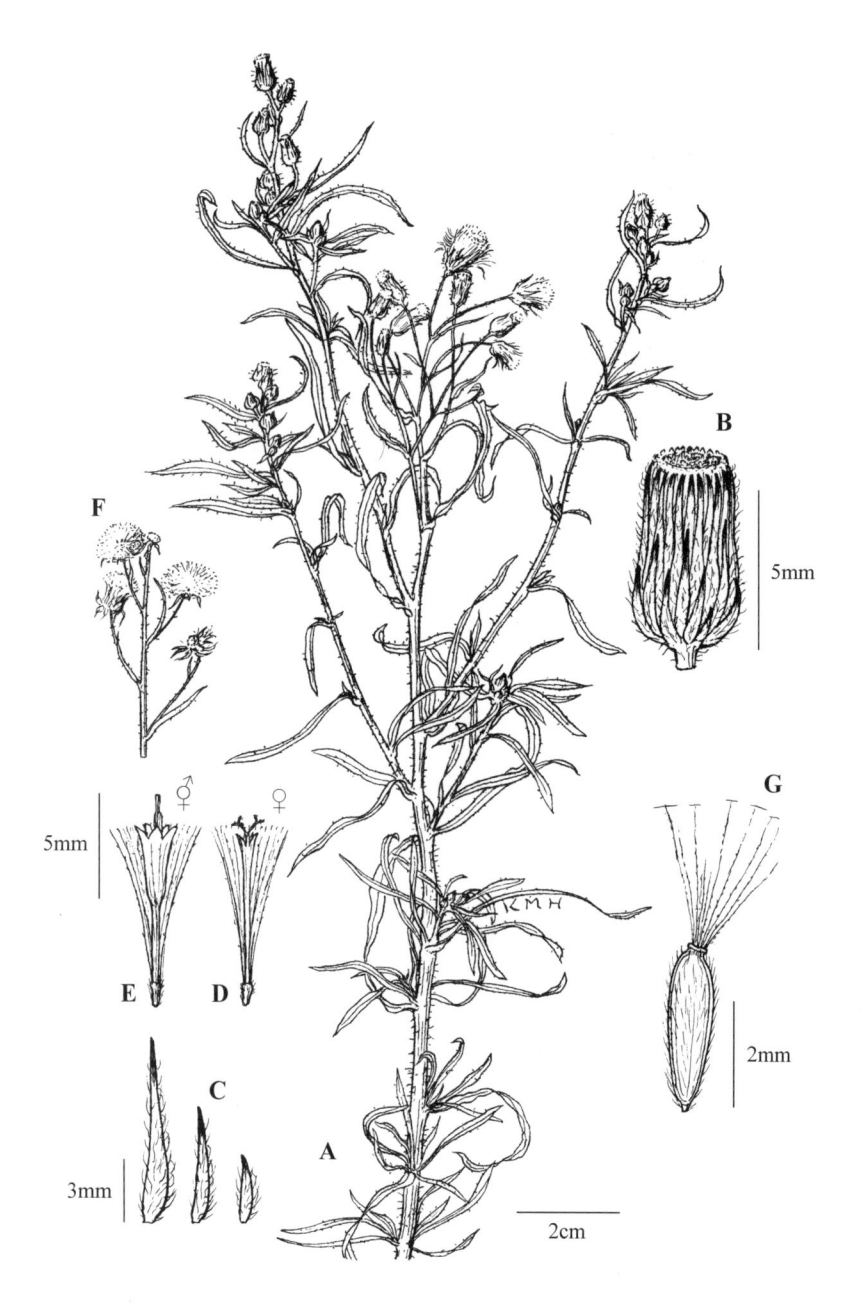

Conyza bonariensis (L.) Cronquist (Argentine Fleabane). A habit; B capitulum; C phyllaries; D outer floret; E inner floret; F fruiting stem; G achene. (Ash, E Kent). STACE 855/725.

139. ASTERACEAE

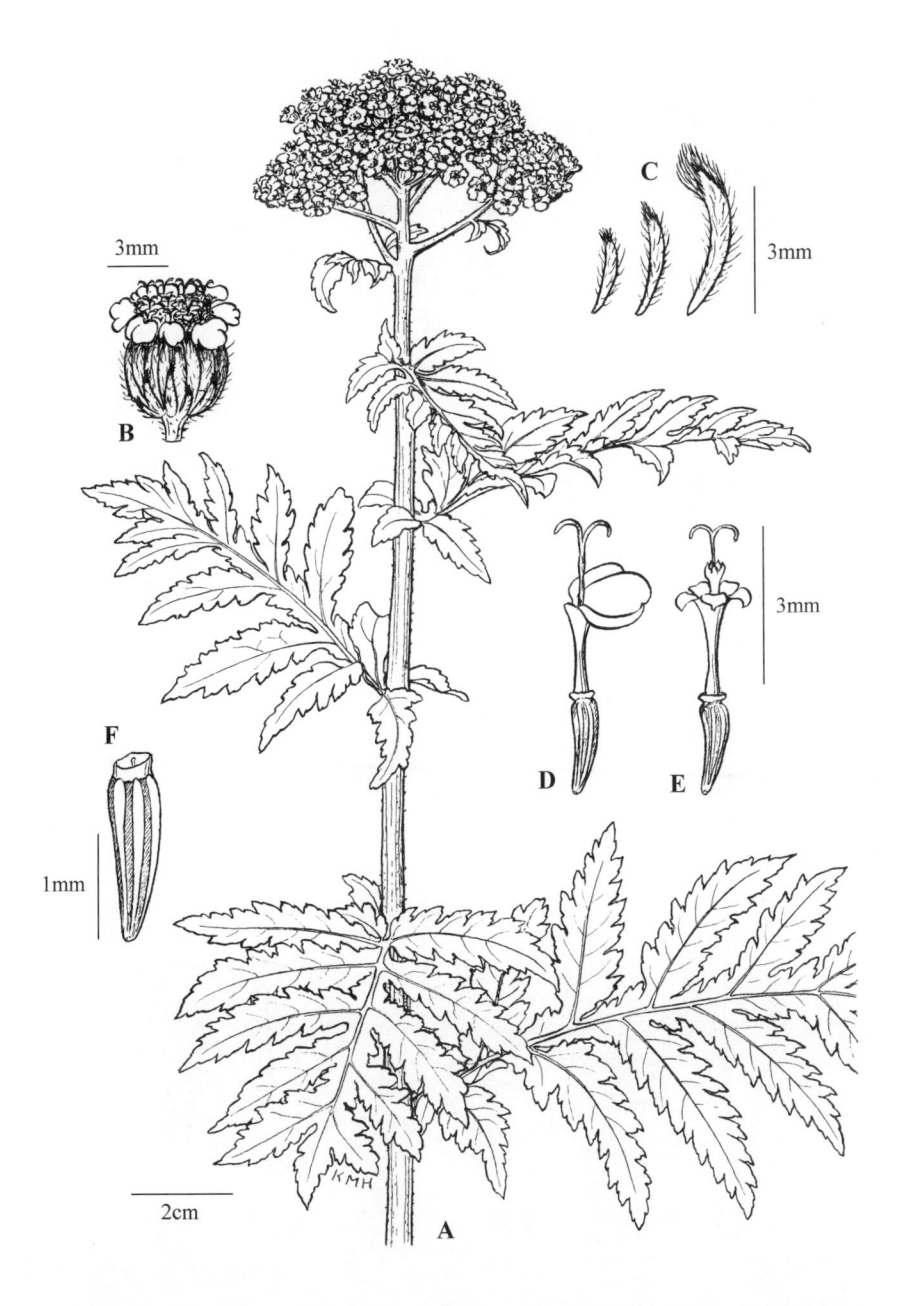

Tanacetum macrophyllum (Waldst. & Kit.) Sch. Bip. (Rayed Tansy). A habit; B capitulum; C phyllaries; D ray floret; E disc floret; F achene. (*Hort.* Ashbourne, Derbys). STACE 858/728.

326   139. ASTERACEAE

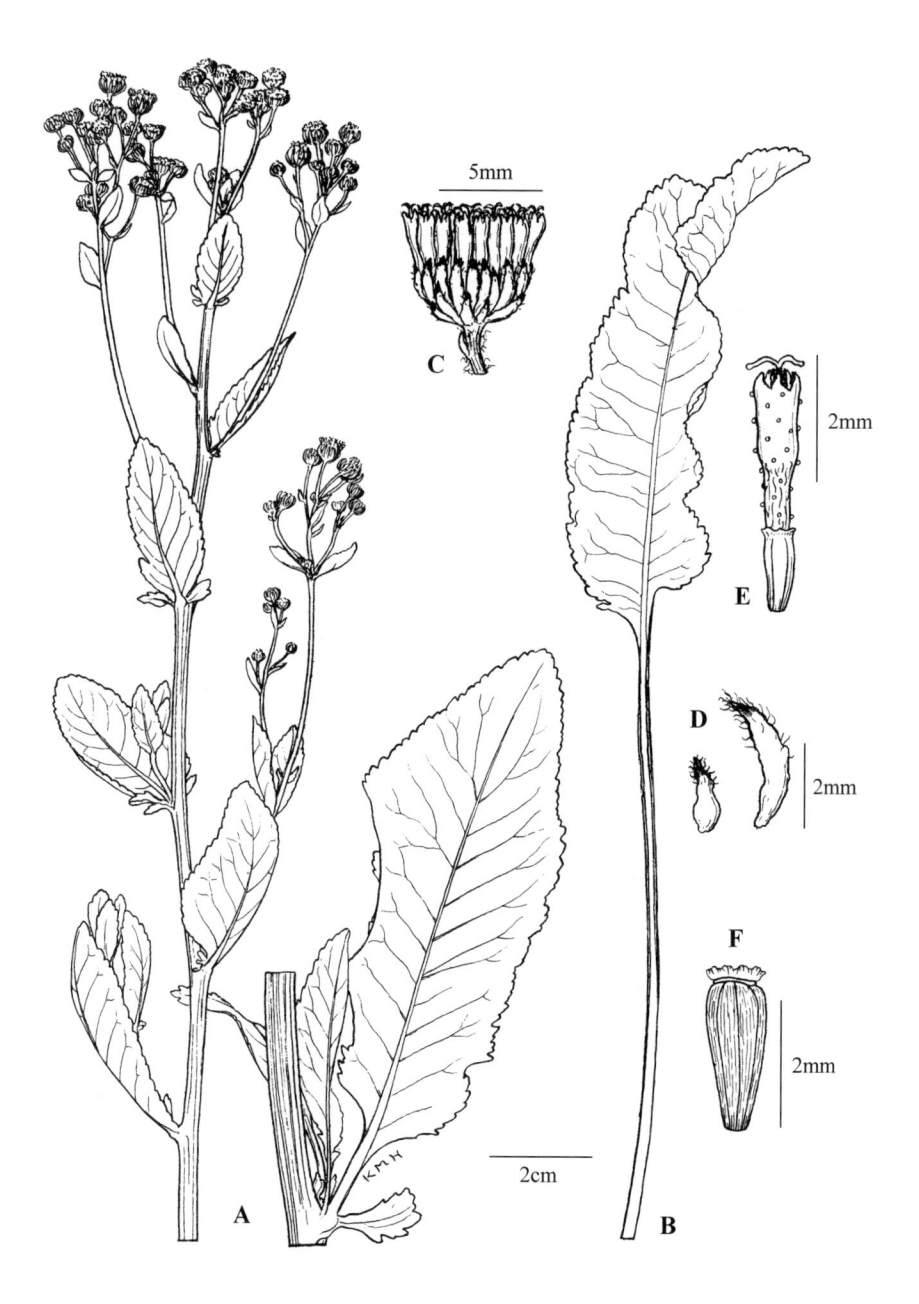

Tanacetum balsamita L. (Costmary). A habit; B basal leaf; C capitulum; D phyllaries; E floret; F achene. (Gosford, E Lothian). STACE 859/728.

139. ASTERACEAE

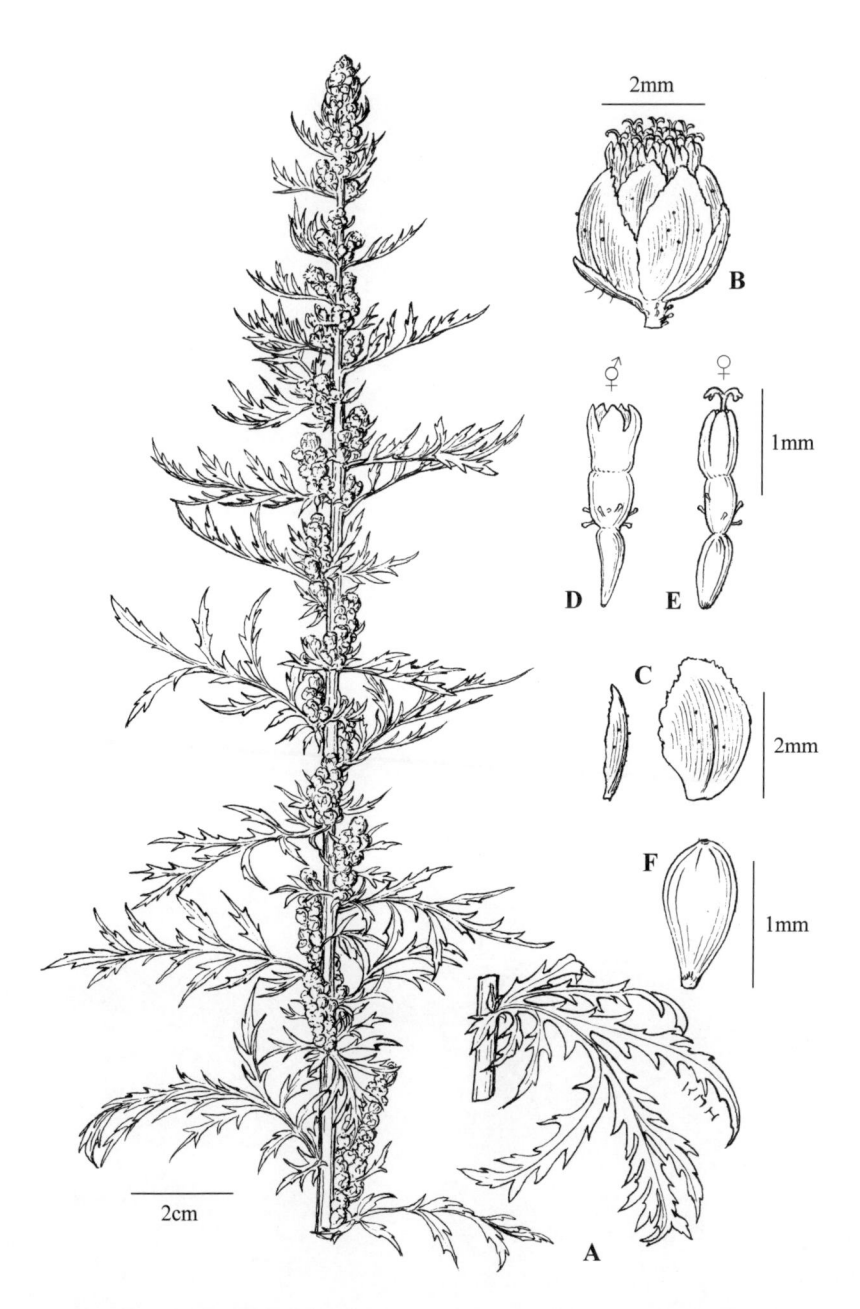

Artemisia biennis Willd. (Slender Mugwort). A habit; B capitulum; C two views of phyllary; D inner floret; E outer floret; F achene. (*Hort.* Bromley, Kent). STACE 861/731.

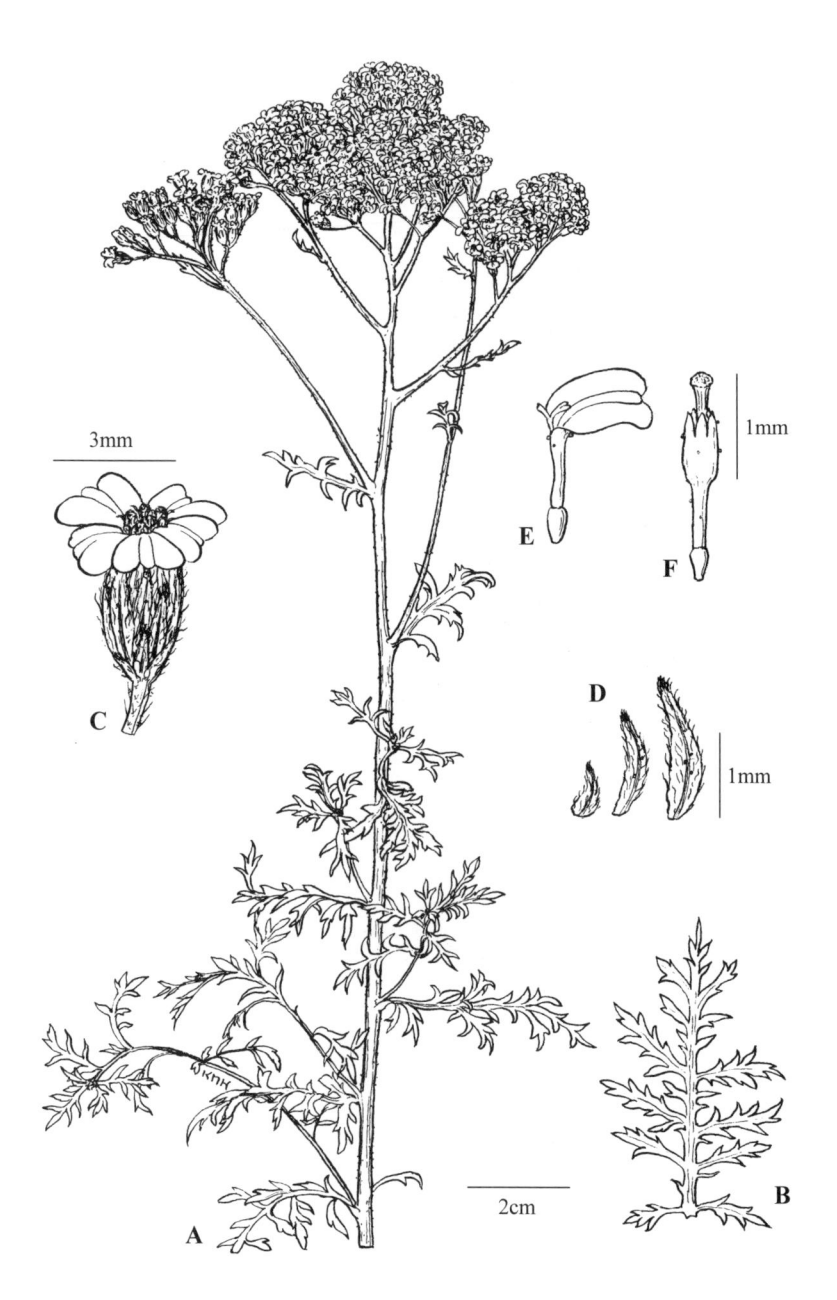

Achillea ligustica All. (Southern Yarrow). A habit; B leaf; C capitulum; D phyllaries; E ray floret; F disc floret. (Newport, Mons). STACE 863/732.

139. ASTERACEAE 329

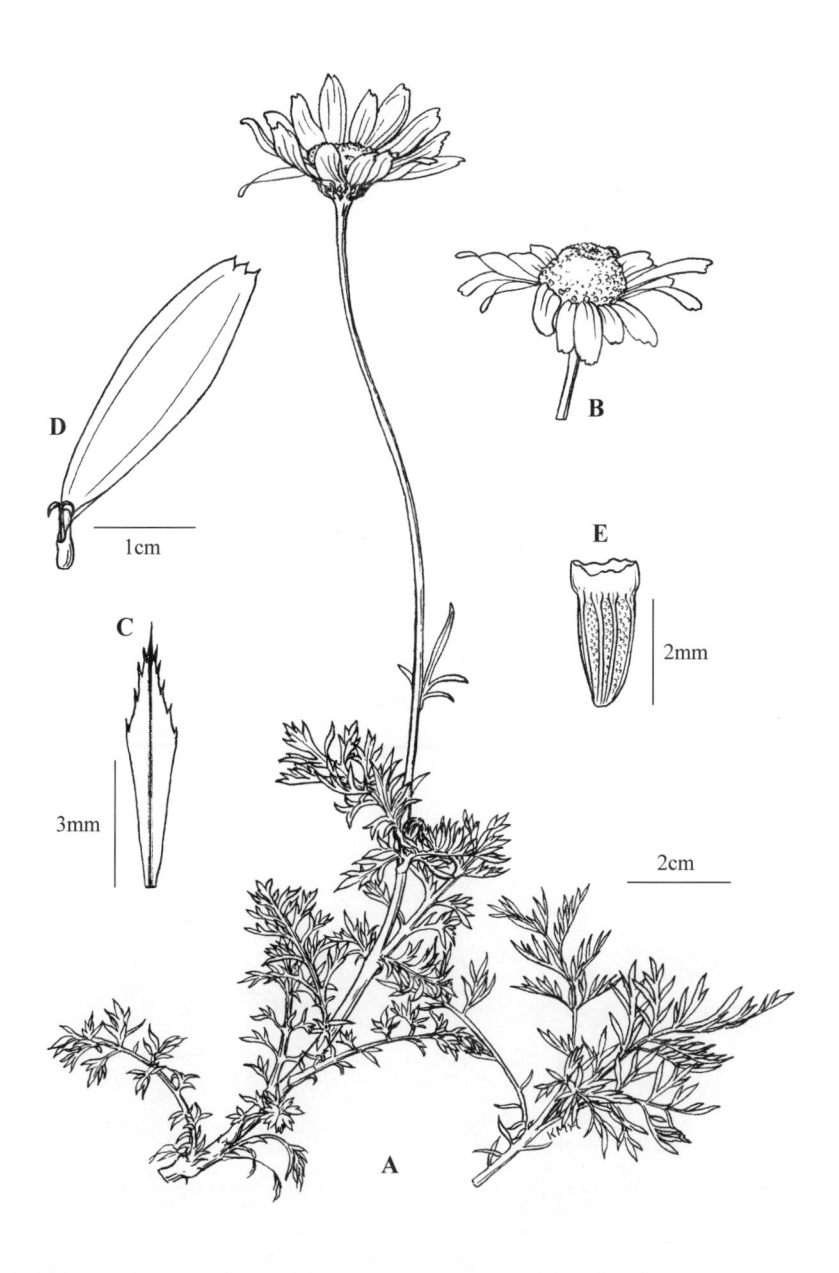

Anthemis punctata Vahl ssp. **cupaniana** (Tod. ex Nyman) R. Fern. (Sicilian Chamomile). A habit; B capitulum; C receptacular scale; D ray floret; E achene. (Fishguard, Pembs). STACE 864/733.

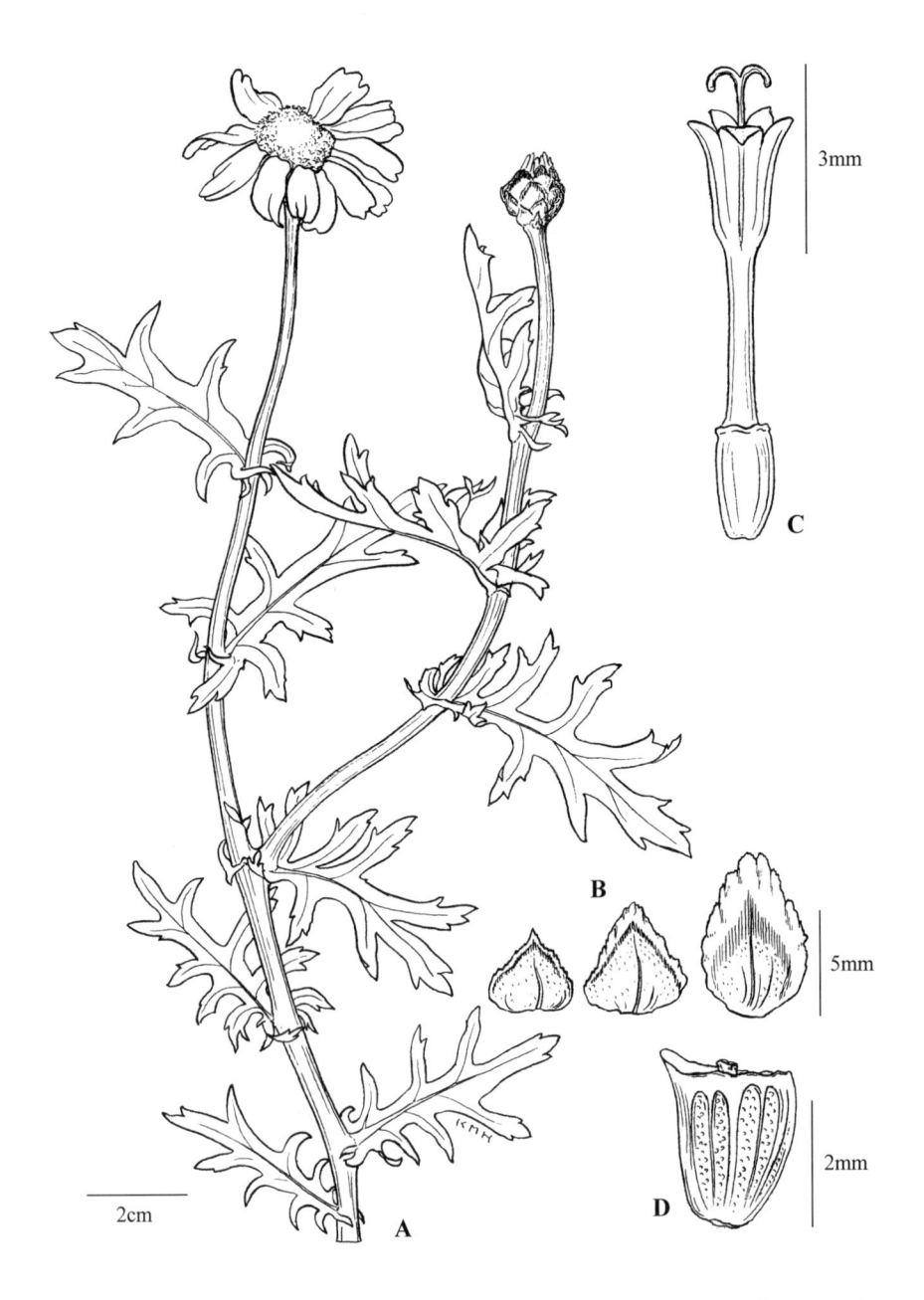

Chrysanthemum coronarium L. (Crown Daisy). A habit; B phyllaries; C disc floret; D achene. (Newton Abbot, S Devon). STACE 865/735.

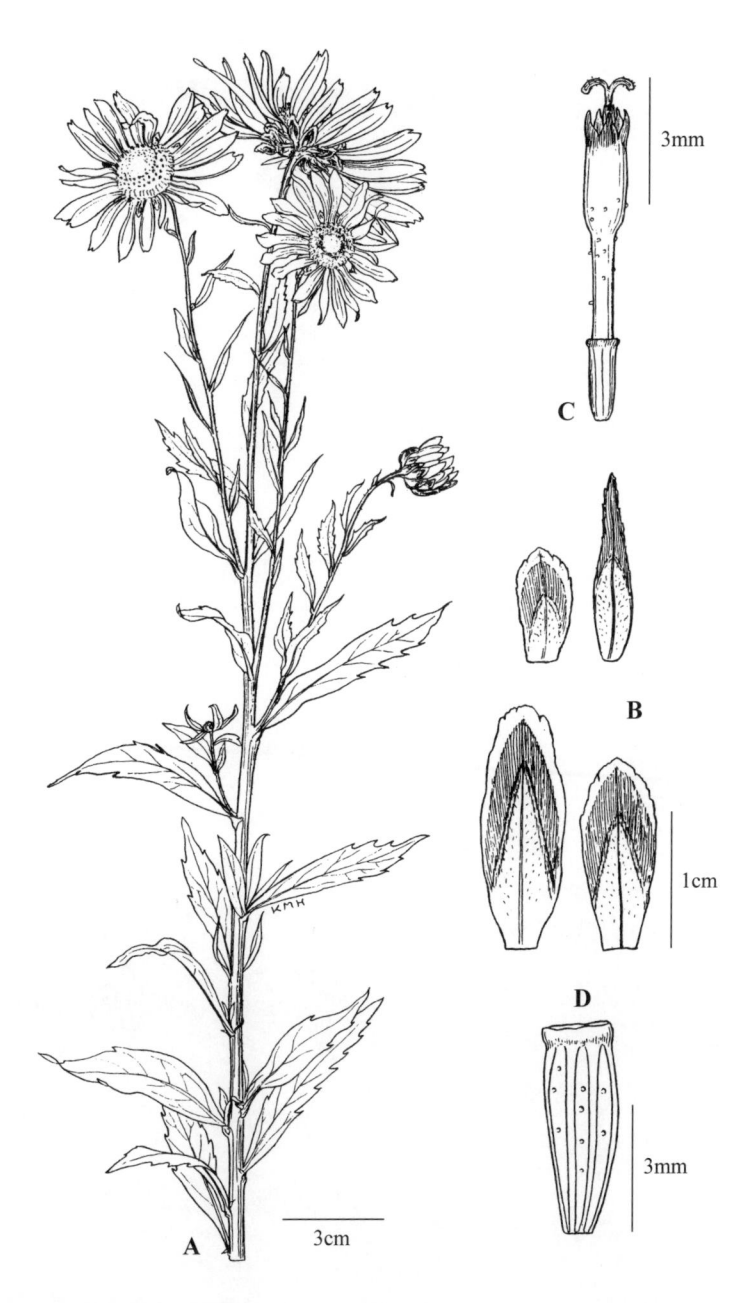

Leucanthemella serotina (L.) Tzvelev (Autumn Oxeye). A habit; B phyllaries; C disc floret; D achene. (*Hort*. Ashbourne, Derbys). STACE 867/735.

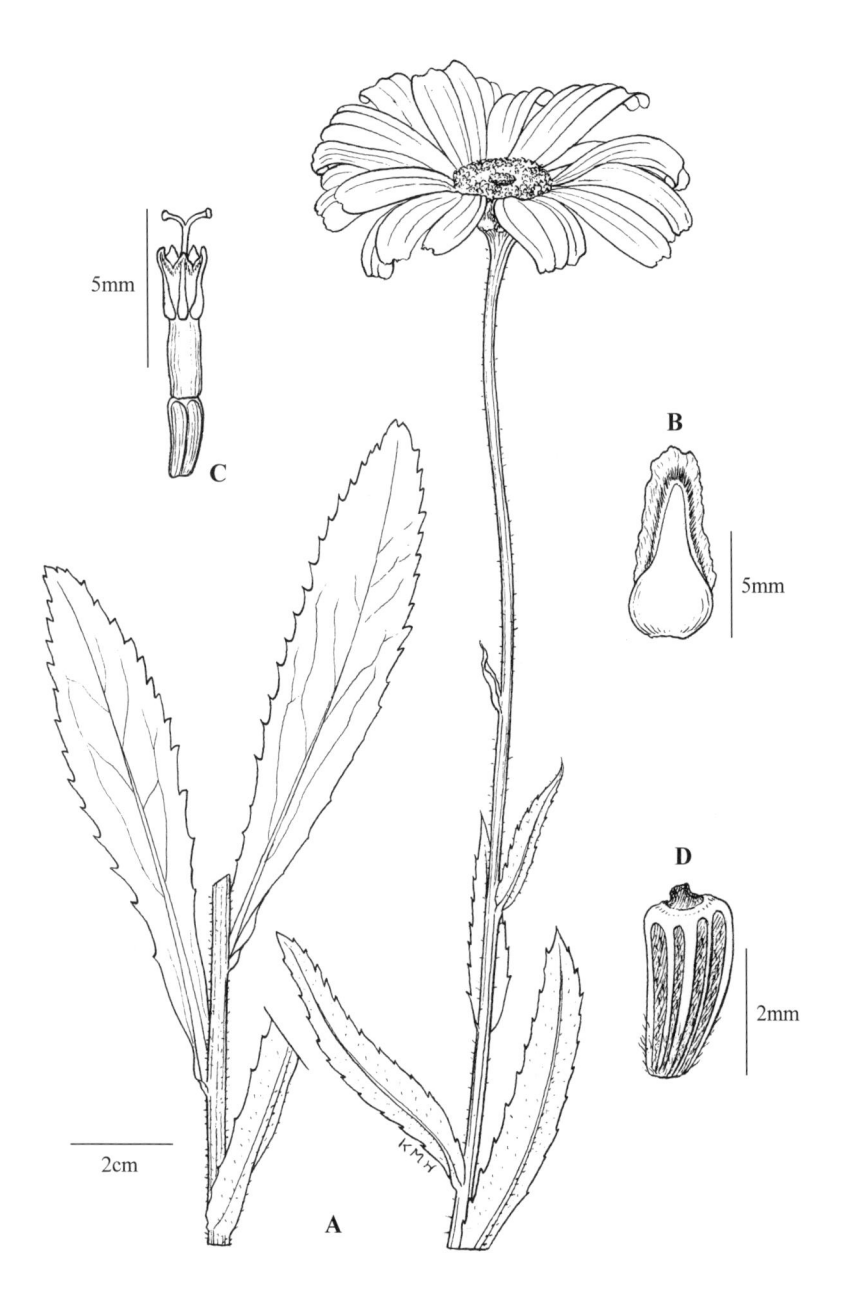

Leucanthemum x superbum (Bergmans ex J. W. Ingram) D. H. Kent (Shasta Daisy). A habit; B phyllary; C disc floret; D achene. (Mells, N Somerset). STACE 867/735.

139. ASTERACEAE

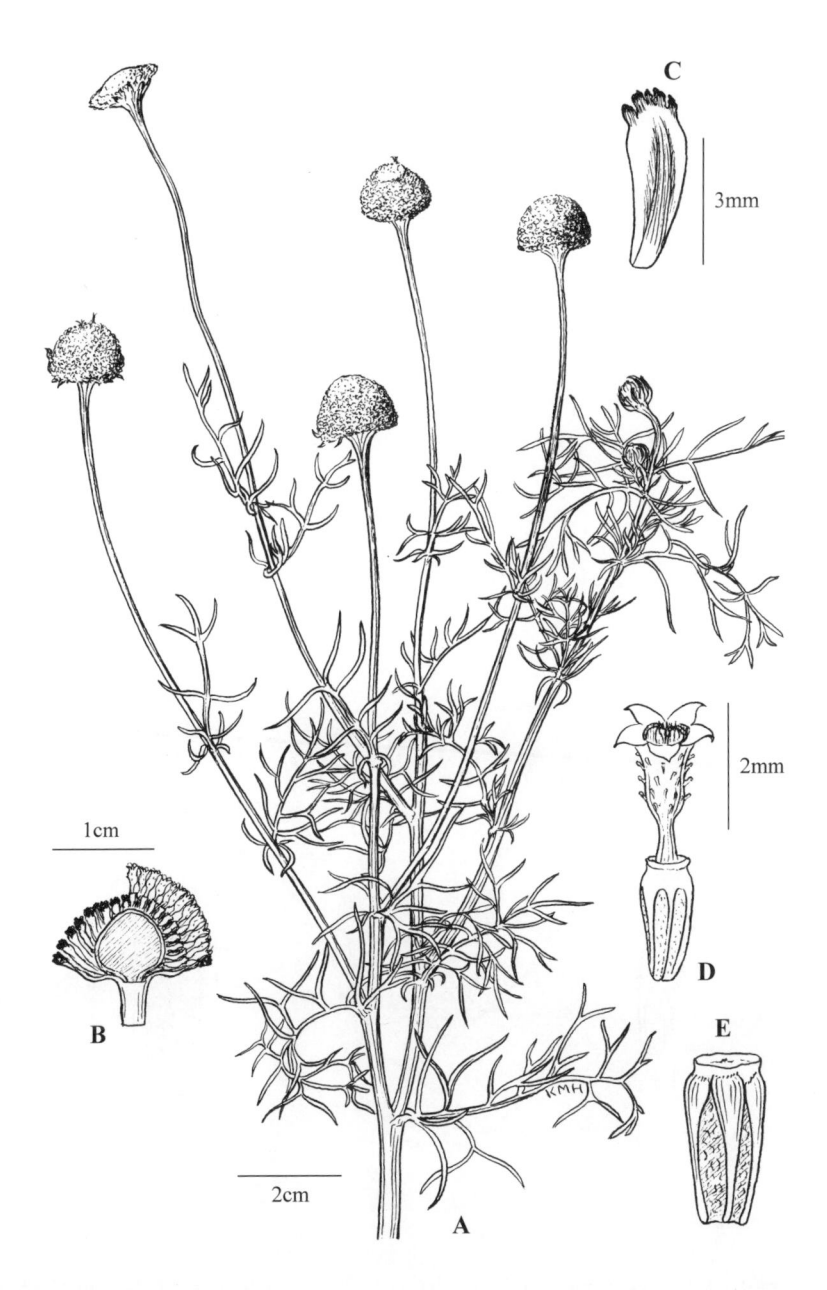

Tripleurospermum decipiens (Fisch. & C. A. Mey.) Bornm. (No English name). A habit; B LS of capitulum; C phyllary; D floret; E achene. (Grimsby, N Lincs). STACE —/736.

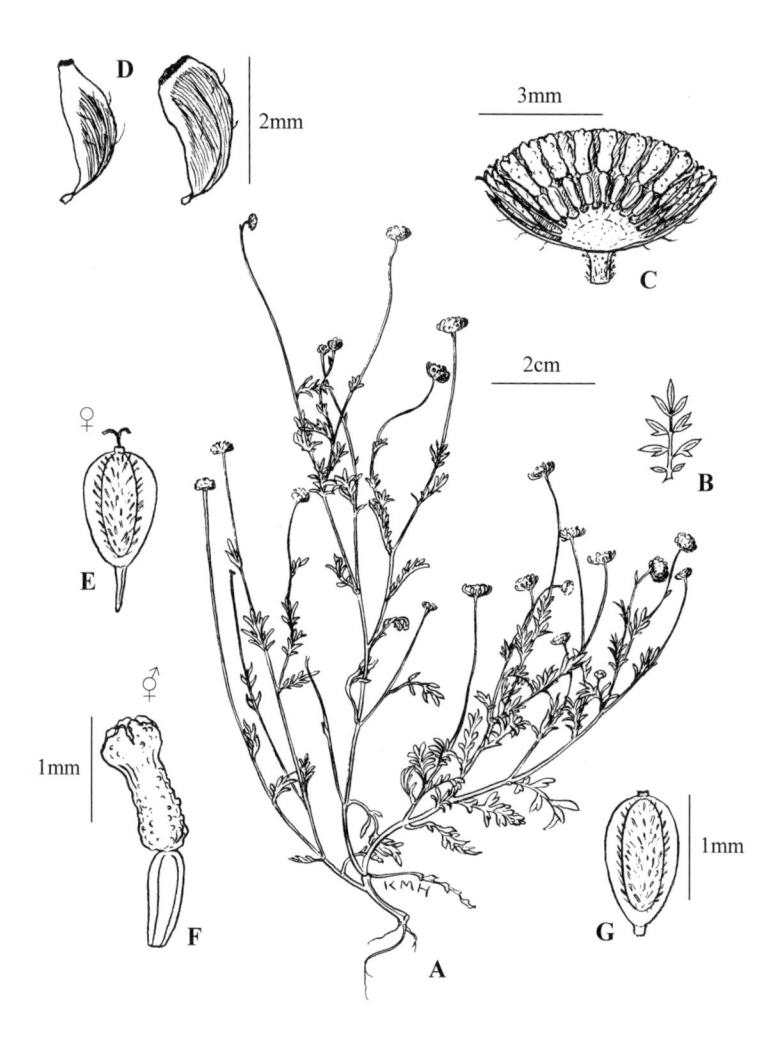

Cotula australis (Sieber ex Spreng.) Hook. f. (Annual Buttonweed). A habit; B leaf; C LS of capitulum; D phyllaries; E outer floret and achene; F inner floret and achene; G outer achene. (Newton Abbot, S Devon). STACE 869/737.

139. ASTERACEAE

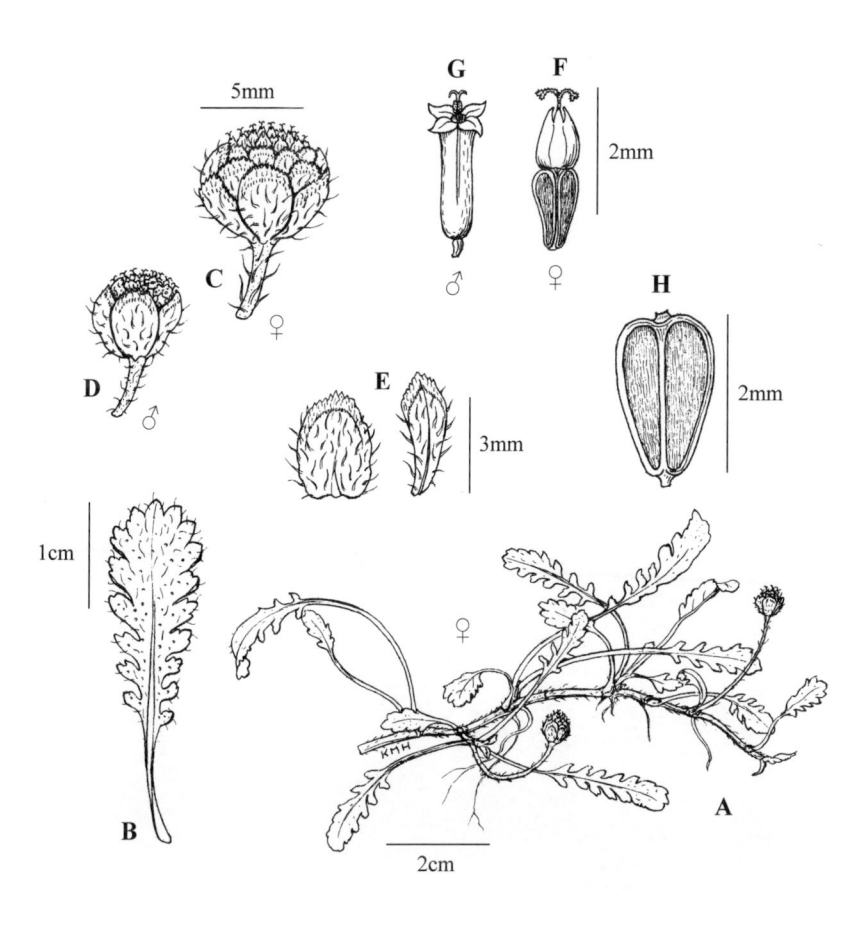

Cotula dioica (Hook. f.) Hook. f. (Hairless Leptinella). A habit of female plant; B leaf; C female capitulum; D male capitulum; E phyllaries; F female floret; G male floret; H achene. (A-C, E-F, H, Pontefract Castle, SW Yorks; D, G, Source unknown). STACE 869/737.

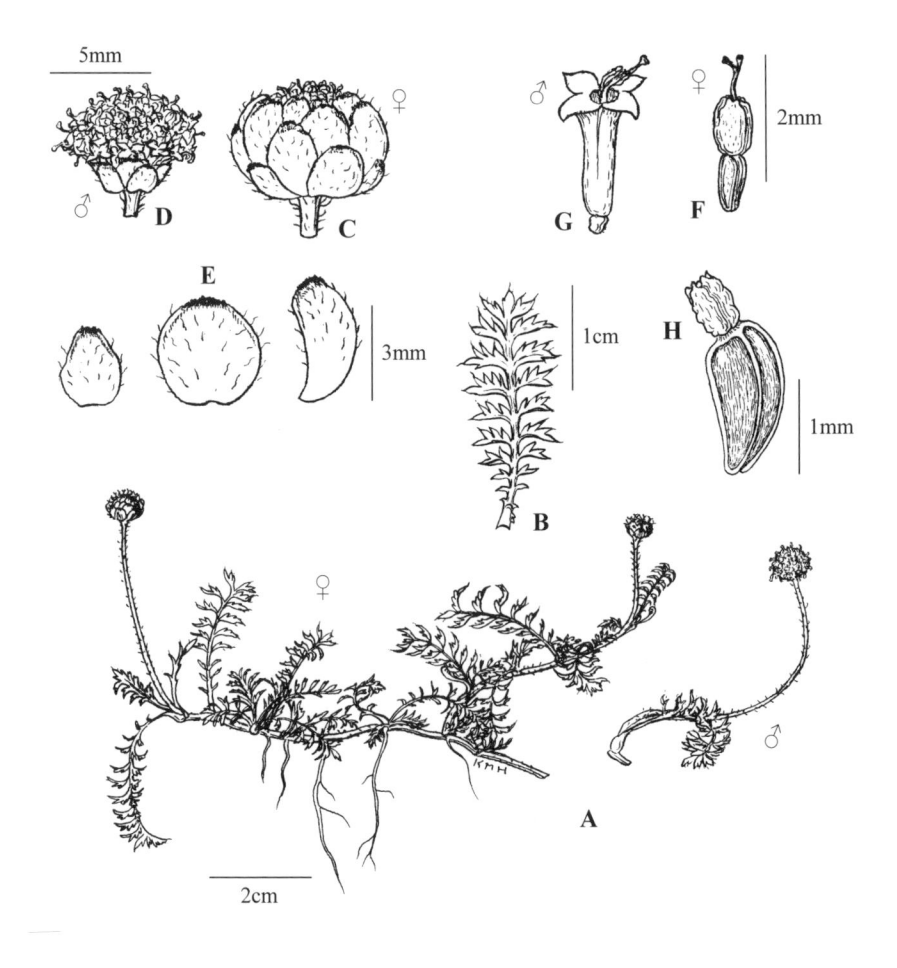

Cotula squalida (Hook. f.) Hook. f. (Leptinella). A habit of male and female plants; B leaf; C female capitulum; D male capitulum; E phyllaries; F female floret; G male floret; H achene. (*Hort*. Leicester University Botanical Garden, Leics). STACE 869/737.

139. ASTERACEAE

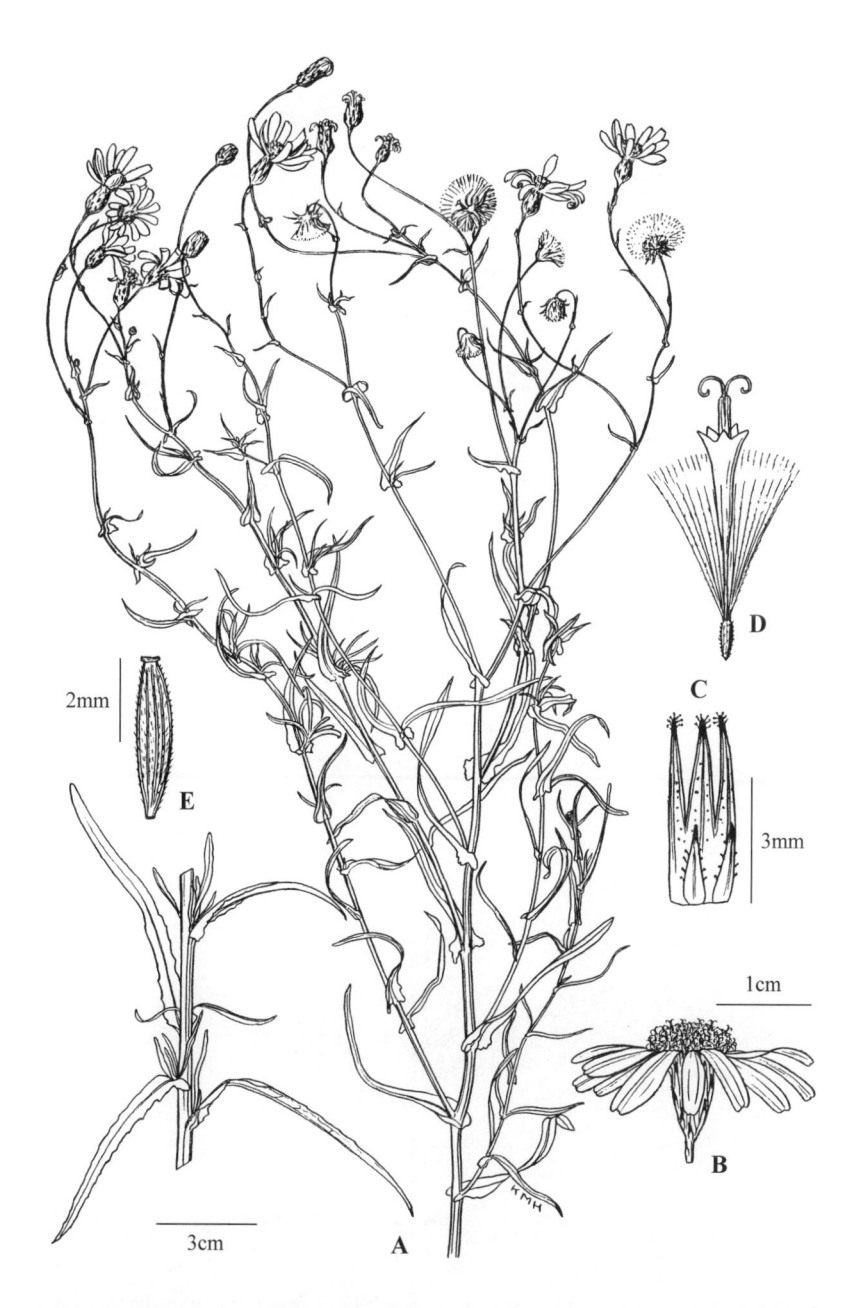

Senecio inaequidens DC. (Narrow-leaved Ragwort). A habit; B capitulum; C phyllaries; D disc floret; E achene. (*Hort.* ex Biggleswade, Beds). STACE 872/739.

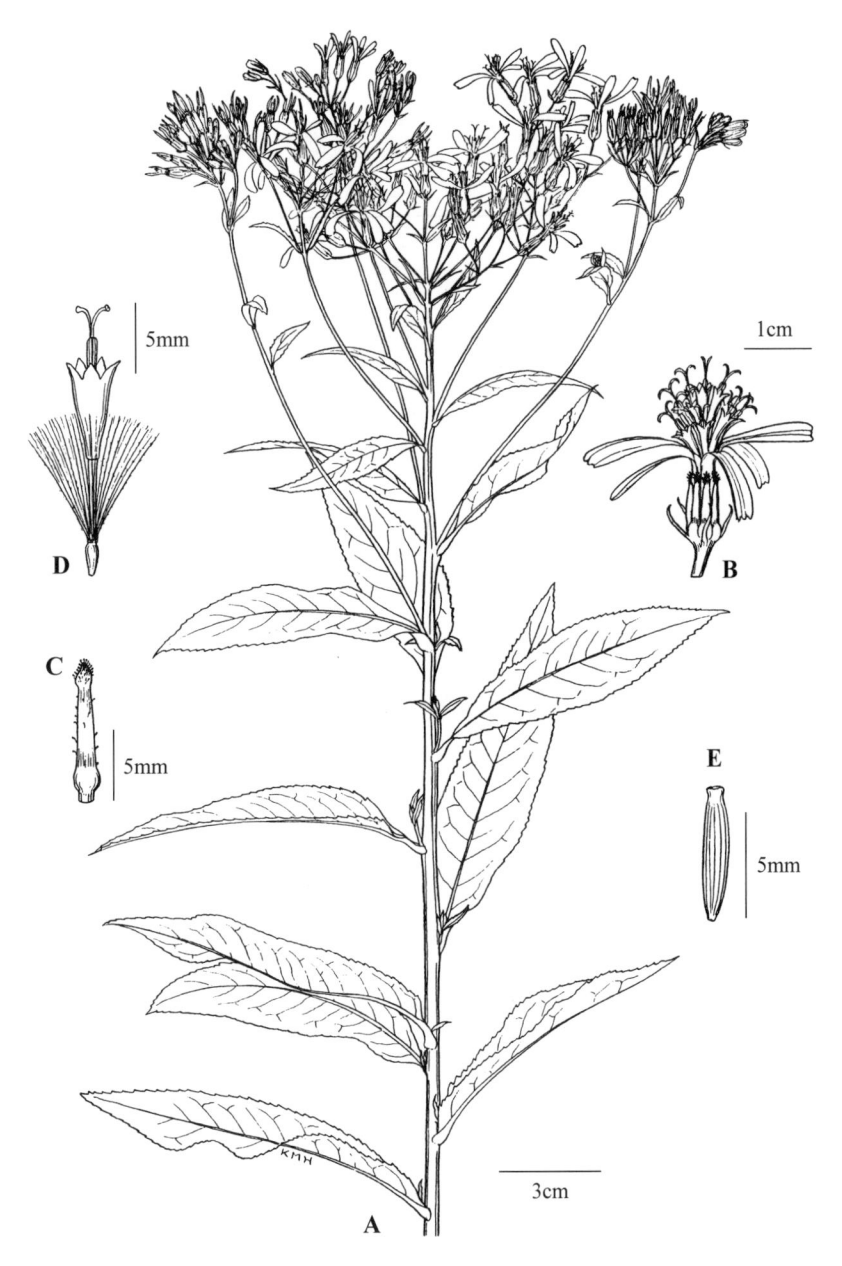

Senecio ovatus (P. Gaertn., B. Mey. & Scherb.) Willd. ssp. **alpestris** (Gaudin) Herborg (Wood Ragwort). A habit; B capitulum; C phyllary; D disc floret; E achene. (*Hort.* ex Trough of Bowland, MW Yorks). STACE 872/739.

139. ASTERACEAE

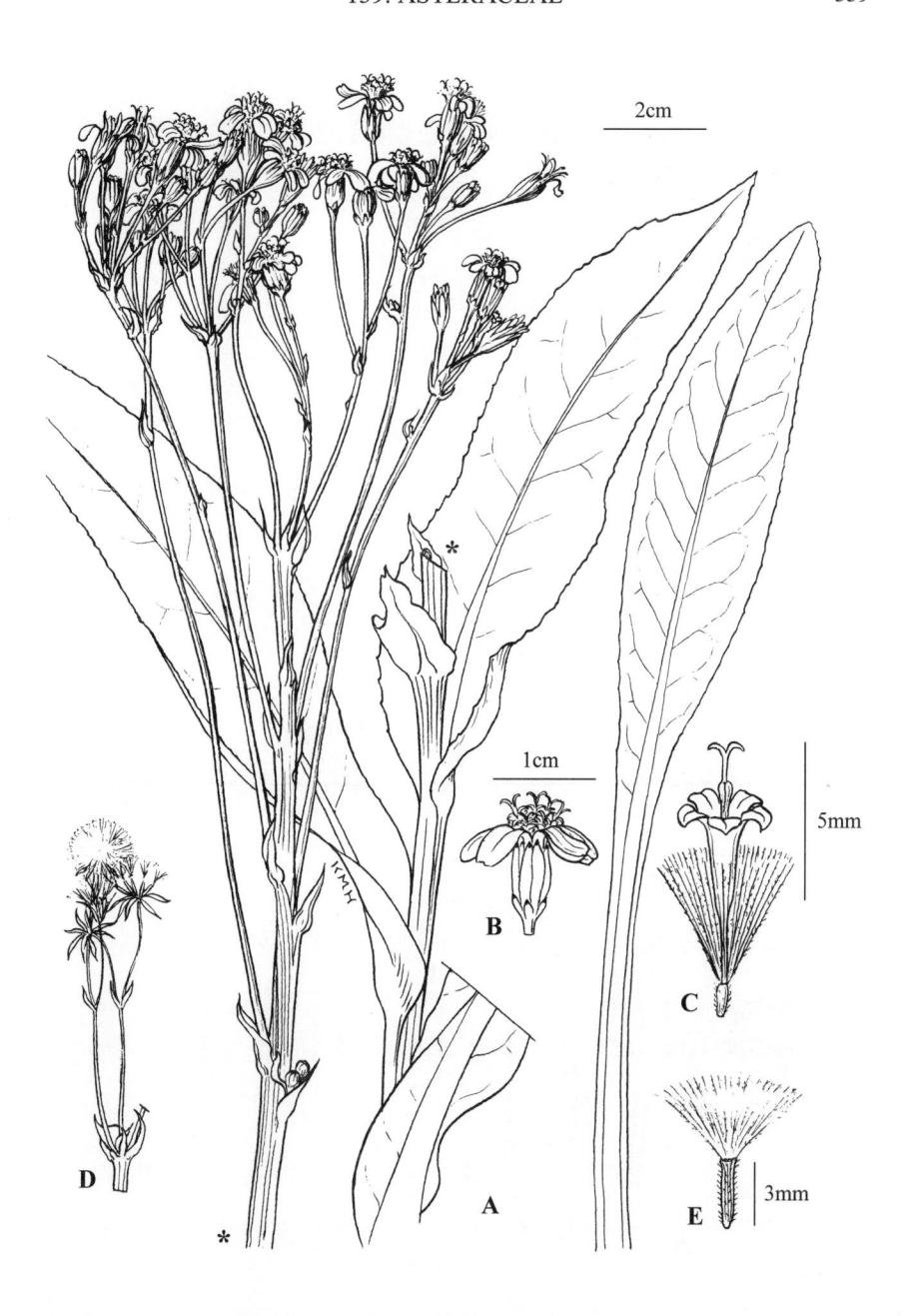

Senecio doria L. (Golden Ragwort). A habit and basal leaf; B capitulum; C disc floret; D fruiting stem; E achene. (*Hort*. Sevenoaks, W Kent). STACE 872/739.

340 139. ASTERACEAE

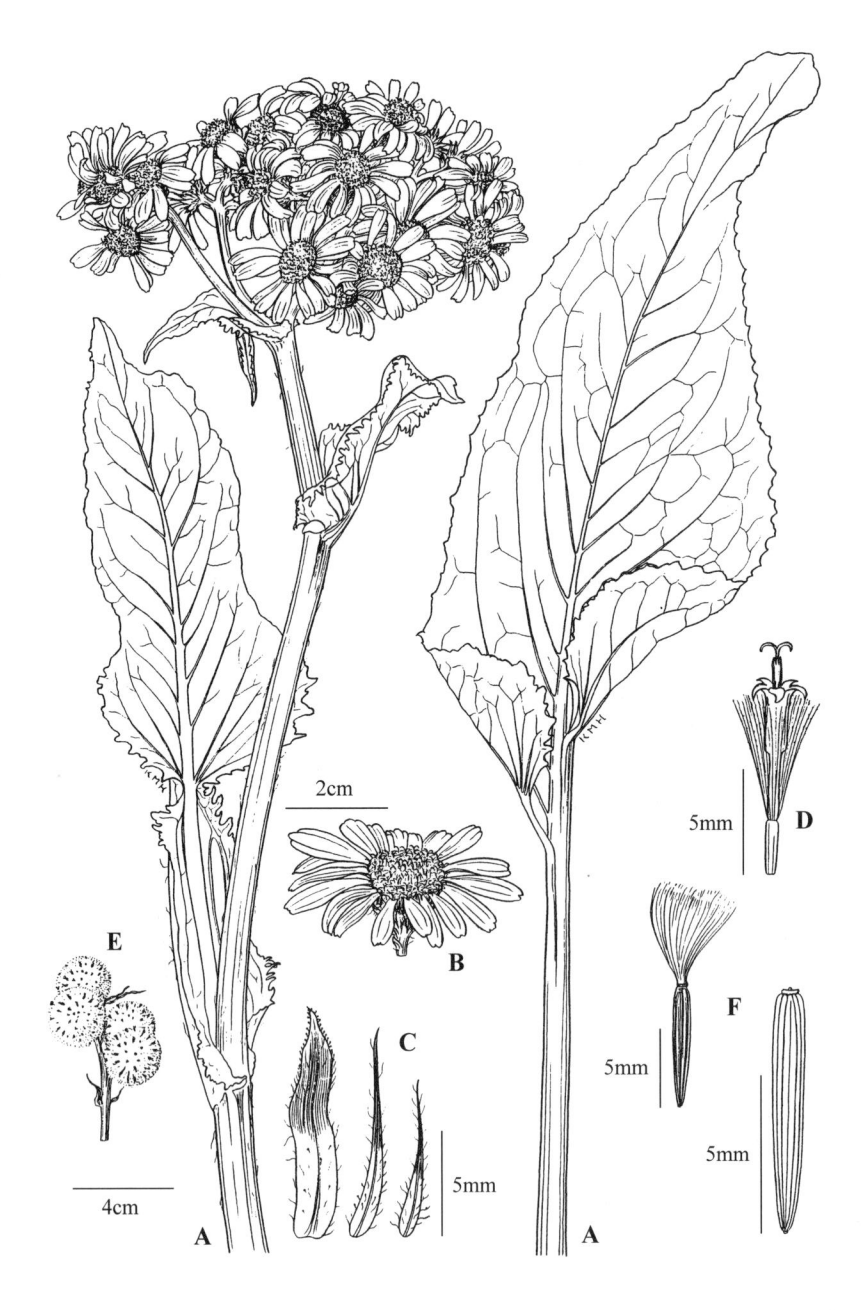

Senecio smithii DC. (Magellan Ragwort). A habit; B capitulum; C phyllaries; D disc floret; E fruiting stem; F achene. (Orkney). STACE 873/740.

139. ASTERACEAE

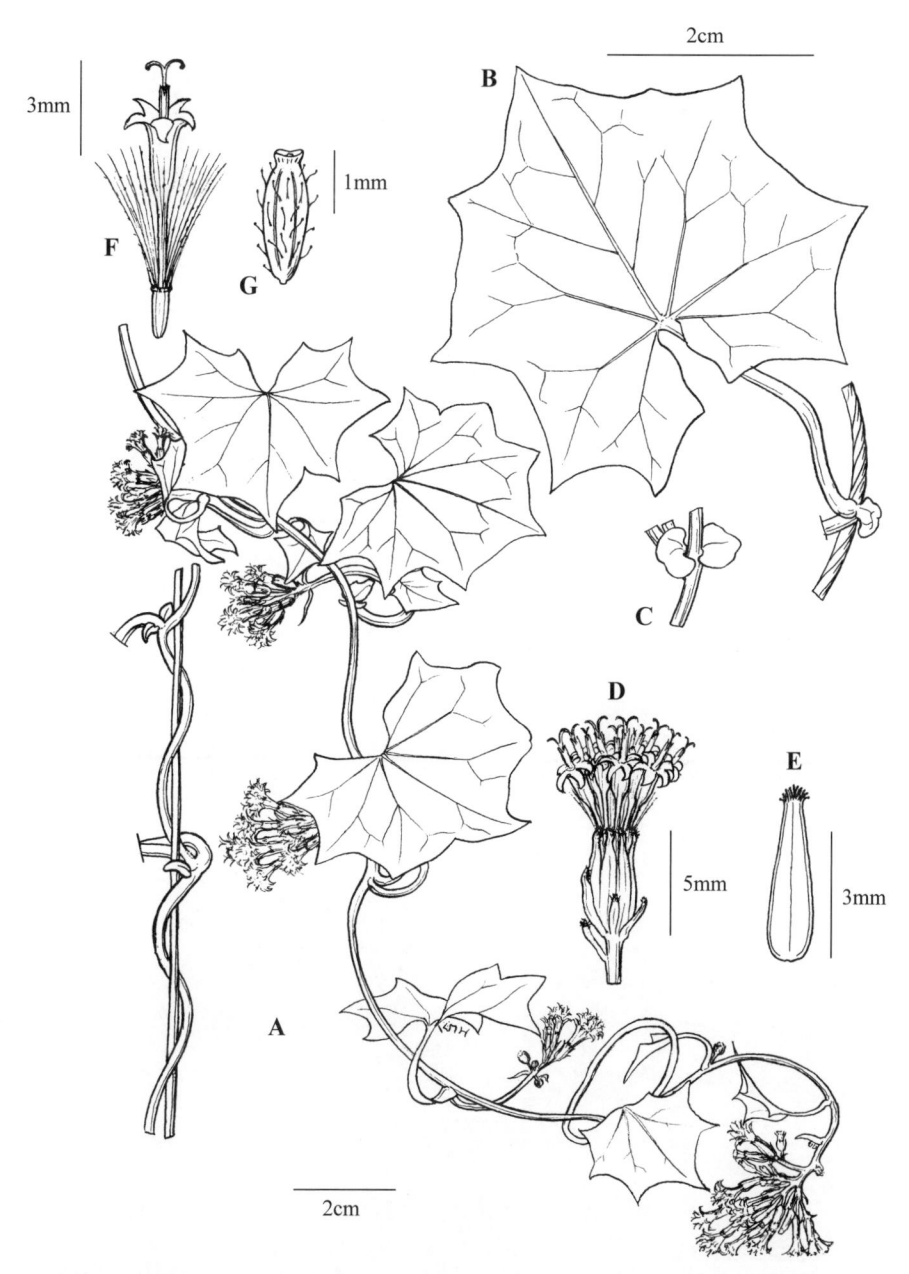

Delairea odorata Lem. (German-ivy). A habit; B lower stem leaf; C auricles at base of leaf on non-flowering shoot; D capitulum; E phyllary; F floret; G achene. (St Aubyn, Jersey). STACE 877/742.

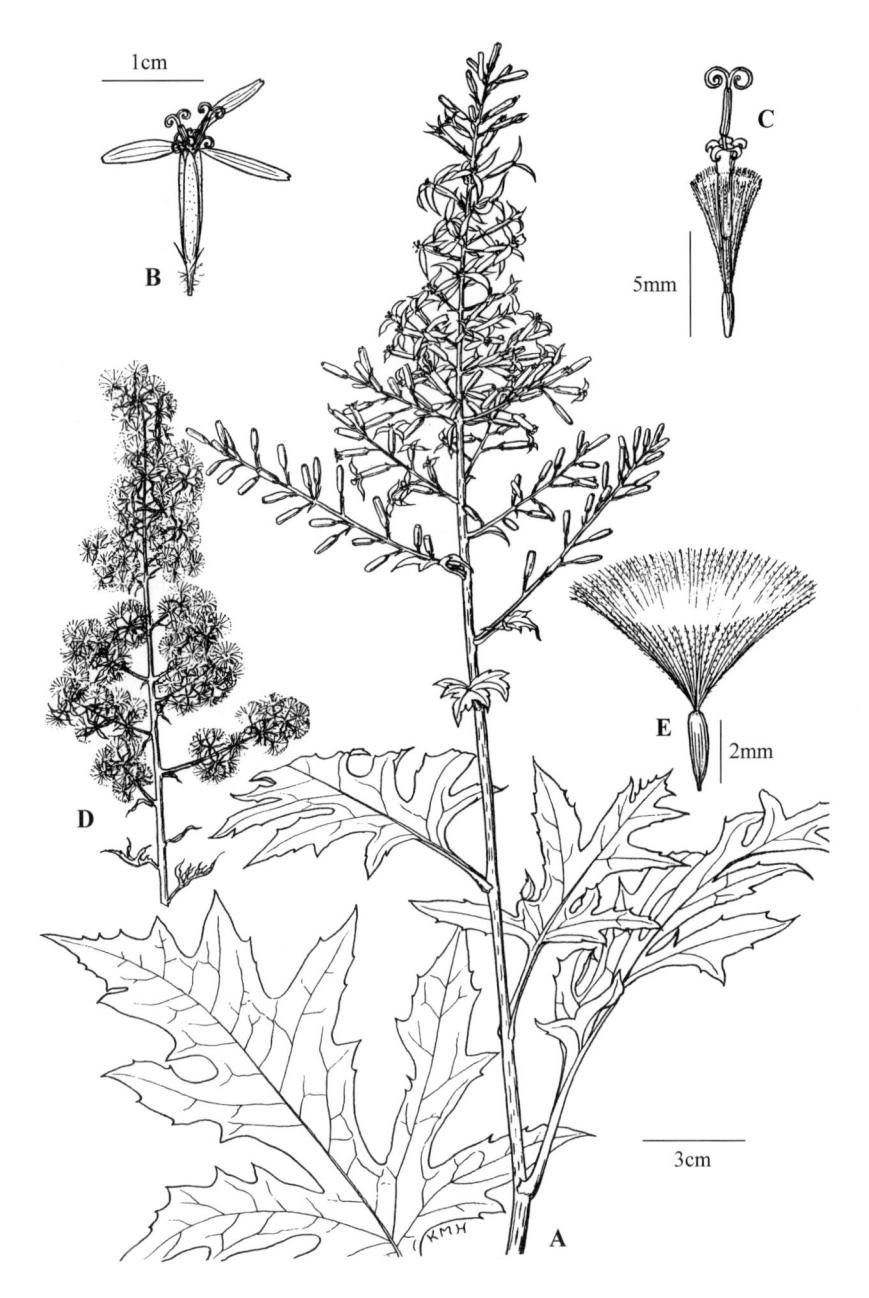

Sinacalia tangutica (Maxim.) B. Nord. (Chinese Ragwort). A habit; B capitulum; C disc floret; D fruiting panicle; E achene. (*Hort*. Ashbourne, Derbys). STACE 879/745.

139. ASTERACEAE

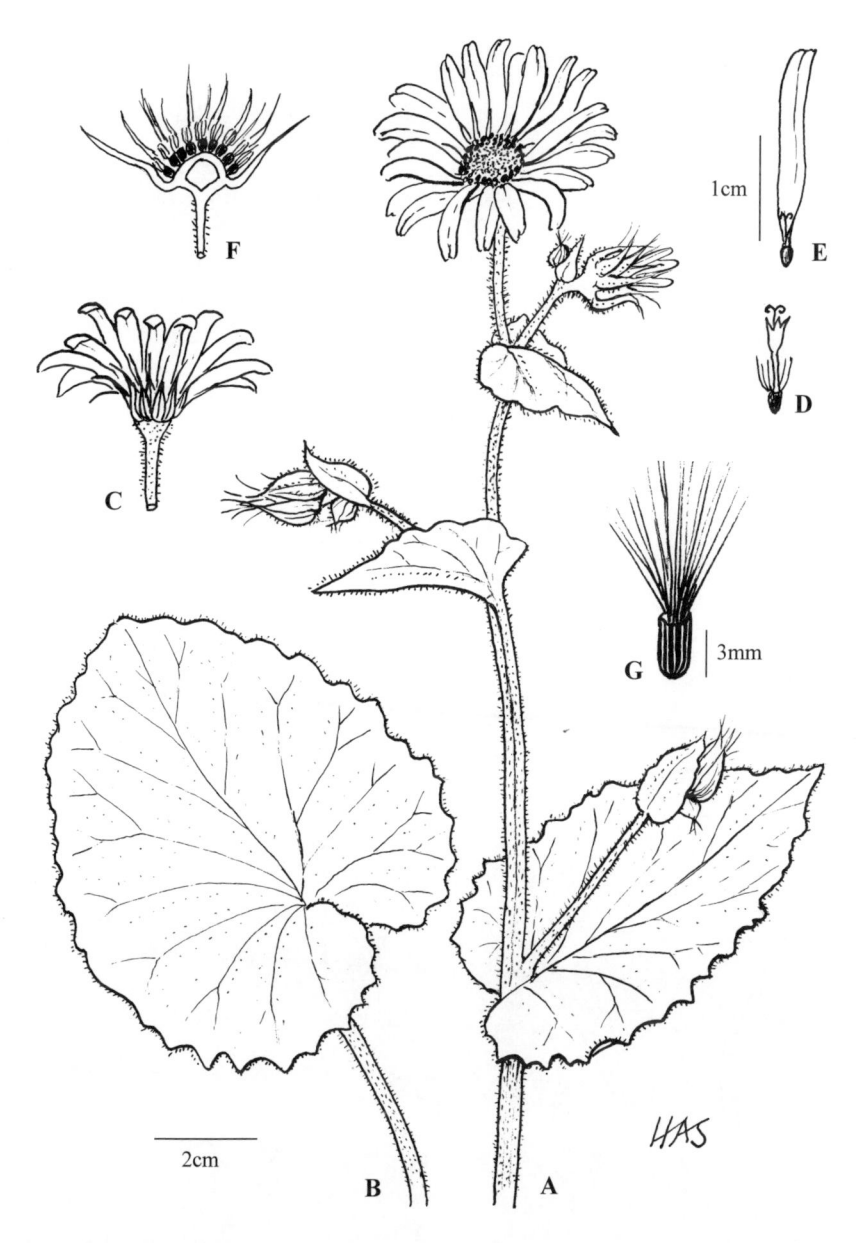

Doronicum pardalianches L. (Leopard's-bane). A flowering stem; B basal leaf; C capitulum; D disc floret; E ray floret; F TS of capitulum G achene. (Aberdeen, S Aberdeen). STACE 880/745.

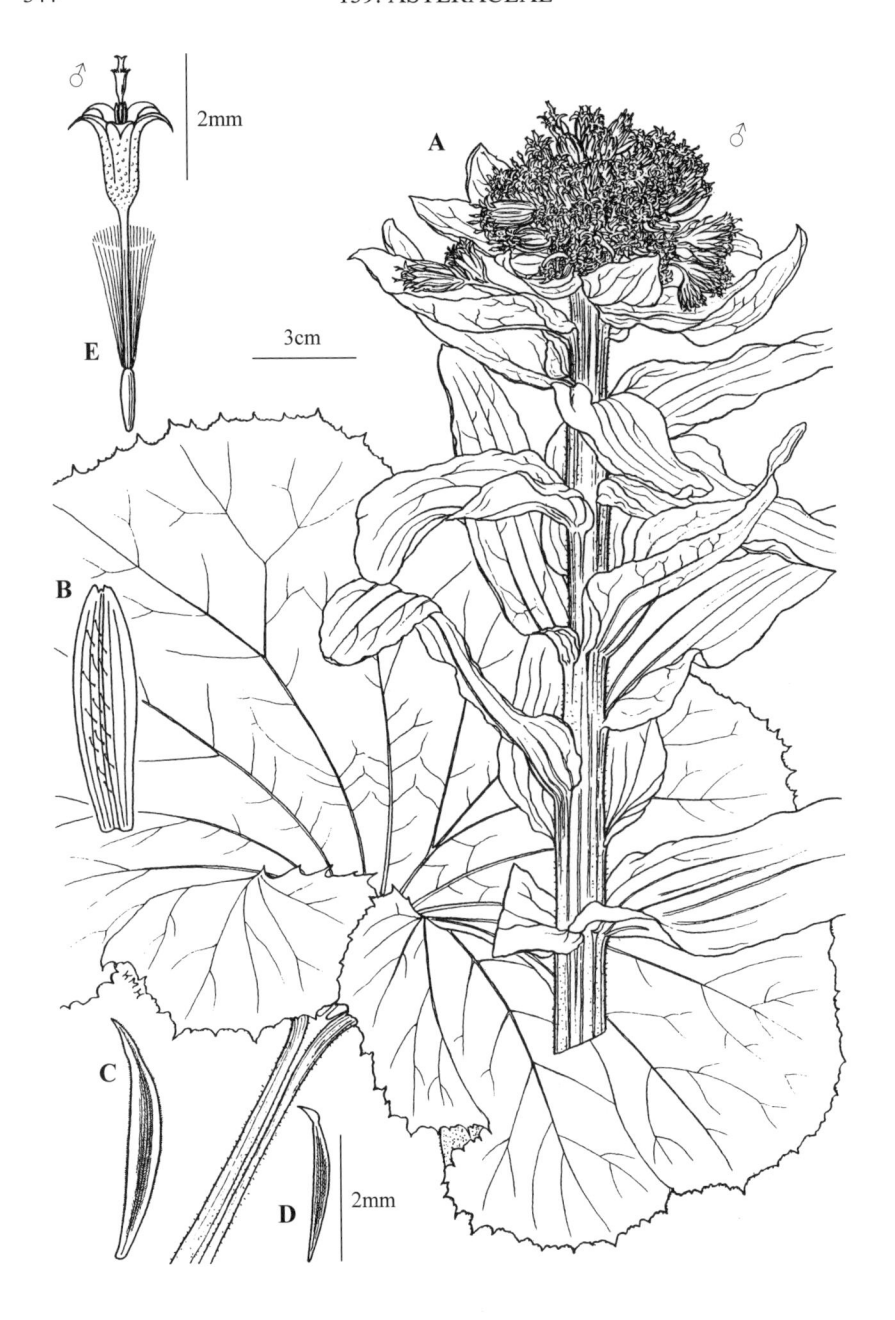

Petasites japonicus (Siebold & Zucc.) Maxim. (Giant Butterbur, male plant). A habit; B bracteole; C outer phyllary; D inner phyllary; E floret. (Groombridge, E Sussex). STACE 881/747.

139. ASTERACEAE 345

Calendula officinalis L. (Pot Marigold). A habit; B capitulum in fruit; C achene. (Greenhithe, W Kent). STACE 882/747.

Calendula arvensis L. (Field Marigold). A habit; B capitulum; C fruiting stem; D outer and inner achenes. (*Hort.* ex Brittany, France). STACE 882/747.

139. ASTERACEAE

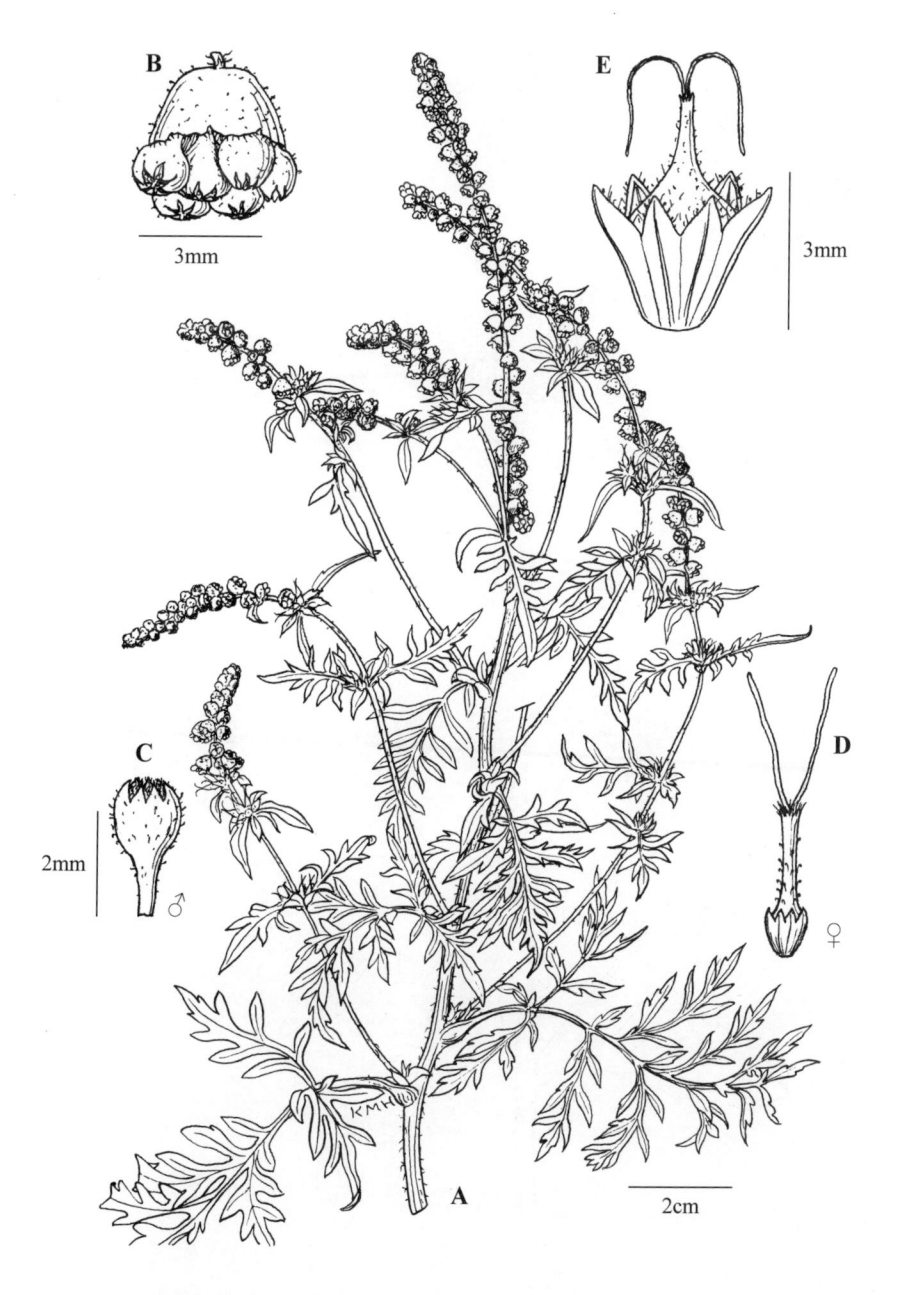

Ambrosia artemisiifolia L. (Ragweed). A habit; B male capitulum; C male floret; D female floret; E young fruit. (Stretford, S Lancs). STACE 883/748.

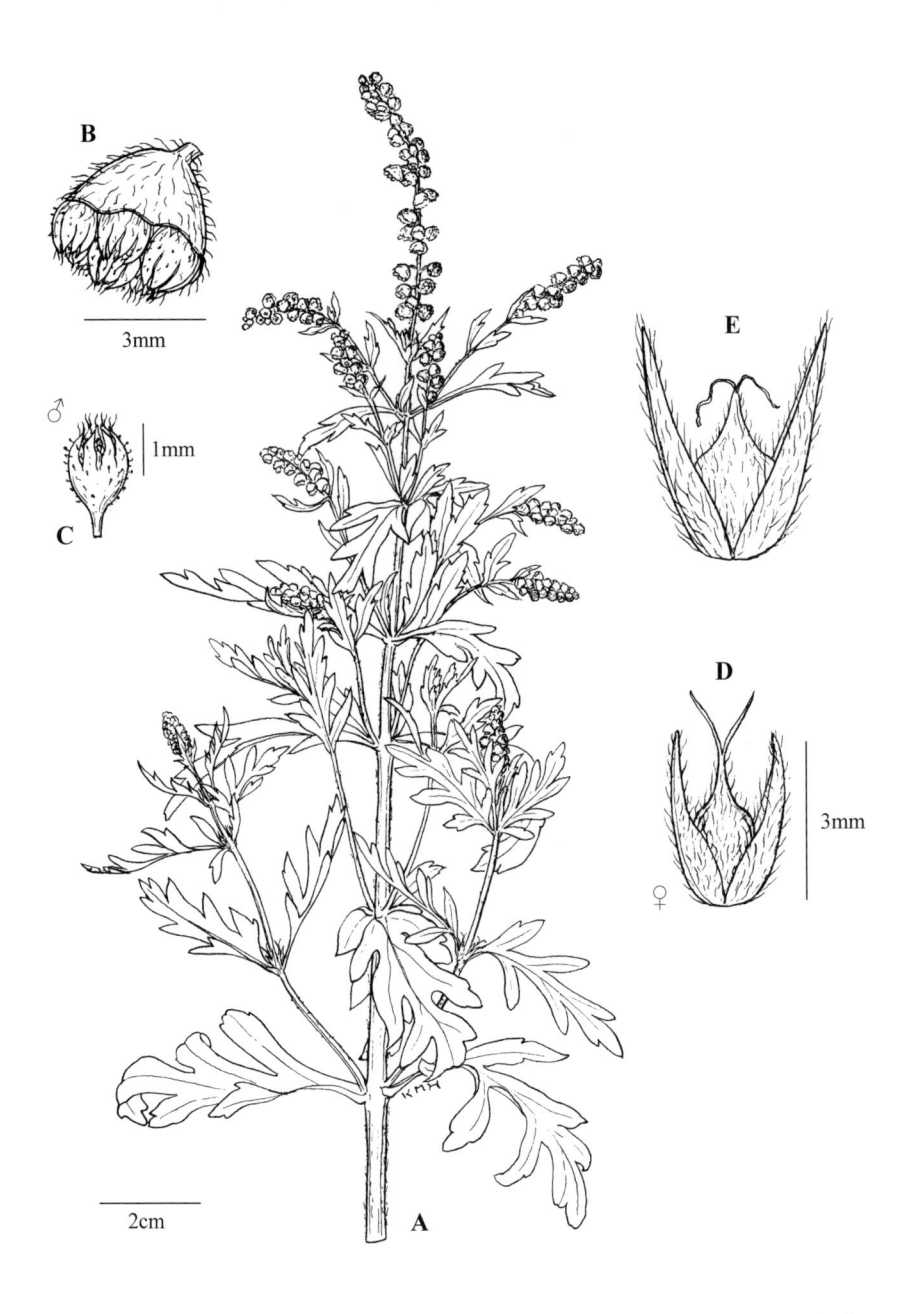

Ambrosia psilostachya DC. (Perennial Ragweed). A habit; B male capitulum; C male floret; D female floret; E young fruit. (St Anne's, W Lancs). STACE 883/748.

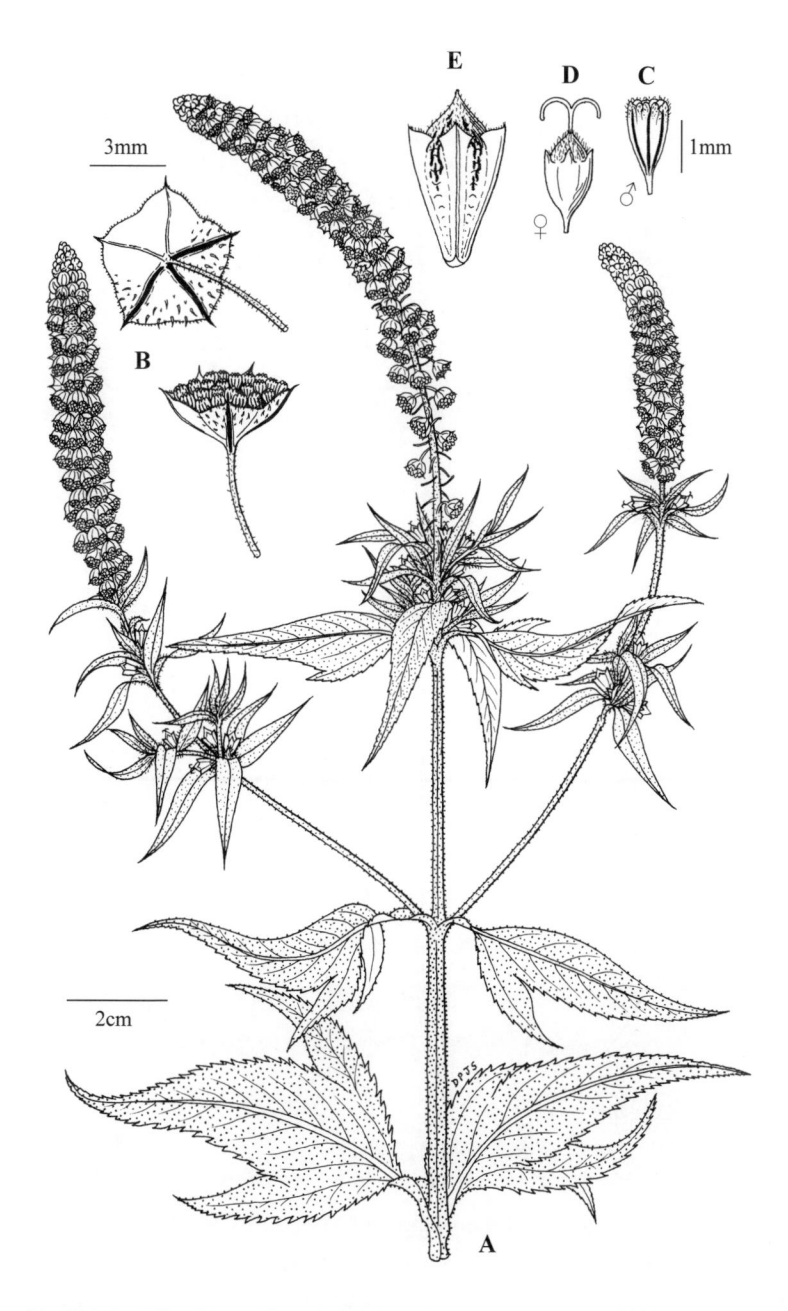

139. ASTERACEAE

Ambrosia trifida L. (Giant Ragweed). A habit; B male capitulum; C male floret; D female floret; E achene. (*Hort*. Ware, Herts). STACE 883/748.

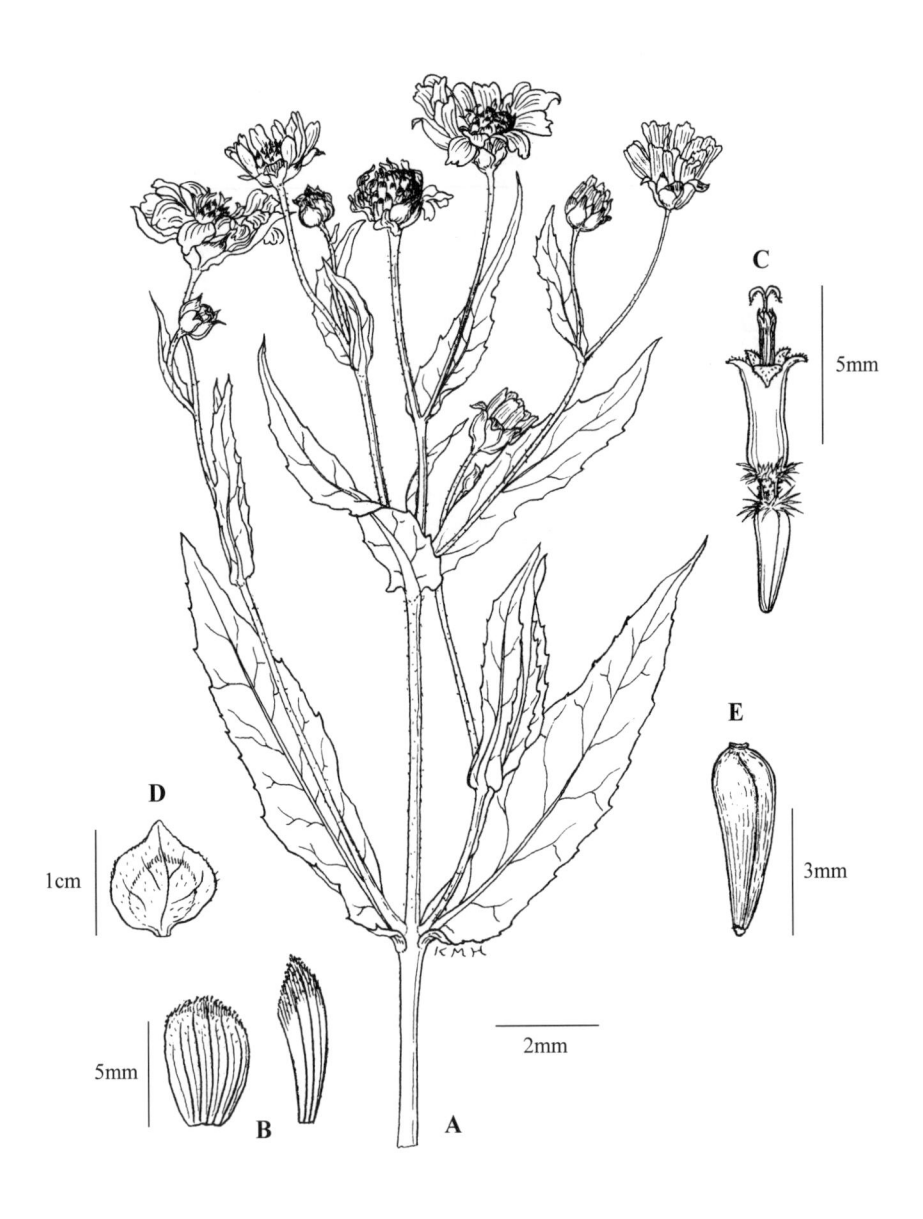

Guizotia abyssinica (L. f.) Cass. (Niger). A habit; B phyllaries; C disc floret; D receptacular scale; E achene. (*Hort.* Ashbourne, Derbys). STACE 886/750.

139. ASTERACEAE

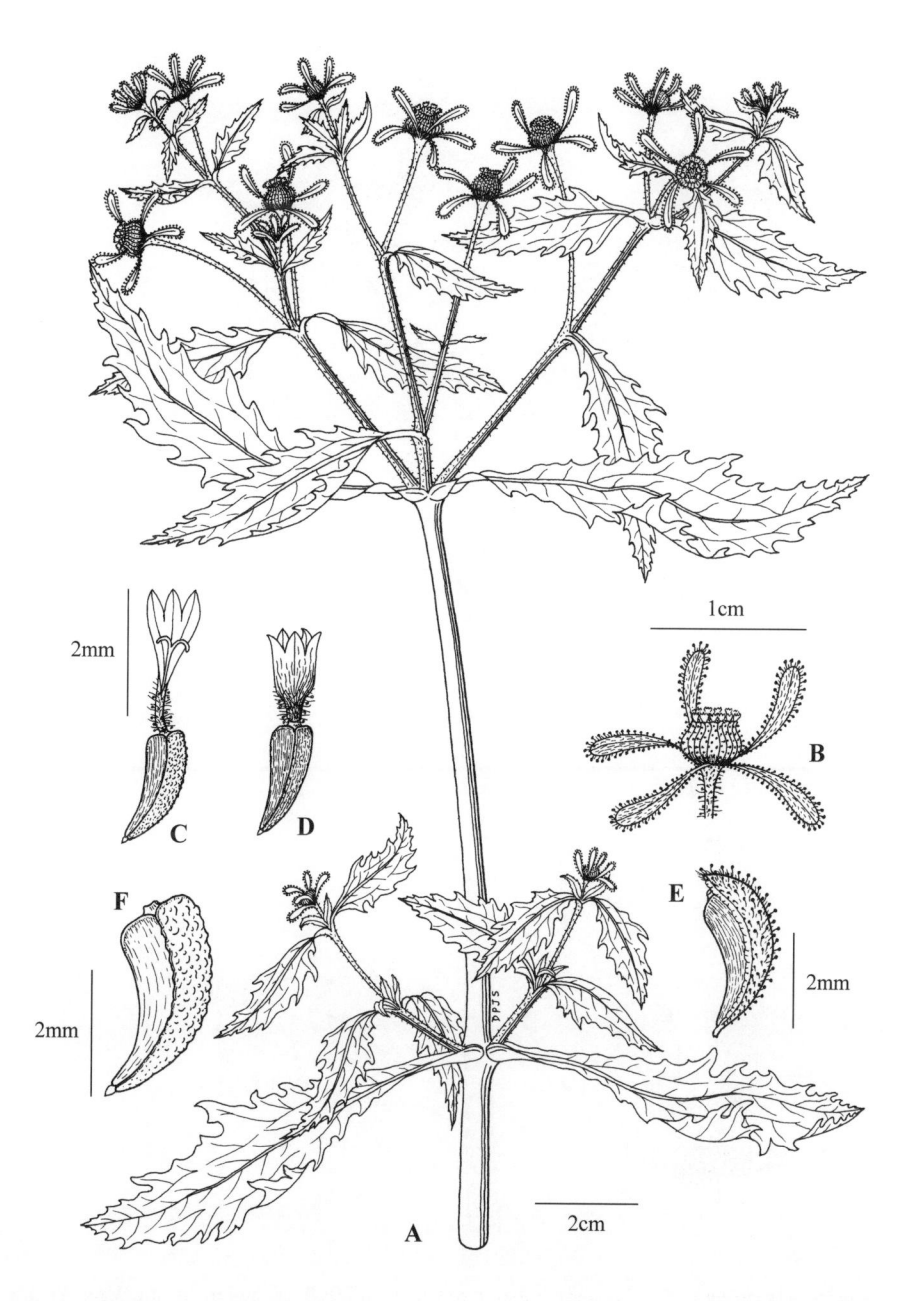

Sigesbeckia orientalis L. (Eastern St Paul's-wort). A habit; B capitulum; C ray floret; D disc floret; E ray achene with enveloping phyllary; F ray achene. (Maulden, Beds). STACE 886/751.

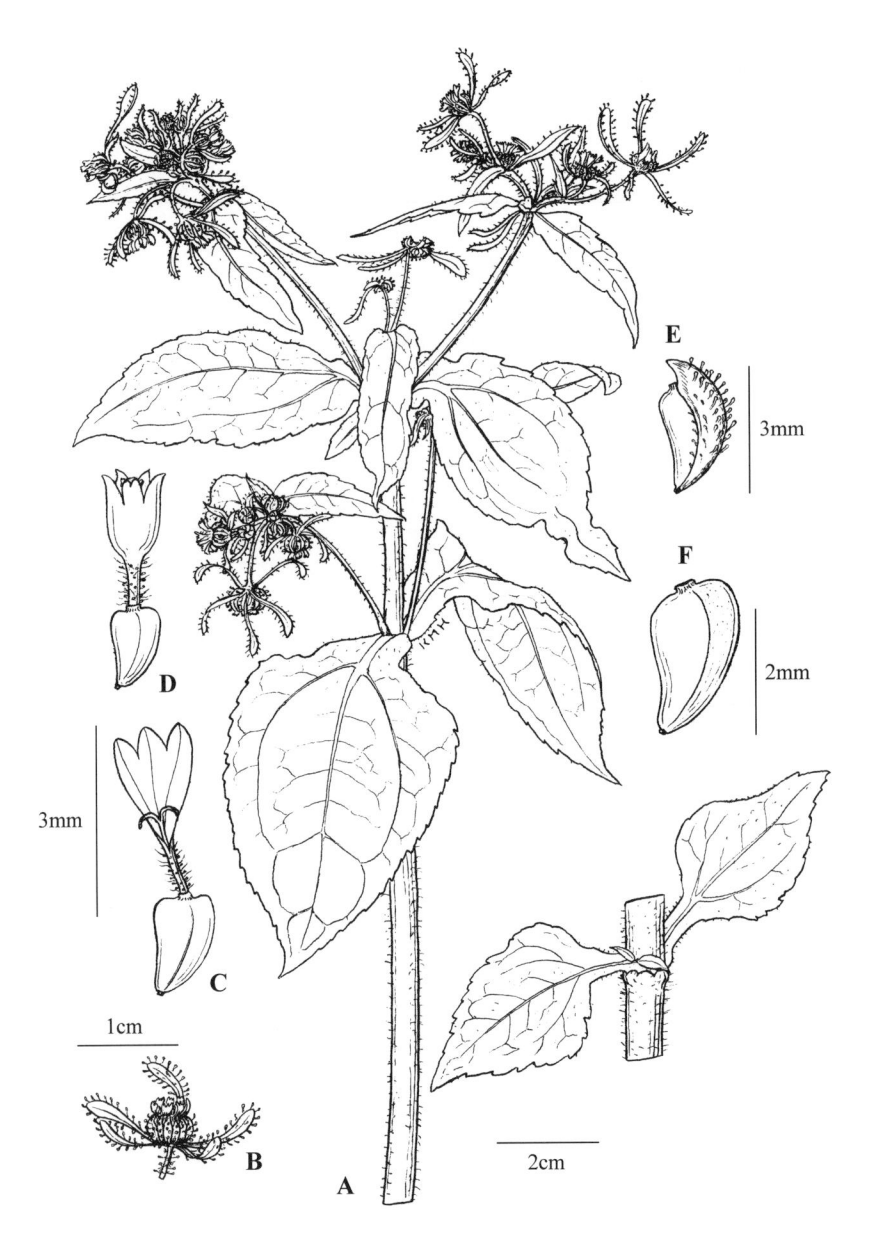

Sigesbeckia serrata DC. (Western St Paul's-wort). A habit; B capitulum; C ray floret; D disc floret; E ray achene with enveloping phyllary; F ray achene. (*Hort.* ex Freshfield, S Lancs). STACE 886/751.

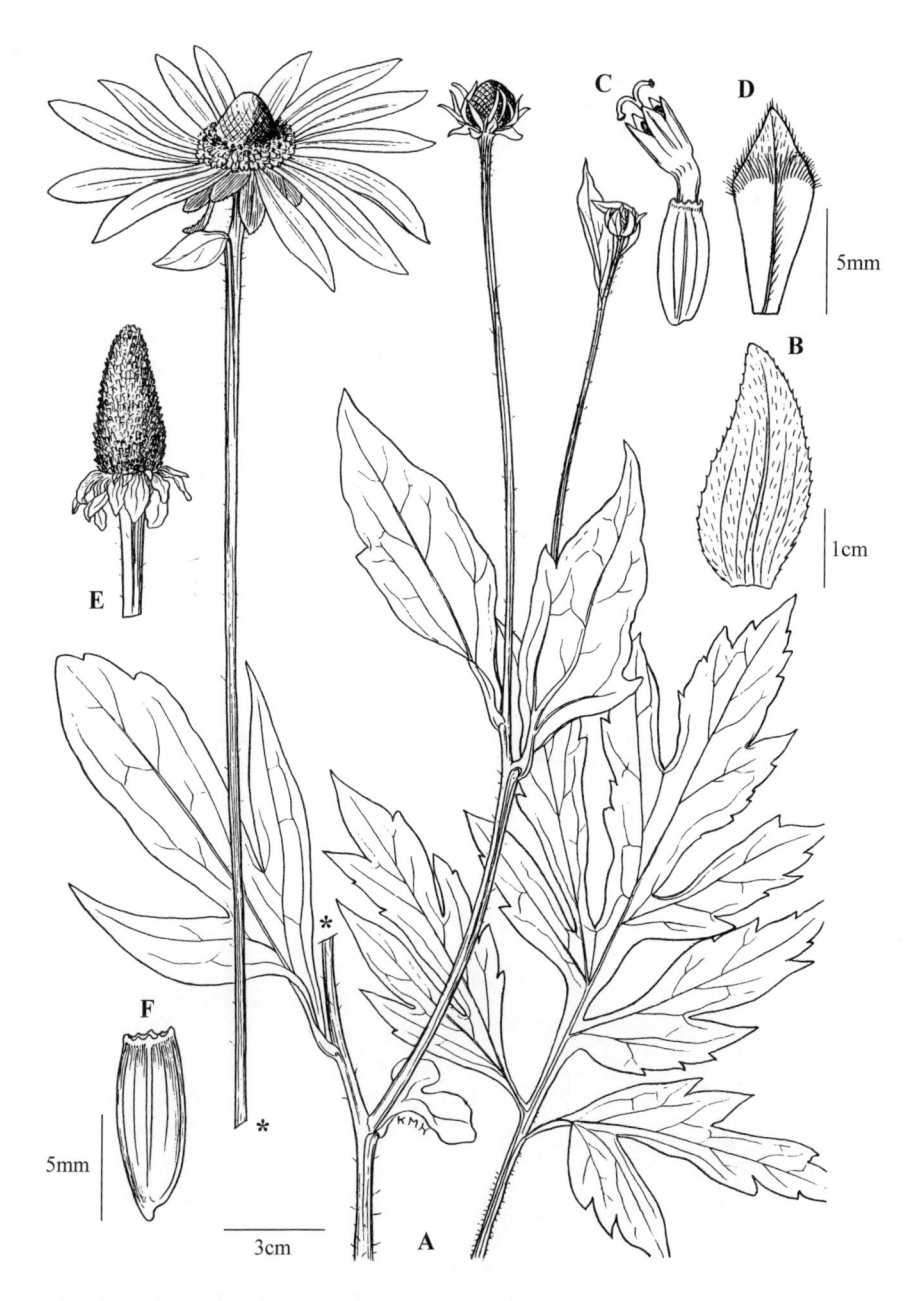

139. ASTERACEAE 353

Rudbeckia laciniata L. (Coneflower). A habit; B phyllary; C disc floret; D receptacular scale; E capitulum in fruit; F achene. (Friockheim, Angus). STACE 887/751.

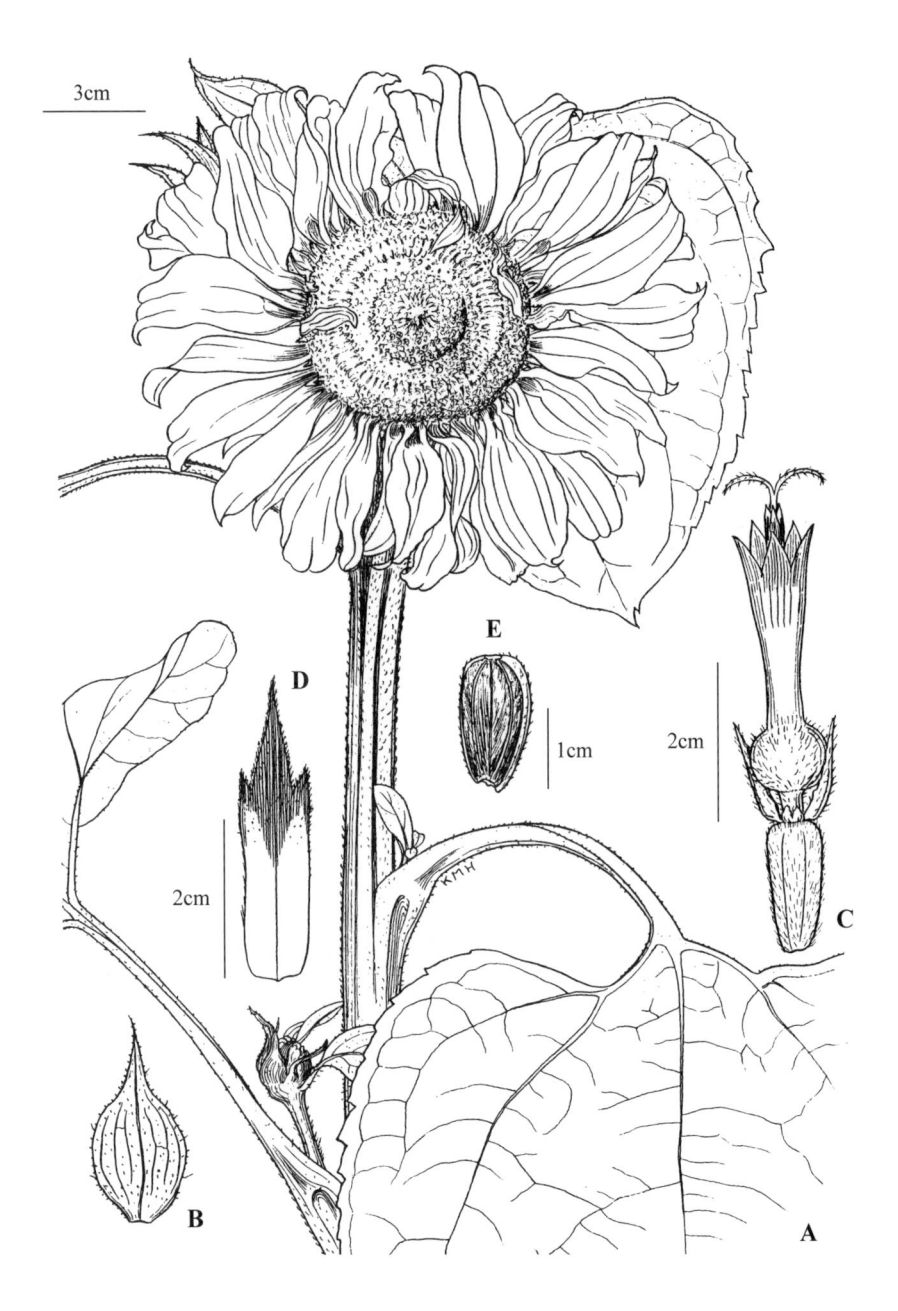

Helianthus annuus L. (Sunflower). A habit; B phyllary; C disc floret; D receptacular scale; E achene. (Tattershall, N Lincs). STACE 887/751.

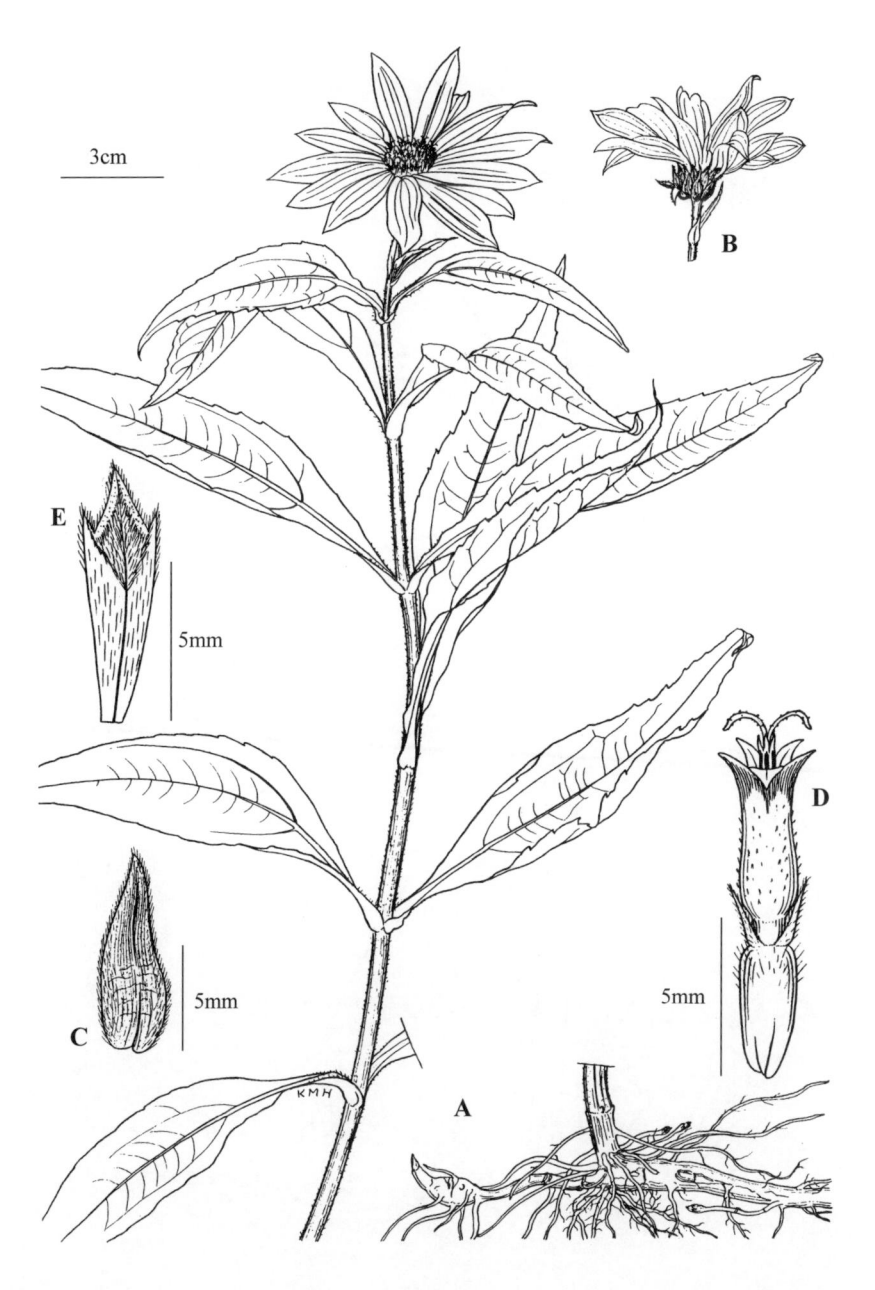

Helianthus x laetiflorus Pers. (Perennial Sunflower). A habit; B capitulum; C phyllary; D disc floret; E receptacular scale. (Banstead Downs, Surrey). STACE 888/753.

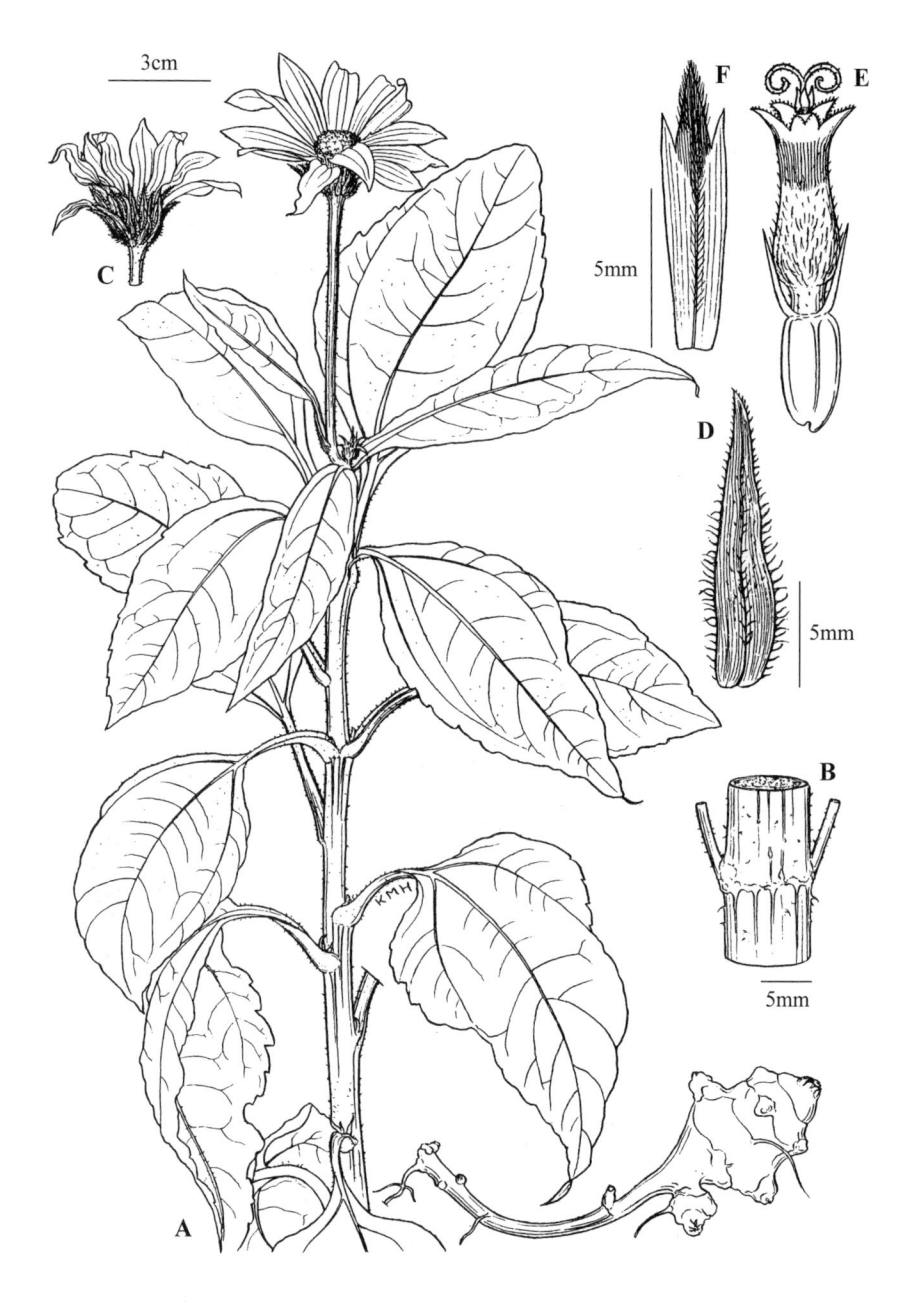

Helianthus tuberosus L. (Jerusalem Artichoke). A habit; B part of stem at node; C capitulum; D phyllary; E disc floret; F receptacular scale. (*Hort*. Ashbourne, Derbys). STACE 888/753.

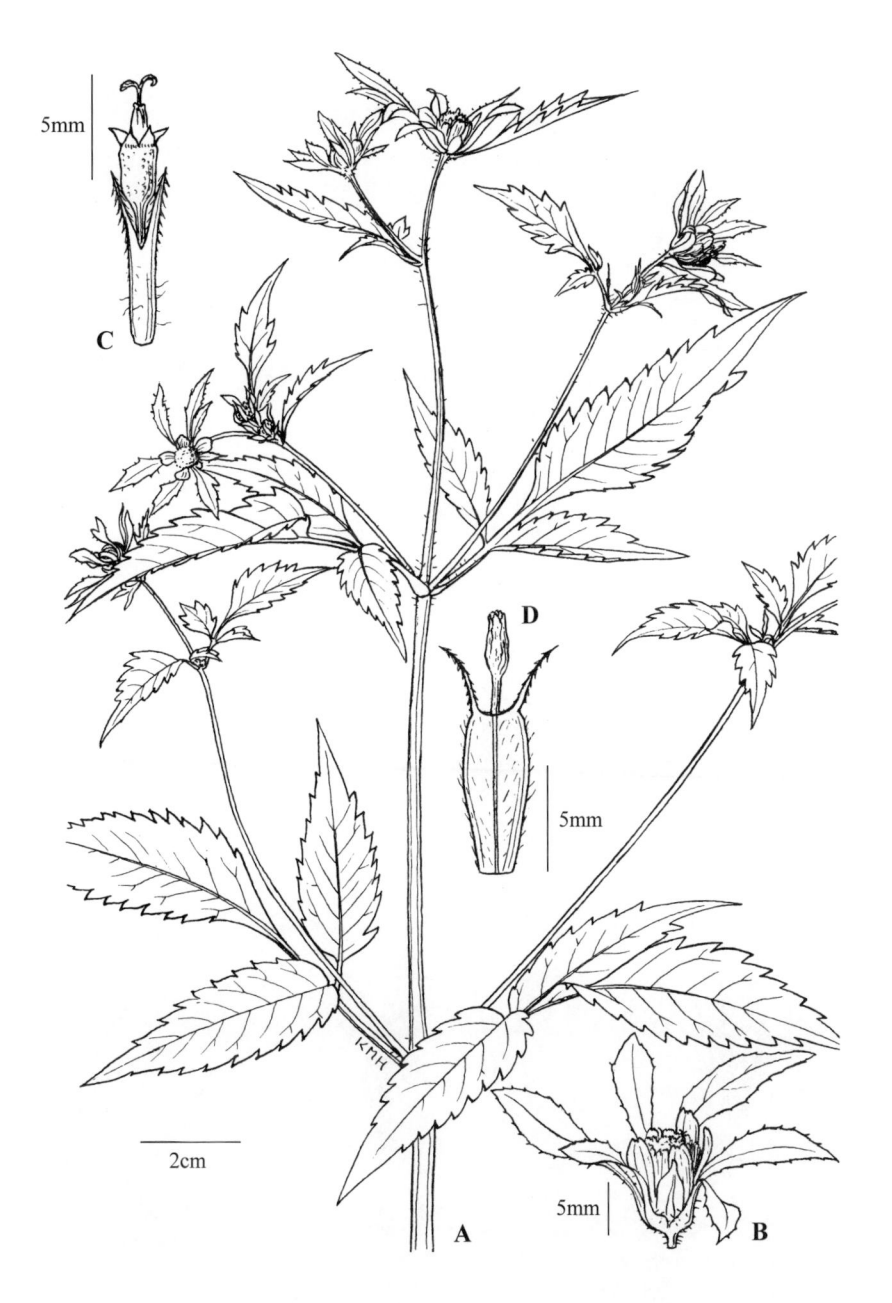

Bidens frondosa L. (Beggarticks). A habit; B capitulum; C disc floret; D achene. (*Hort.* ex Newport, Mons). STACE 890/754.

Bidens pilosa L. (Black-jack). A habit; B capitulum; C capitulum of variant without rays; D disc floret; E capitulum in fruit; F achene. (Ash, E Kent). STACE 890/754.

139. ASTERACEAE

Bidens bipinnata L. (Spanish-needles). A habit; B capitulum; C disc floret; D capitulum in fruit; E achenes. (Birchington, E Kent). STACE 890/754.

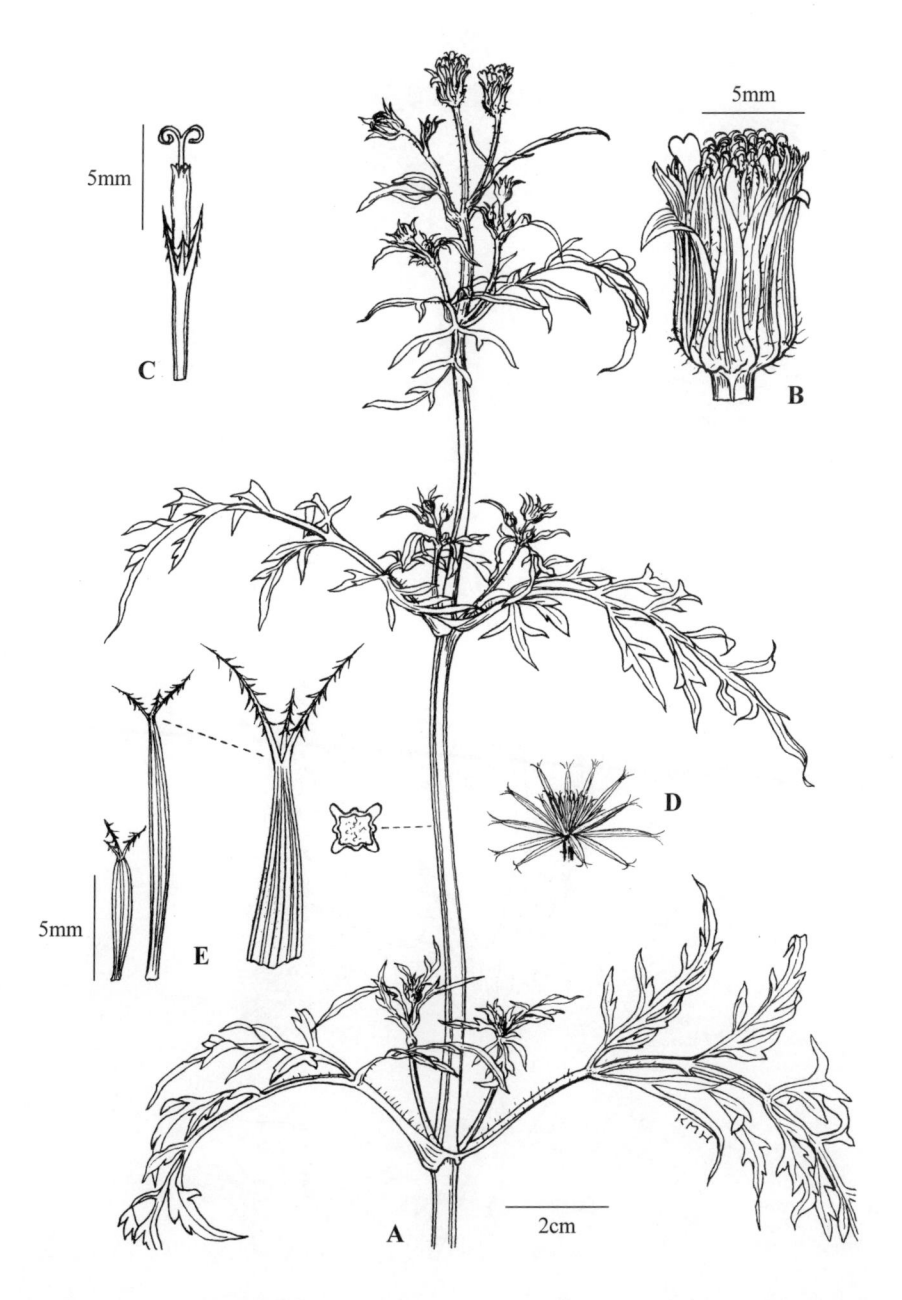

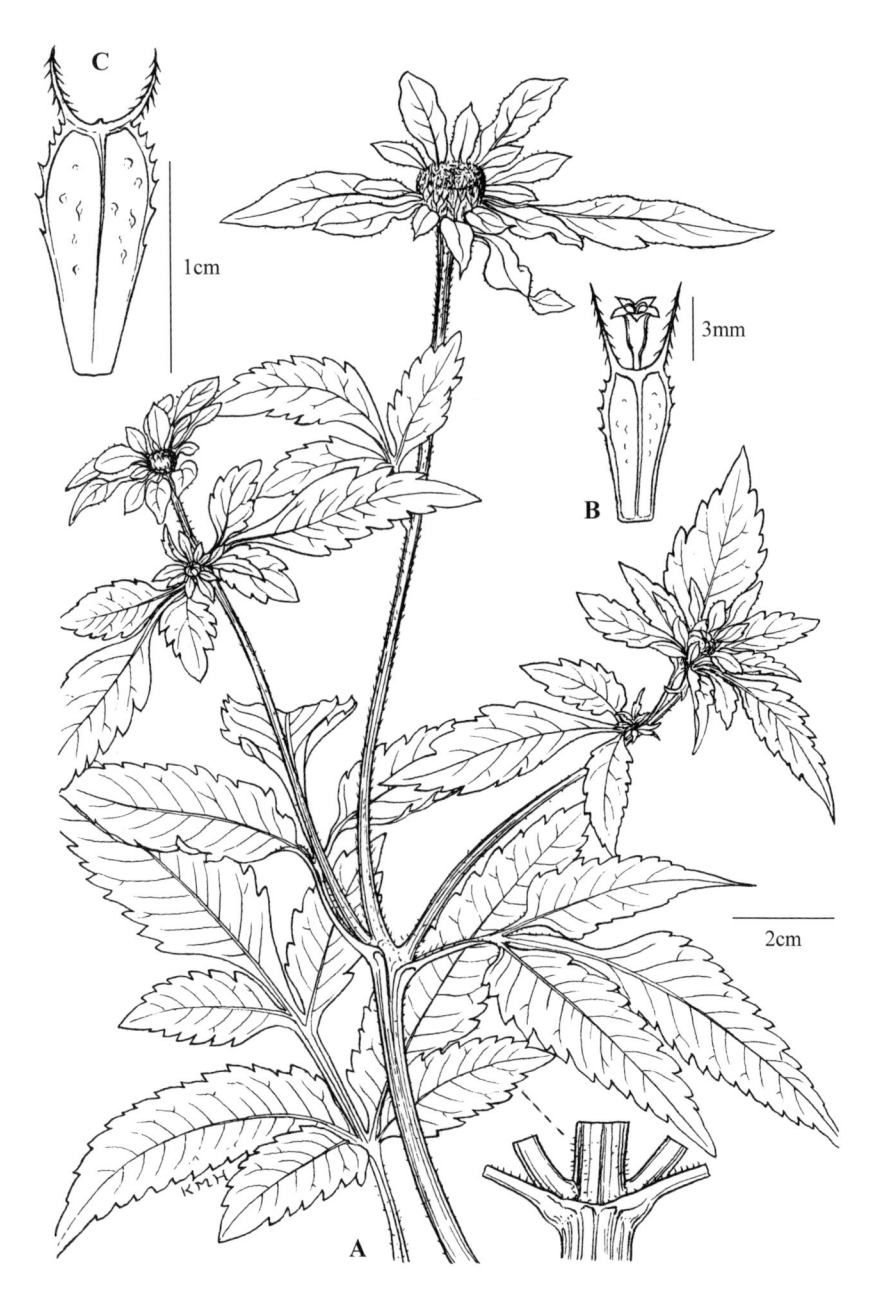

Bidens vulgata Greene (Tall Beggarticks). A habit; B disc floret; C achene. (*Hort.* ex Croydon, Surrey). STACE 890/754.

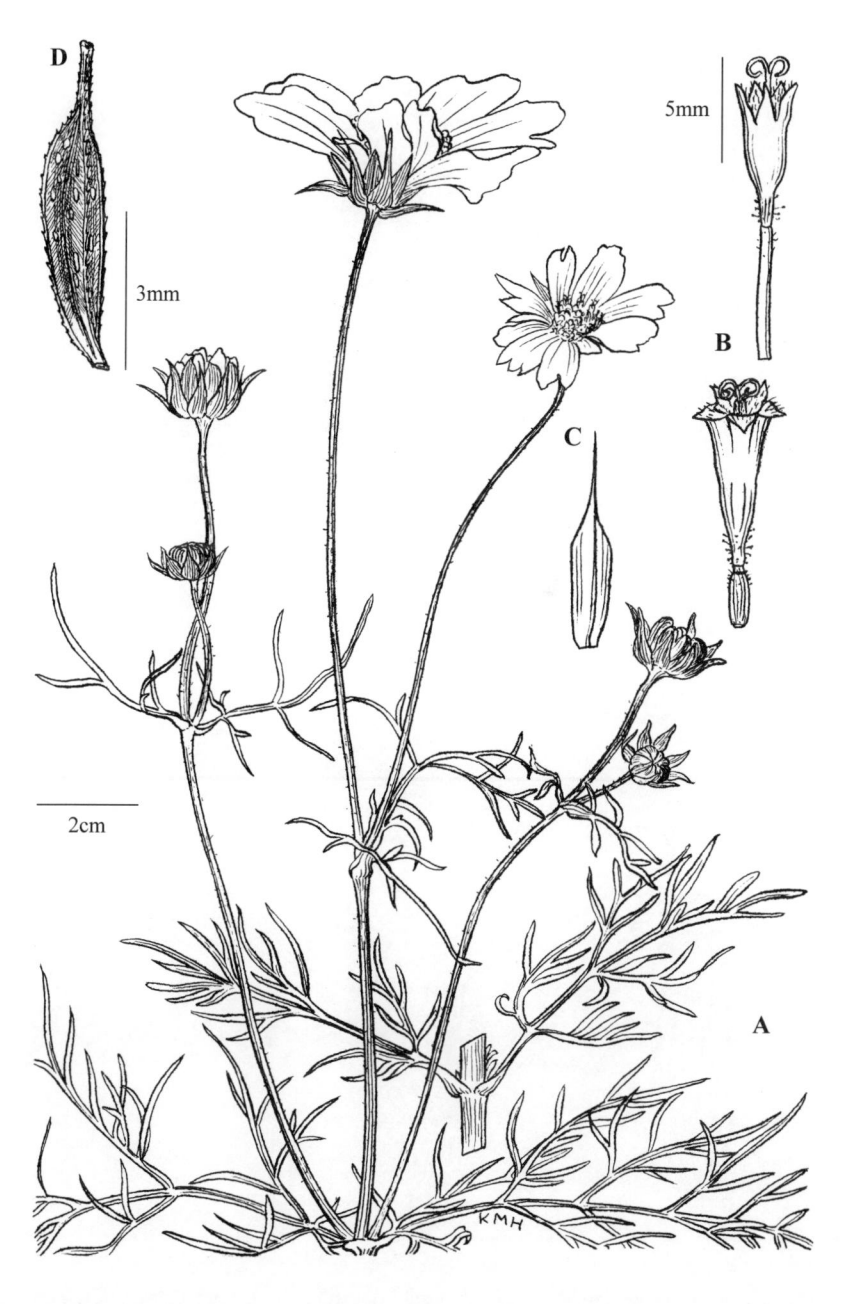

Cosmos bipinnatus Cav. (Mexican Aster). A habit; B disc florets; C receptacular scale; D disc achene. (Baildon, MW Yorks). STACE 891/755.

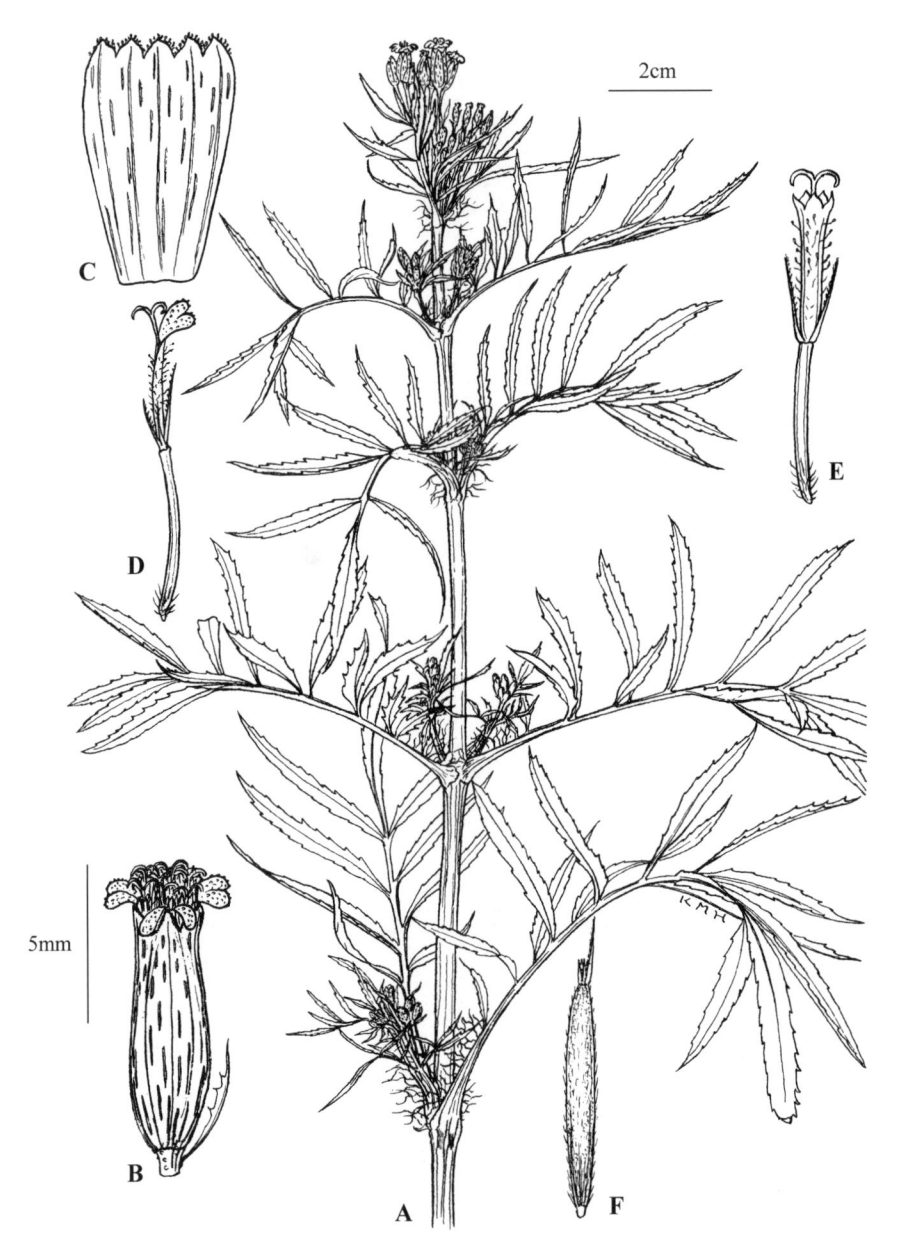

Tagetes minuta L. (Southern Marigold). A habit; B capitulum; C united phyllaries opened out; D ray floret showing pappus bristles; E disc floret showing pappus bristles; F achene. (Sandwich, E Kent). STACE 892/756.

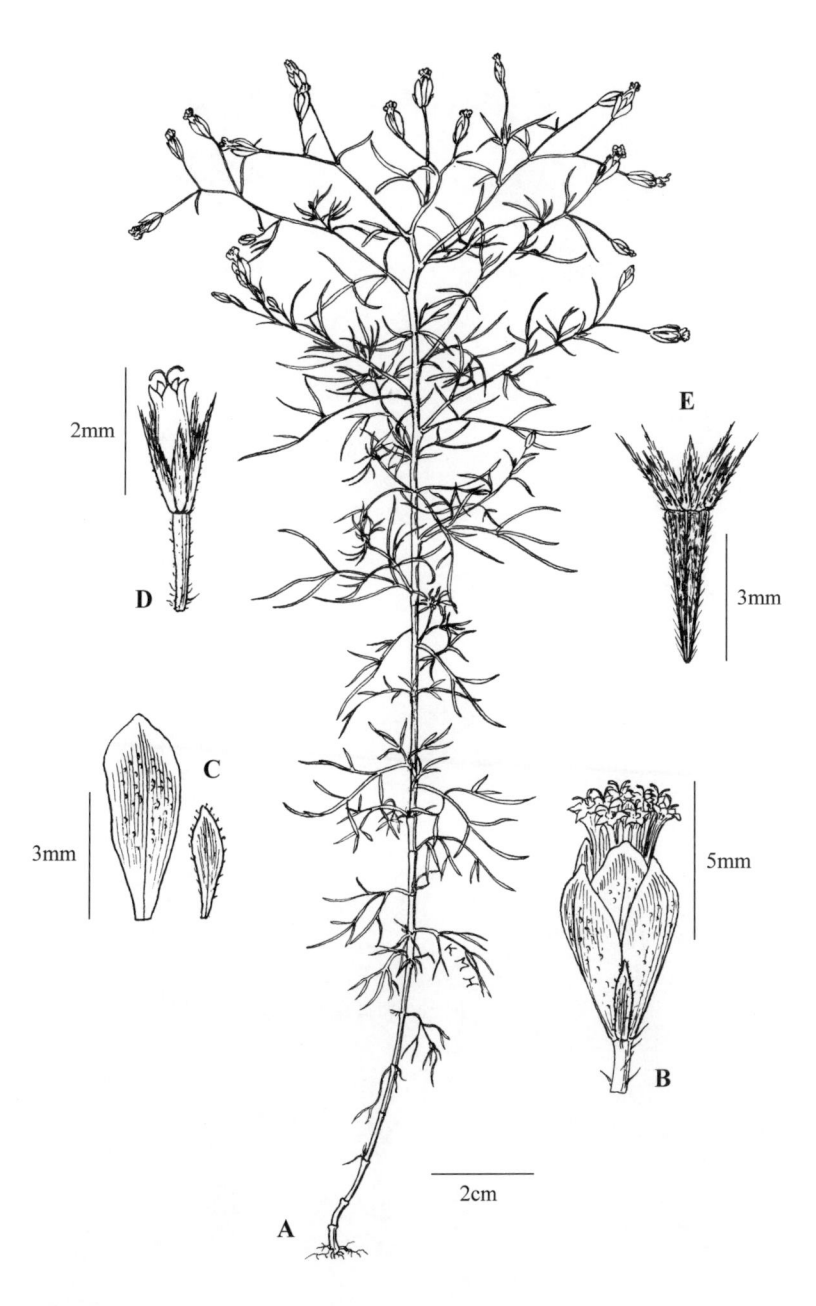

139. ASTERACEAE

Schkuhria pinnata (Lam.) Kuntze (Dwarf Marigold). A habit; B capitulum; C phyllaries; D floret; E achene. (Sandwich, E Kent). STACE 892/756.

Gaillardia x grandiflora Van Houtte (Blanketflower). A habit; B lower stem leaf; C phyllaries; D disc floret; E capitulum in fruit; F disc achene. (Dungeness, E Kent). STACE 894/756.

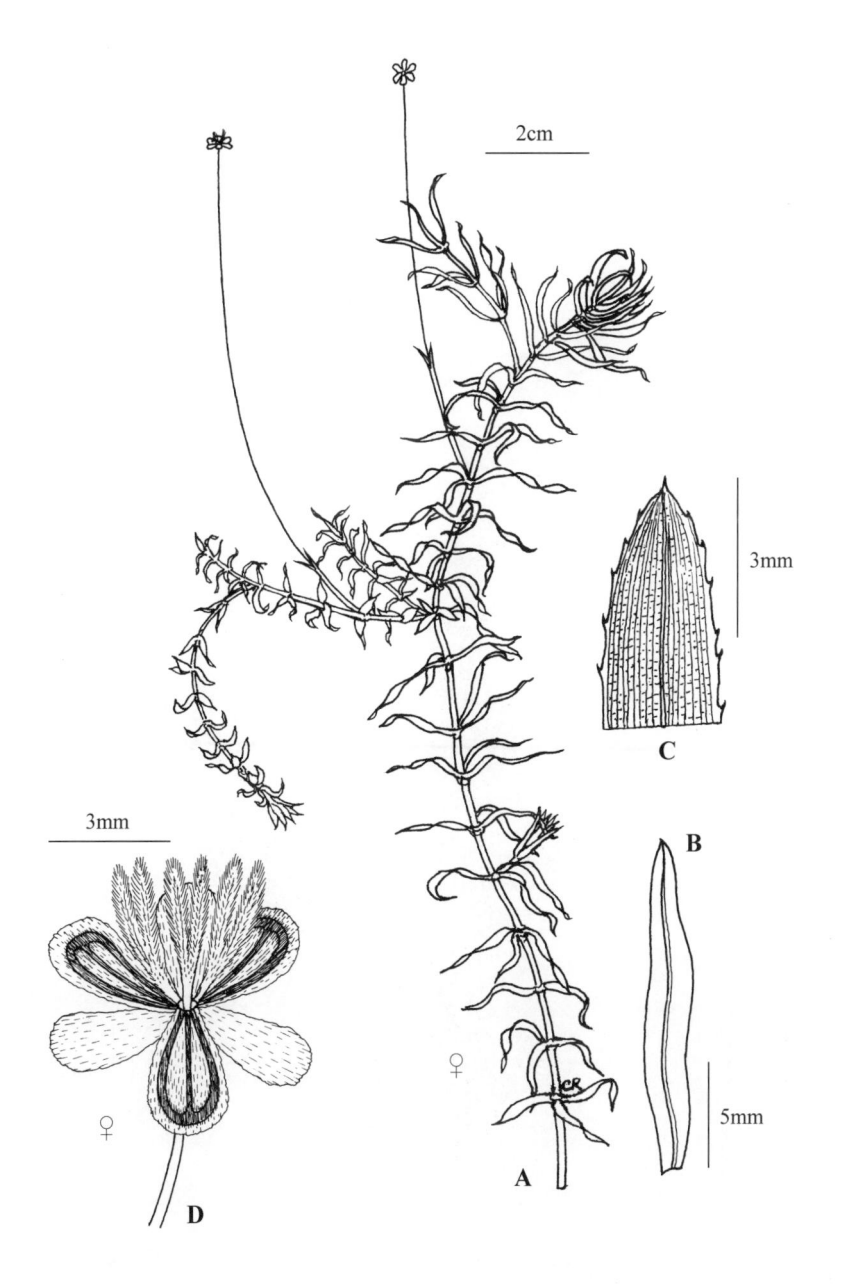

Elodea nuttallii (Planch.) H. St. John (Nuttall's Waterweed). A habit of female plant; B leaf; C apex of leaf; D female flower. (Deadwater, N Hants). STACE 901/762.

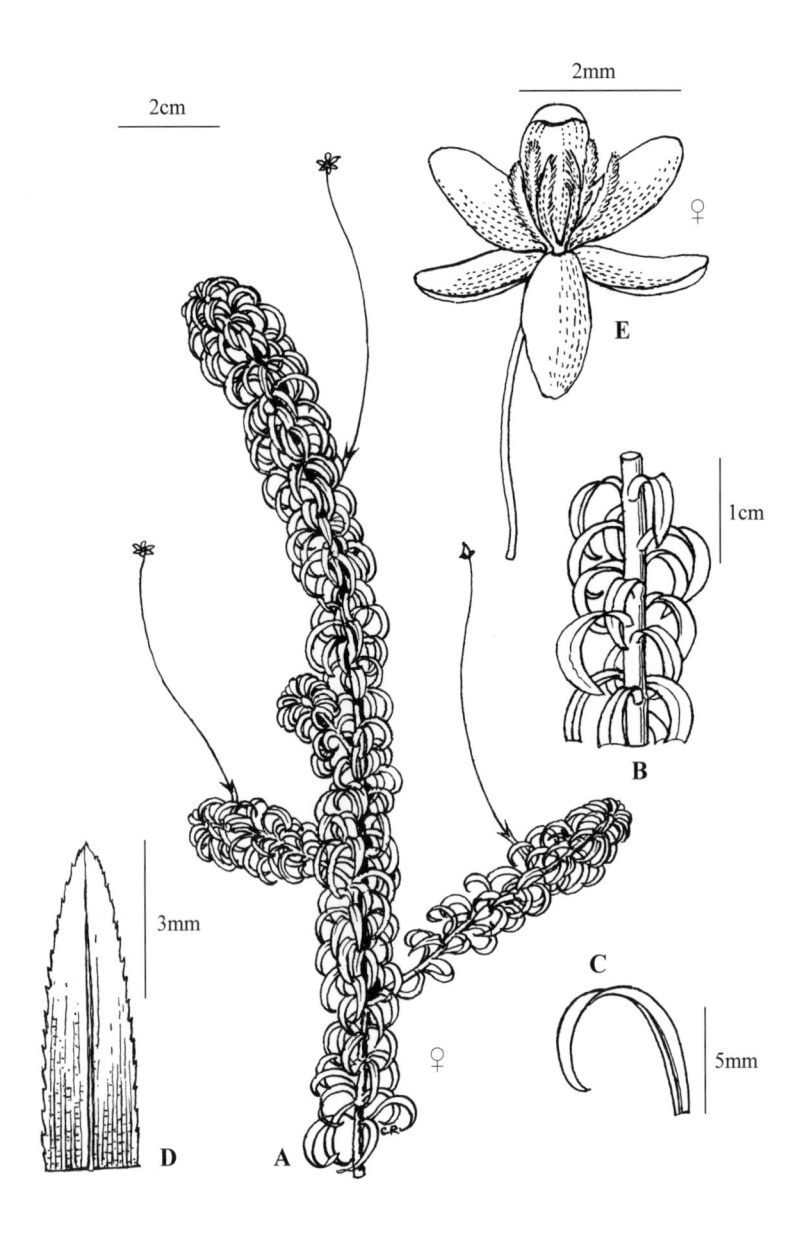

Lagarosiphon major (Ridl.) Moss ex Wager (Curly Waterweed). A habit of female plant; B part of stem; C leaf; D leaf apex; E female flower. (Source unknown). STACE 902/763.

151. ARACEAE

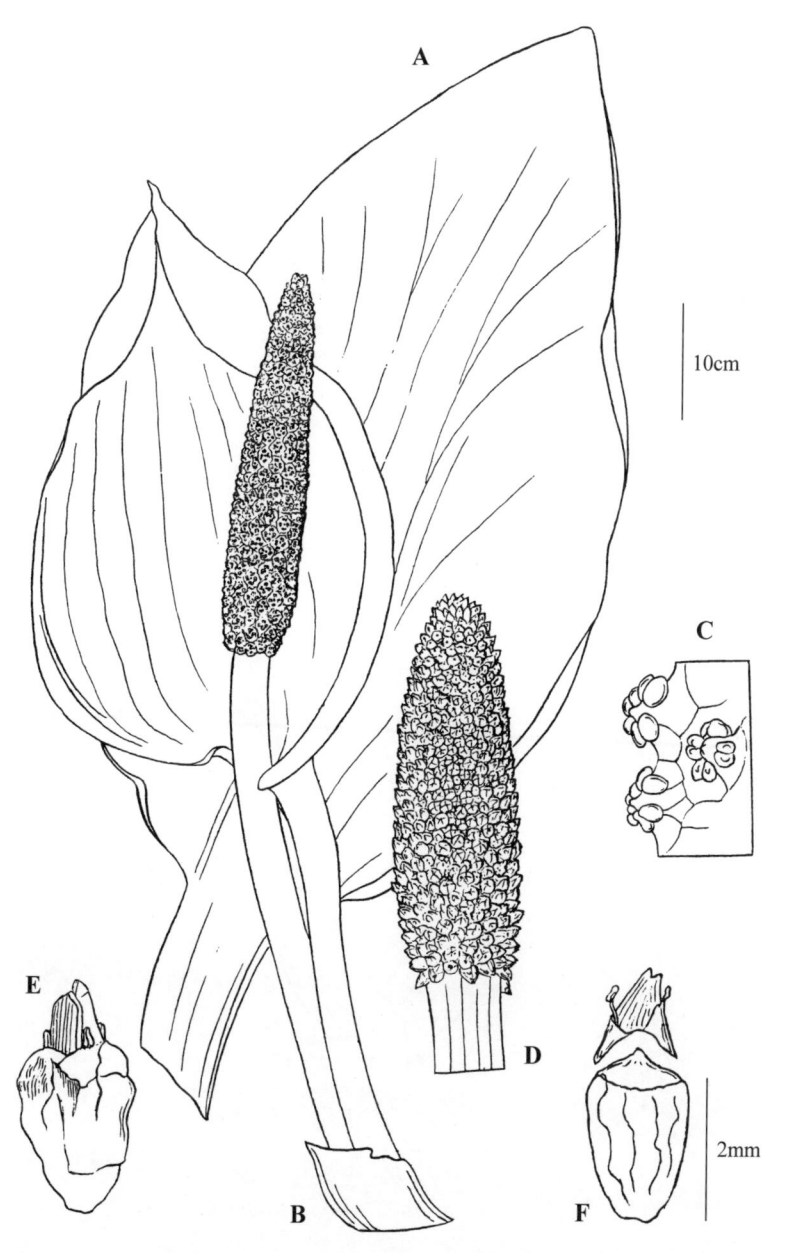

Lysichiton americanus Hultén & H. St. John (American Skunk-cabbage). A basal leaf; B spathe and spadix; C flowers on spadix; D fruiting spadix; E immature fruit; F mature fruit with distal portion detaching from berry. (Bedgebury, W Kent). STACE 919/777.

151. ARACEAE

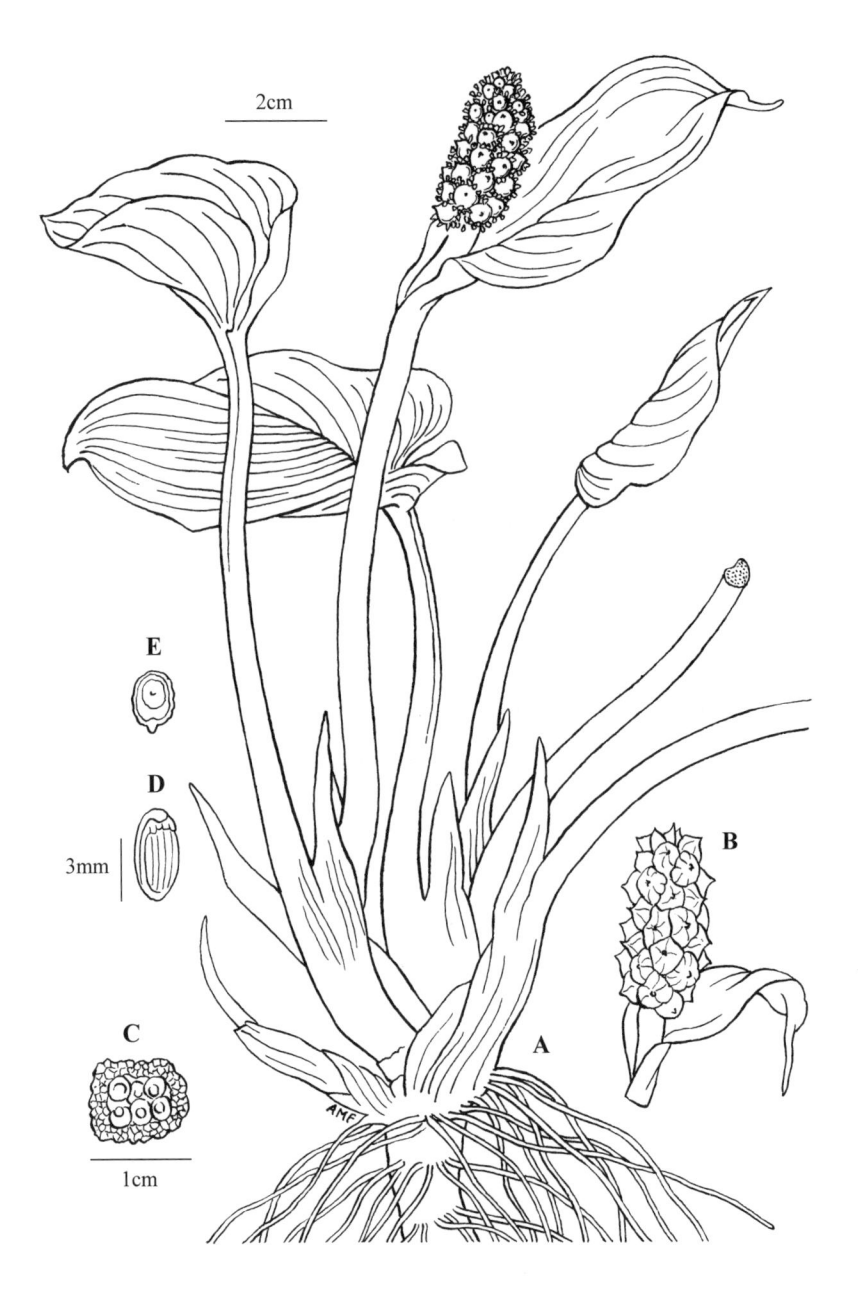

Calla palustris L. (Bog Arum). A habit; B fruiting spadix; C TS of fruit; D seed; E TS of seed. (Beckley, E Sussex). STACE 920/778.

151. ARACEAE

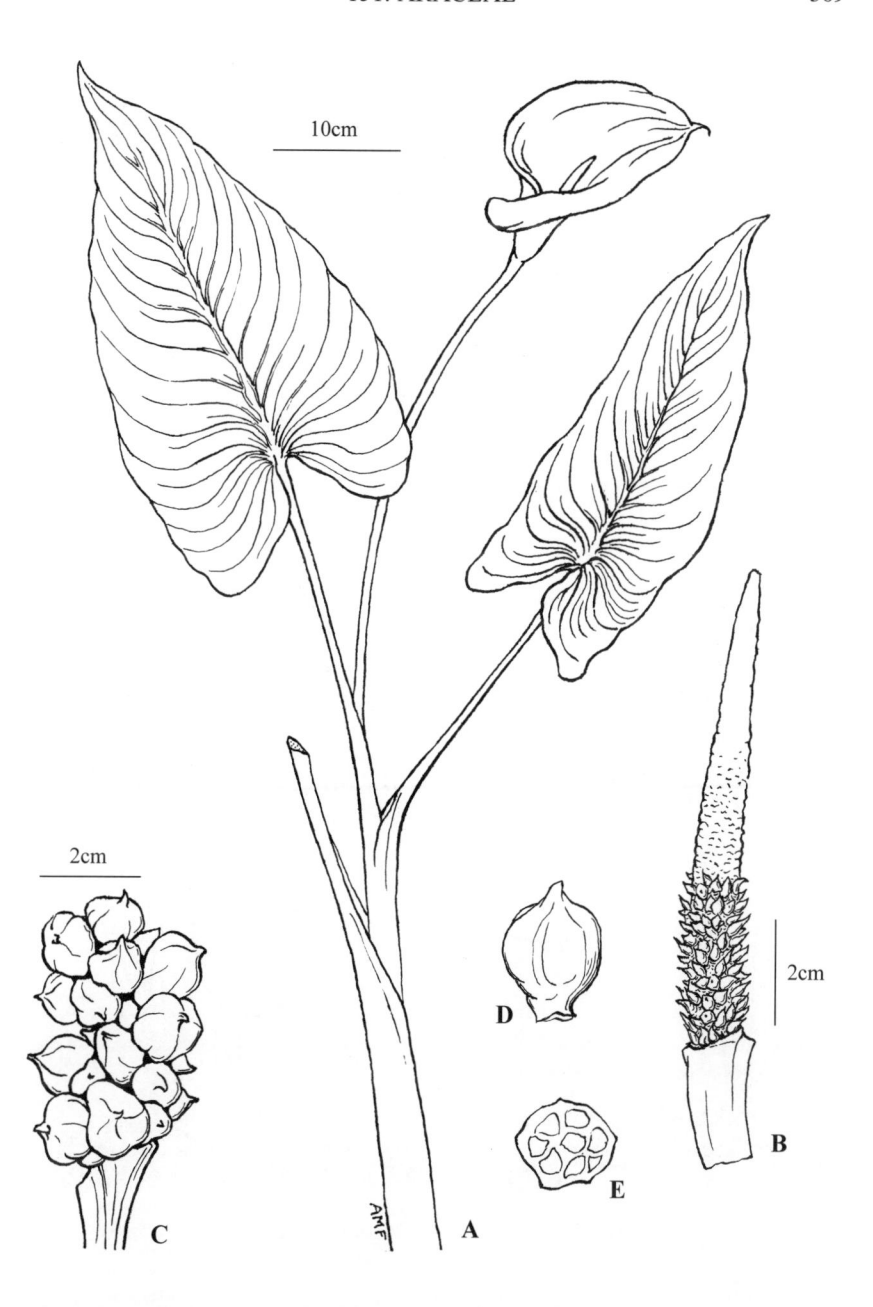

Zantedeschia aethiopica (L.) Spreng. (Altar-lily). A habit; B spadix; C fruiting spadix; D berry; E TS of berry. (Source unknown). STACE 920/778.

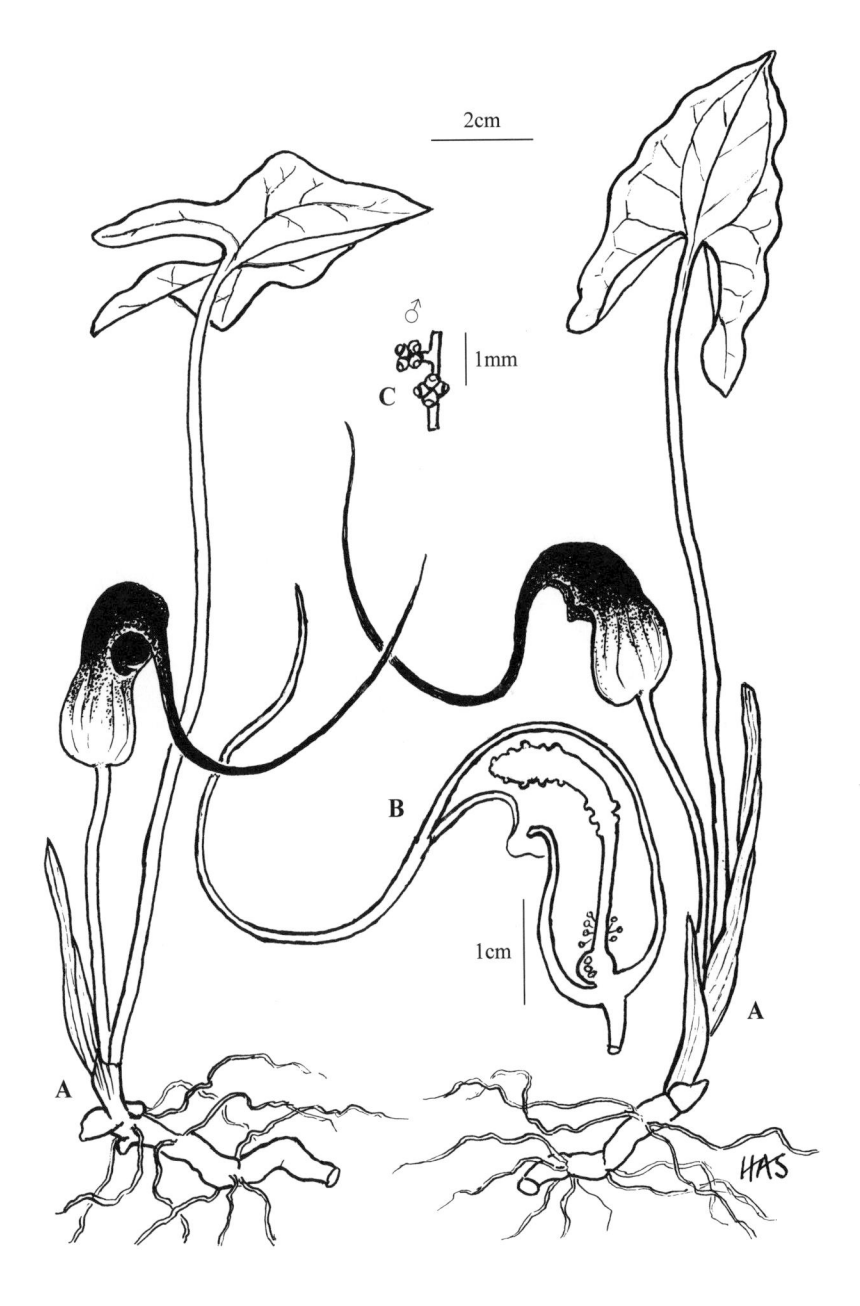

Arisarum proboscideum (L.) Savi (Mousetailplant). A habit; B LS of flower; C male flowers. (*Hort.* Aberdeen, S Aberdeen). STACE 921/779.

155. JUNCACEAE

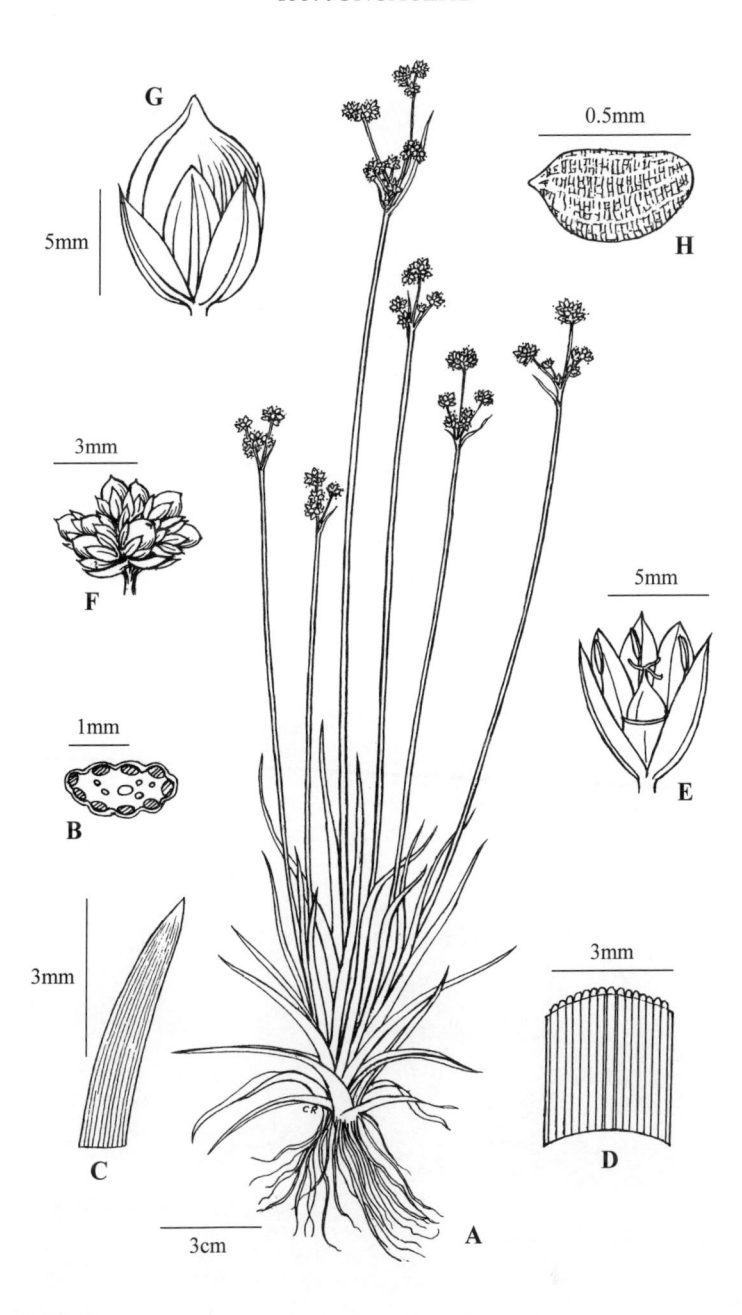

Juncus planifolius R. Br. (Broad-leaved Rush). A habit; B TS of stem; C apex of leaf; D part of leaf; E flower; F cluster of fruits; G fruit; H seed. (Ireland). STACE 932/785.

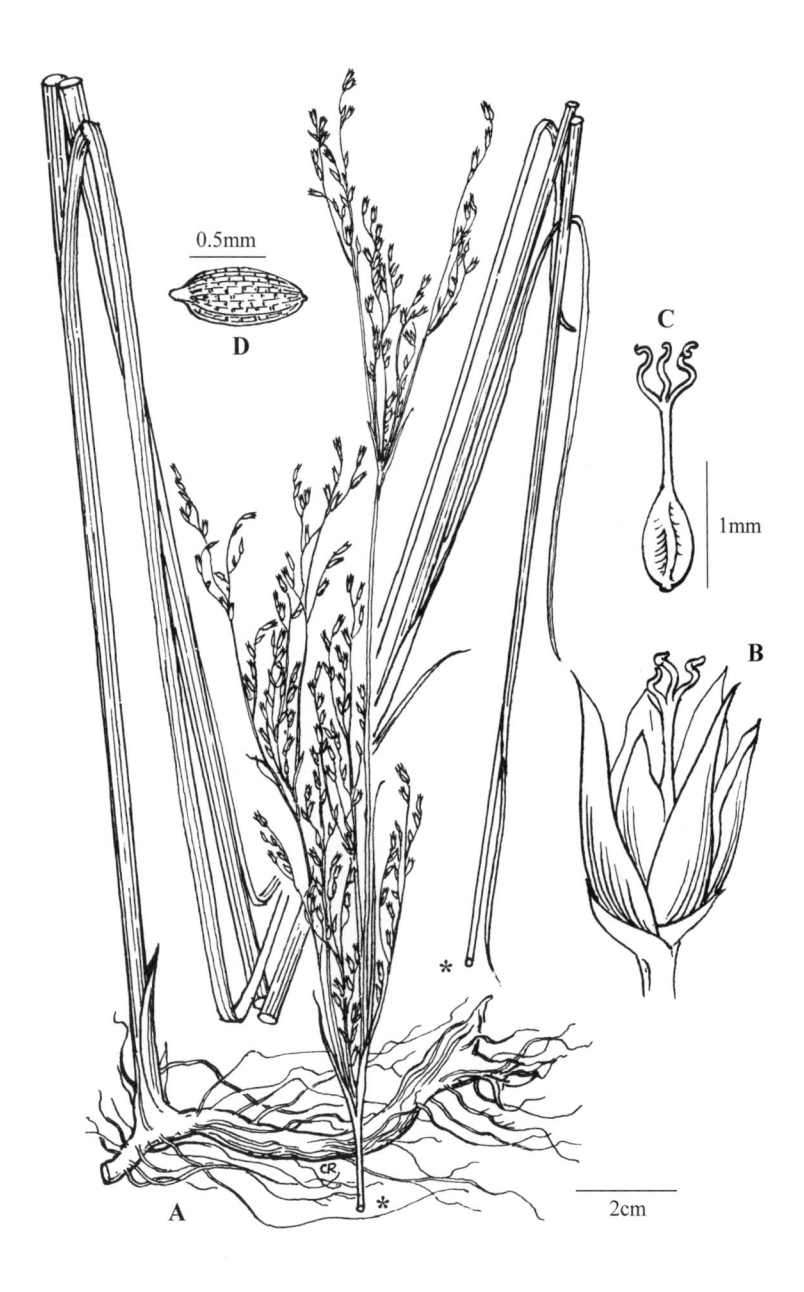

Juncus subulatus Forssk. (Somerset Rush). A habit; B flower; C gynoecium; D seed. (Berrow, N Somerset). STACE 935/789.

155. JUNCACEAE 373

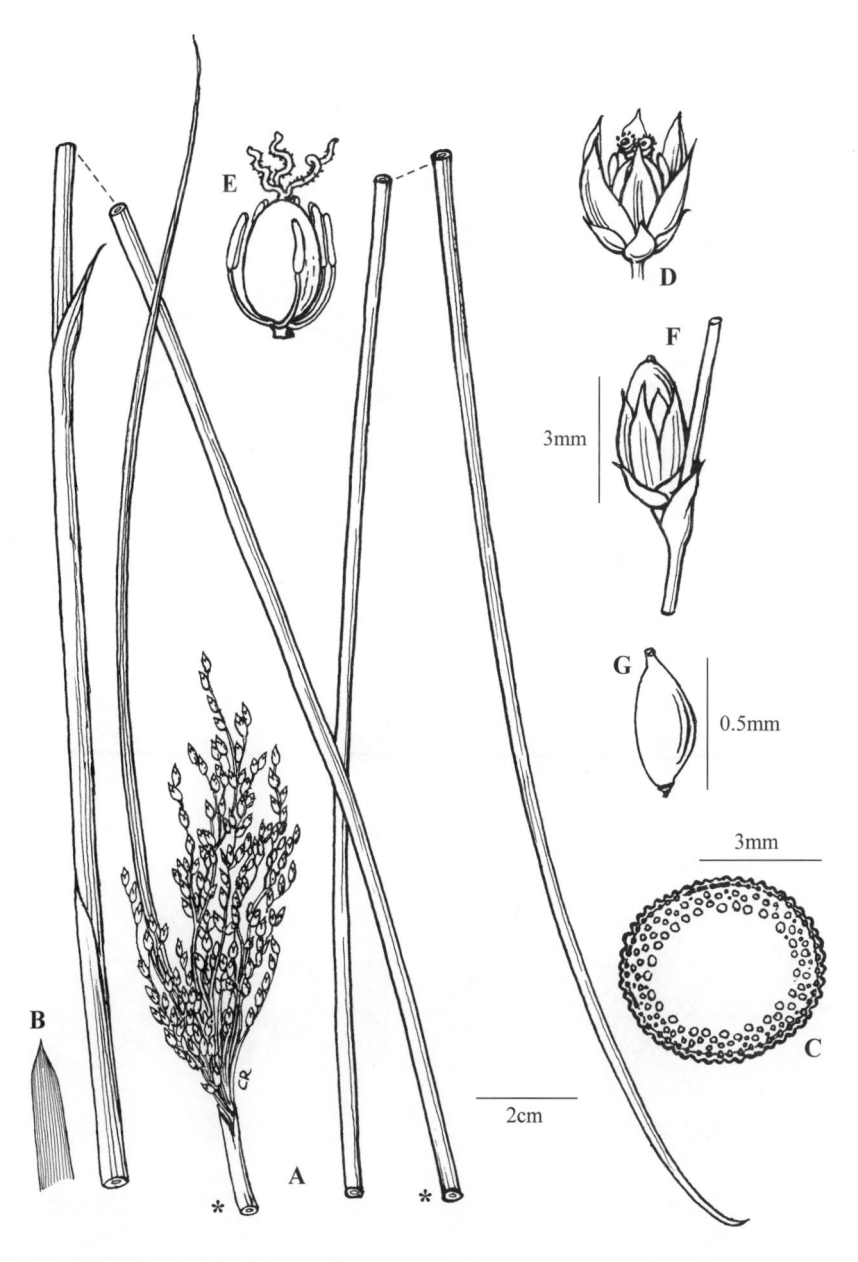

Juncus pallidus R. Br. (Great Soft-rush). A flowering stem and leaf; B apex of basal leaf-sheath; C TS of stem; D flower; E flower minus perianth; F fruit; G seed. (Source unknown). STACE 936/790.

374 156. CYPERACEAE

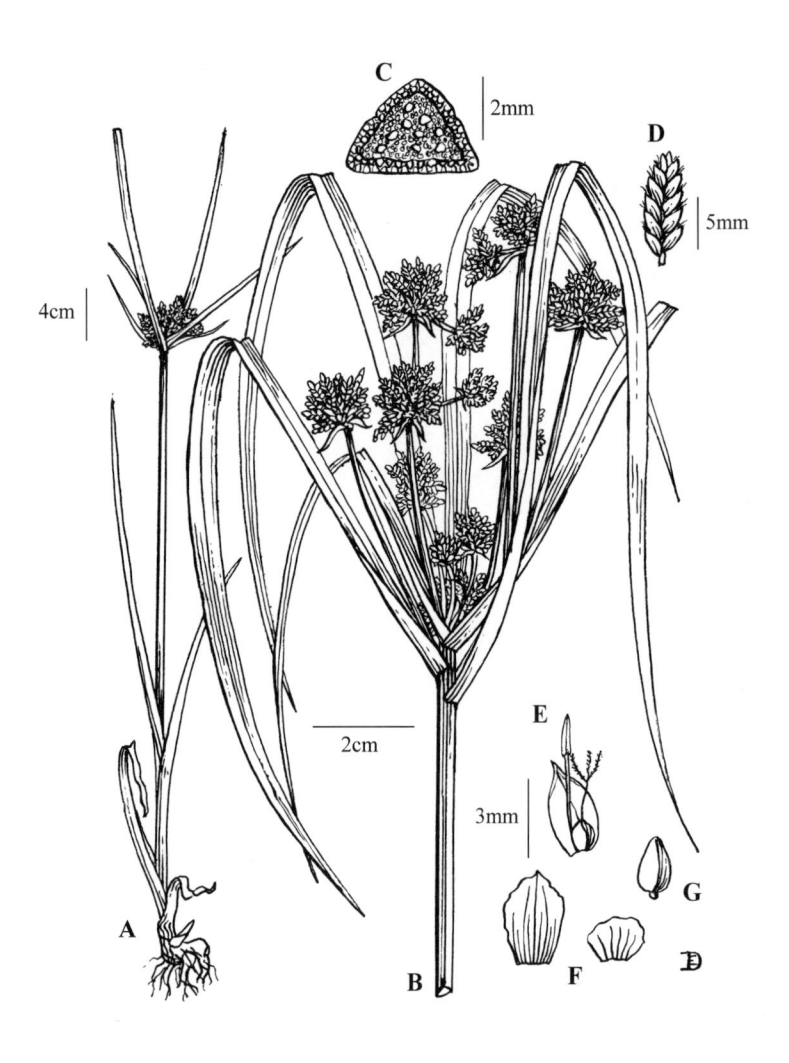

Cyperus eragrostis Lam. (Pale Galingale). A habit; B inflorescence; C TS of stem; D spikelet; E flower; F glumes; G nutlet. (Source unknown). STACE 950/801.

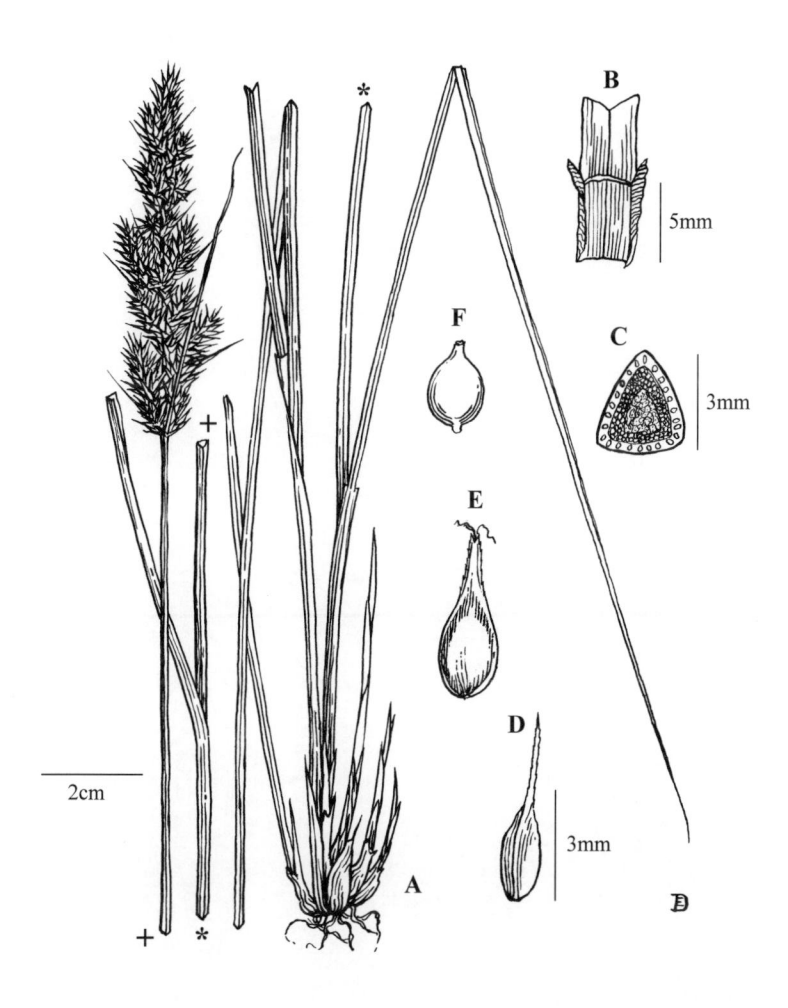

Carex vulpinoidea Michx. (American Fox-sedge). A habit; B ligule; C TS of stem; D female glume; E utricle; F nutlet. (Source unknown). STACE 963/811.

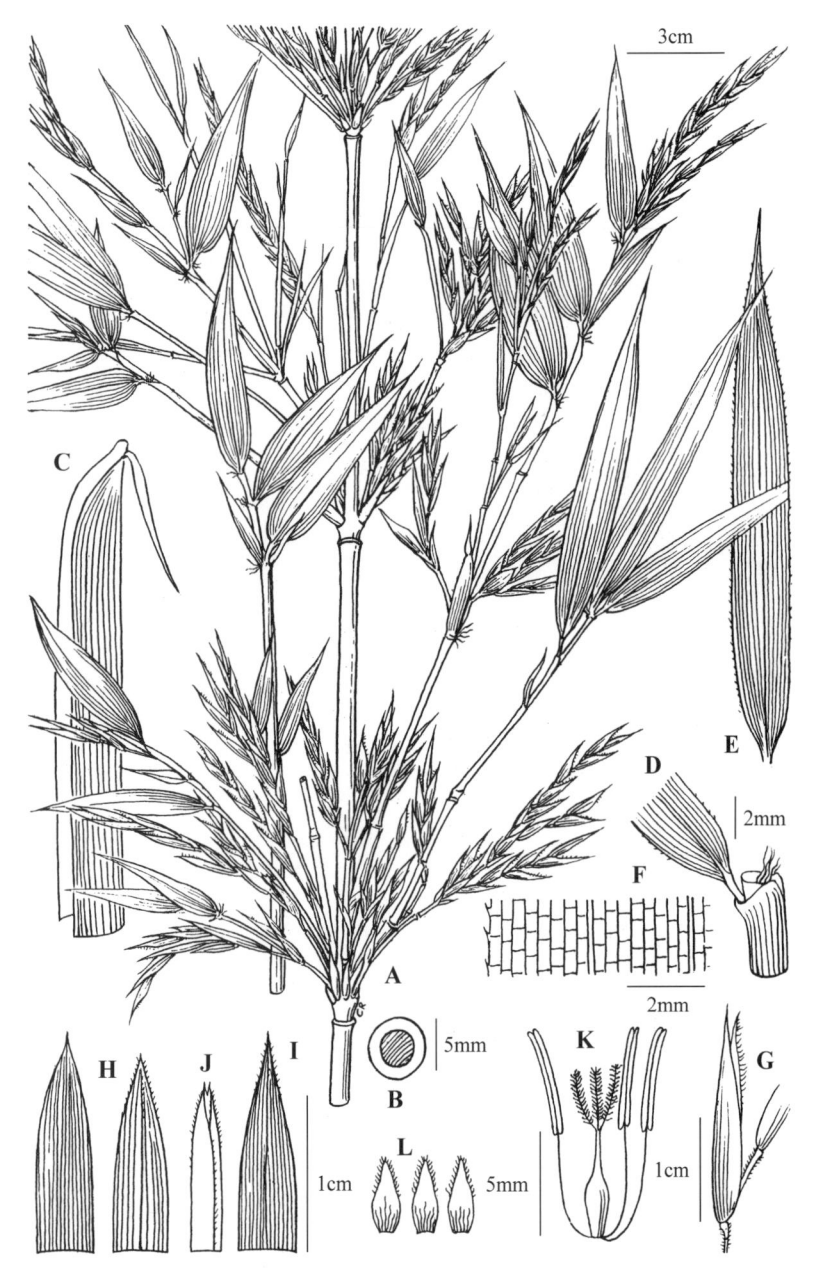

Semiarundinaria fastuosa (Lat.-Marl. ex Mitford) Makino ex Nakai (Narihira Bamboo). A flowering branch; B TS of main stem; C stem sheath; D ligule; E leaf blade; F tessellation of leaf blade; G part of spikelet; H lower and upper glumes; I lemma; J palea; K flower; L lodicules. (Source unknown). STACE 994/837.

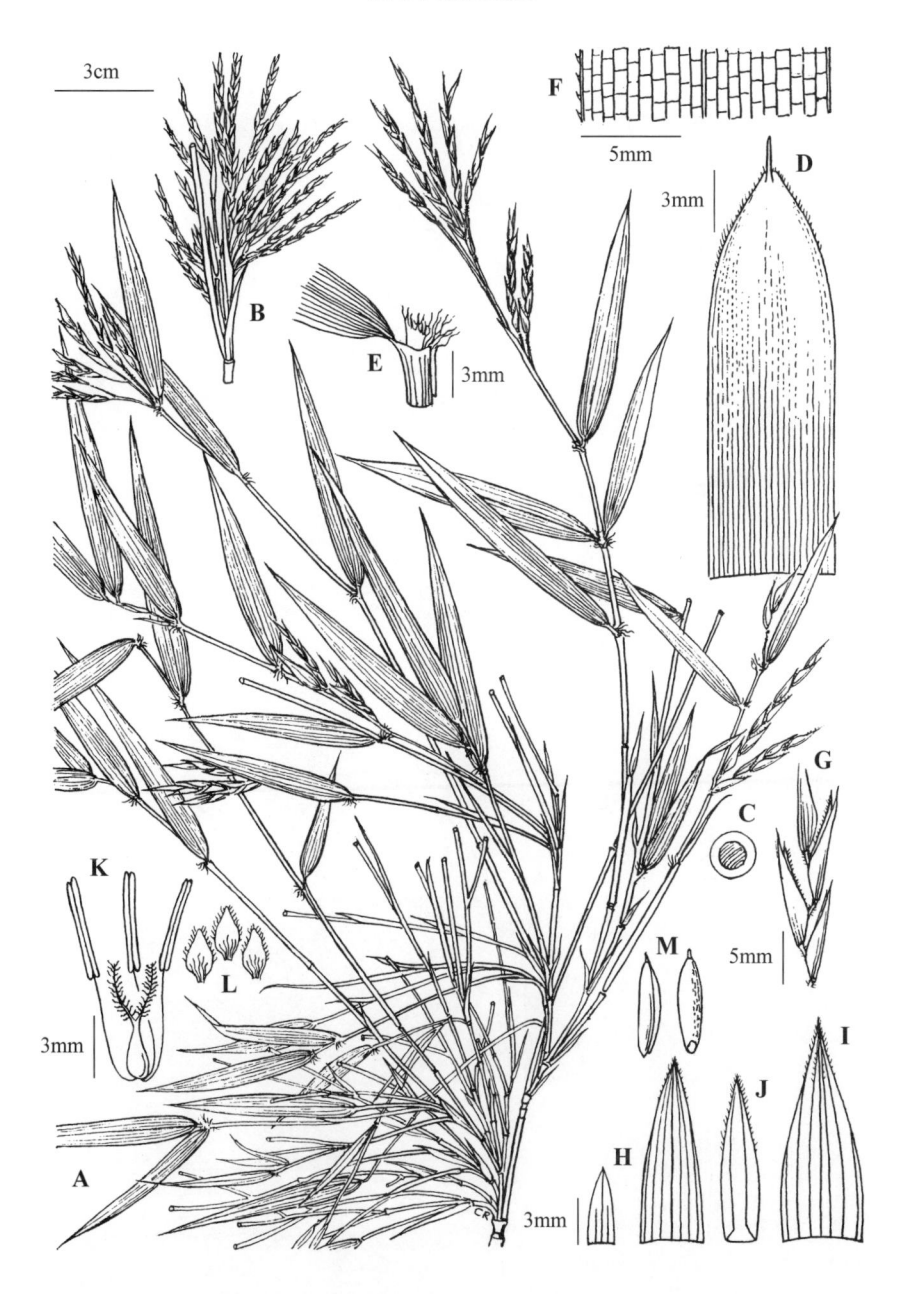

157. POACEAE

Yushania anceps (Mitford) W.C. Lin (Indian Fountain-bamboo). A flowering branch; B variant with denser spikelets; C TS of main stem; D stem sheath; E ligule; F tessellation of leaf blade; G part of spikelet; H lower and upper glumes; I lemma; J palea; K flower; L lodicules; M caryopsis. (Source unknown). STACE 995/837.

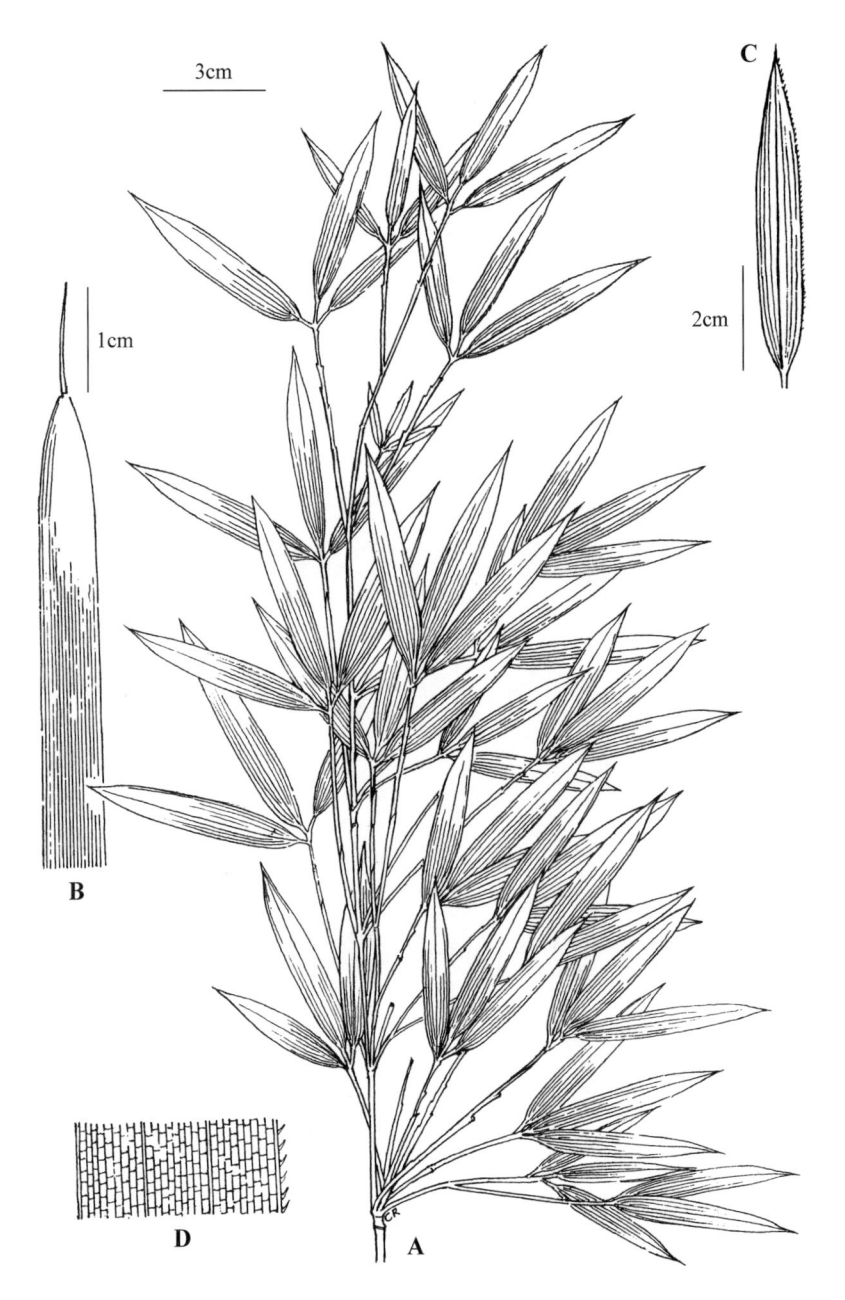

Fargesia spathacea Franch. (Chinese Fountain-bamboo). A non-flowering branch; B stem sheath; C leaf blade; D tessellation of leaf blade. (Source unknown). STACE 995/837.

157. POACEAE

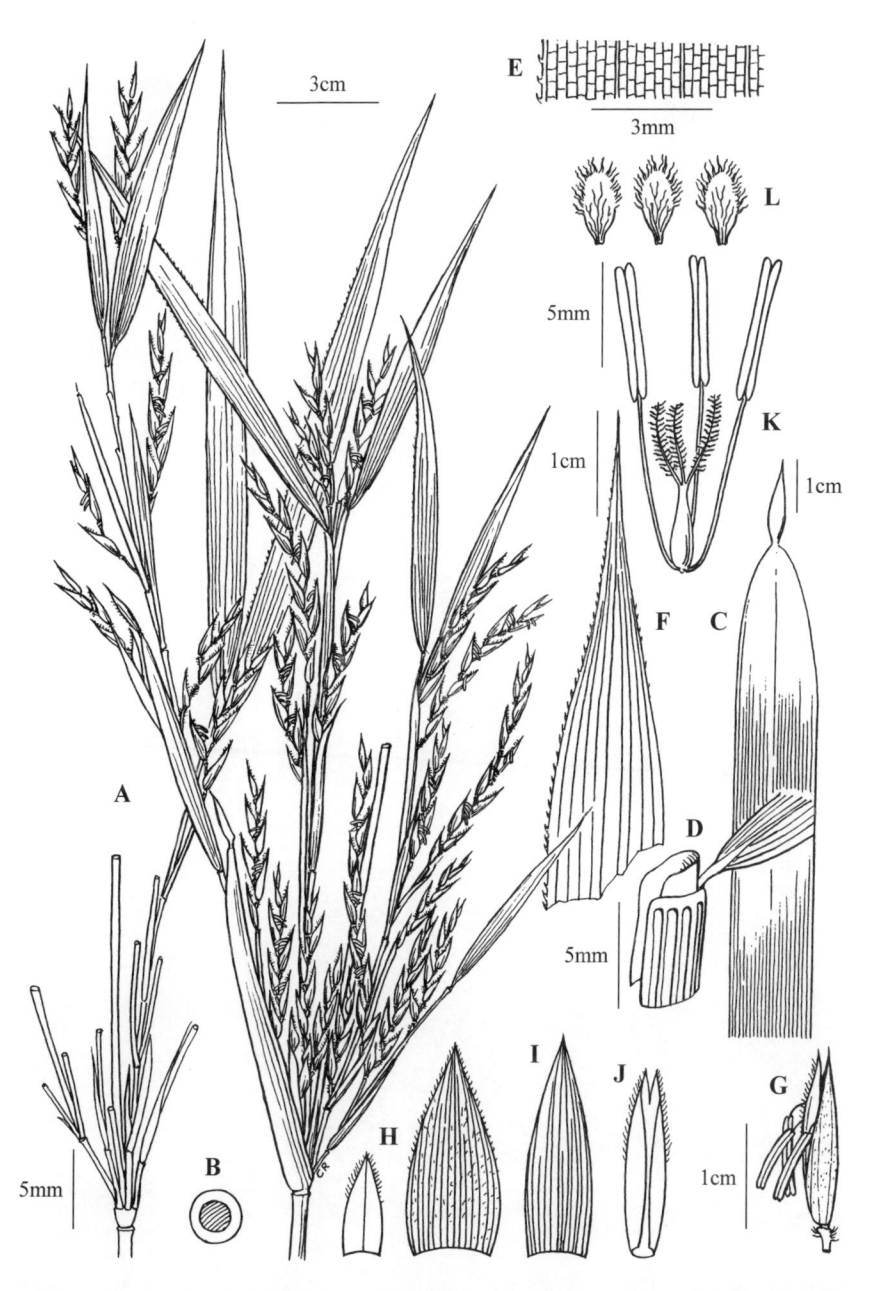

Pleioblastus simonii (Carrière) Nakai (Simon's Bamboo). A flowering branch; B TS of main stem; C stem sheath; D ligule; E tessellation of leaf blade; F apex of leaf; G floret; H lower and upper glumes; I lemma; J palea; K flower; L lodicules. (Source unknown). STACE 996/838.

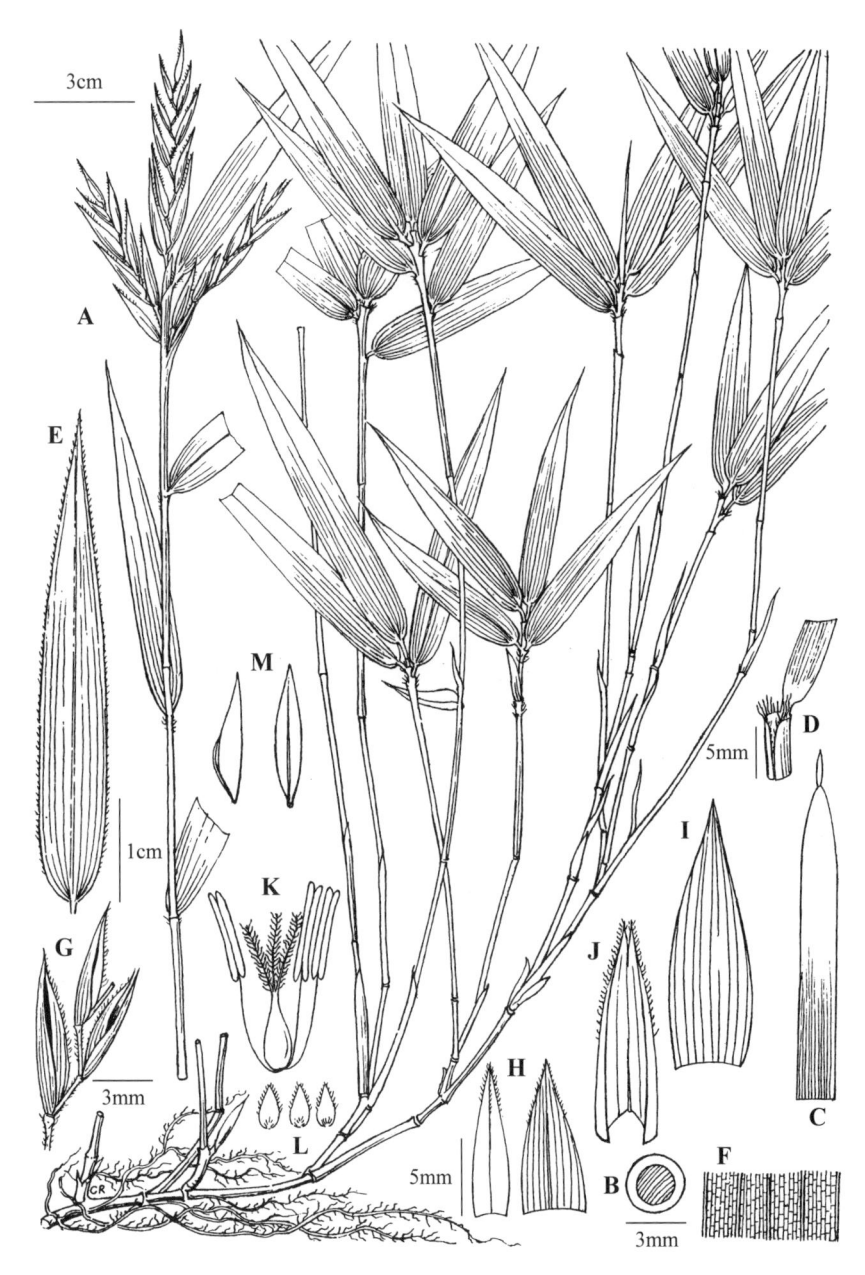

Pleioblastus humilis (Mitford) Nakai (No English name). A flowering branch; B TS of main stem; C stem sheath; D ligule; E leaf blade; F tessellation of leaf blade; G part of spikelet; H lower and upper glumes; I lemma; J palea; K flower; L lodicules; M caryopsis. (Source unknown). STACE 996/838.

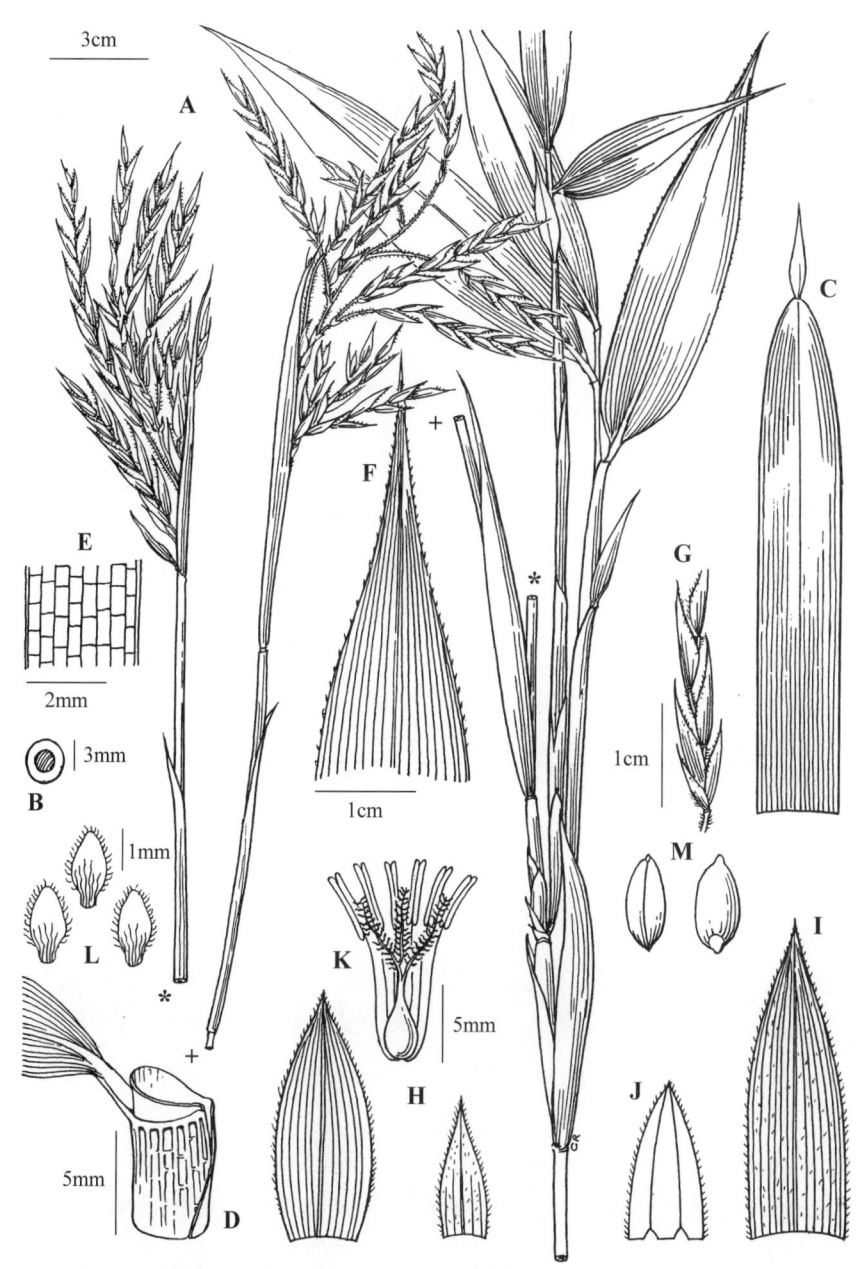

157. POACEAE

Sasa palmata (Burb.) E.G. Camus (Broad-leaved Bamboo). A flowering branch; B TS of main stem; C stem sheath; D ligule; E tessellation of leaf blade; F apex of leaf; G part of spikelet; H lower and upper glumes; I lemma; J palea; K flower; L lodicules; M caryopsis. (Source unknown). STACE 996/838.

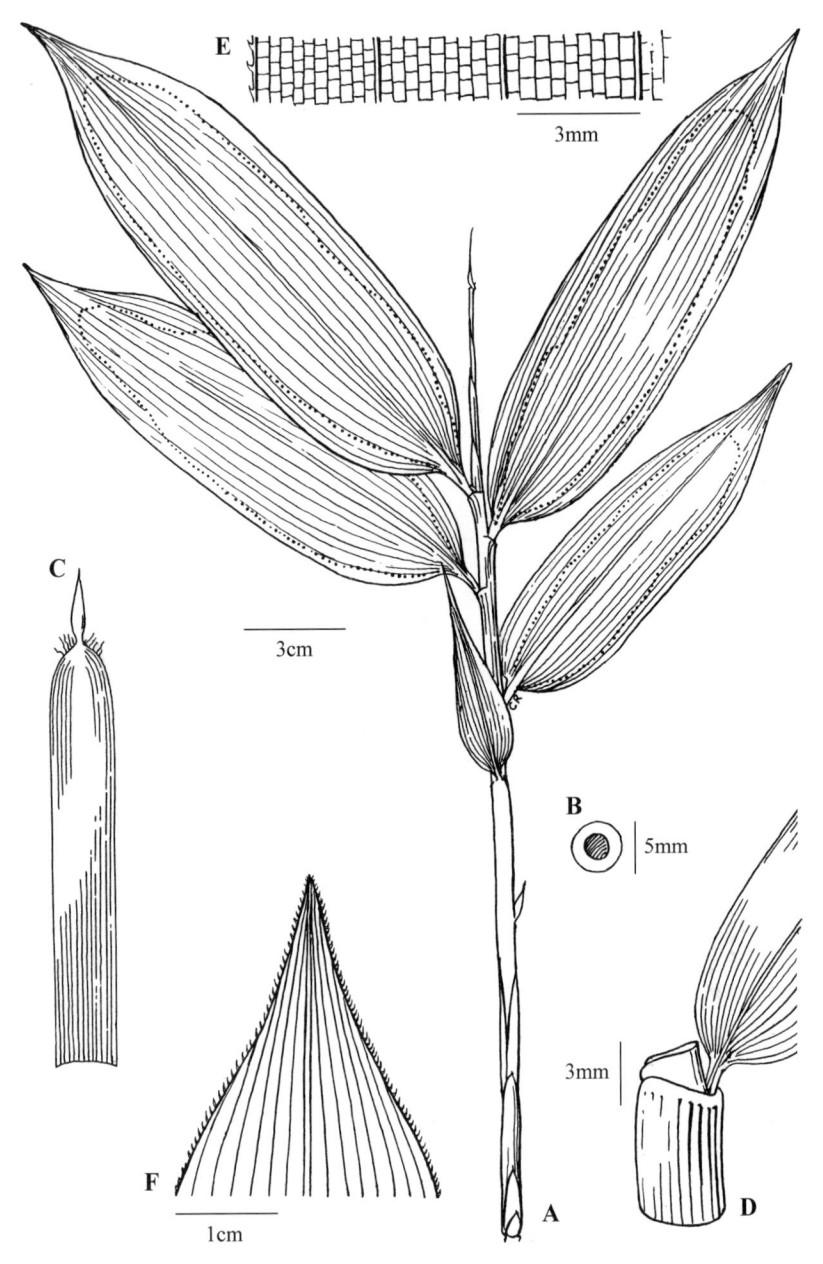

Sasa veitchii (Carrière) Rehder (Veitch's Bamboo). A non-flowering branch; B TS of main stem; C stem sheath; D ligule; E tessellation of leaf blade; F apex of leaf. (Source unknown). STACE 996/838.

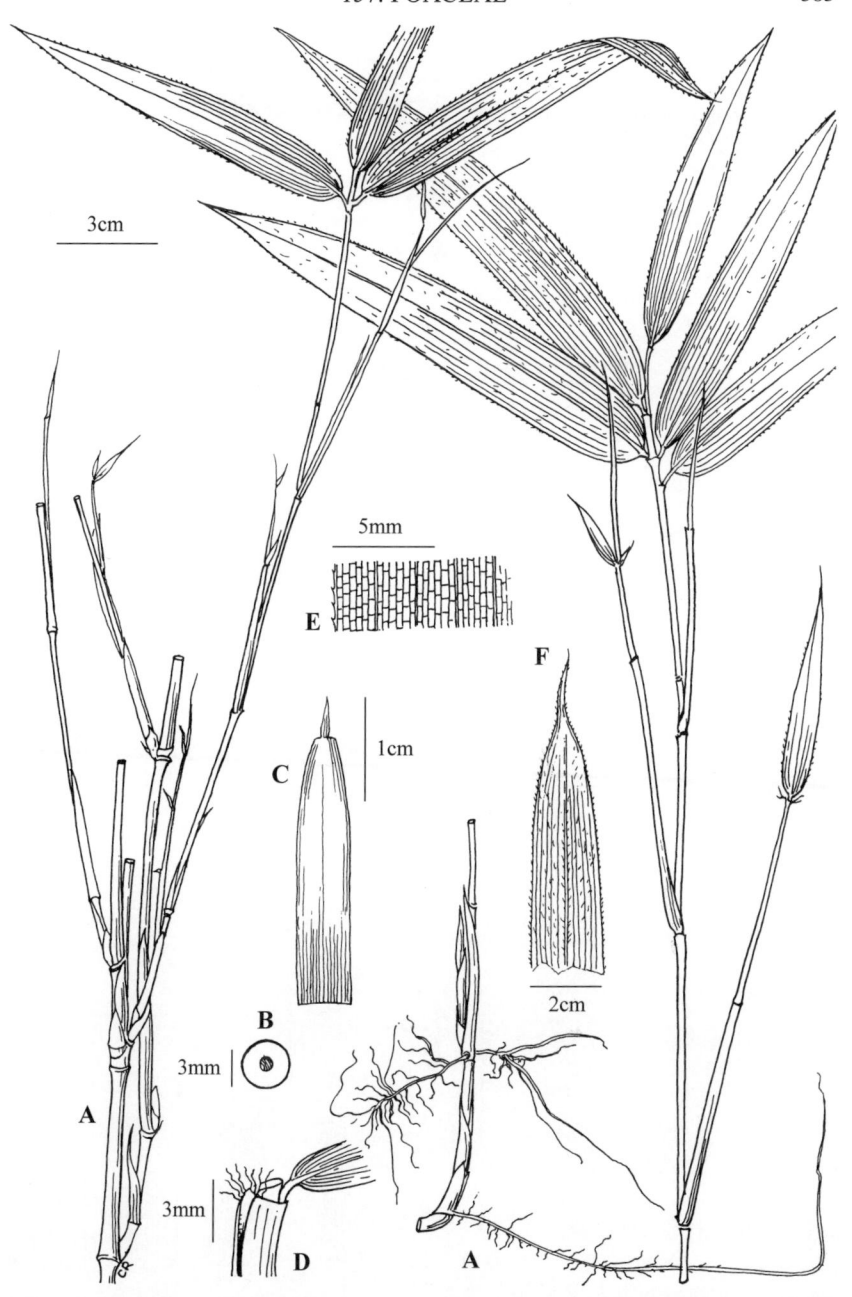

Sasaella ramosa (Makino) Makino (Hairy Bamboo). A non-flowering branch; B TS of main stem; C stem sheath; D ligule; E tessellation of leaf blade; F underside of leaf apex. No flower spikes known. (Source unknown). STACE 996/838.

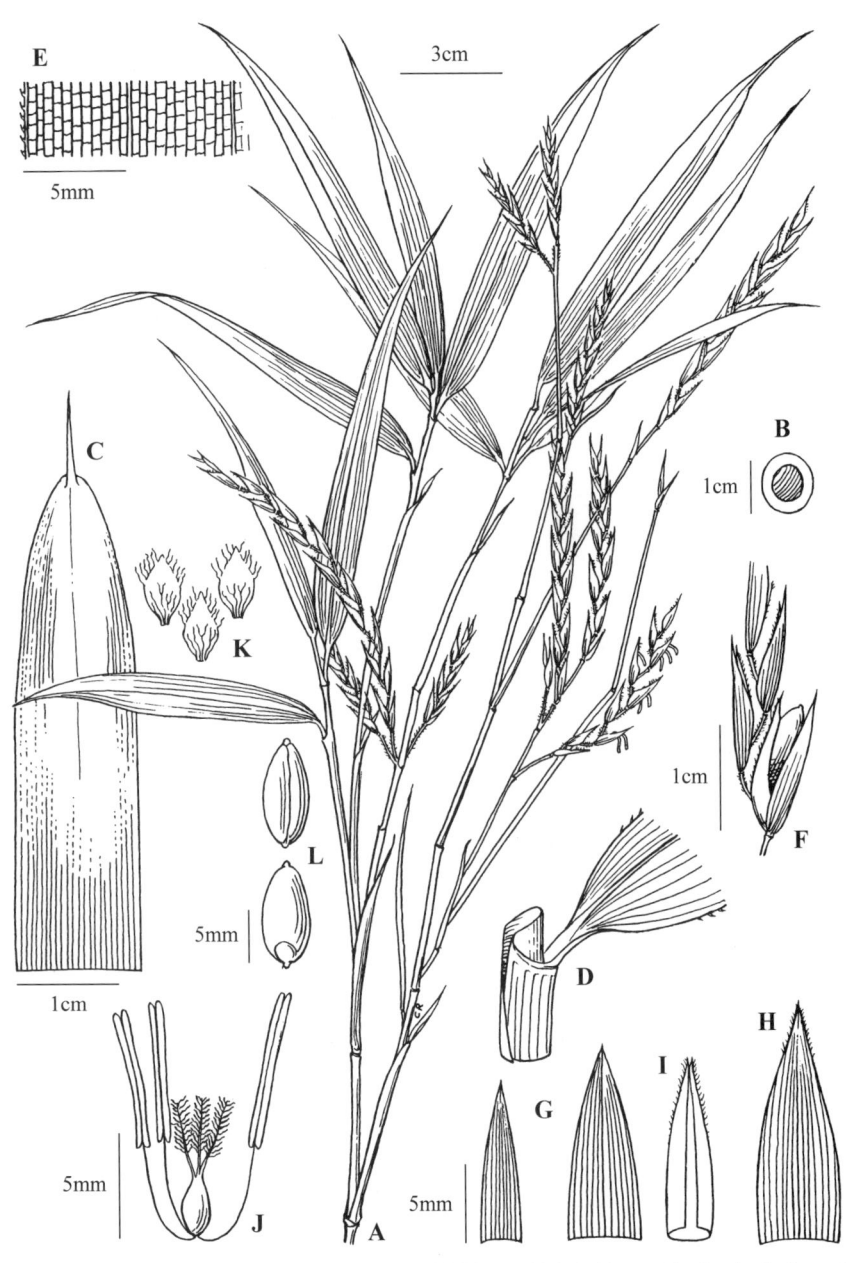

Pseudosasa japonica (Siebold & Zucc. ex Steud.) Makino ex Nakai (Arrow Bamboo). A flowering branch; B TS of main stem; C stem sheath; D ligule; E tessellation of leaf blade; F part of spikelet; G lower and upper glumes; H lemma; I palea; J flower; K lodicules; L caryopsis. (Source unknown). STACE 997/839.

Oryzopsis miliacea (L.) Benth. & Hook. f. ex Asch. & Schweinf. (Smilo-grass). A habit; B ligule; C part of panicle; D lower and upper glumes; E lemma; F dorsal view of lemma; G palea; H flower; I caryopsis. (Jersey). STACE 1000/842.

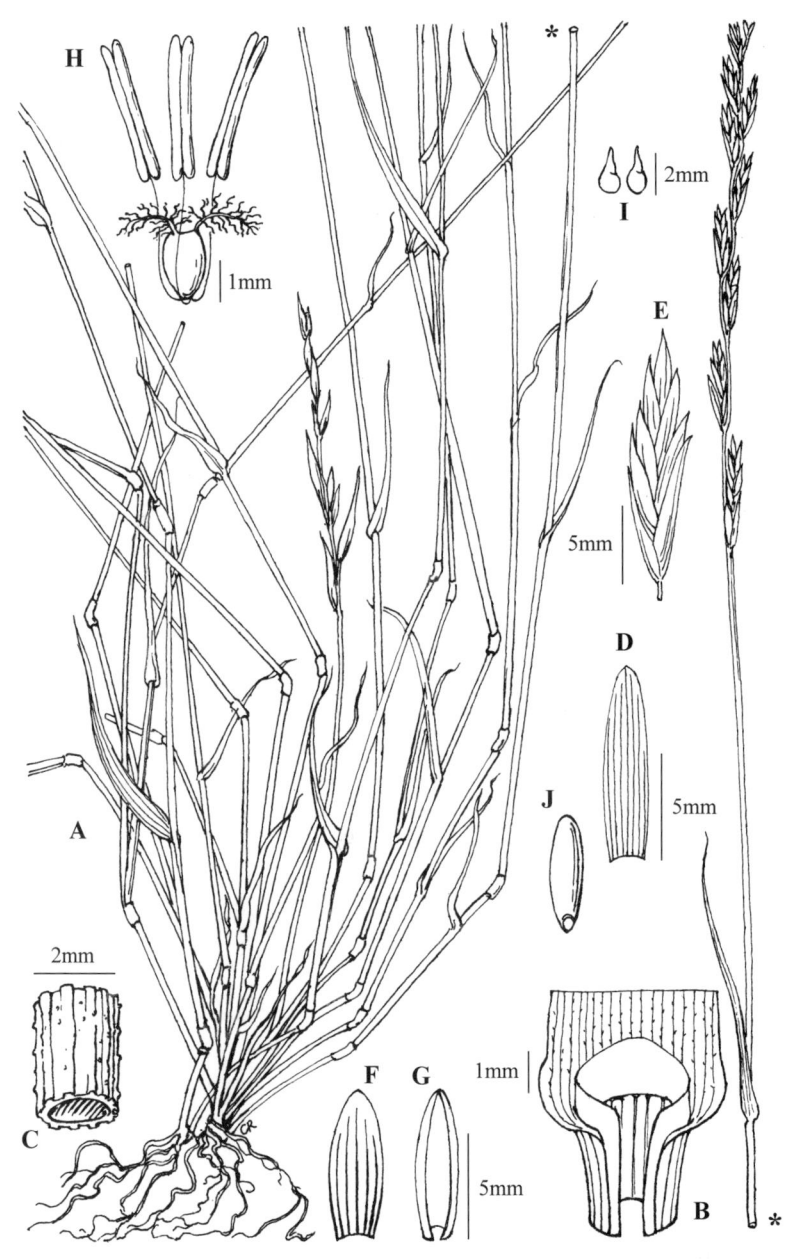

Lolium rigidum Gaudin (Mediterranean Rye-grass). A habit; B ligule; C part of culm below inflorescence; D lower glume; E terminal spikelet; F lemma; G palea; H flower; I lodicules; J caryopsis. (Source unknown). STACE 1014/852.

157. POACEAE

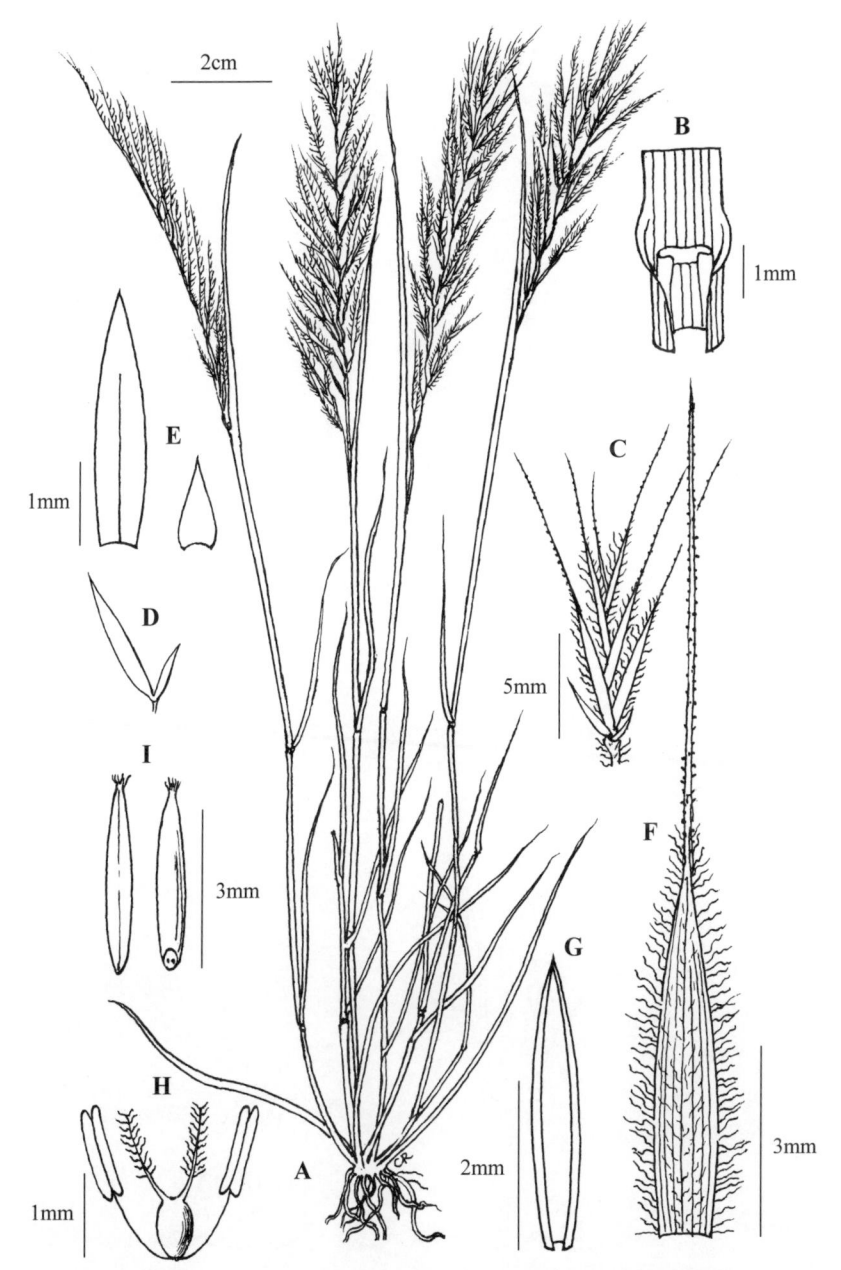

Vulpia ciliata Dumort. ssp. **ciliata** (Bearded Fescue subspecies). A habit; B ligule; C spikelet; D side view of glumes; E upper and lower glumes; F lemma; G palea; H flower; I caryopsis. (Ardingly, E Sussex). STACE 1016/854.

388   157. POACEAE

Lamarckia aurea (L.) Moench (Golden Dog's-tail). A habit; B ligule; C typical cluster of three sterile spikelets and two fertile spikelets; D two fertile spikelets; E lemma; F palea; G flower; H caryopsis. (Blackmoor, N Hants). STACE 1017/854.

157. POACEAE

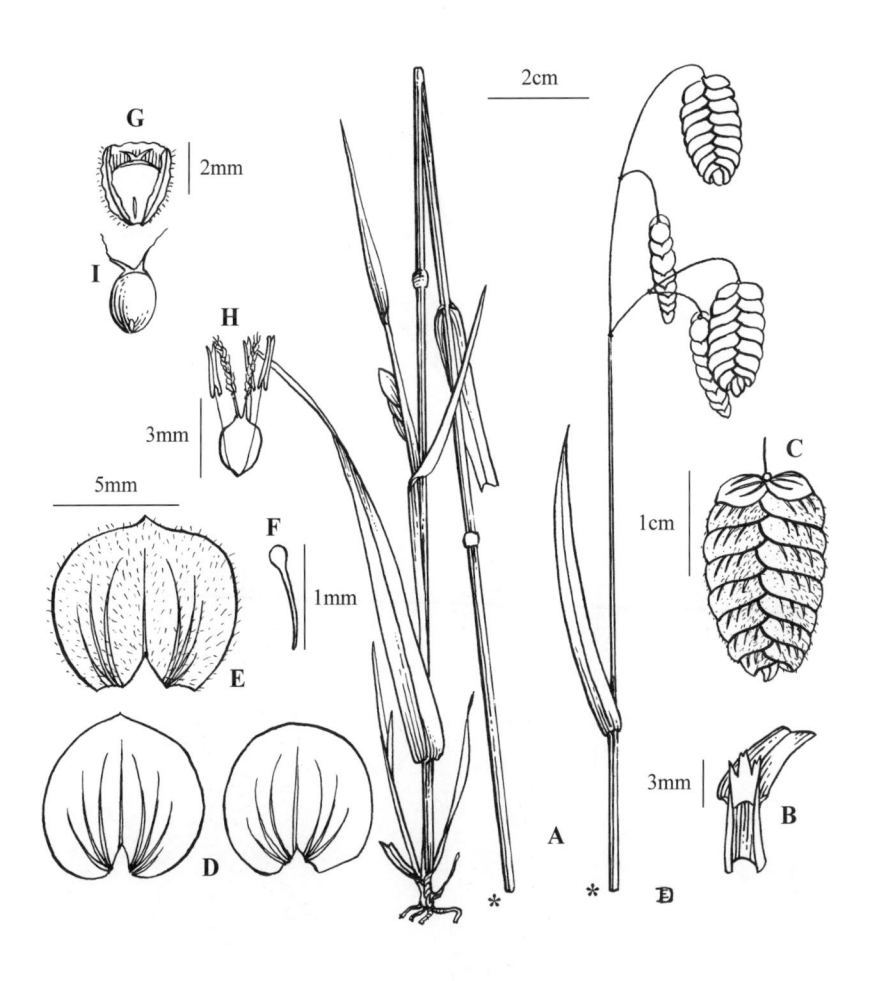

Briza maxima L. (Greater Quaking-grass). A habit; B ligule; C spikelet; D lower and upper glumes; E lemma; F clavate hair from back of mature lemma; G palea enclosing ovary; H flower; I caryopsis. (*Hort*. Ipswich, E Suffolk). STACE 1019/856.

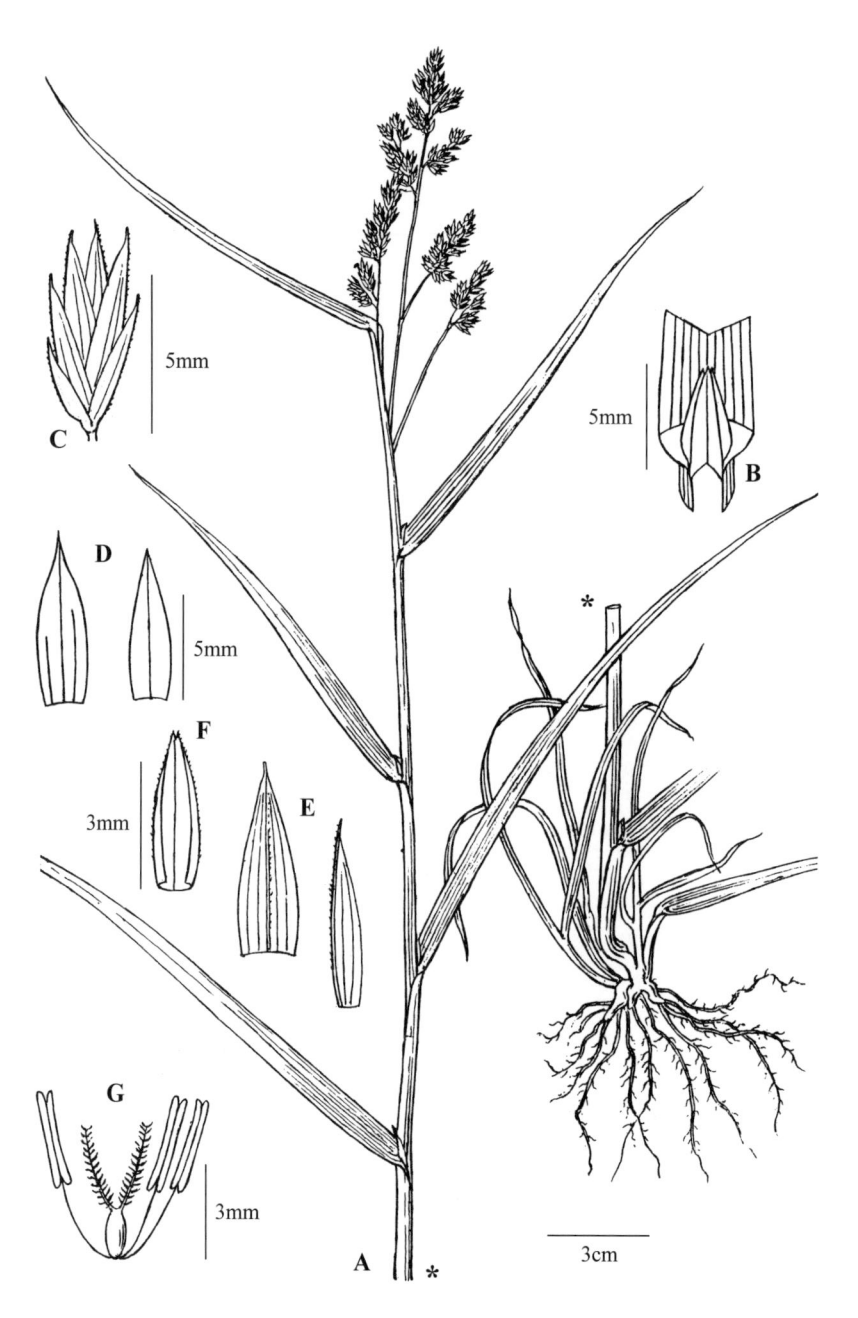

Dactylis polygama Horv. (Slender Cock's-foot). A habit; B ligule; C spikelet; D upper and lower glumes; E dorsal and side view of lemma; F palea; G flower. (Bucks). STACE 1024/859.

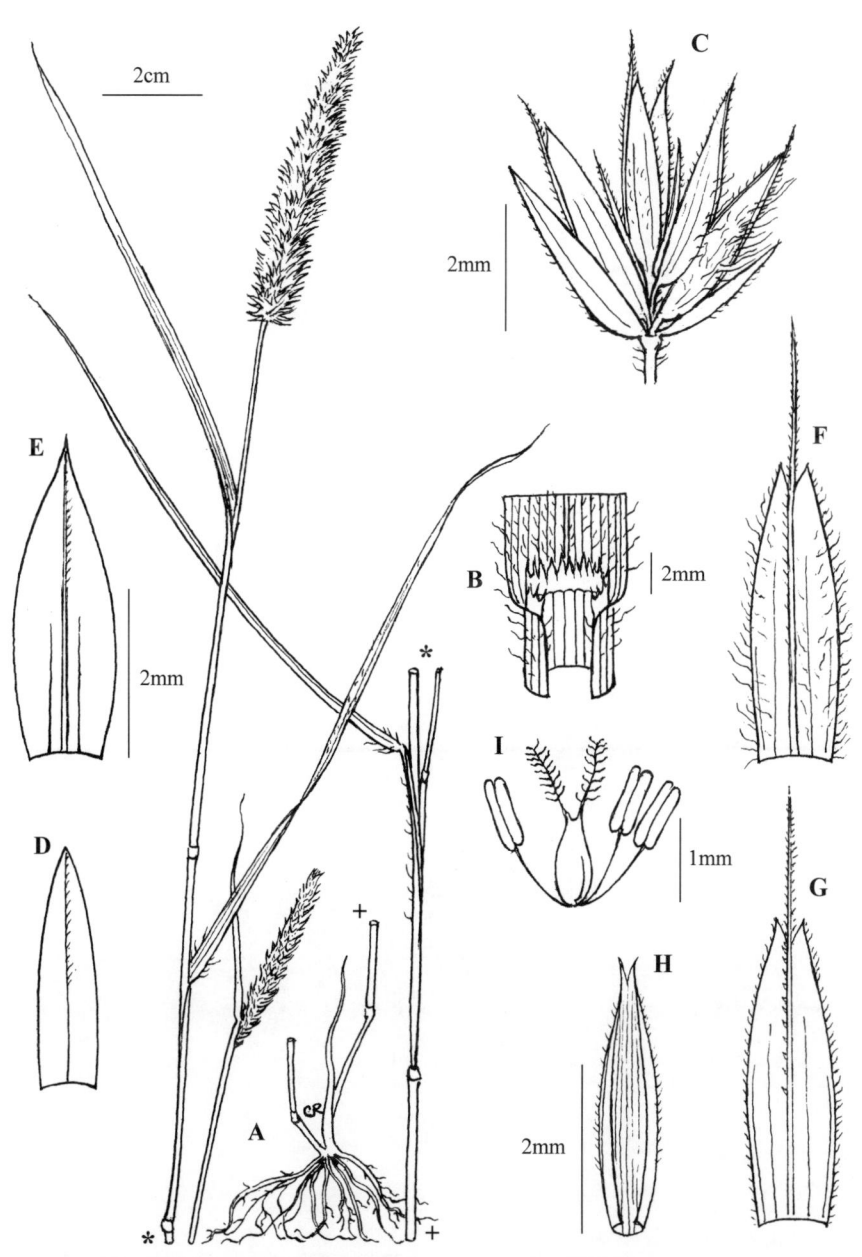

Rostraria cristata (L.) Tzvelev (Mediterranean Hair-grass). A habit; B ligule; C spikelet; D lower glume; E upper glume; F lemma of lowest floret; G lemma of middle floret; H palea; I flower. (*Hort.* Kew, Surrey). STACE 1033/867.

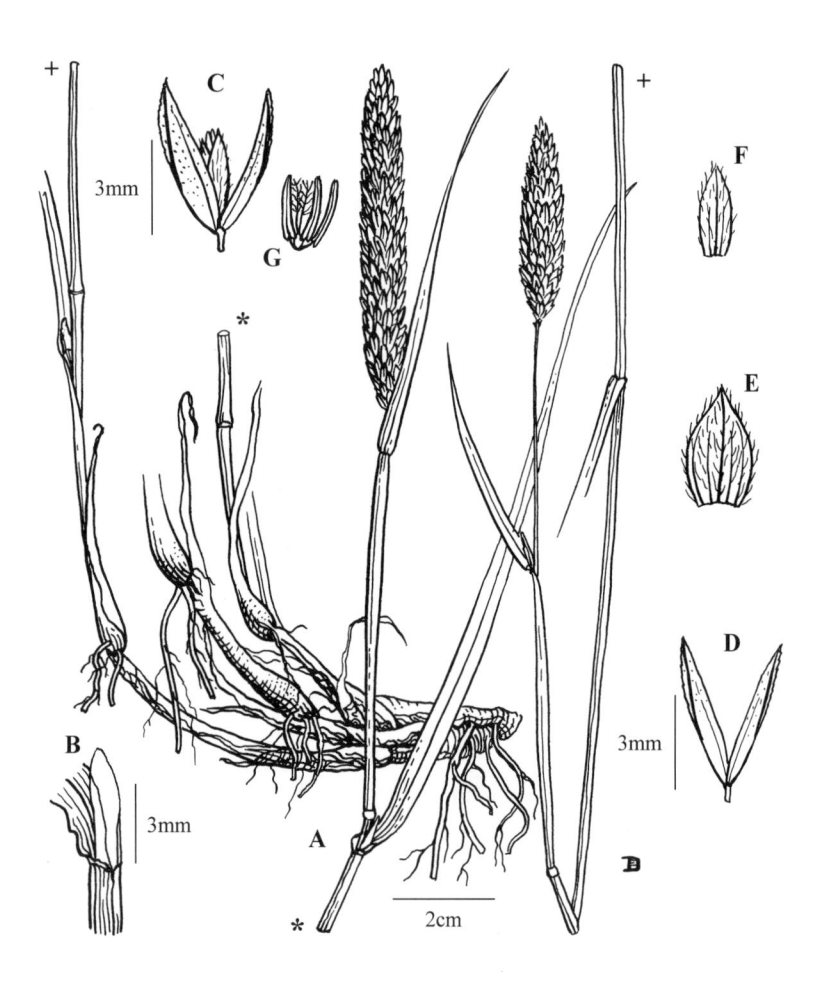

Phalaris aquatica L. (Bulbous Canary-grass). A habit; B ligule; C spikelet; D glumes; E lemma of fertile floret; F palea of fertile floret; G flower. (Source unknown). STACE 1039/872.

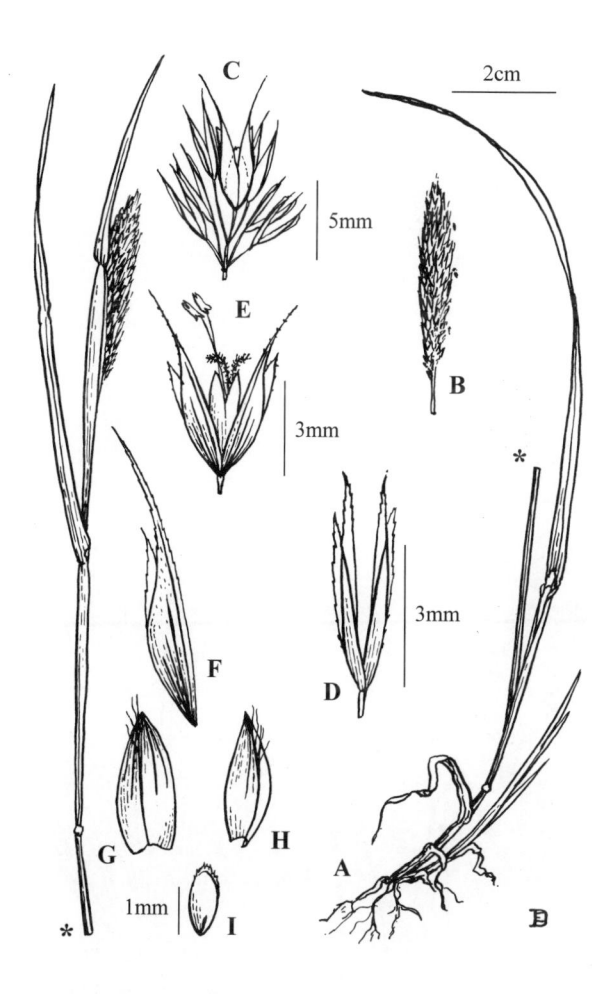

Phalaris paradoxa L. (Awned Canary-grass). A habit; B panicle; C cluster of sterile and fertile spikelets; D sterile spikelet; E fertile spikelet; F glume; G lemma; H palea; I caryopsis. (Source unknown). STACE 1040/872.

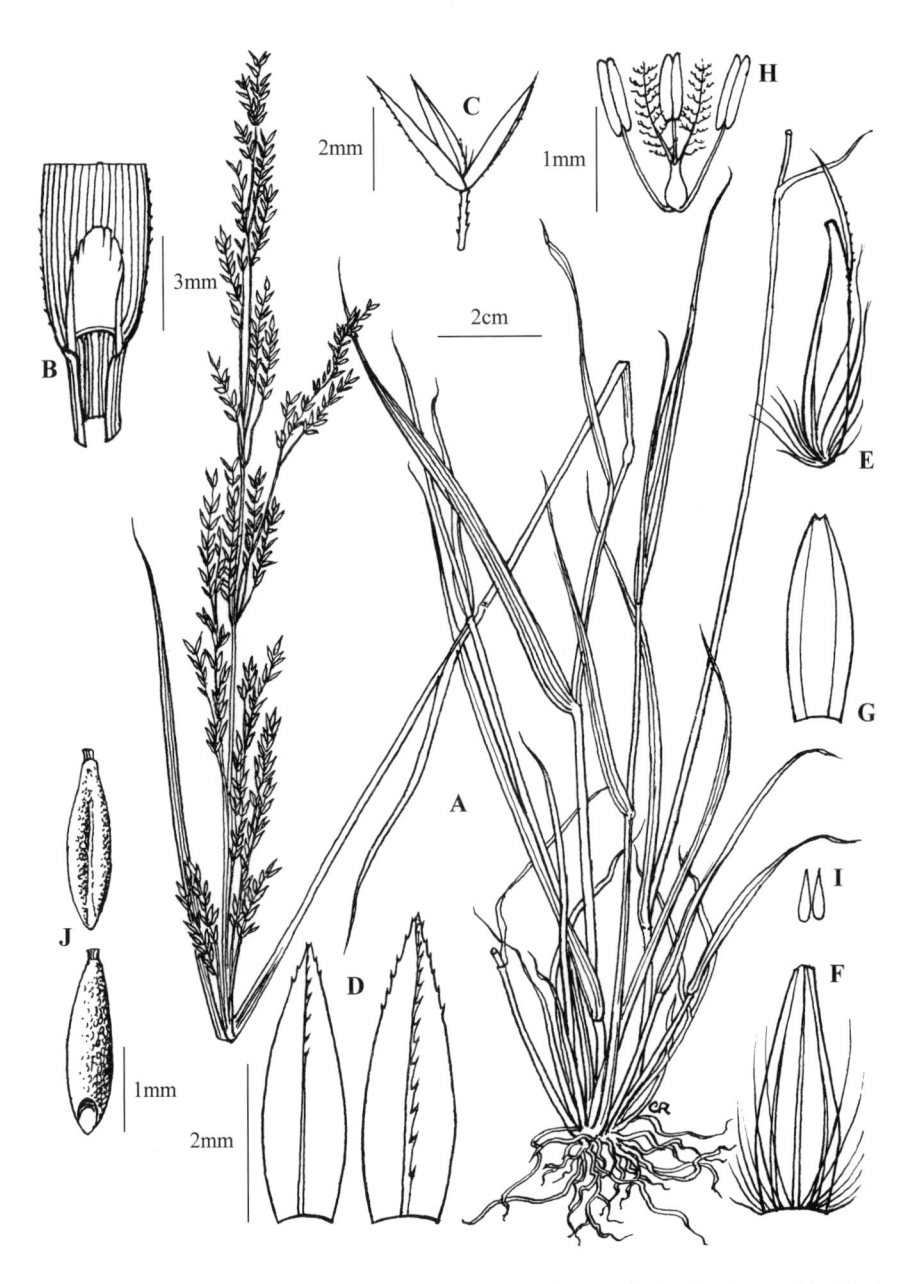

157. POACEAE

Agrostis lachnantha Nees (African Bent). A habit; B ligule; C spikelet; D glumes; E side view of lemma; F dorsal view of lemma; G palea; H flower; I lodicules; J caryopsis. (Source unknown). STACE 1044/874.

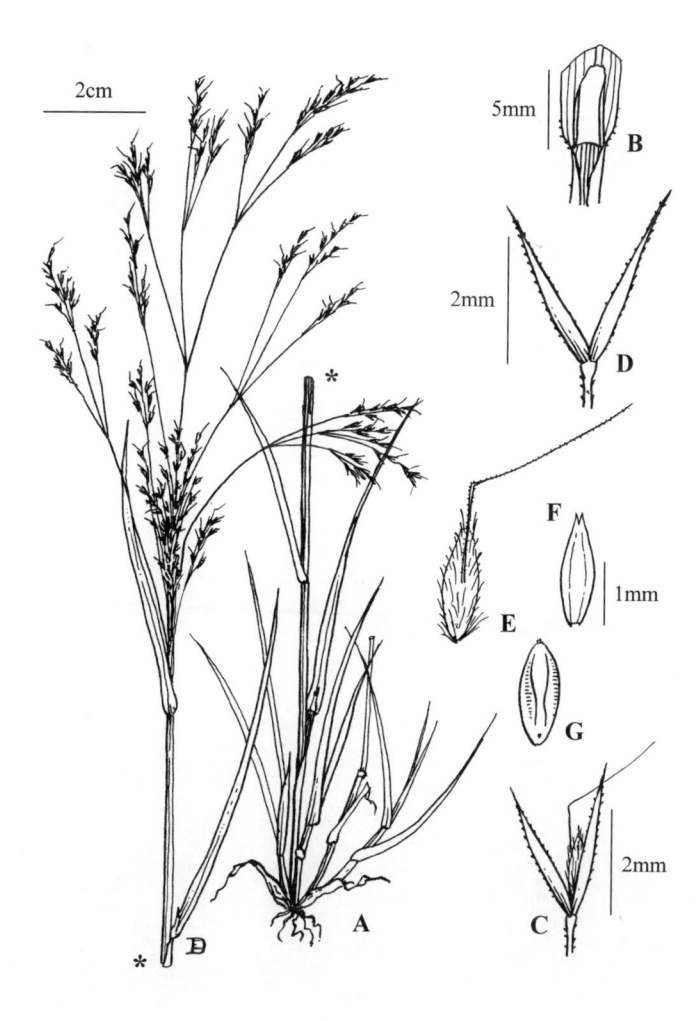

Agrostis avenacea J.F. Gmel. (Blown-grass). A habit; B ligule; C spikelet; D glumes; E lemma; F palea; G caryopsis. (Source unknown). STACE 1044/874.

Agrostis scabra Willd. (Rough Bent). A habit and mature panicle; B ligule; C spikelet; D glume; E lemma; F palea; G caryopsis. (Source unknown). STACE 1044/876.

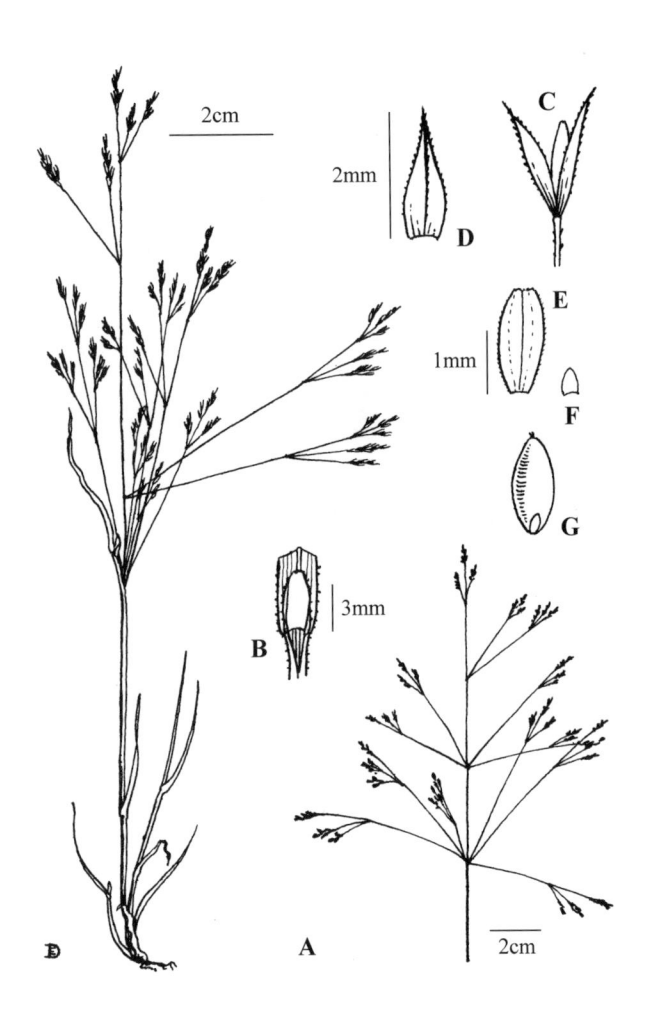

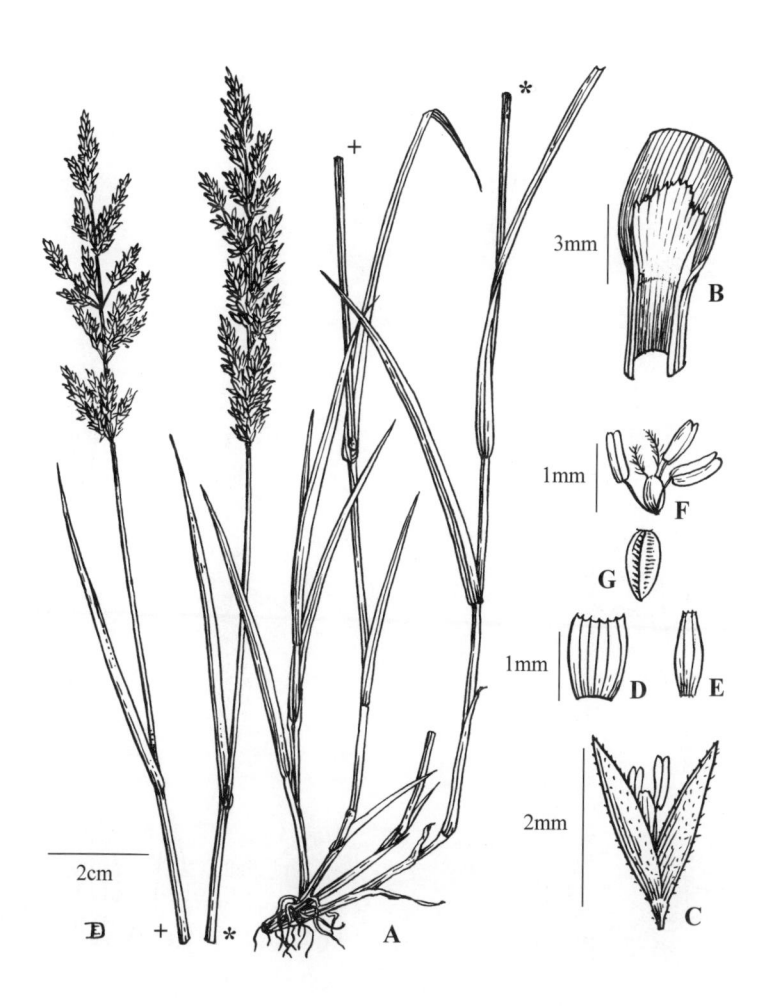

157. POACEAE

Polypogon viridis (Gouan) Breistr. (Water Bent). A habit; B ligule; C spikelet; D lemma; E palea; F flower; G caryopsis. (Source unknown). STACE 1049/879.

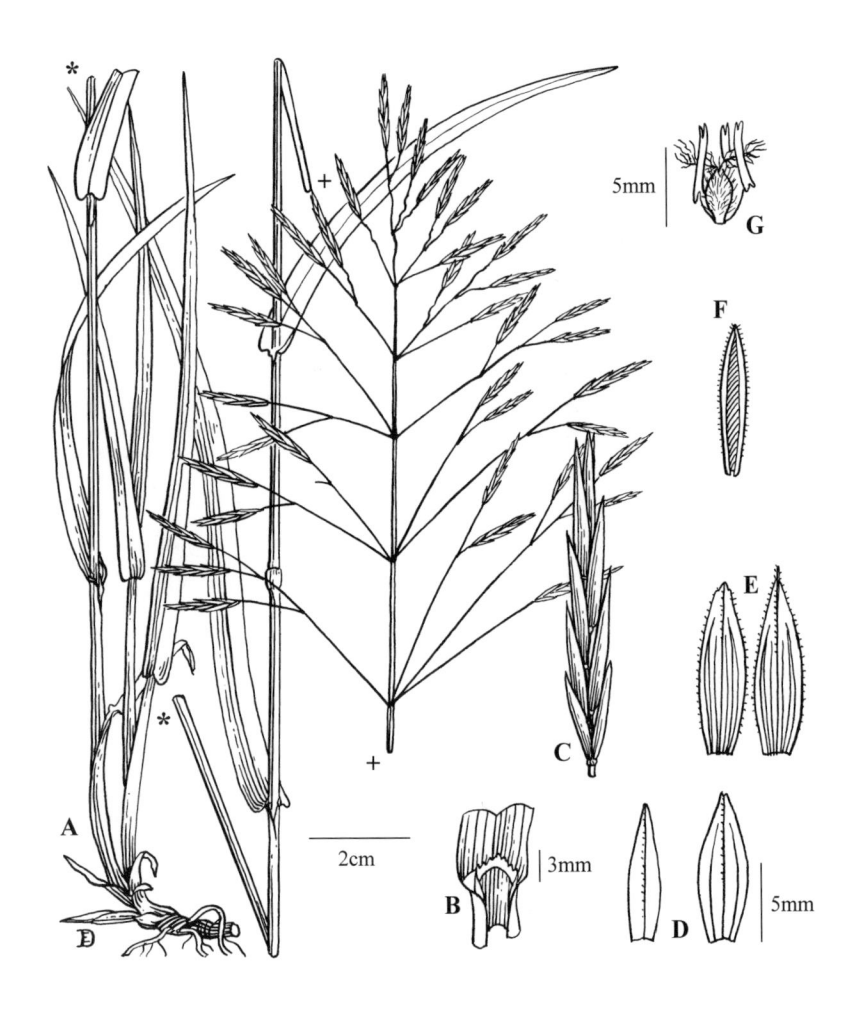

Bromopsis inermis (Leyss.) Holub ssp. **inermis** (Hungarian Brome). A habit; B ligule; C spikelet; D lower and upper glumes; E awnless and awned lemmas; F palea; G flower. (*Hort*. Ipswich, E Suffolk). STACE 1059/887.

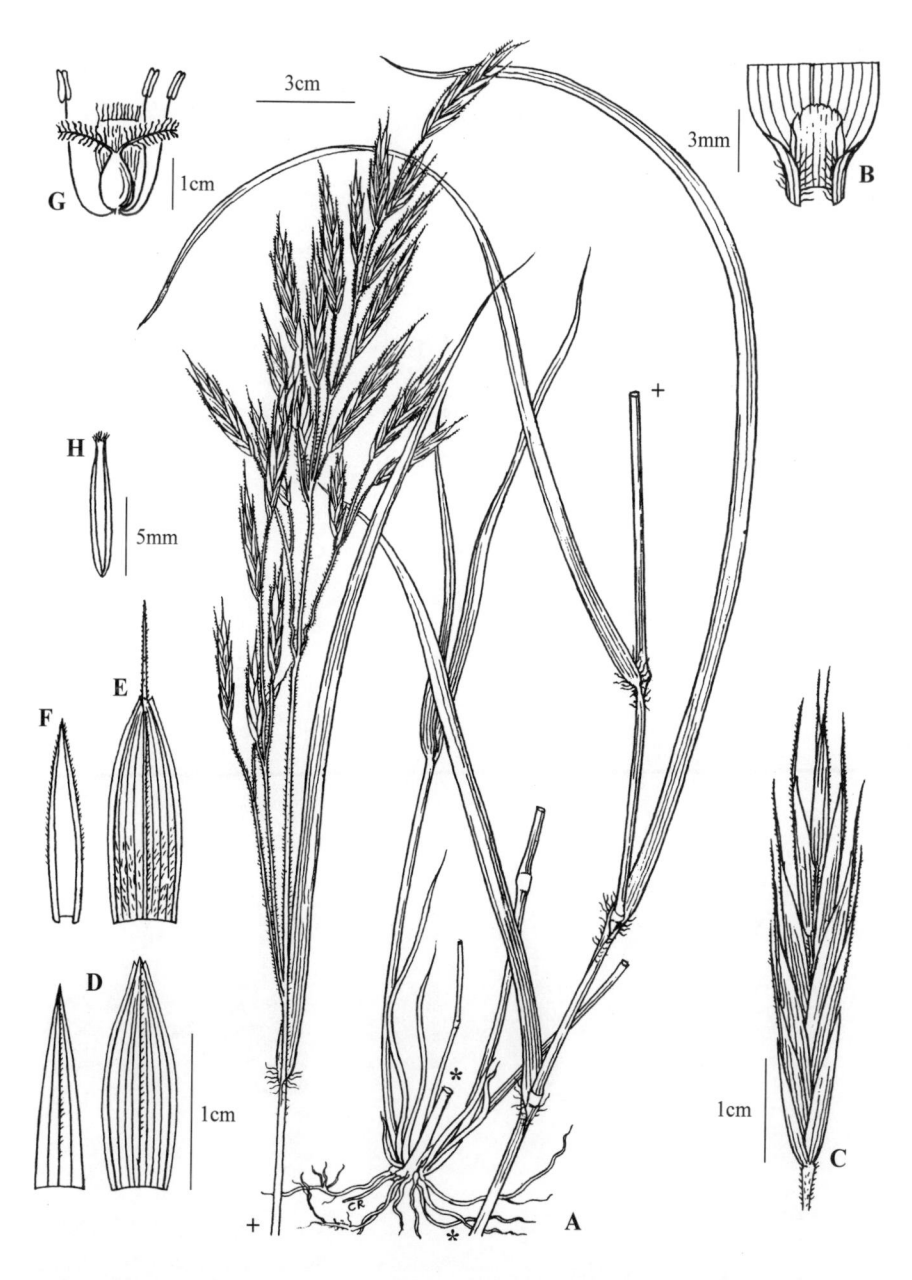

157. POACEAE

Ceratochloa carinata (Hook. & Arn.) Tutin (California Brome). A habit; B ligule; C spikelet; D lower and upper glumes; E lemma; F palea; G flower; H caryopsis. (Kew, Surrey). STACE 1061/890.

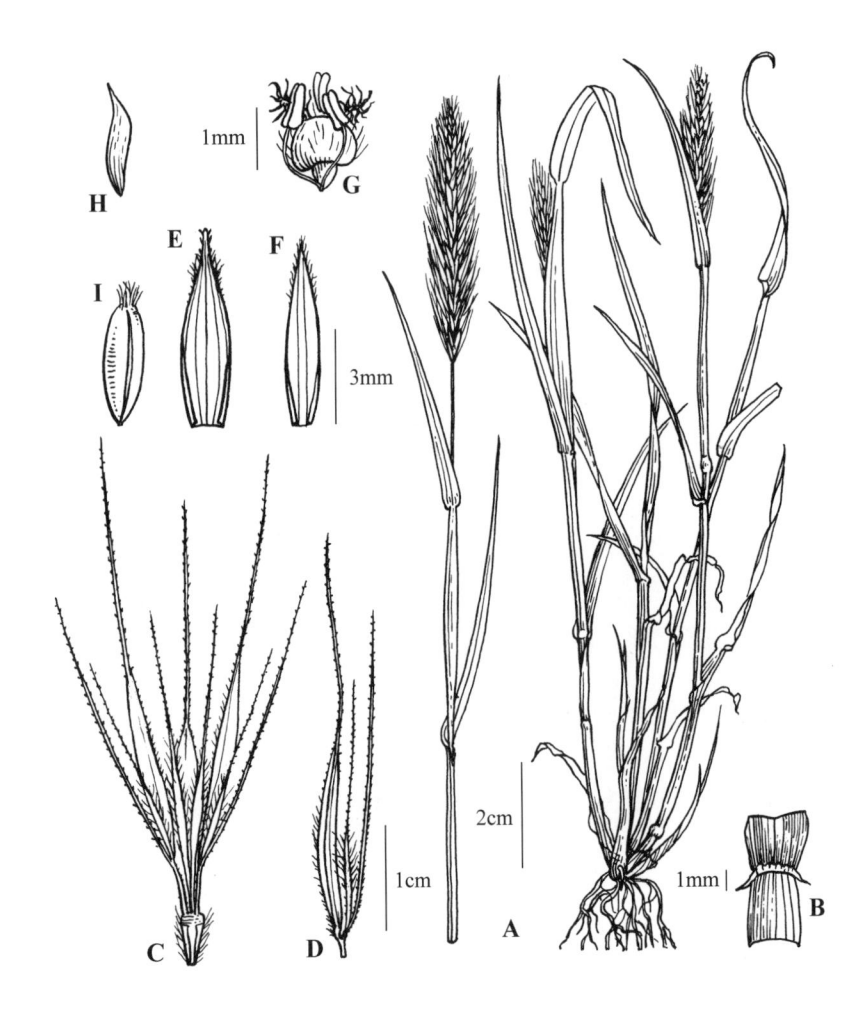

Hordeum murinum L. ssp. **glaucum** (Steud.) Tzvelev (Wall Barley subspecies). A habit and flowering shoot; B ligule; C triplet of spikelets; D sterile lateral spikelet; E lemma of fertile spikelet, minus awn; F palea; G flower; H lodicule; I caryopsis. (Source unknown). STACE 1071/897.

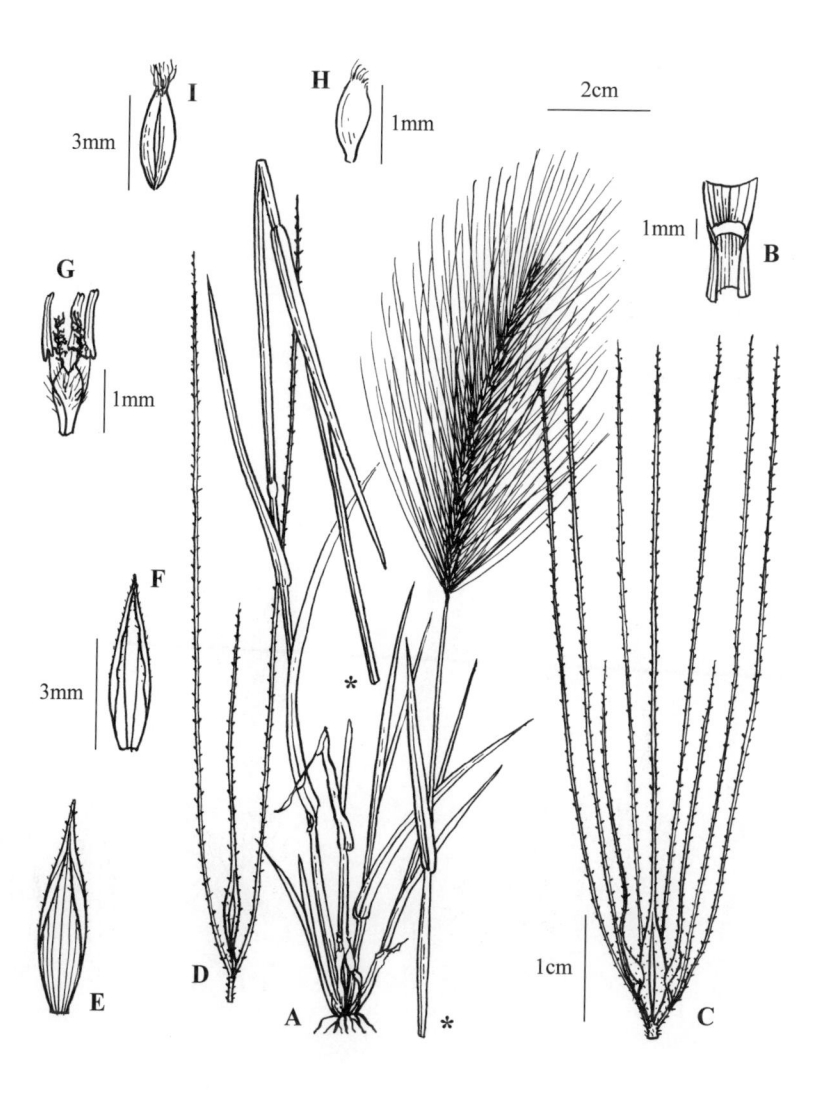

157. POACEAE

Hordeum jubatum L. (Foxtail Barley). A habit; B ligule; C triplet of spikelets; D sterile lateral spikelet; E lemma of fertile spikelet, minus awn; F palea; G flower; H lodicule; I caryopsis. (Source unknown). STACE 1071/897.

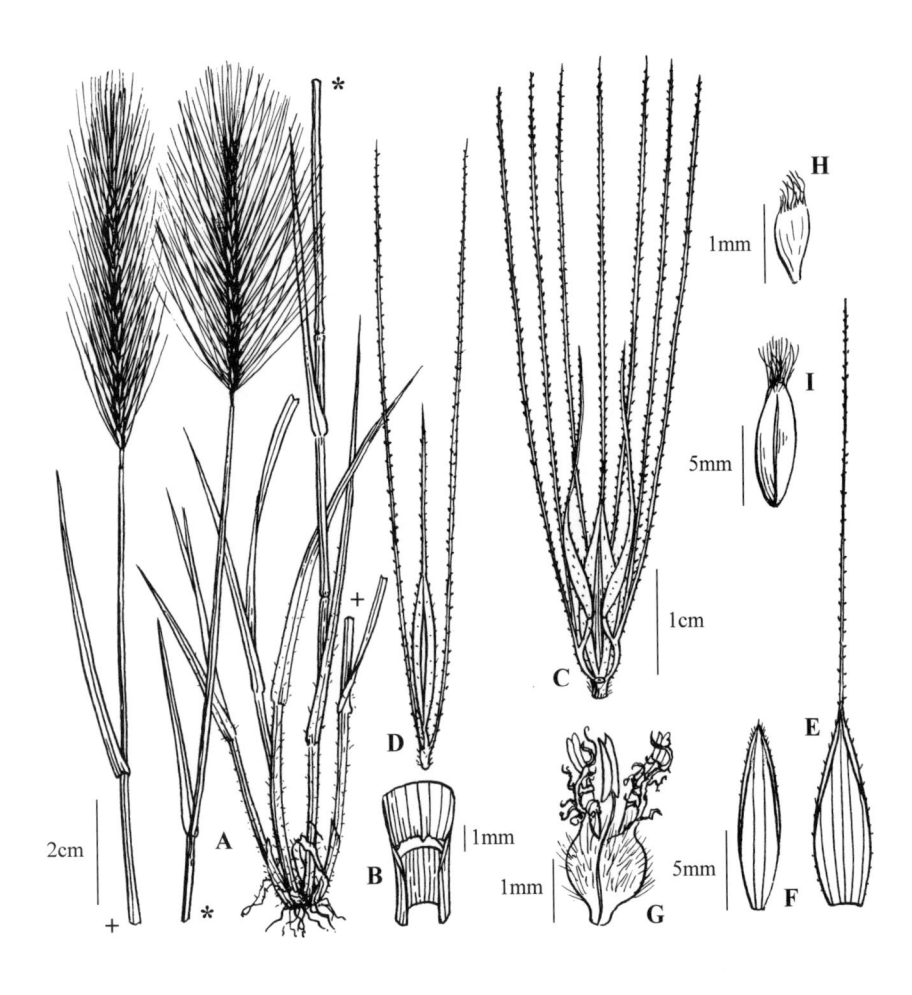

Hordeum pubiflorum Hook.f. (Antarctic Barley). A habit; B ligule; C triplet of spikelets; D sterile lateral spikelet; E lemma of fertile spikelet; F palea of fertile spikelet; G flower; H lodicule; I caryopsis. (Source unknown). STACE 1072/897.

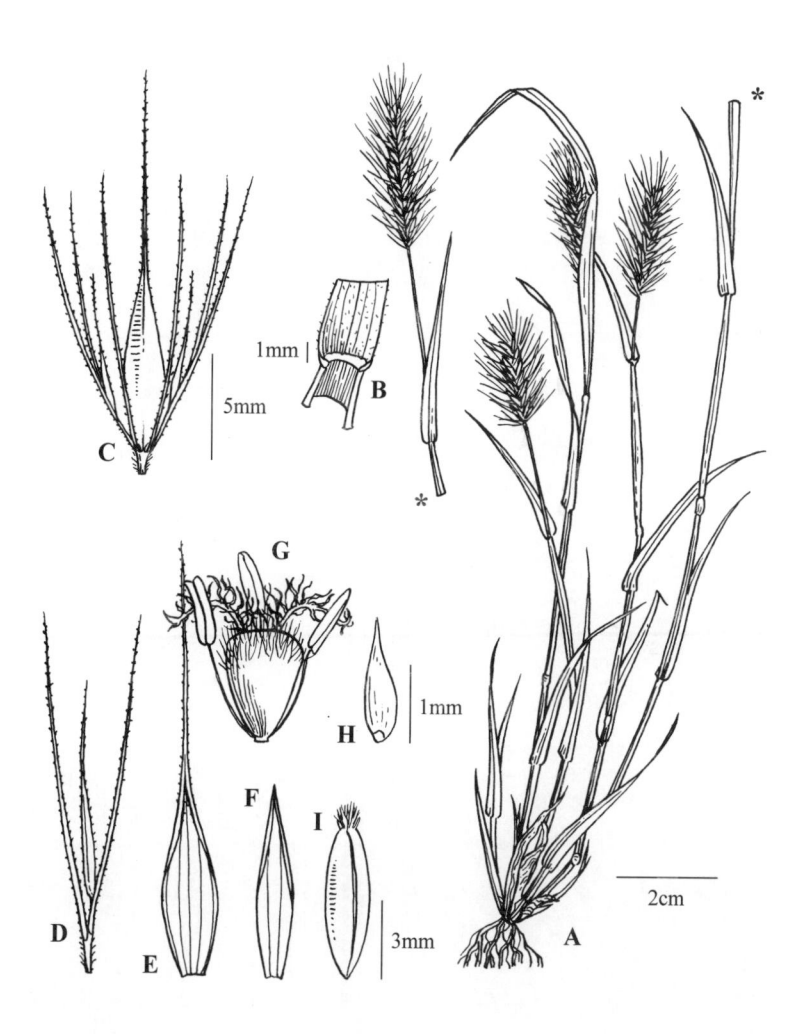

Hordeum geniculatum All. (Mediterranean Barley). A habit; B ligule; C triplet of spikelets; D sterile lateral spikelet; E lemma of fertile spikelet; F palea; G flower; H lodicule; I caryopsis. (Source unknown). STACE 1072/897.

404
157. POACEAE

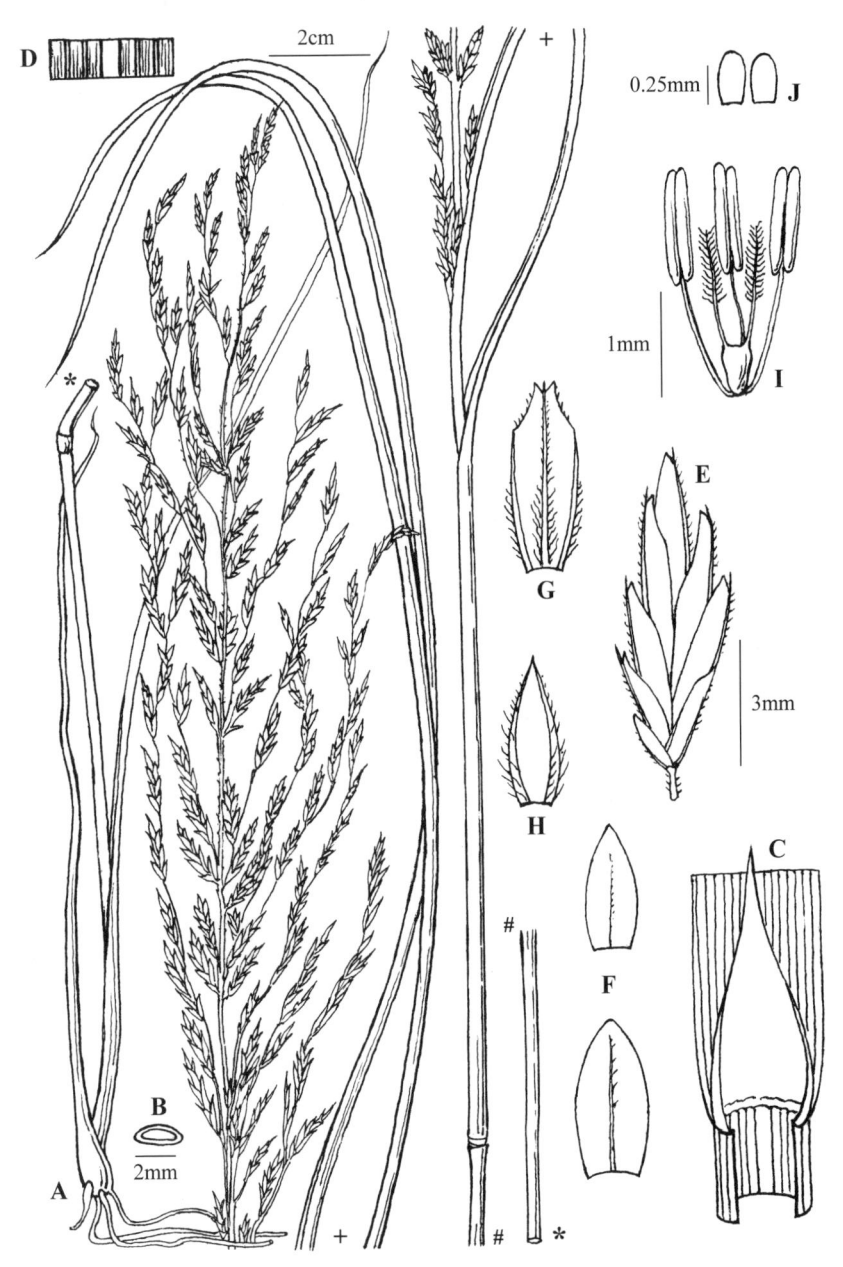

Leptochloa fusca (L.) Kunth (Brown Beetle-grass). A habit; B TS of culm above node; C ligule; D part of leaf blade showing broad midrib; E spikelet; F upper and lower glumes; G lemma; H palea; I flower; J lodicules. (Blackmoor, N Hants). STACE 1076/901.

157. POACEAE

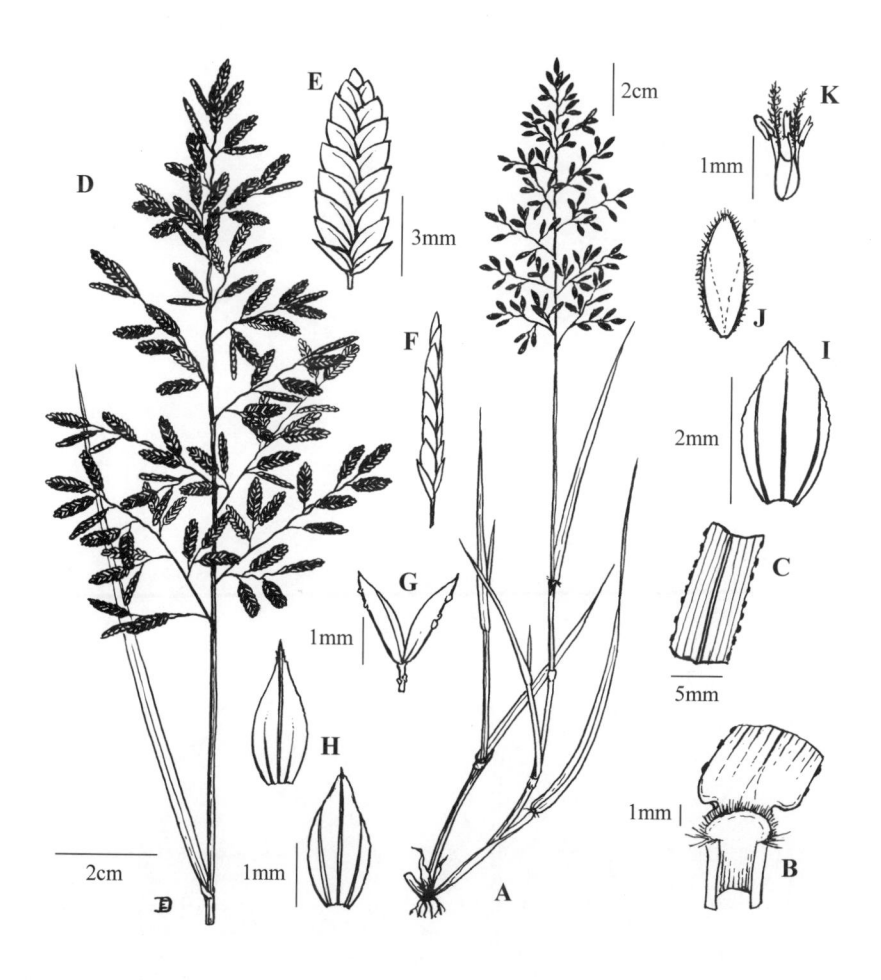

Eragrostis cilianensis (All.) Vignolo ex Janch. (Stink-grass). A habit; B ligule; C part of leaf showing marginal glands; D panicle; E spikelet; F dorsal view of spikelet; G glumes; H inner view of upper and lower glumes; I lemma; J palea; K flower. (Source unknown). STACE 1079/903.

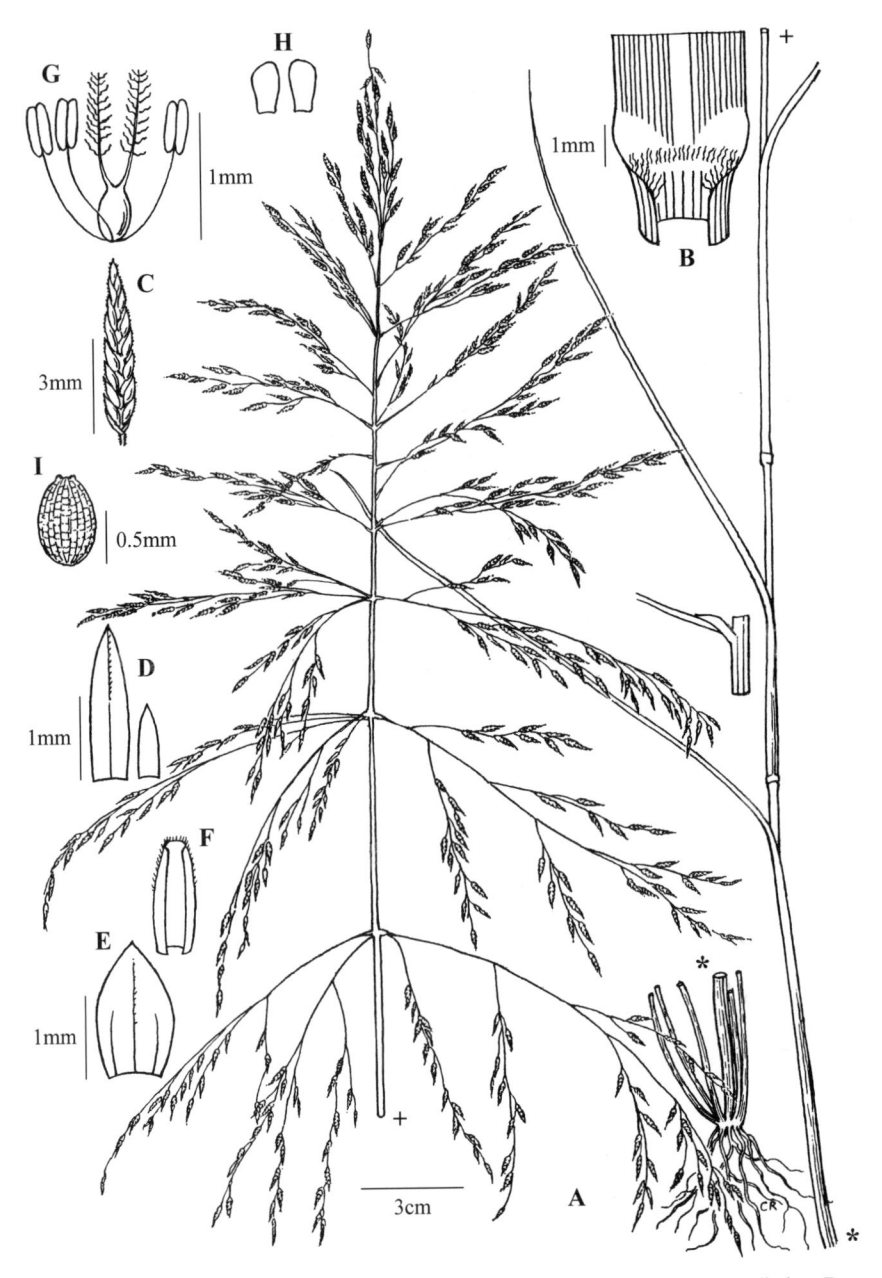

Eragrostis parviflora (R. Br.) Trin. (Weeping Love-grass). A habit; B ligule; C spikelet; D upper and lower glumes; E lemma; F palea; G flower; H lodicules; I caryopsis. (Source unknown). STACE 1079/903.

157. POACEAE

Eleusine indica (L.) Gaertn. ssp. **africana** (Kenn.-O'Byrne) S.M. Phillips (Yard-grass subspecies). A habit; B ligule; C spikelet; D lower and upper glumes; E lemma; F palea; G flower; H lodicules; I caryopsis. (Source unknown). STACE 1080/904.

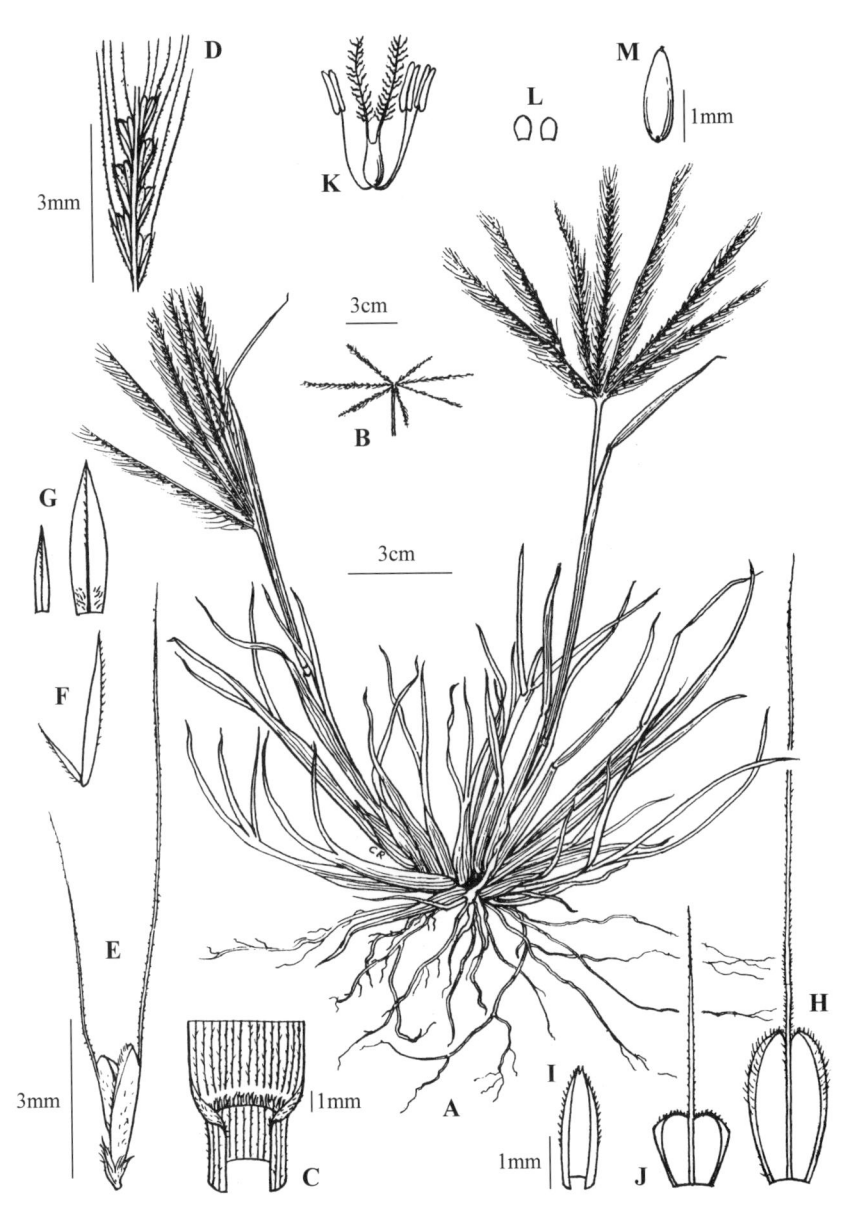

Chloris truncata R. Br. (Windmill-grass). A habit; B inflorescence at maturity; C ligule; D part of spike; E spikelet minus glumes; F side view of glumes; G lower and upper glumes; H lemma of lower floret; I palea of lower floret; J reduced lemma of upper floret; K flower; L lodicules; M caryopsis. (Blackmoor, N Hants). STACE 1083/906.

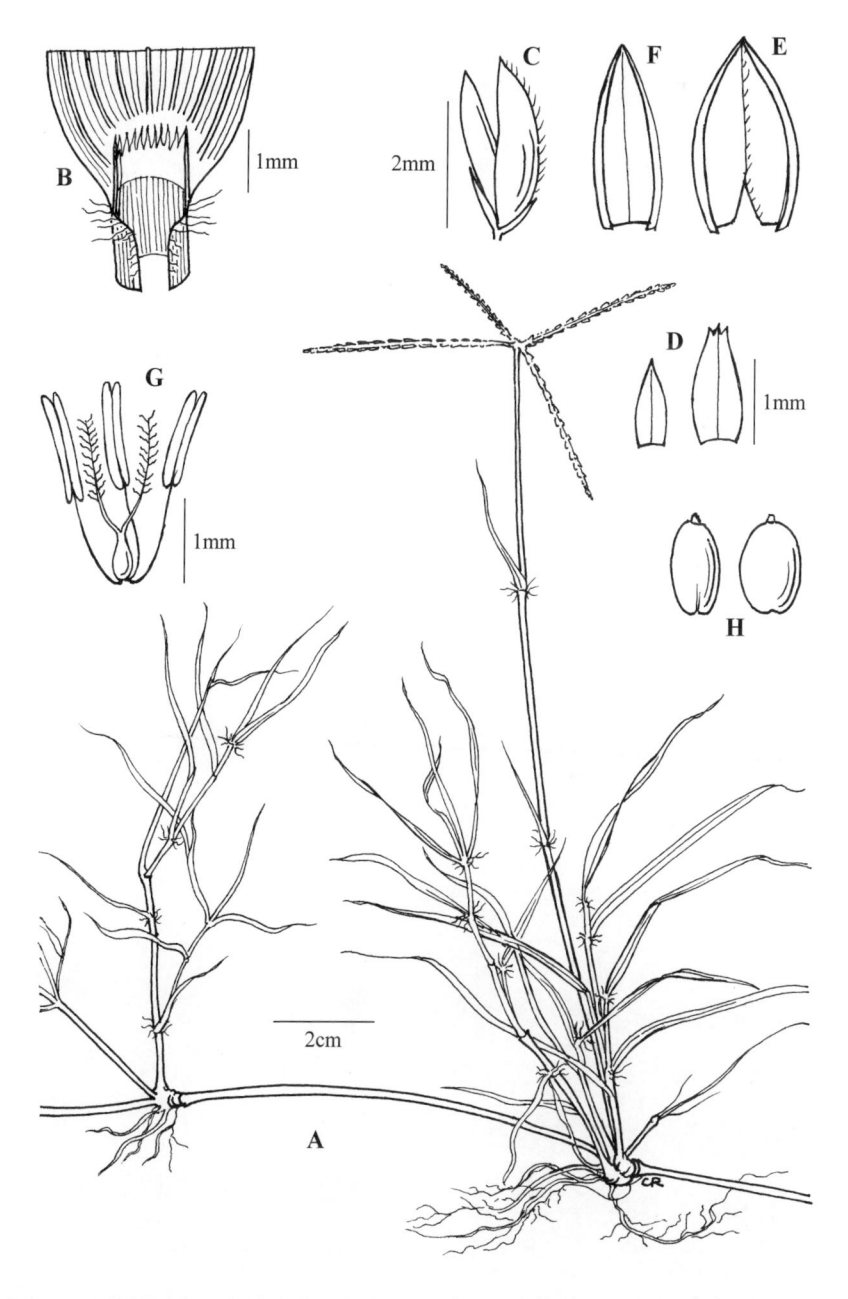

157. POACEAE

Cynodon incompletus Nees (African Bermuda-grass). A habit; B ligule; C spikelet; D lower and upper glumes; E lemma; F palea; G flower; H caryopsis. (Blackmoor, N Hants). STACE 1083/907.

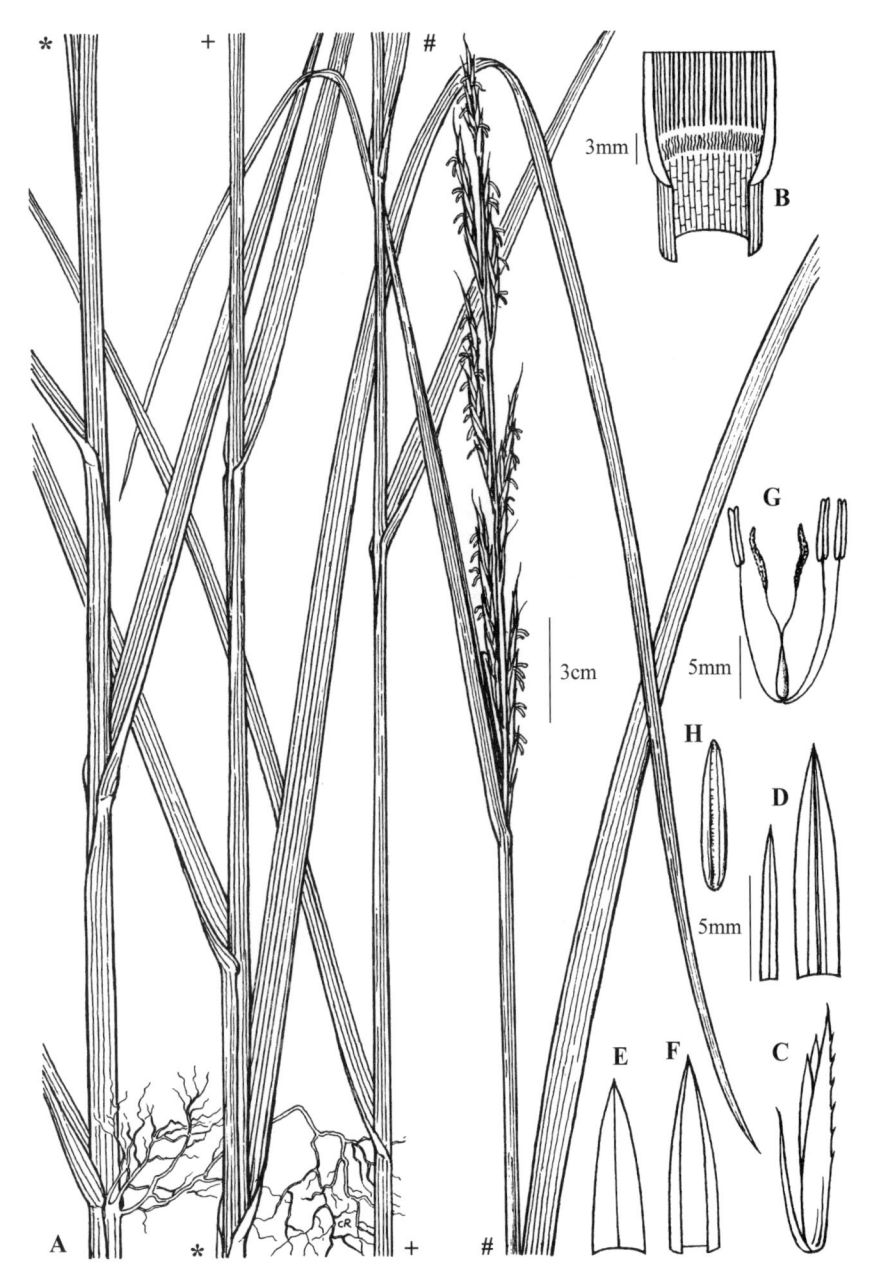

Spartina alterniflora Loisel. var. **glabra** (Muhl. ex Bigelow) Fernald (Smooth Cord-grass variety). A habit; B ligule; C spikelet; D lower and upper glumes; E lemma; F palea; G flower; H caryopsis. (Eling, S Hants). STACE 1085/907.

Tragus racemosus (L.) All. (European Bur-grass). A habit; B ligule; C group of spikelets at one node; D lower and upper glumes; E lemma; F palea. (Source unknown). STACE 1085/908.

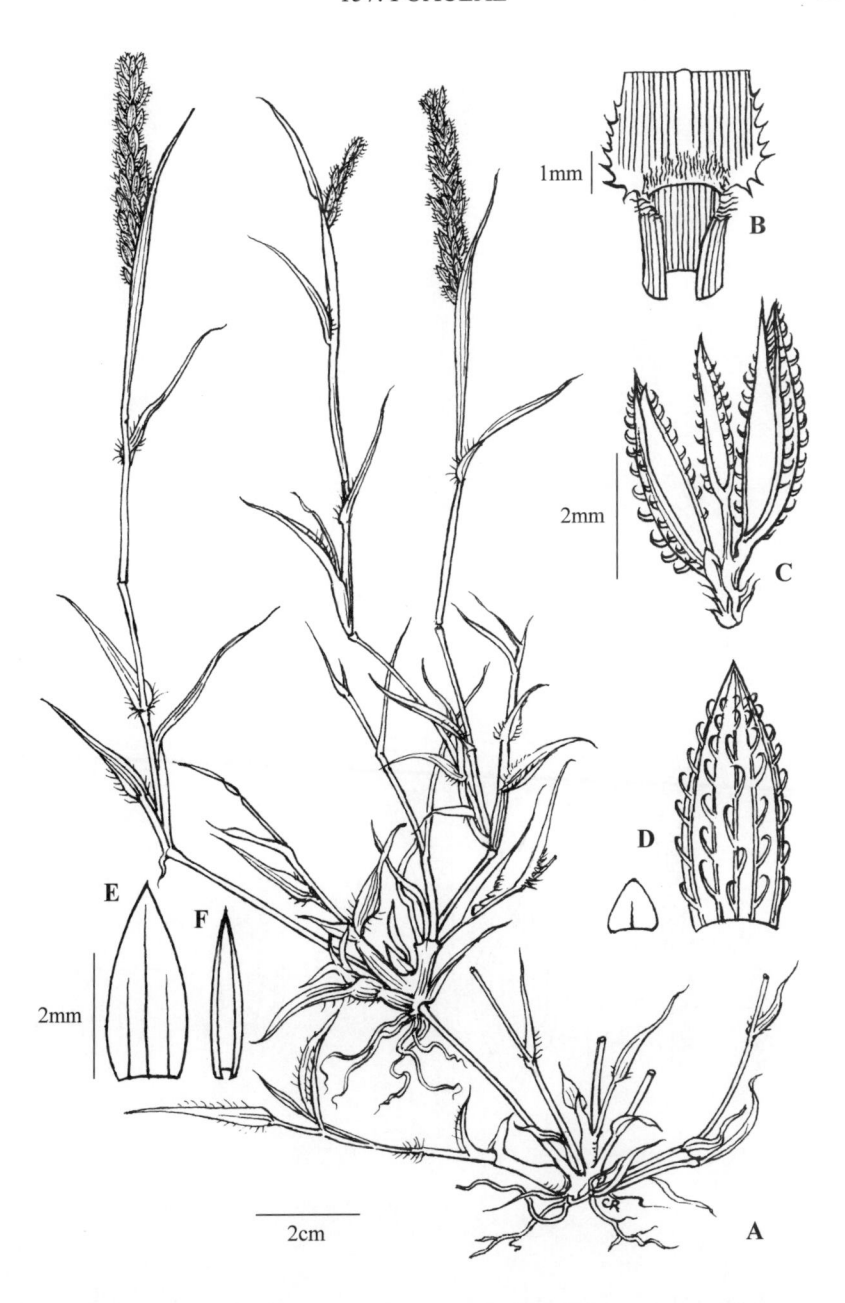

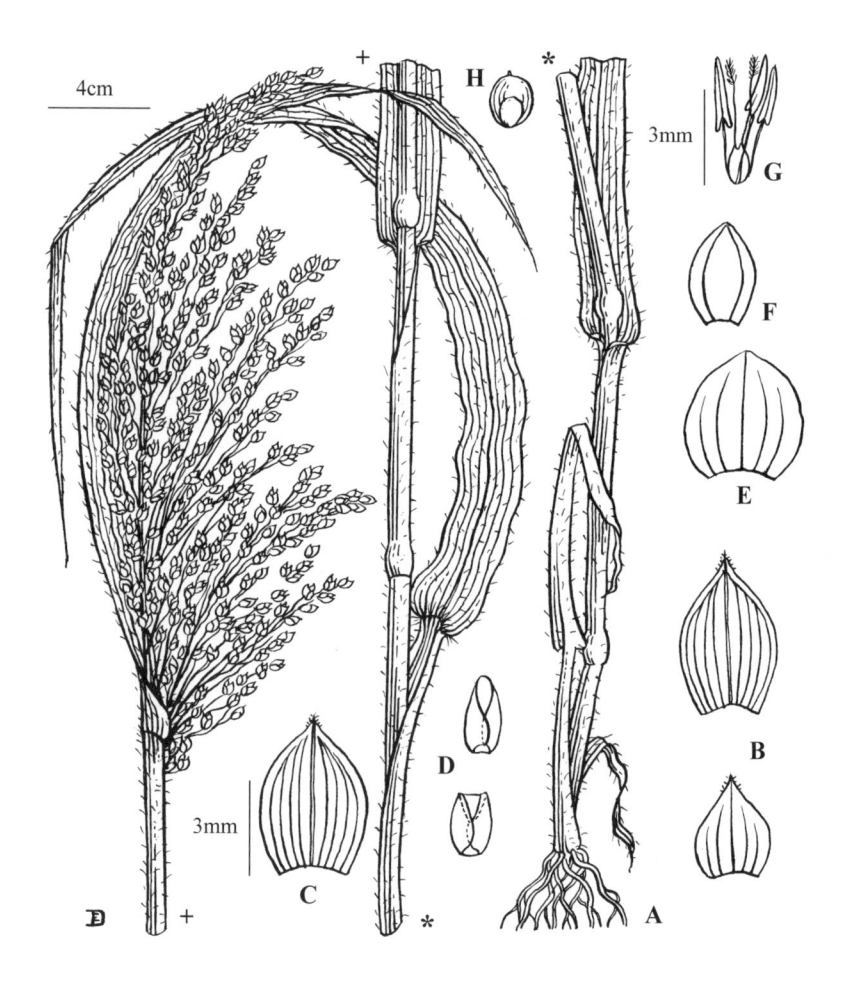

Panicum miliaceum L. (Common Millet). A habit; B lower and upper glumes; C lemma of lower floret; D palea of lower floret, sometimes with apex folded over; E lemma of upper floret; F palea of upper floret; G flower; H caryopsis. (*Hort*. Sevenoaks, W Kent). STACE 1088/910.

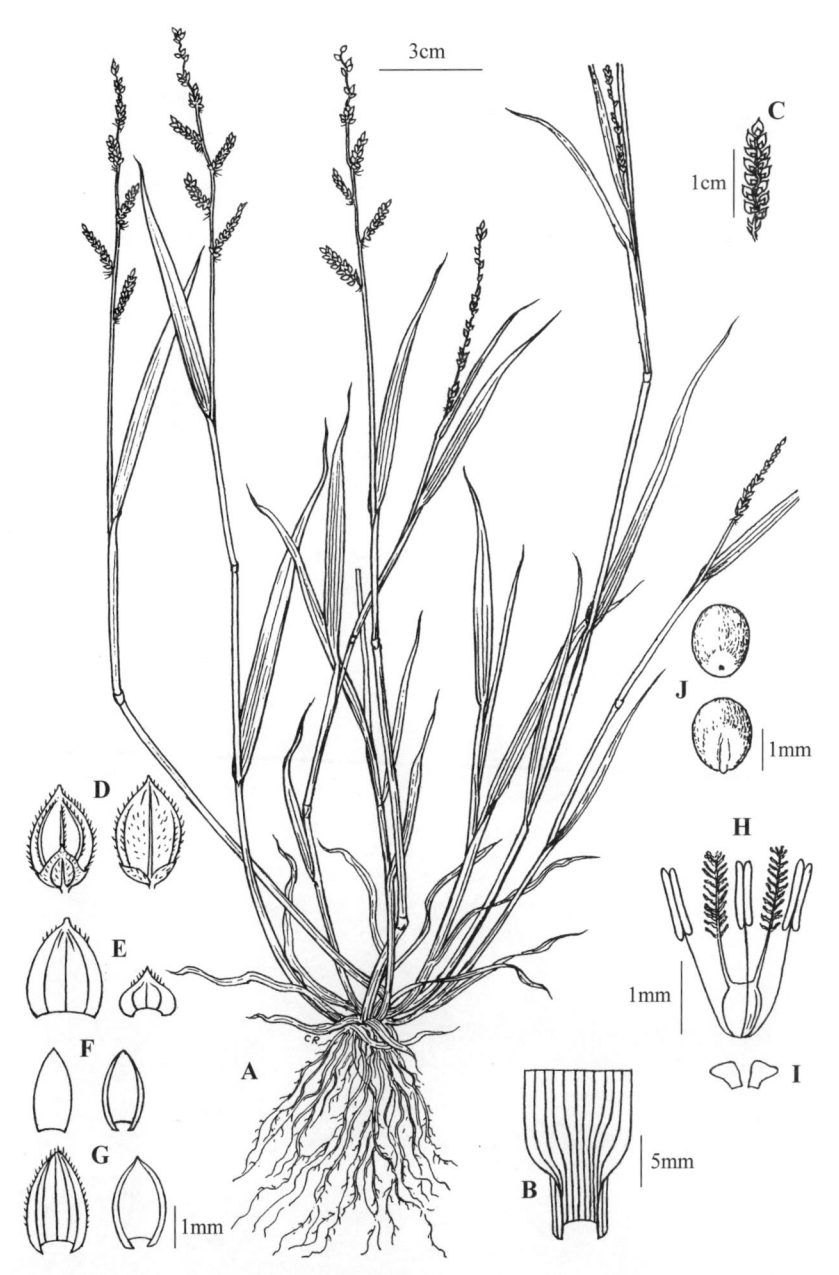

Echinochloa colona (L.) Link (Shama Millet). A habit; B ligule (absent); C flower spike; D front and back of spikelet; E upper and lower glumes; F lemma and palea of lower floret; G lemma and palea of upper floret; H flower; I lodicules; J caryopsis. (*Hort.* Ware, Herts). STACE 1090/911.

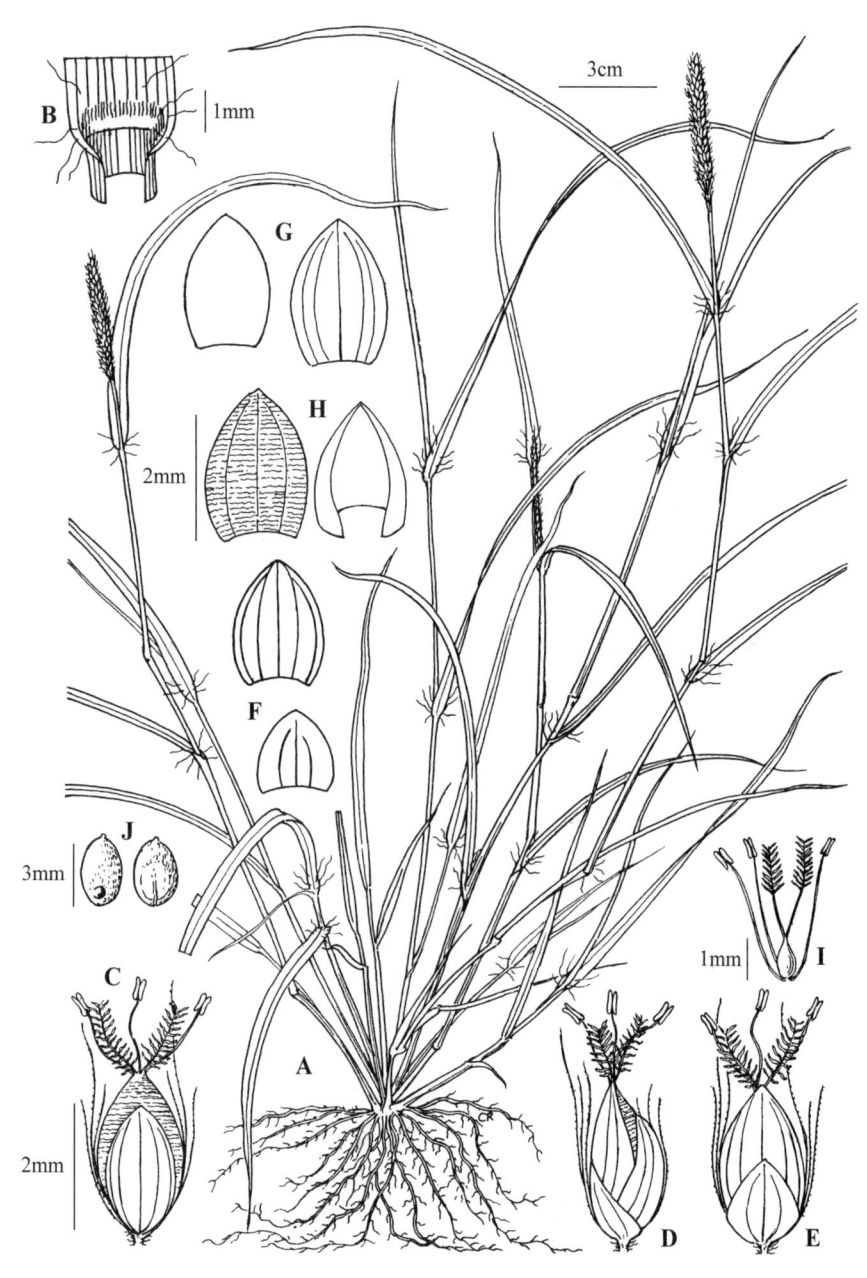

Setaria parviflora (Poir.) Kerguélen (Knotroot Bristle-grass). A habit; B ligule; C, D, E three views of spikelet; F upper and lower glumes; G lemma and palea of lower floret; H lemma and palea of upper floret; I flower; J caryopsis. (*Hort.* Ware, Herts). STACE 1095/915.

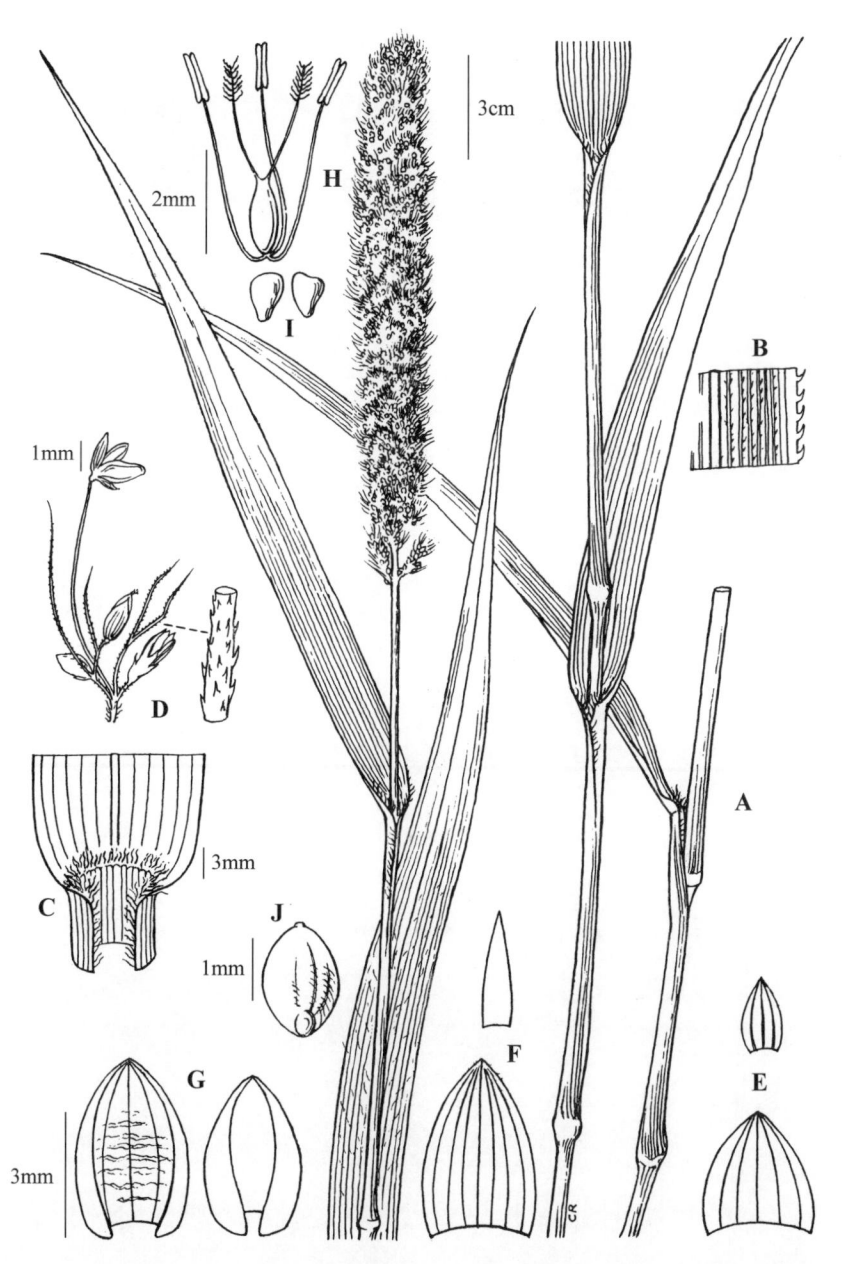

157. POACEAE

Setaria italica (L.) P. Beauv. (Foxtail Bristle-grass). A habit; B part of leaf; C ligule; D cluster of spikelets with barbed bristles; E upper and lower glumes; F lemma and palea of lower floret; G lemma and palea of upper floret; H flower; I lodicules; J caryopsis. (Source unknown). STACE 1096/915.

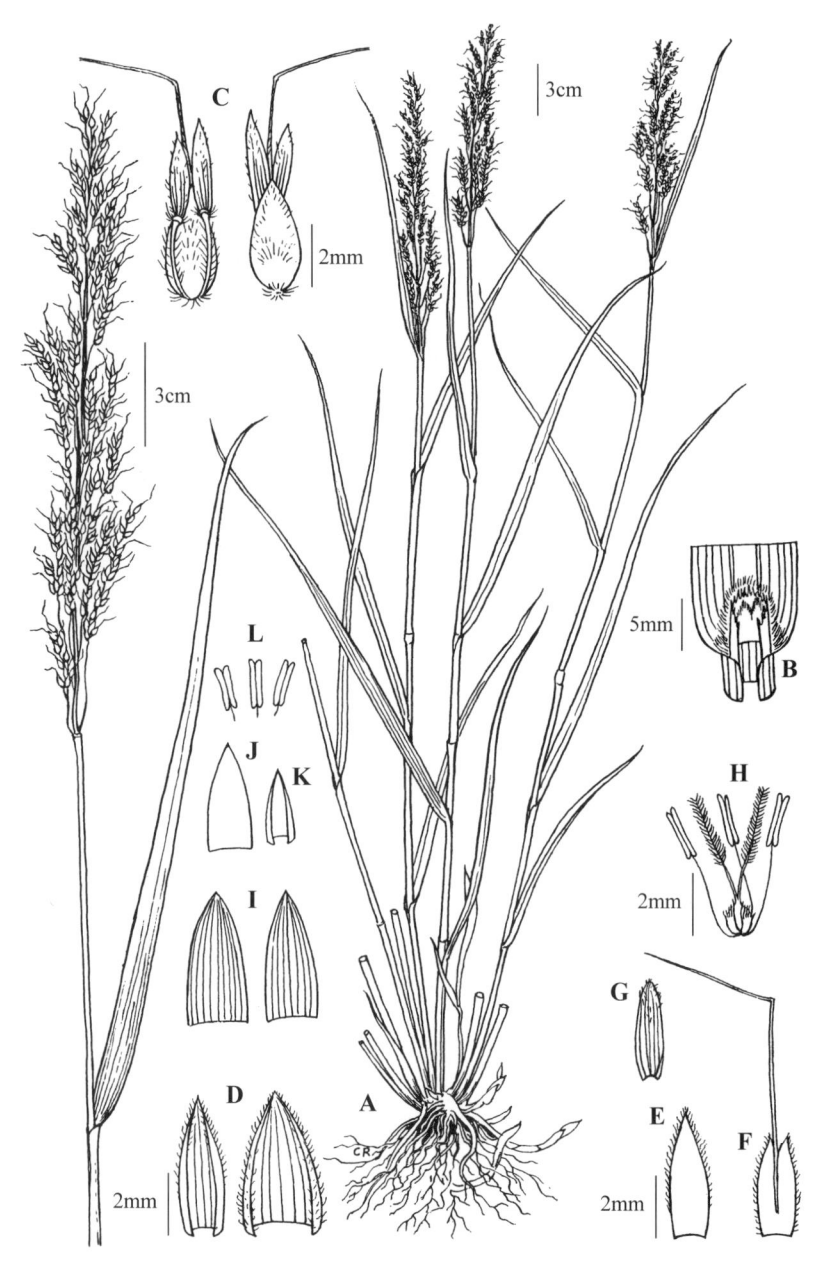

Sorghum halepense (L.) Pers. (Johnson-grass). A habit; B ligule; C terminal triplet of one sessile and two stalked spikelets; D upper and lower glumes of sessile spikelet; E, F, G, H, lower lemma, upper lemma, palea and flower of sessile spikelet; I, J, K, L glumes, lemma, palea and anthers of stalked spikelet. (*Hort.* Ware, Herts). STACE 1097/918.

157. POACEAE 417

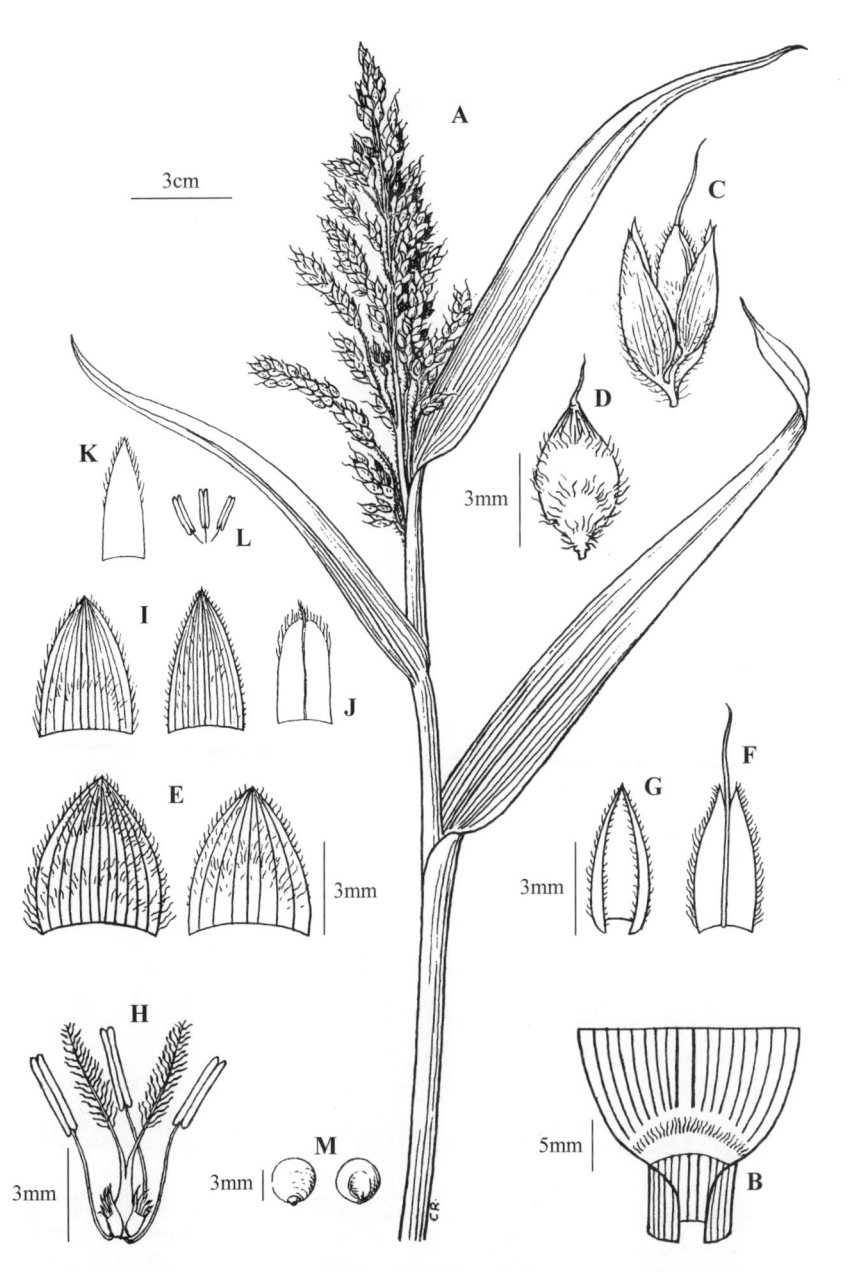

Sorghum bicolor (L.) Moench (Great Millet variant 1). A habit; B ligule; C terminal triplet of spikelets; D sessile spikelet; E, F, G, H upper and lower glumes, upper lemma, palea and flower of sessile spikelet; I, J, K, L upper and lower glumes, upper lemma, palea and anthers of stalked spikelet; M caryopsis. Lower lemmas not observed. (Source unknown). STACE 1098/918.

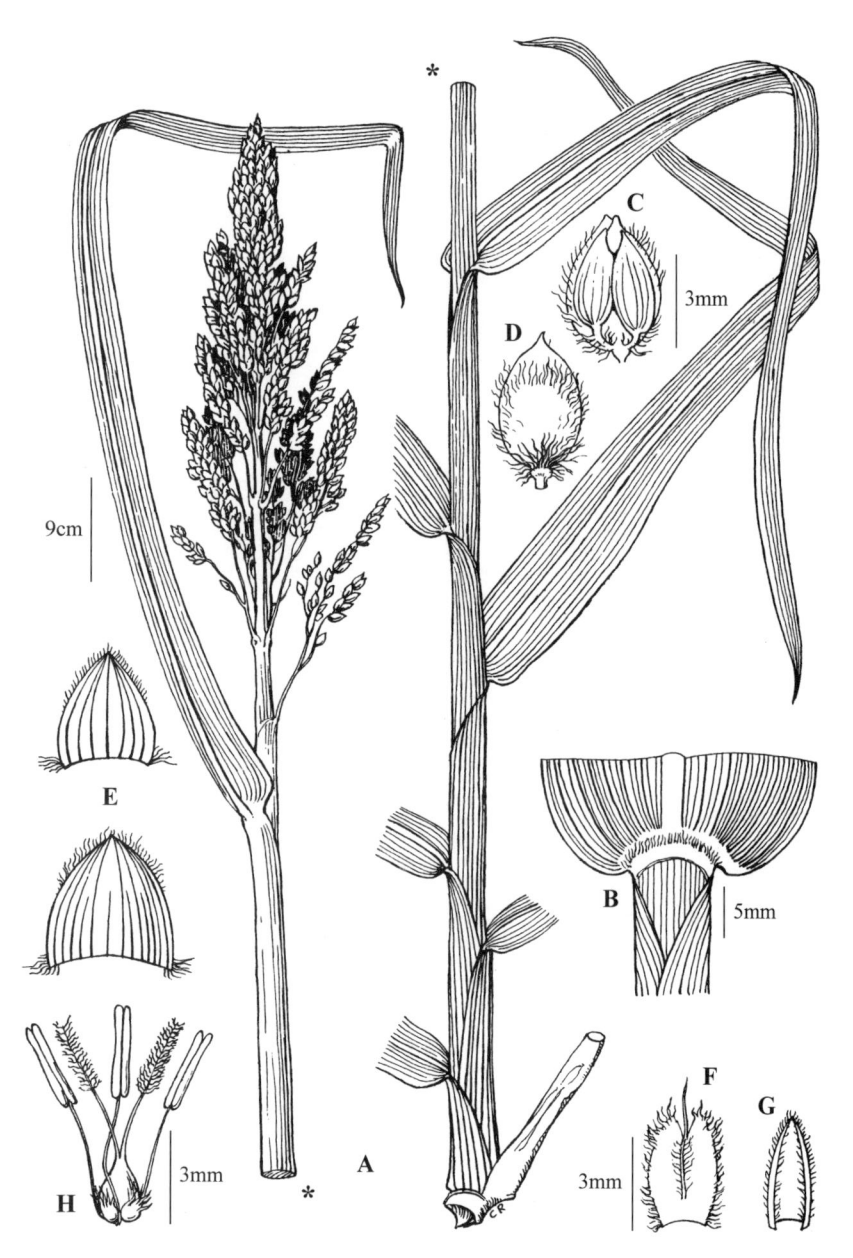

Sorghum bicolor (L.) Moench (Great Millet variant 2). A habit; B ligule; C terminal triplet of one sessile and two stalked spikelets; D sessile spikelet; E upper and lower glumes of sessile spikelet; F upper lemma, G palea; H flower of sessile spikelet. Lower lemma not observed. (Source unknown). STACE 1098/918.

162. LILIACEAE

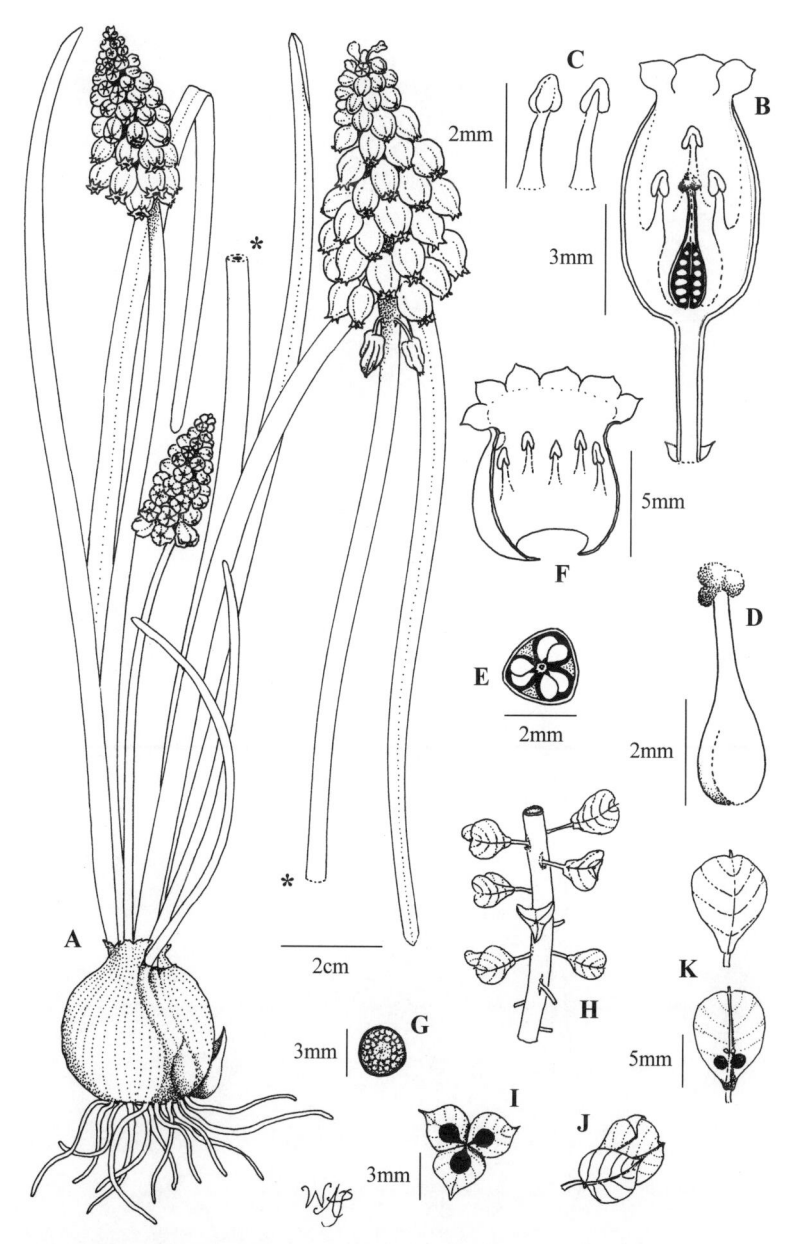

Muscari armeniacum Lechtlin ex Baker (Garden Grape-hyacinth). A habit; B LS of flower; C stamen; D gynoecium; E TS of ovary; F corolla opened out; G TS of scape; H part of fruiting raceme; I dehisced fruit with seeds; J fruit; K one valve of the fruit. (St. Mary's, Isles of Scilly). STACE 1118/935.

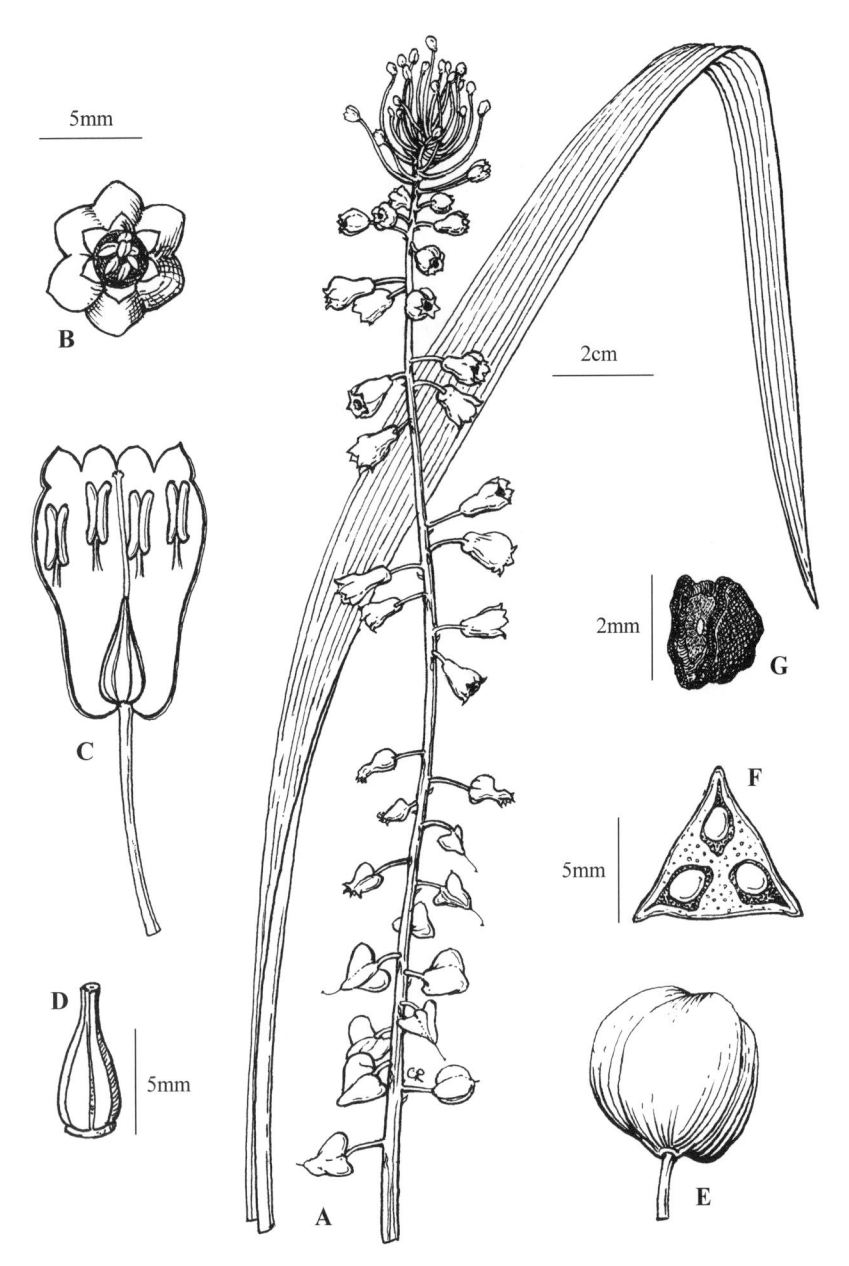

Muscari comosum (L.) Mill. (Tassel Hyacinth). A inflorescence and basal leaf; B flower; C flower minus part of corolla; D gynoecium minus style; E fruit; F TS of fruit; G seed. (St. Anne's, W Lancs). STACE 1118/935.

162. LILIACEAE

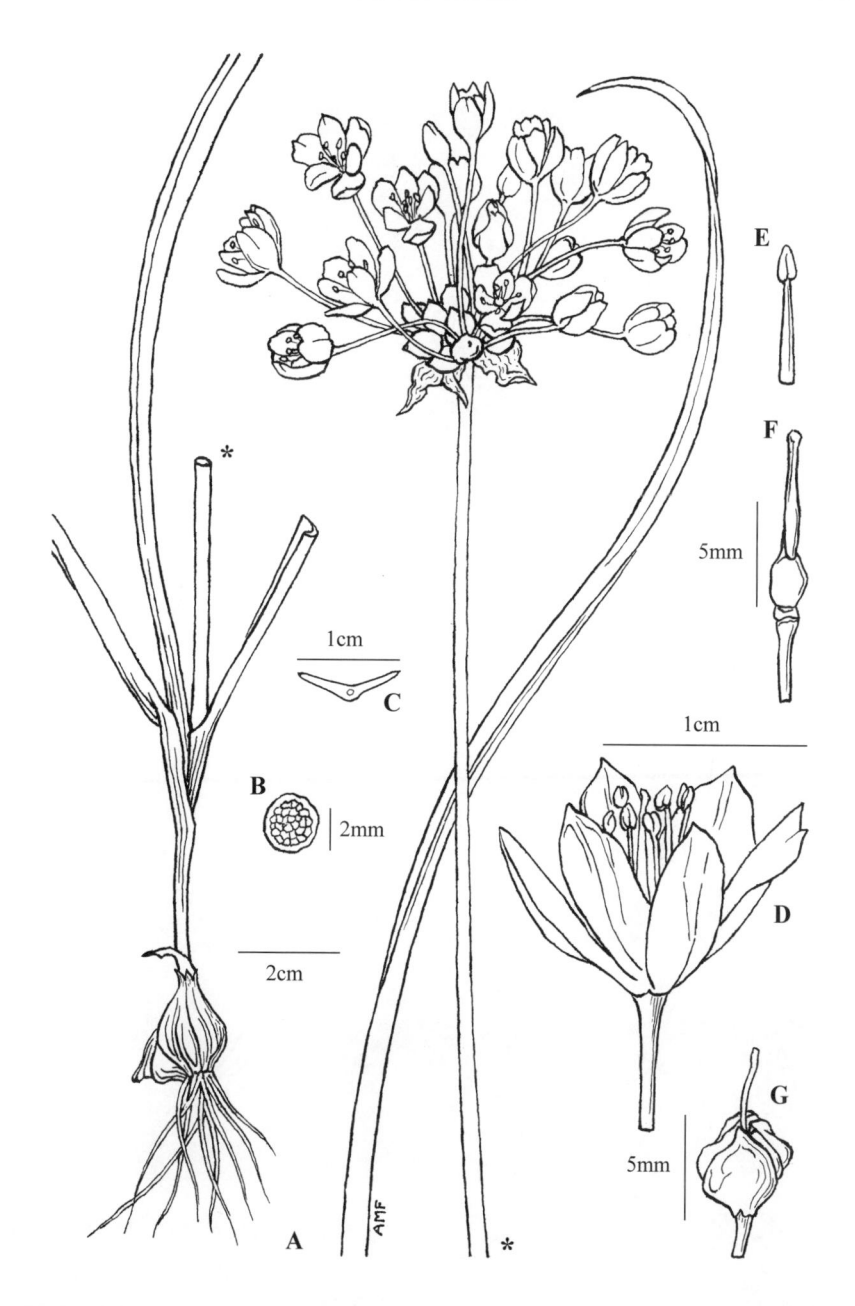

Allium roseum L. (Rosy Garlic). A habit; B TS of stem; C TS of leaf; D flower; E stamen; F gynoecium; G capsule. (Source unknown). STACE 1122/938.

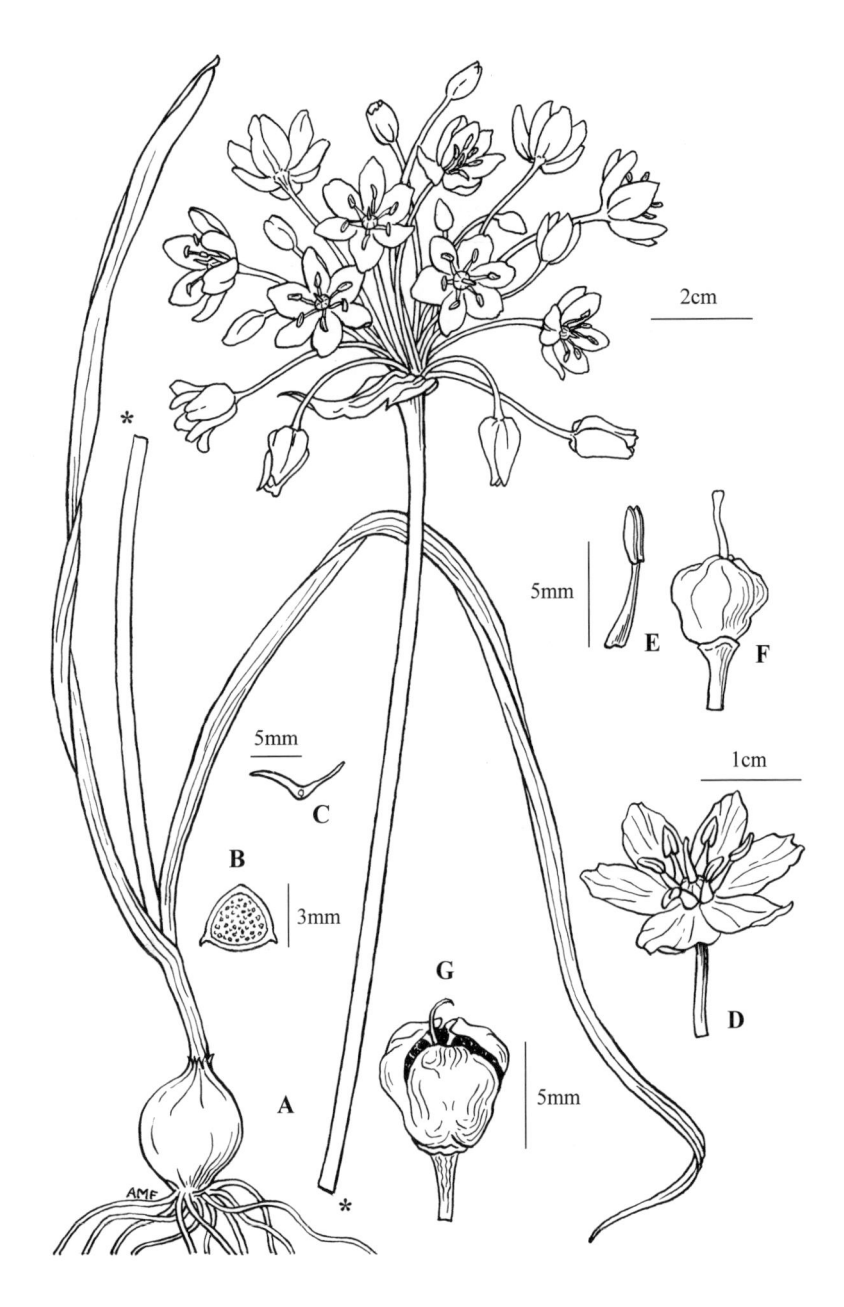

Allium neapolitanum Cirillo (Neapolitan Garlic). A habit; B TS of stem; C TS of leaf; D flower; E stamen; F gynoecium; G capsule. (Guernsey). STACE 1122/938.

162. LILIACEAE 423

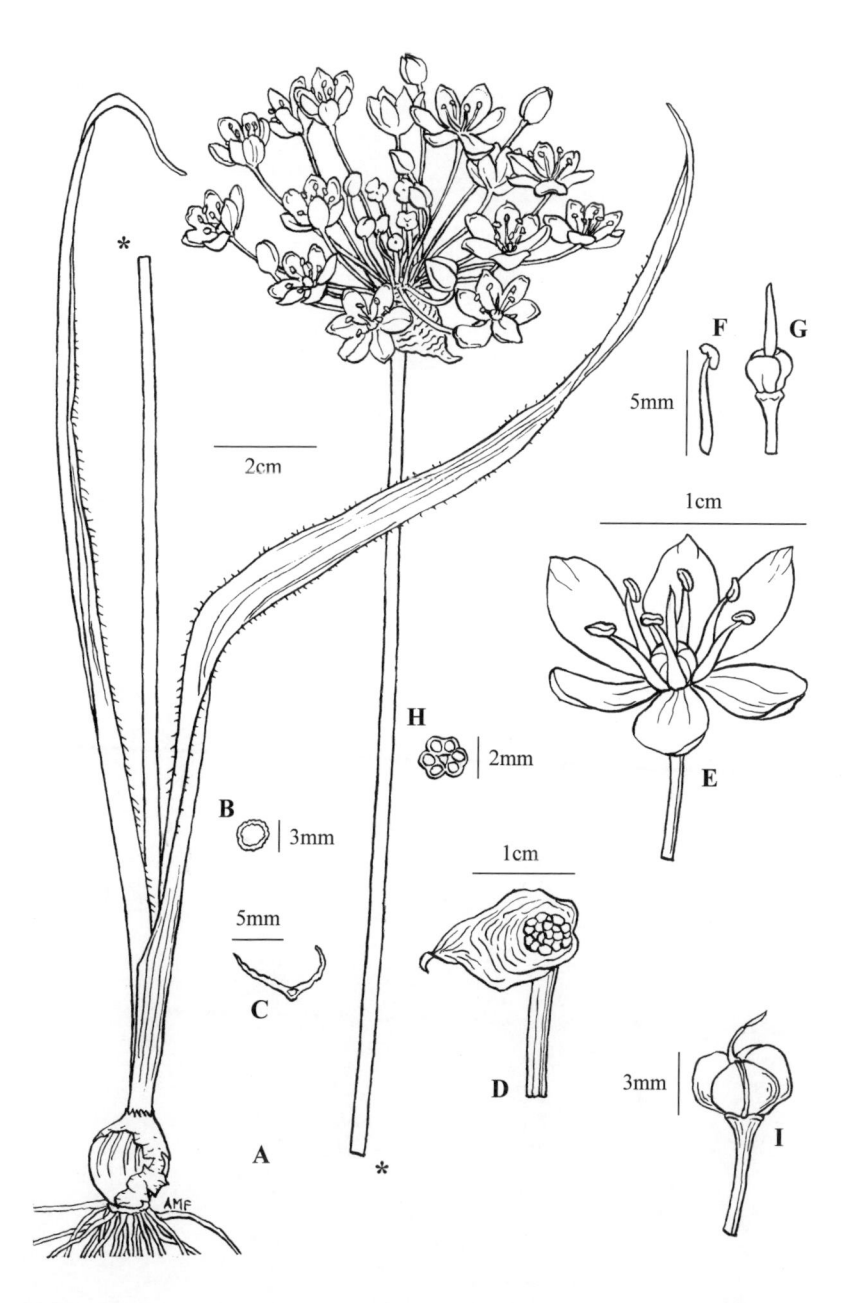

Allium subhirsutum L. (Hairy Garlic). A habit; B TS of stem; C TS of leaf; D spathe; E flower; F stamen; G gynoecium; H TS of ovary; I capsule. (Source unknown). STACE 1122/938.

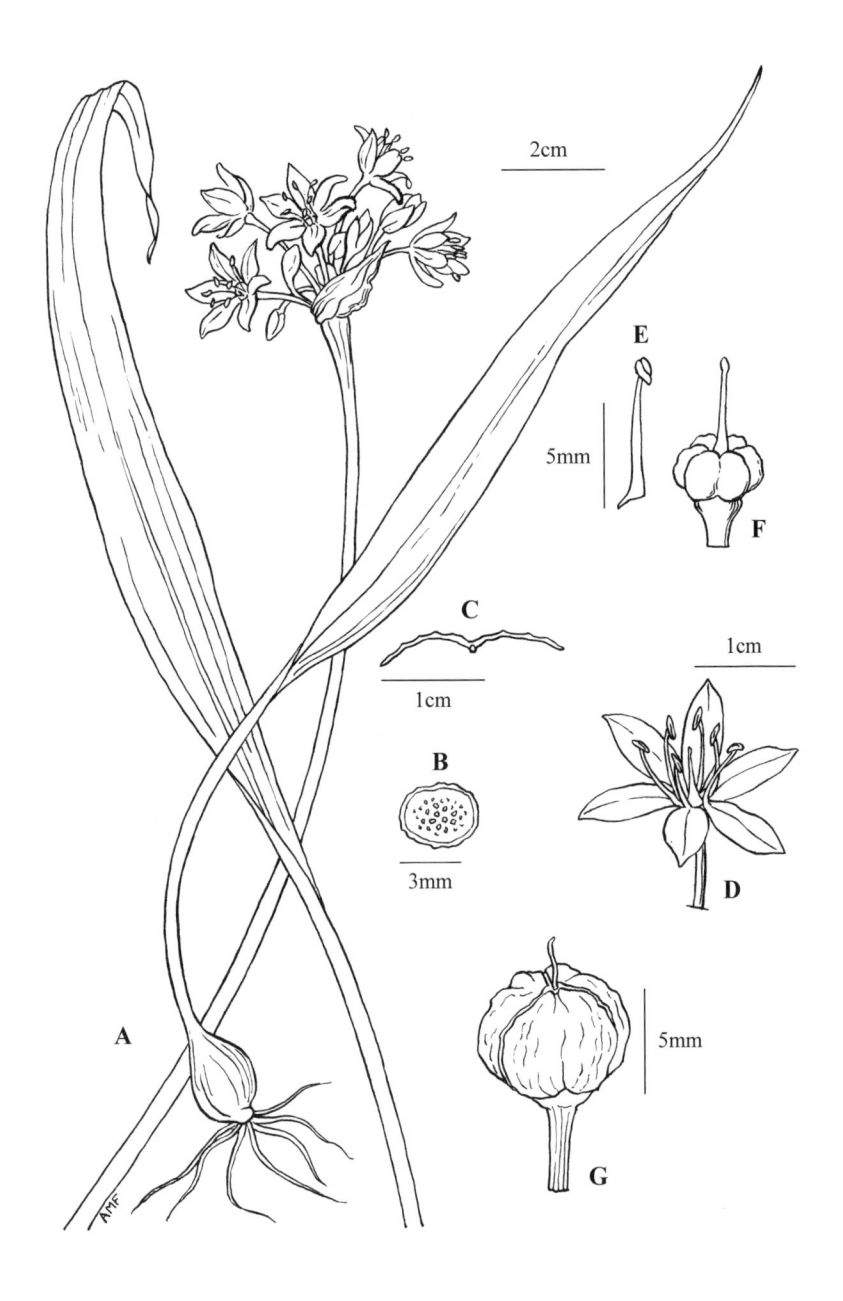

Allium moly L. (Yellow Garlic). A habit; B TS of stem; C TS of leaf; D flower; E stamen; F gynoecium; G capsule. (Source unknown). STACE 1124/938.

162. LILIACEAE

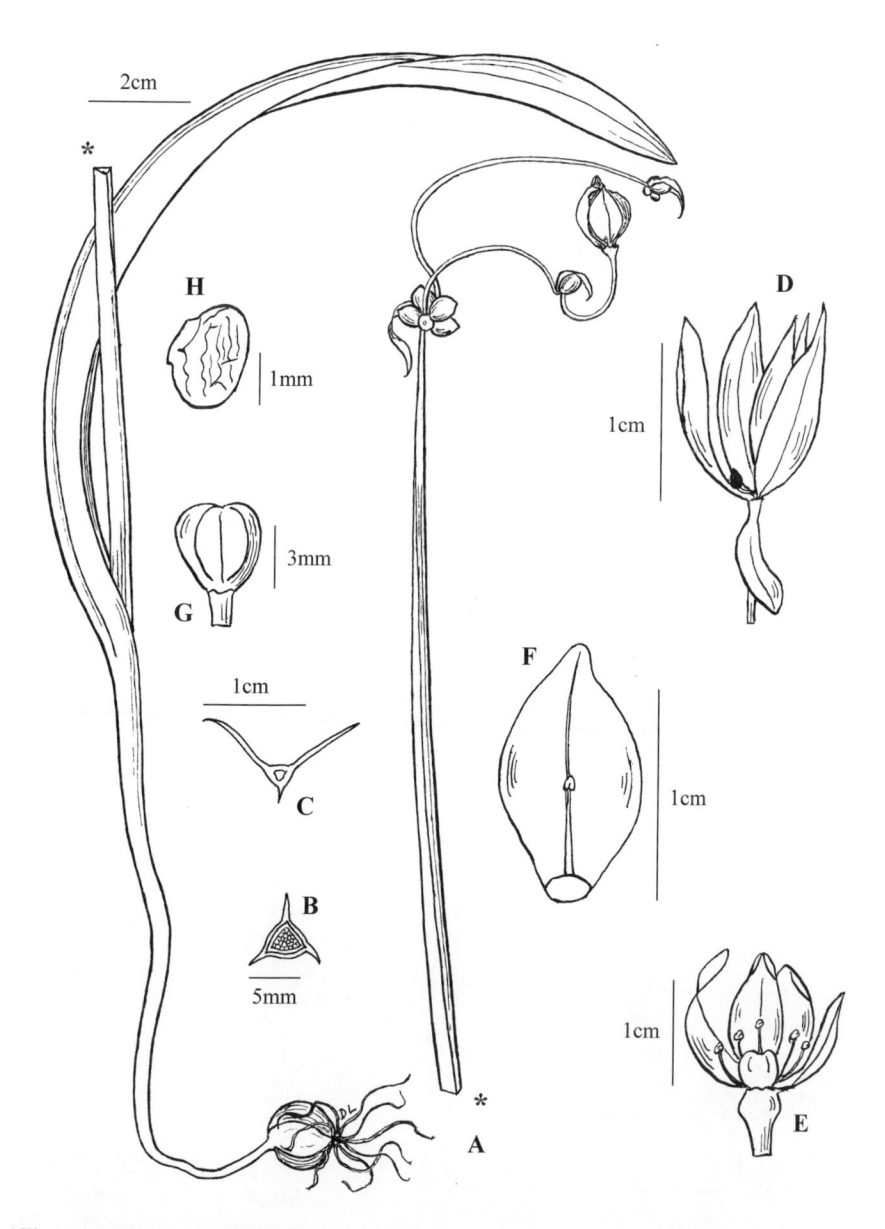

Allium paradoxum (M. Bieb.) G. Don (Few-flowered Garlic). A habit; B TS of stem; C TS of leaf; D flower; E flower opened out, minus part of perianth; F petal and stamen; G capsule; H seed. (A–B *Hort*. Sevenoaks, W Kent; C–F Royal Agricultural College, Cirencester, E Gloucs). STACE 1124/940.

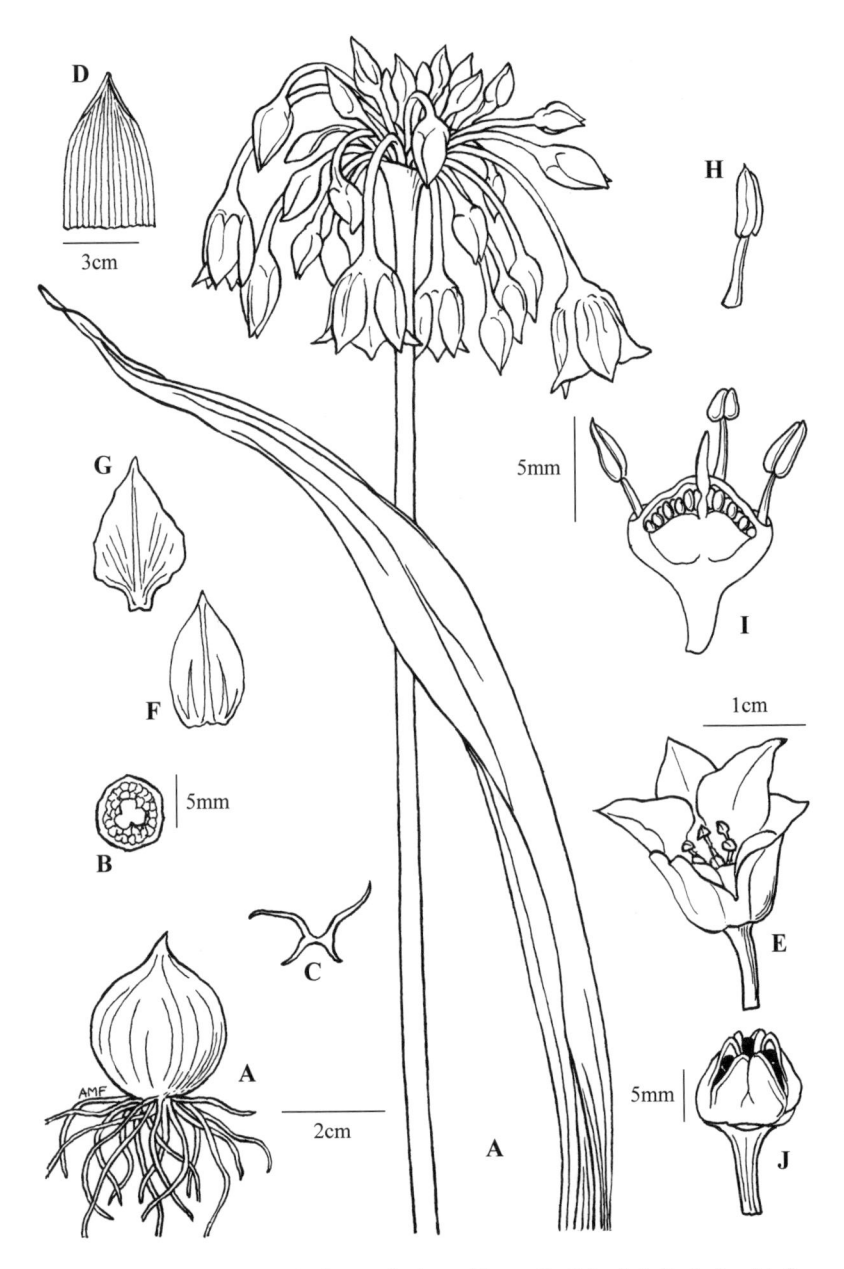

Nectaroscordum siculum (Ucria) Lindl. ssp. **siculum** (Honey Garlic). A bulb, leaf and inflorescence; B TS of stem; C TS of leaf; D spathe; E flower; F outer tepal; G inner tepal; H stamen; I LS of gynoecium; J capsule. (*Hort*. Saffron Walden, N Essex). STACE 1126/941.

162. LILIACEAE

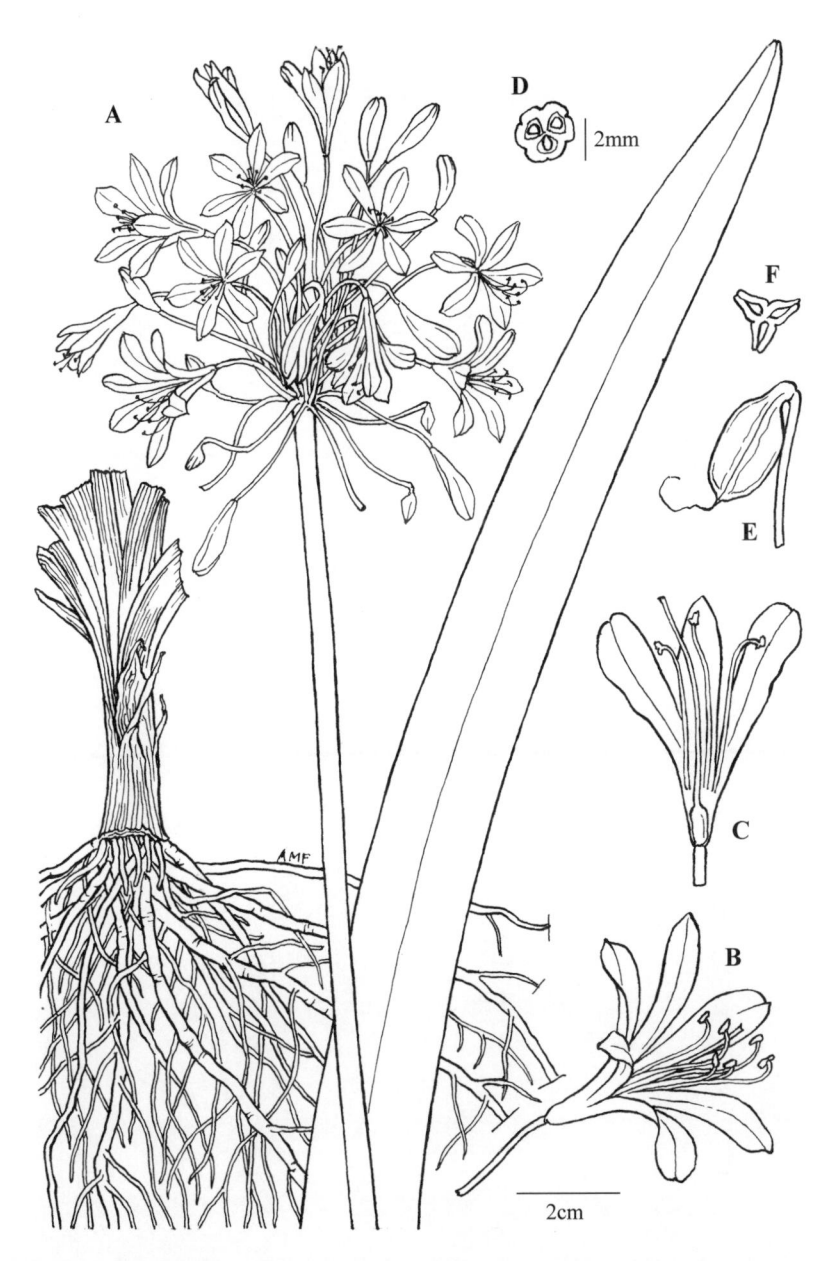

Agapanthus praecox Willd. ssp. **orientalis** (F.M. Leight.) F.M. Leight. (African Lily). A habit; B flower; C half of flower opened out; D TS of ovary; E immature fruit; F TS of immature fruit. (Tresco, Scilly). STACE 1126/941.

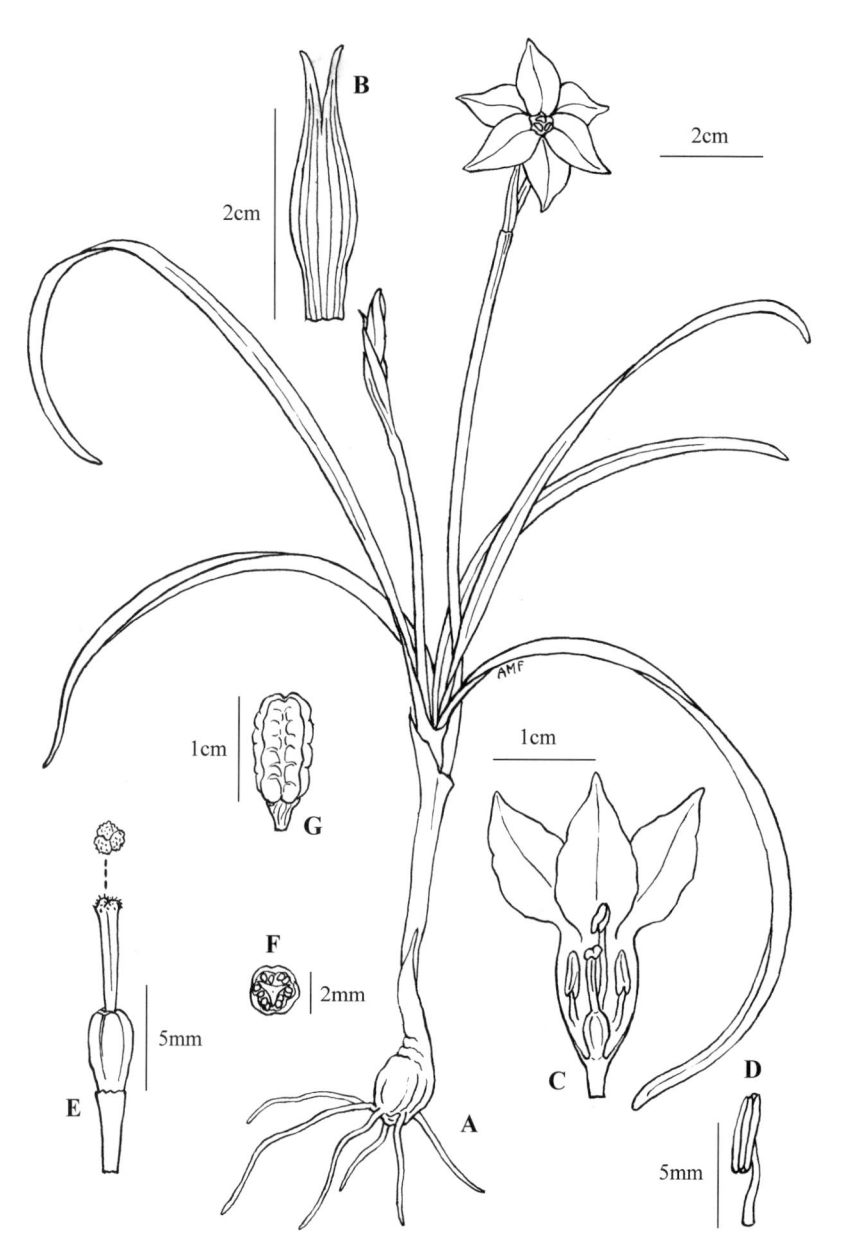

Tristagma uniflorum (Lindl.) Traub (Spring Starflower). A habit; B spathe flattened out; C half of flower opened out; D stamen; E gynoecium; F TS of ovary; G capsule. (*Hort*. Saffron Walden, N Essex). STACE 1126/941.

162. LILIACEAE

Amaryllis belladonna L. (Jersey Lily). A habit; B TS of stem; C stamen; D LS of gynoecium; E TS of ovary. (Channel Islands). STACE 1127/942.

430 162. LILIACEAE

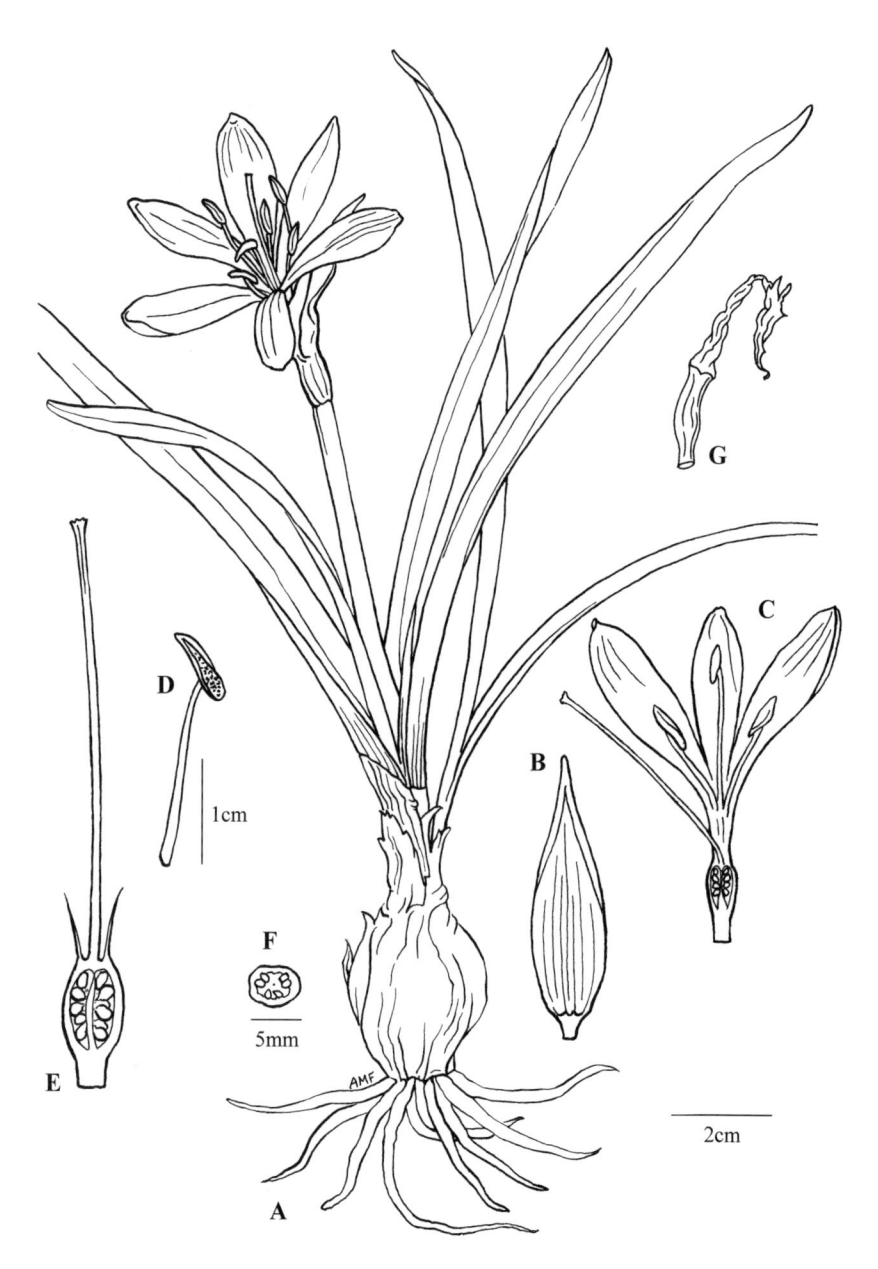

Sternbergia lutea (L.) Ker-Gawl. ex Spreng. (Winter Daffodil). A habit; B spathe; C LS of flower; D stamen from shorter whorl; E LS of gynoecium; F TS of ovary; G sterile fruit. (Source unknown). STACE 1127/942.

162. LILIACEAE

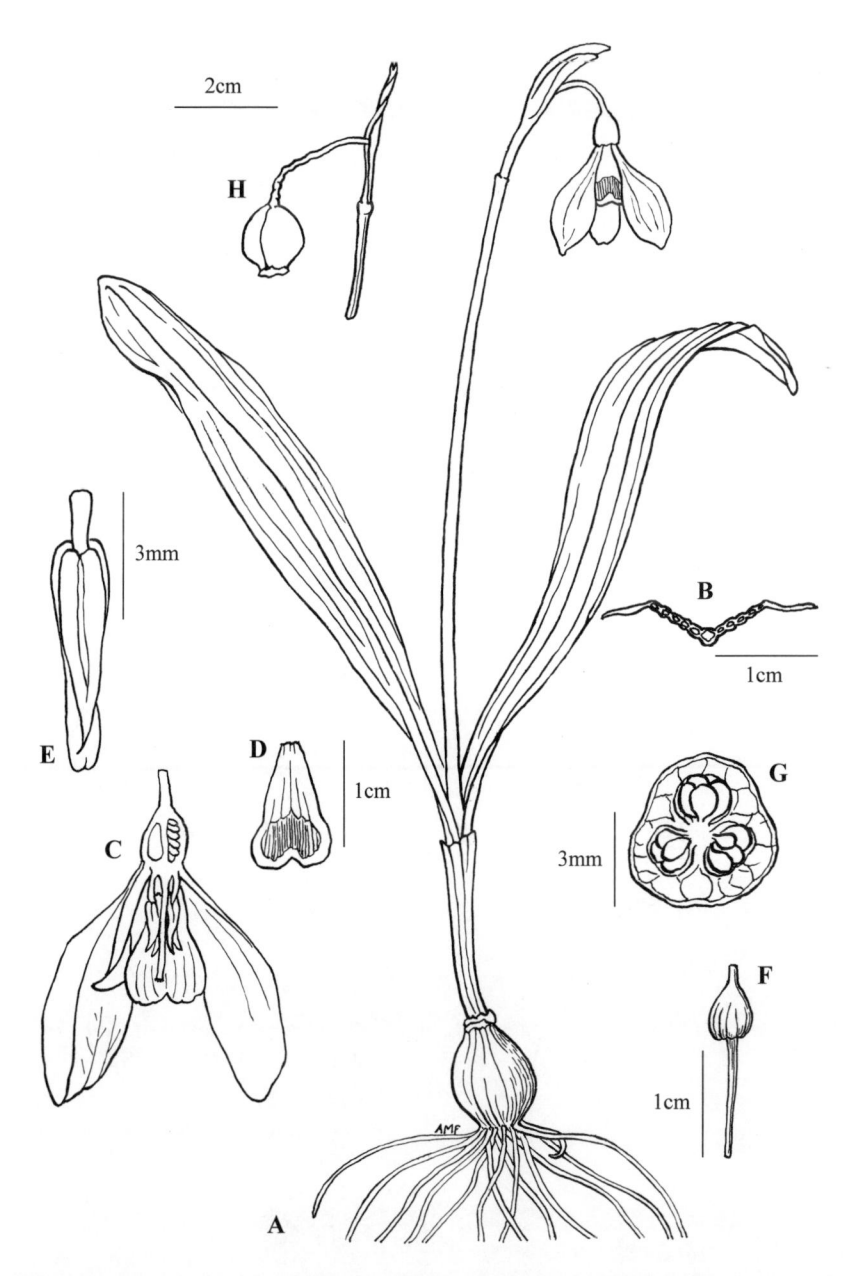

Galanthus plicatus M. Bieb. ssp. **plicatus** (Pleated Snowdrop). A habit; B TS of leaf; C LS of flower; D inner tepal; E stamen; F gynoecium; G TS of ovary; H fruit. (Source unknown). STACE 1130/945.

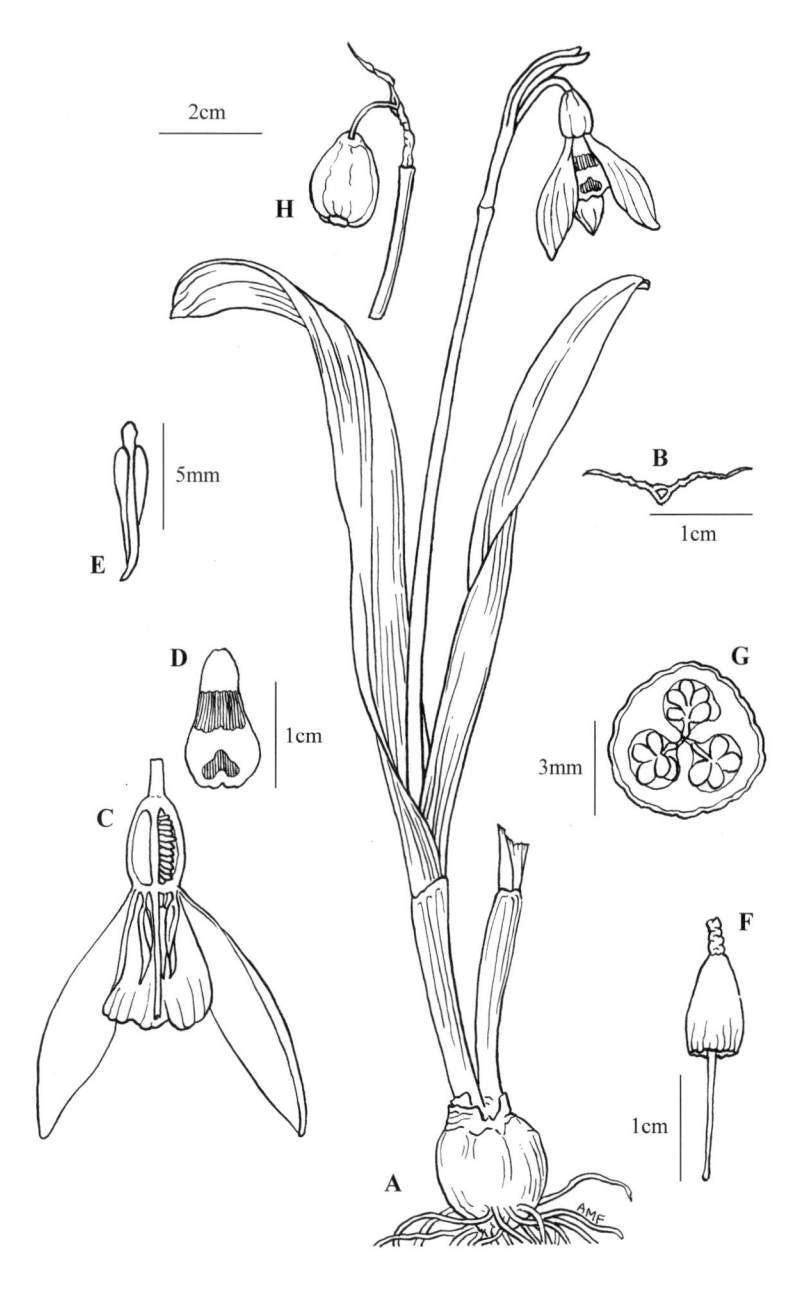

Galanthus elwesii Hook. f. (Greater Snowdrop). A habit; B TS of leaf; C LS of flower; D inner tepal; E stamen; F gynoecium; G TS of ovary; H fruit. (Source unknown). STACE 1129/945.

162. LILIACEAE

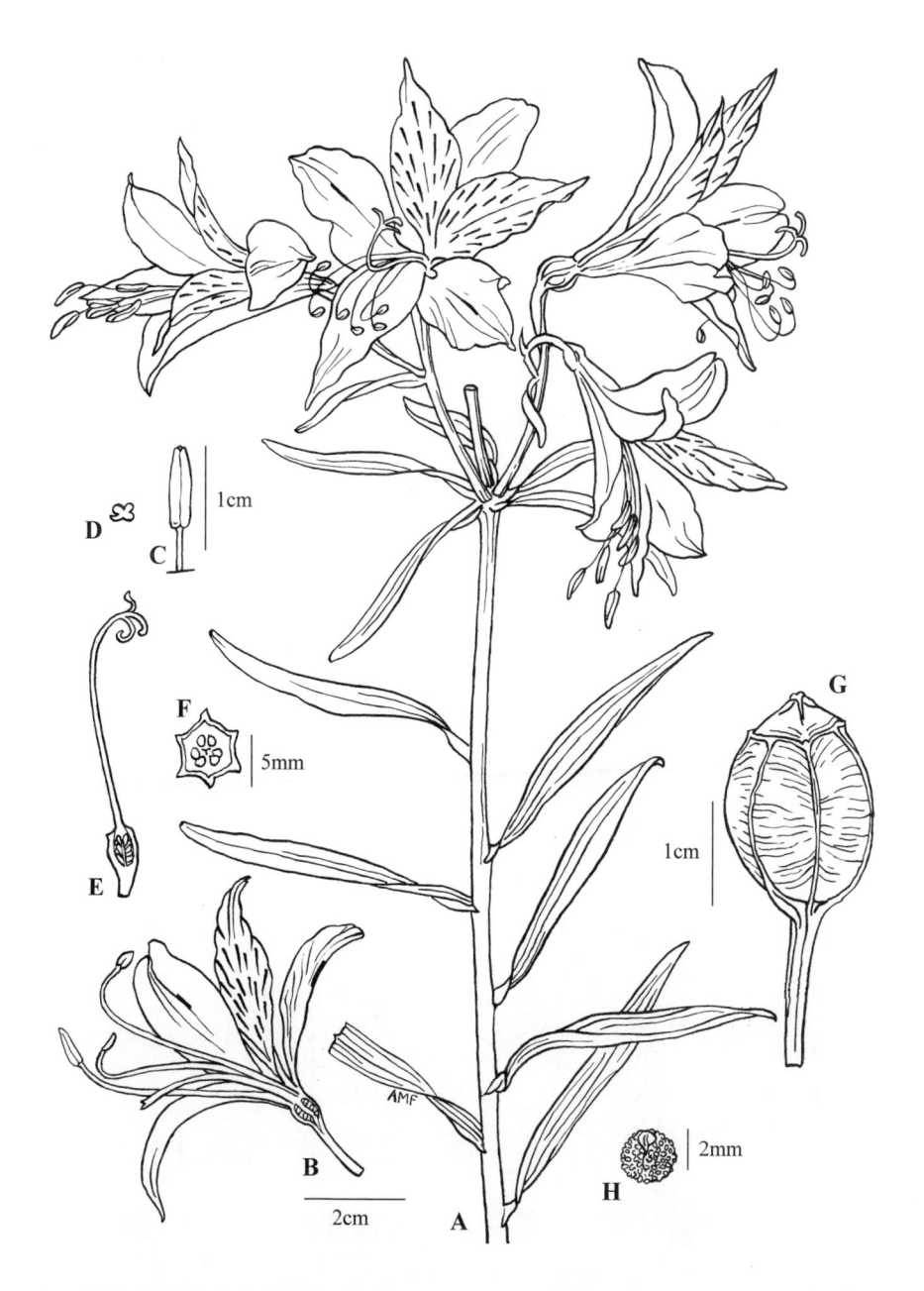

Alstroemeria aurea Graham (Peruvian Lily). A habit; B flower; C anther; D TS of anther; E LS of gynoecium; F TS of ovary; G capsule; H seed. (Source unknown). STACE 1136/950.

Libertia formosa Graham (Chilean-iris). A habit; B flower; C flower minus perianth; D TS of ovary; E fruit; F TS of fruit. (St. Mary's, Isles of Scilly). STACE 1138/951.

163. IRIDACEAE

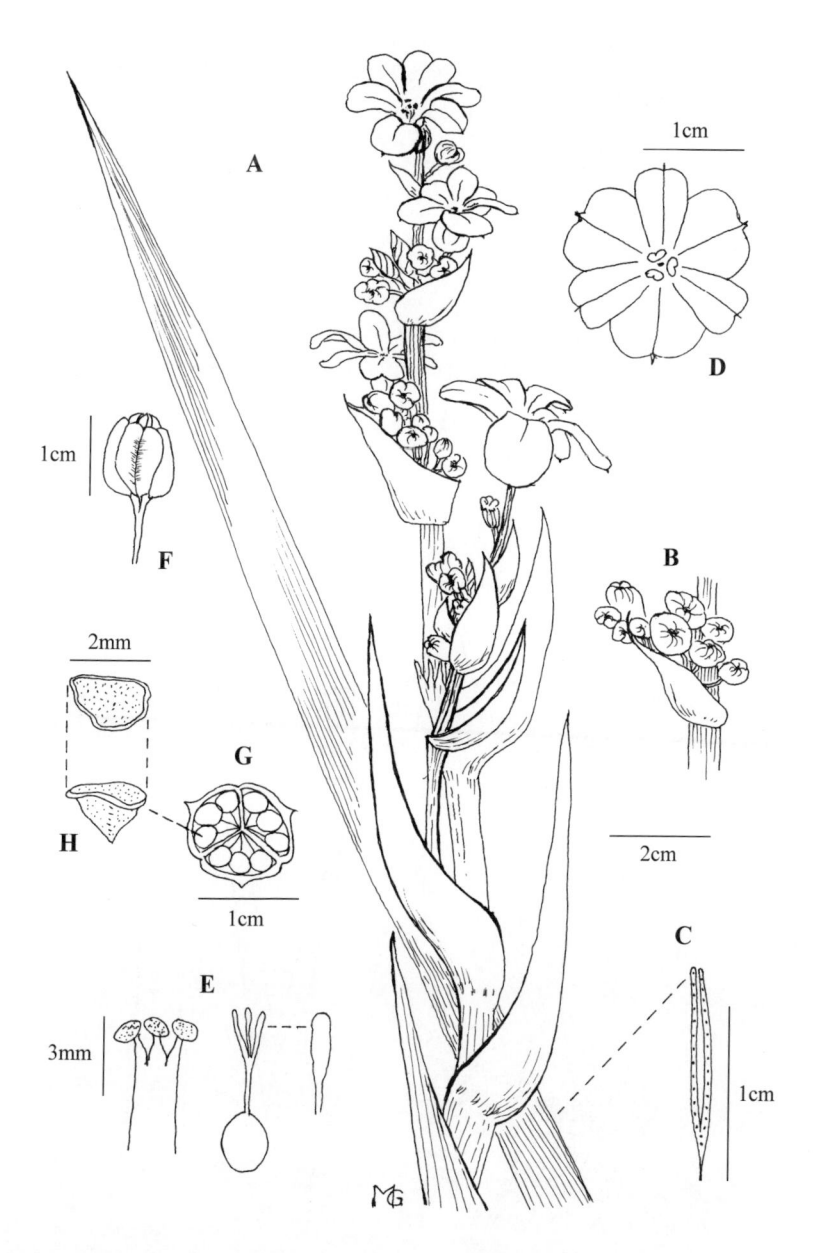

Sisyrinchium striatum Sm. (Pale Yellow-eyed-grass). A inflorescence and lower leaf; B flowers at node after anthesis; C TS of leaf base; D flower; E staminal sheath and gynoecium; F fruit; G TS of fruit; H seed. (*Hort.* Bexley, W Kent). STACE 1139/952.

436   163. IRIDACEAE

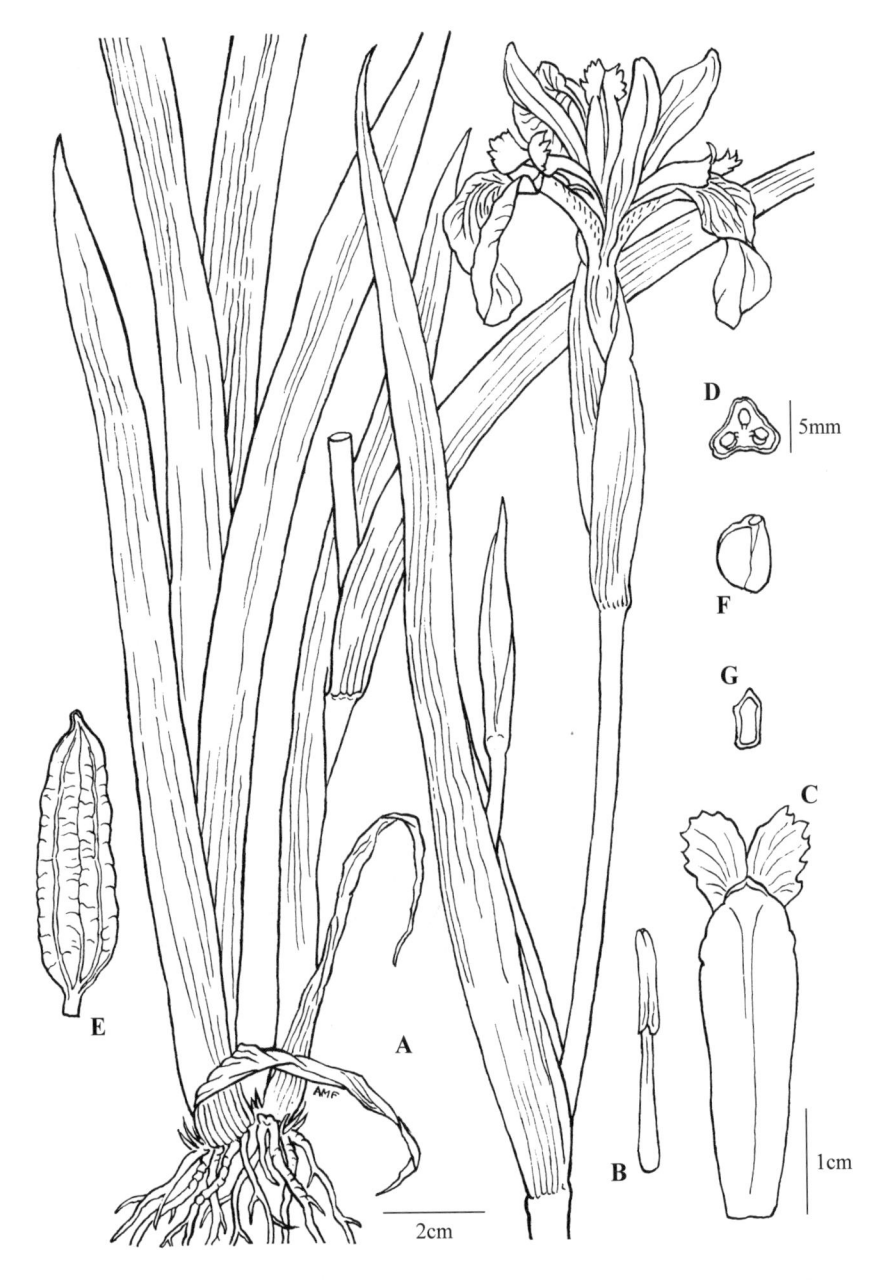

Iris versicolor L. (Purple Iris). A habit; B stamen; C one branch of style; D TS of ovary; E capsule; F seed; G TS of seed. (Epping Forest, S Essex). STACE 1141/954.

163. IRIDACEAE

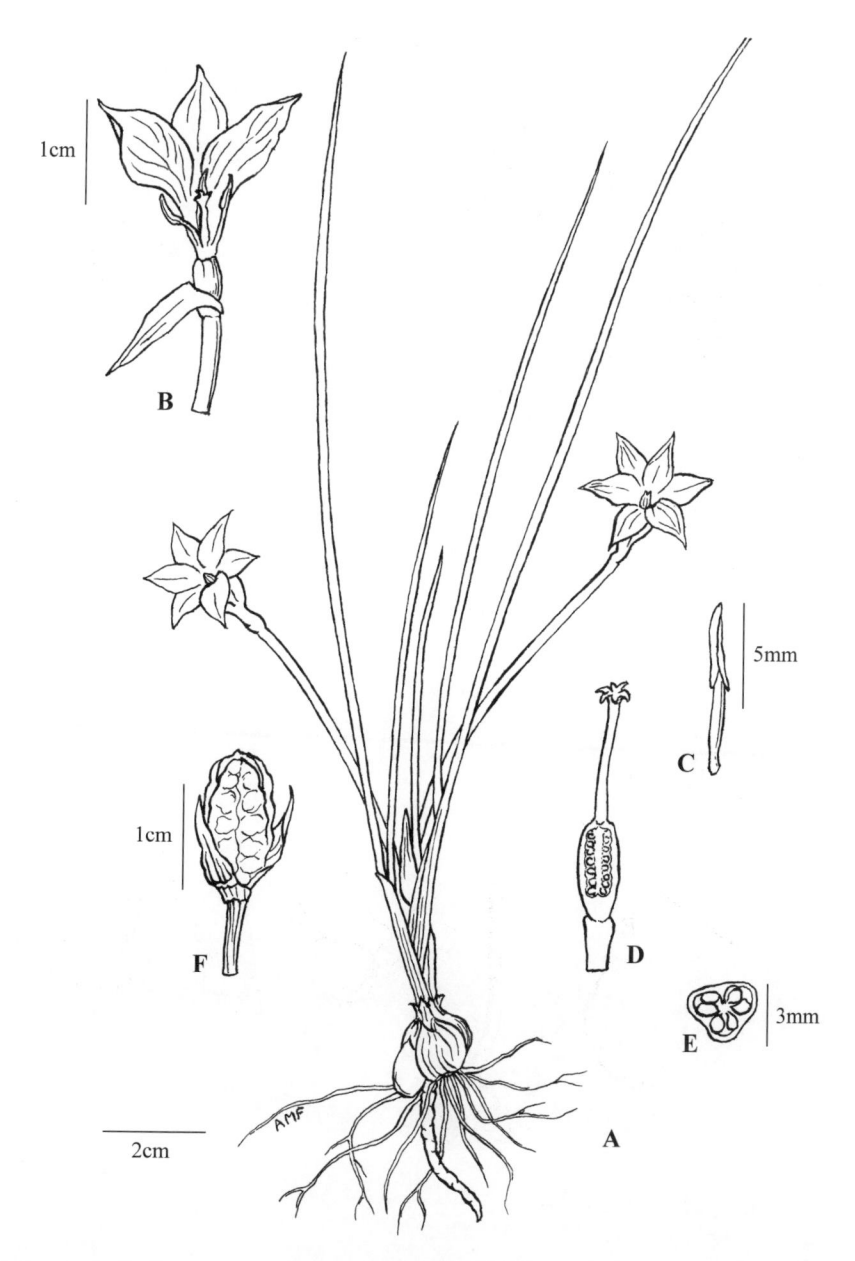

Romulea rosea (L.) Eckl. var. **australis** (Ewart) M.P. de Vos (Oniongrass). A habit; B part of flower opened out, stamens spread; C stamen; D LS of gynoecium; E TS of ovary; F capsule. (Guernsey). STACE 1143/955.

438   163. IRIDACEAE

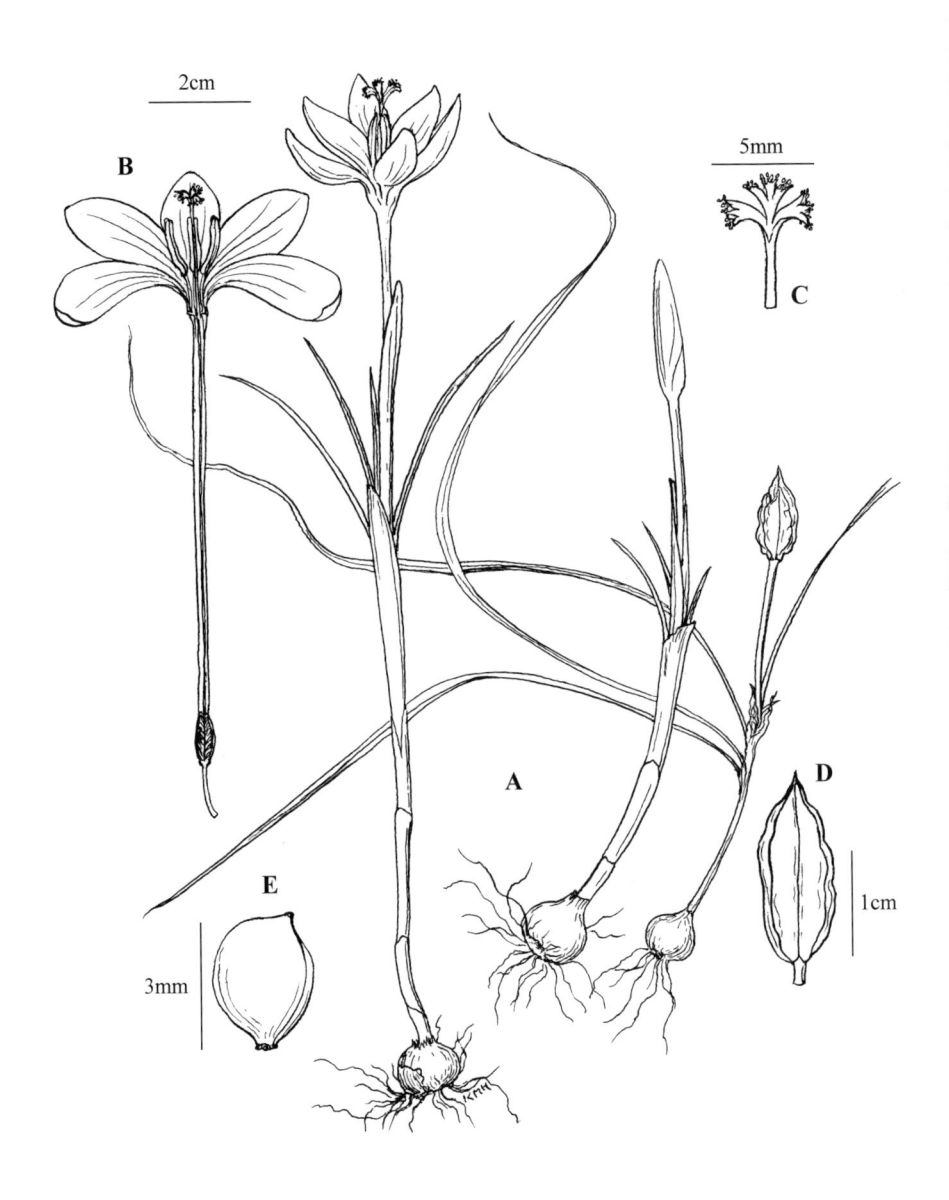

Crocus tommasinianus Herb. (Early Crocus). A habit, before, during and after anthesis; B flower minus one tepal; C stigma; D fruit; E seed. (*Hort*. Ashbourne, Derbys). STACE 1145/958.

163. IRIDACEAE

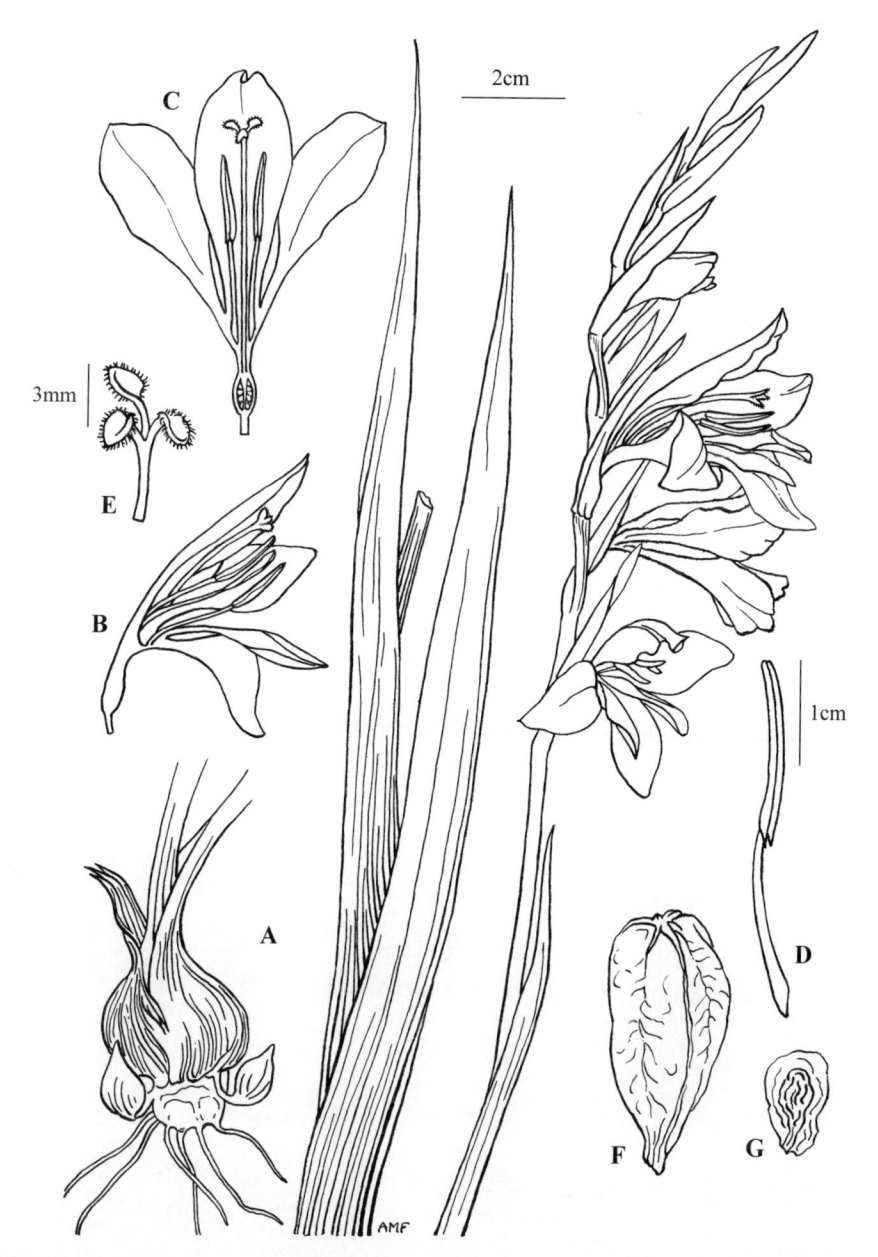

Gladiolus communis L. ssp. **byzantinus** (Mill.) R.C.V. Douin (Eastern Gladiolus). A habit with corm and flowering stem; B flower minus part of perianth; C LS of flower; D stamen; E stigma; F capsule; G seed. (Source unknown). STACE 1146/959.

440　　163. IRIDACEAE

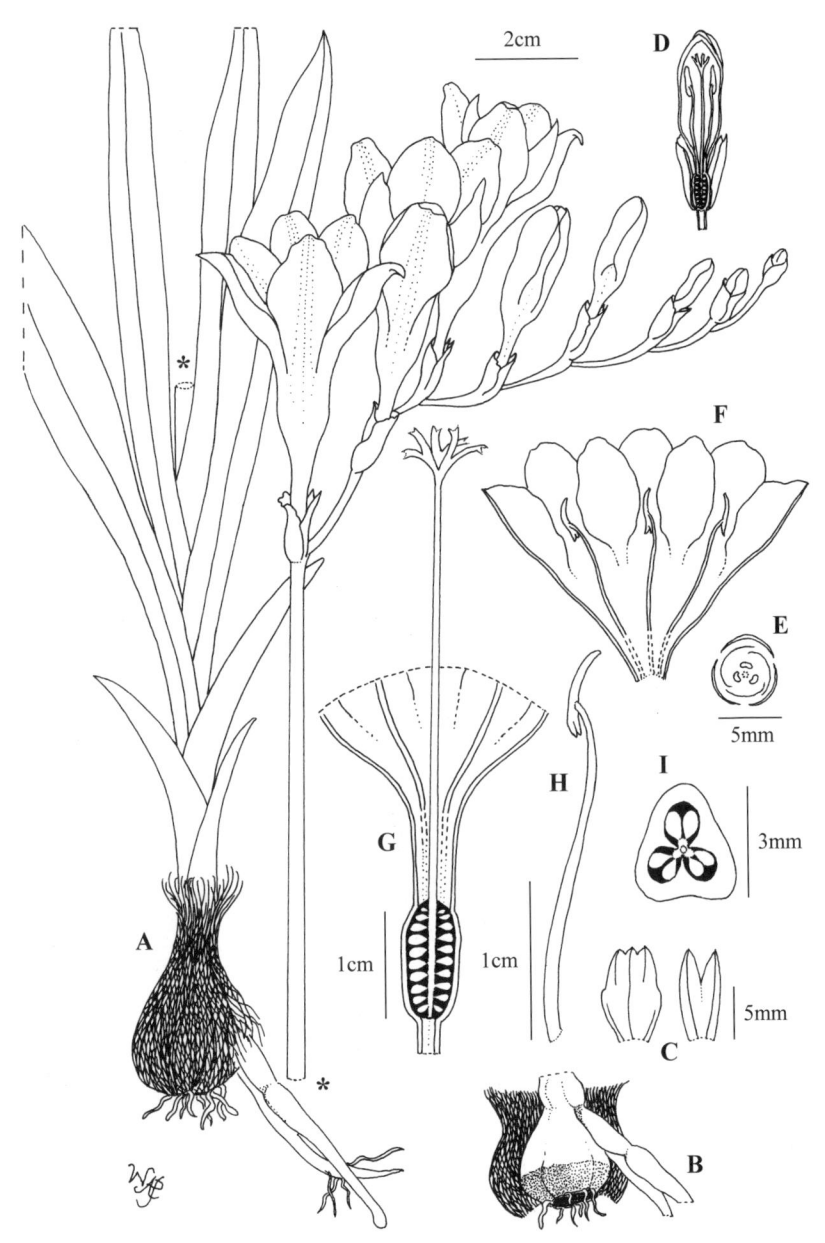

Freesia × hybrida L.H. Bailey (Freesia). A habit; B rootstock with tunic opened out; C bracts; D LS of flower bud; E TS of flower bud; F flower opened out; G LS of gynoecium; H stamen; I TS of ovary. (*Hort*. Horsham, W Sussex). STACE 1147/960.

163. IRIDACEAE 441

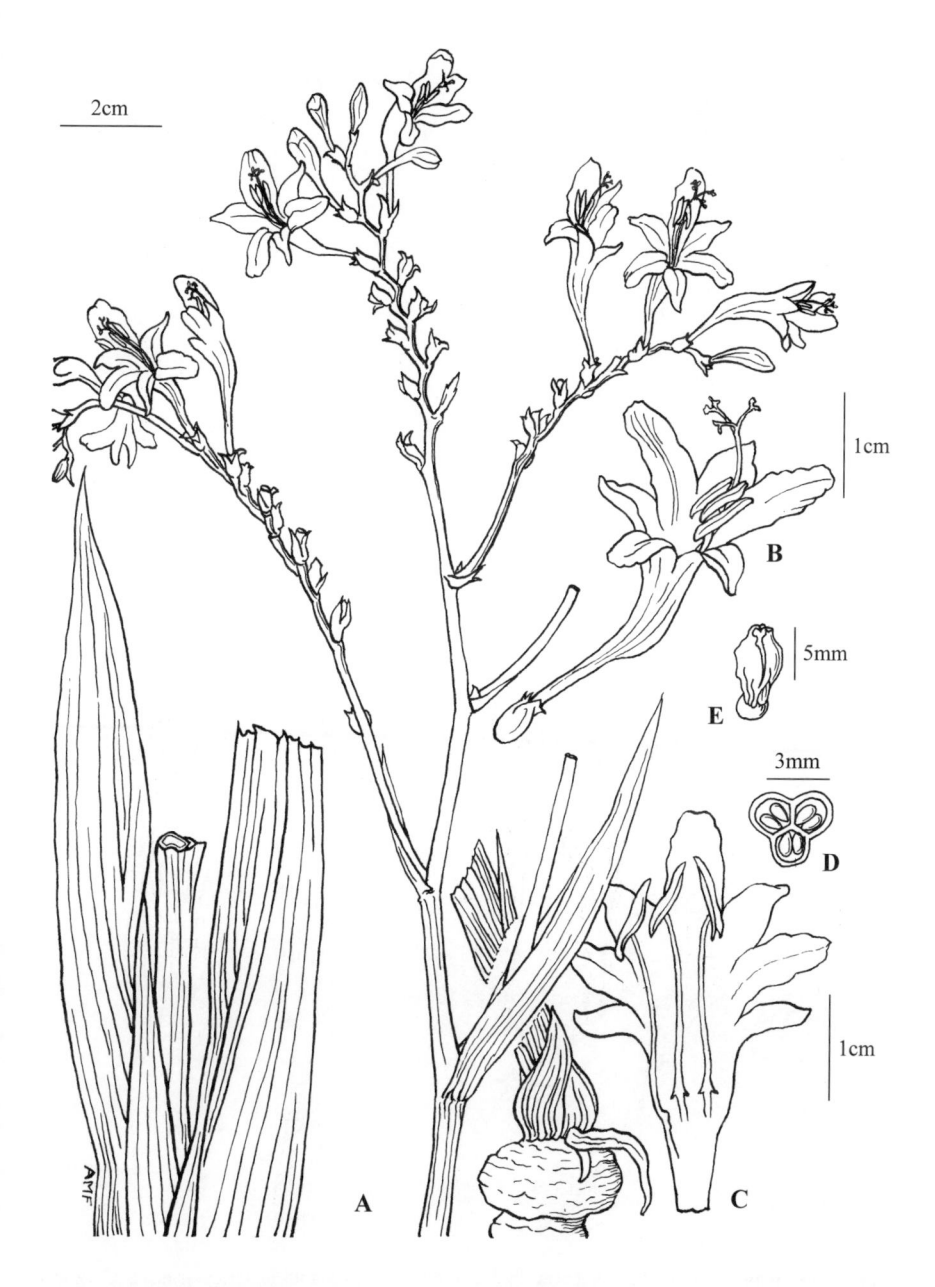

Crocosmia x crocosmiiflora (Lemoine) N.E. Br. (Montbretia). A habit; B flower; C corolla opened out; D TS of ovary; E capsule. (Source unknown). STACE 1148/960.

442 164. AGAVACEAE

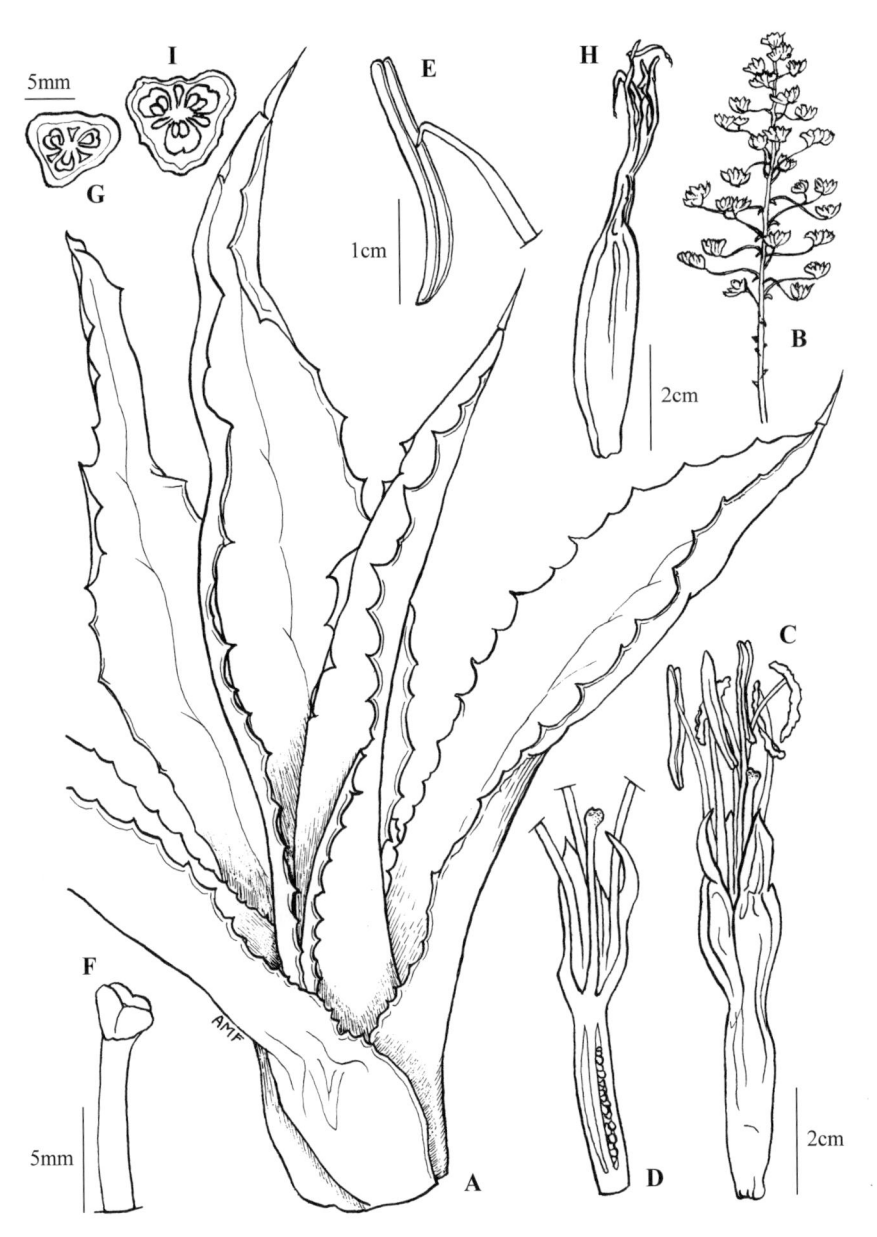

Agave americana L. (Centuryplant). A leaf rosette of young plant; B inflorescence; C flower; D LS of flower; E stamen; F stigma; G TS of ovary; H fruit; I TS of fruit. (Guernsey). STACE 1150/962.

166. ORCHIDACEAE

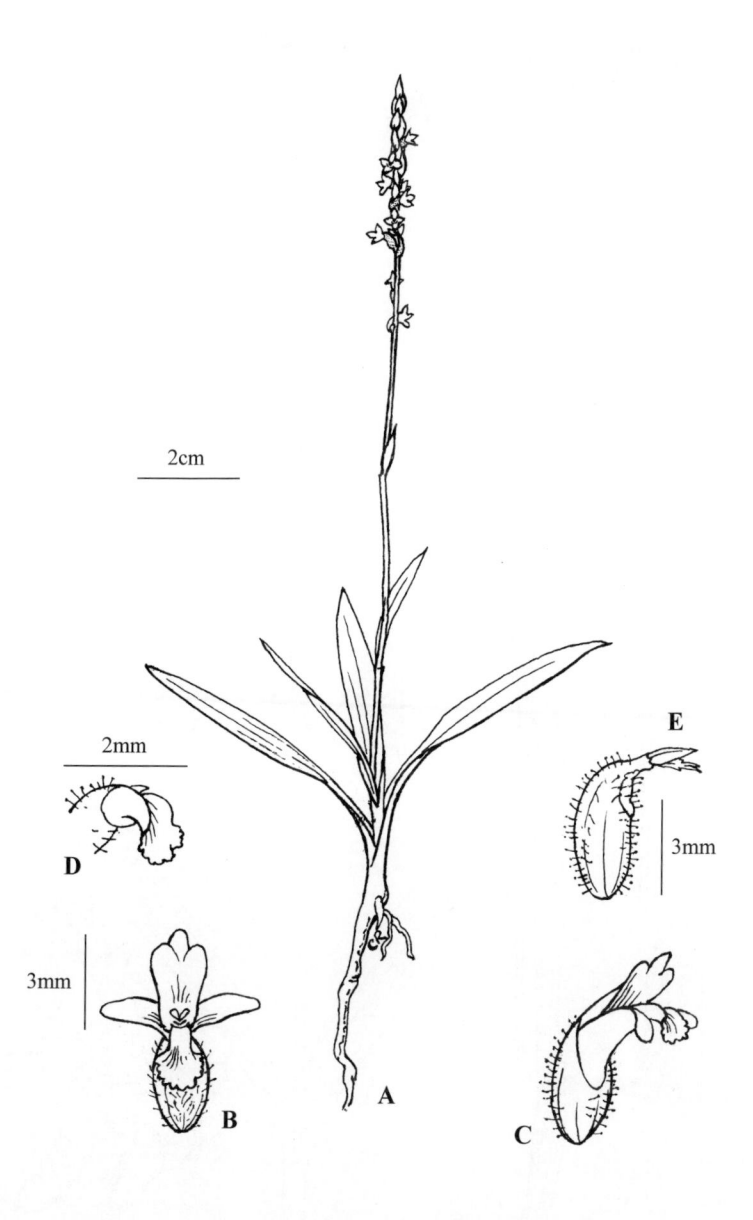

Spiranthes lancea (Thunb. ex Sw.) Backer, Bakh. f. & v. Steenis (No English name). A habit; B front view of flower; C side view of flower; D lower lip of corolla; E ovary. (Greenhouse weed; Source unknown). STACE —/—.

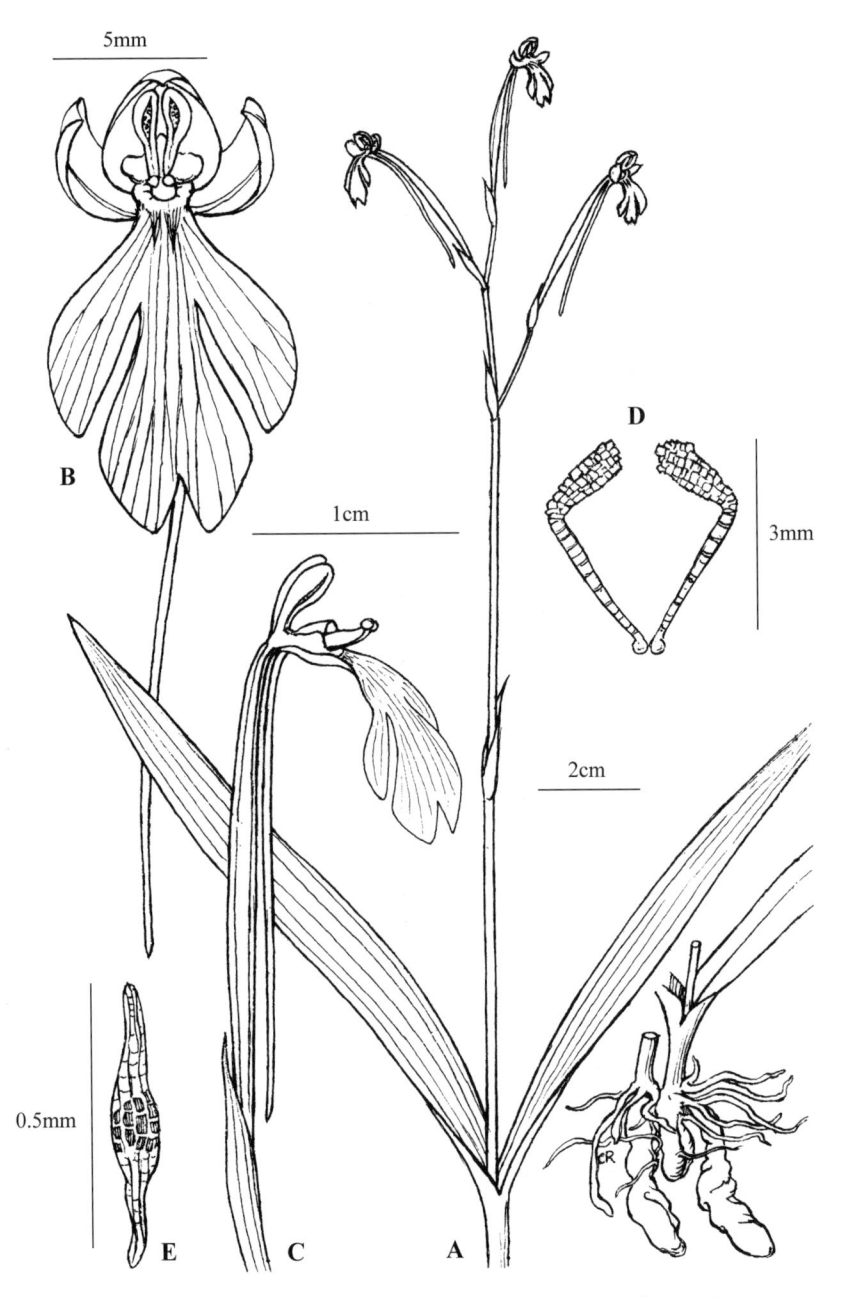

Cynorkis fastigiata Thouars (No English name). A habit; B flower; C flower minus upper tepals; D pollinia; E seed. (Greenhouse weed, Glasgow University Garden, Lanarks). STACE —/—.

SELECTED BIBLIOGRAPHY

Brummitt, R.K. & Powell, C.E., eds. (1992). *Authors of plant names*. Royal Botanic Gardens, Kew.

Butcher, R.W. (1961). *A new illustrated British Flora*, vols. 1-2. Leonard Hill, London.

Clapham, A.R., Tutin, T.G. & Moore, D.M. (1987). *Flora of the British Isles*, 3rd ed. Cambridge University Press.

Clapham, A.R., Tutin, T.G. & Warburg, E.F. (1957-1965). *Flora of the British Isles. Illustrations*, Parts 1-4. Cambridge University Press, London.

Clement, E.J. & Foster, M.C. (1994). *Alien plants of the British Isles. A provisional catalogue of vascular plants (excluding grasses)*. Botanical Society of the British Isles, London.

Griffiths, M. (1994). *The New Royal Horticultural Society Dictionary. Index of garden plants*. Macmillan Press, London and Basingstoke.

Hickey, M. & King, C. (2000). *The Cambridge illustrated glossary of botanical terms*. Cambridge University Press, Cambridge.

[Meikle, R.D.] (1980). *Draft index of author abbreviations compiled at the Herbarium, Royal Botanic Gardens, Kew*. HMSO, Basildon.

Ross-Craig, Stella (1948-1973). *Drawings of British Plants*, parts 1-31. (Also issued as vols. 1-8). G. Bell and Sons, London.

Ryves, T.B., Clement, E.J. & Foster, M.C. (1996). *Alien Grasses of the British Isles*. Botanical Society of the British Isles, London.

Sell, P. & Murrell, G. (1996). *Flora of Great Britain and Ireland*, vol. 5. Cambridge University Press, Cambridge.

Stace, C. (1991). *New Flora of the British Isles*. Cambridge University Press, Cambridge.

Stace, C. (1997). *New Flora of the British Isles*, 2nd ed. Cambridge University Press, Cambridge.

Stace, C. (1999). *Field Flora of the British Isles*. Cambridge University Press, Cambridge.

Tutin, T.G. *et al.*, eds. (1964-1980). *Flora Europaea*, vols. 1-5. Cambridge University Press, London.

Tutin, T.G. *et al.*, eds. (1993). *Flora Europaea*, 2nd ed., vol. 1. Cambridge University Press, London.

Walters, S.M. *et al.*, eds. (1986-2000). *The European Garden Flora*, vols. 1-6. Cambridge University Press, Cambridge.

INDEX

Compiled by I. R. Thirlwell

INDEX

Abraham-Isaac-Jacob, 235
Abutilon
 theophrasti, 96
Acaena
 anserinifolia, 157
 microphylla, 159
 ovalifolia, 158
Acaena
 Two-spined, 158
ACANTHACEAE, 257
Acanthus
 mollis, 257
 spinosus, 258
Acer
 platanoides, 184
ACERACEAE, 184
Achillea
 ligustica, 328
Aconitum
 x cammarum, 10
Acroptilon
 repens, 293
Aesculus
 hippocastanum, 183
Aetheorhiza
 bulbosa, 302
Agapanthus
 praecox ssp. **orientalis**, 427
AGAVACEAE, 442
Agave
 americana, 442
Agrostis
 avenacea, 395
 lacnantha, 394
 scabra, 396
Ailanthus
 altissima, 185
AIZOACEAE, 39
Alcea
 rosea, 95
Alchemilla
 tytthantha, 160
Alder
 Grey, 35
 Italian, 36
Alexanders
 Perfoliate, 201
Alkanet, 232
 Yellow, 233
Allium
 moly, 424
 neapolitanum, 422

 paradoxum, 425
 roseum, 421
 subhirtum, 423
Alnus
 cordata, 36
 incana, 35
Alstroemeria
 aurea, 433
Altar-lily, 369
Amaranth
 Common, 60
 Green, 61
 Mitchell's, 67
AMARANTHACEAE, 60
Amaranthus
 albus, 64
 blitum, 63
 capensis ssp. **uncinatus**, 66
 deflexus, 62
 hybridus, 61
 mitchellii, 67
 retroflexus, 60
 thunbergii, 65
Amaryllis
 belladonna, 429
Ambrosia
 artemisiifolia, 347
 psilostachya, 348
 trifida, 349
Amelanchier
 lamarckii, 163
Ammi
 majus, 204
 visnaga, 205
Anchusa
 ochroleuca, 233
 officinalis, 232
Anemone
 apennina, 12
 ranunculoides, 13
Anemone
 Blue, 12
 Yellow, 13
Anthemis
 punctata ssp. **cupaniana**, 329
APIACEAE, 198
Apple-of-Peru, 209
APOCYNACEAE, 208
Aptenia
 cordifolia, 39
Arabis
 collina, 107
 turrita, 106

INDEX 449

ARACEAE, 367
ARALIACEAE, 197
Arctotheca
  calendula, 308
Arenaria
  balearica, 69
Argemone
  mexicana, 19
Argentine-pear, 210
Arisarum
  proboscideum, 370
Aristolochia
  rotunda, 7
ARISTOLOCHIACEAE, 7
Artemisia
  biennis, 327
Artichoke
  Jerusalem, 356
Arum
  Bog, 368
Asperula
  arvensis, 272
  taurina, 271
Aster
  lanceolatus, 319
  novae-angliae, 317
  schreberi, 316
  x versicolor, 318
Aster
  Mexican, 361
ASTERACEAE, 286
Atriplex
  halimus, 57
Aubrieta
  deltoidea, 108
Aubretia, 108
Avens
  Large-leaved, 156
Azalea
  Yellow, 121
Bamboo
  Arrow, 384
  Broadleaved, 381
  Hairy, 383
  Narihira, 376
  Simon's, 379
  Veitch's, 382
Barberry
  Great, 16
Barley
  Antarctic, 402
  Foxtail, 401

Mediterranean, 403
Wall, 400
Bartsia
  French, 255
Bayberry, 31
Beadplant, 269
Bear's-breech, 257
  Spiny, 258
Beet
  Caucasian, 58
Beetle-grass
  Brown, 404
Beggarticks, 357
  Tall, 360
Bellflower
  Adria, 264
  Chimney, 263
  Cornish, 262
  Milky, 259
  Peach-leaved, 260
  Trailing, 265
Bent
  African, 394
  Rough, 396
  Water, 397
BERBERIDACEAE, 16
Berberis
  glaucocarpa, 16
Bermuda-buttercup, 188
Bermuda-grass
  African, 409
Beta
  trigyna, 58
BETULACEAE, 35
Bidens
  bipinnata, 359
  frondosa, 357
  pilosa, 358
  vulgata, 360
Bindweed
  Hairy, 223
Bird-in-a-bush, 23
Bistort
  Red, 81
Black-jack, 358
Blanketflower, 364
BLECHNACEAE, 6
Blechnum
  cordatum, 6
Bleeding-heart, 22
Blown-grass, 395

Blue-sow-thistle
  Common, 304
  Pontic, 305
Bog-laurel, 122
Borage
  Slender, 234
BORAGINACEAE, 228
Borago
  pygmaea, 234
Bramble, 152
Brassica
  juncea, 116
  tournefortii, 115
BRASSICACEAE, 101
Bridewort
  Pale, 149
Bristle-grass
  Foxtail, 415
  Knotroot, 414
Briza
  maxima, 389
Brome
  California, 399
  Hungarian, 398
Bromopsis
  inermis ssp. **inermis**, 398
Buckwheat
  Tall, 83
Buddleja
  davidii, 239
BUDDLEJACEAE, 239
Buffalo-bur, 221
  Red, 220
Bugseed, 56
Bullwort, 204
Bupleurum
  fruticosum, 202
  subovatum, 203
Bur-grass
  European, 411
Burr
  New Zealand, 159
Butterbur
  Giant, 344
Buttercup
  Rough-fruited, 15
  St Martin's, 14
Butterfly-bush, 239
Buttonweed
  Annual, 334
Cabbage
  Pale, 115
  Steppe, 118

Calceolaria
  chelidonioides, 244
Calendula
  arvensis, 346
  officinalis, 345
Calla
  palustris, 368
Calotis
  cuneifolia, 312
Calystegia
  pulchra, 223
Campanula
  alliariifolia, 262
  lactiflora, 259
  medium, 261
  persicifolia, 260
  portenschlagiana, 264
  poscharskyana, 265
  pyramidalis, 263
CAMPANULACEAE, 259
Canary-grass
  Awned, 393
  Bulbous, 392
Candytuft
  Garden, 112
CANNABACEAE, 27
Cannabis
  sativa, 27
Canterbury-bells, 261
CAPRIFOLIACEAE, 273
Cardamine
  raphanifolia, 105
  trifolia, 104
Carduus
  pycnocephalus, 291
Carex
  vulpinoidea, 375
Carpobrotus
  acinaciformis, 45
  edulis, 46
Carrot
  Australian, 207
Carthamus
  lanatus, 299
  tinctorius, 298
CARYOPHYLLACEAE, 69
Catchfly
  Forked, 74
  Fringed, 75
  Sweet-William, 76
Caucasian-stonecrop, 139

INDEX 451

Centaurea
  diluta, 296
  melitensis, 295
  montana, 294
  paniculata, 297
Centuryplant, 442
Cephalaria
  gigantea, 284
Cerastium
  tomentosum, 70
Ceratochloa
  carinata, 399
Chaenomeles
  × superba, 162
Chaenorhinum
  origanifolium ssp. crassifolium, 245
Chamomile
  Sicilian, 329
Checkerberry, 125
CHENOPODIACEAE, 48
Chenopodium
  × bontei, 50
  carinatum, 49
  cristatum, 51
  giganteum, 54
  probstii, 53
  pumilio, 48
  schraderianum, 55
  suecicum, 52
Chilean-iris, 434
Chloris
  truncata, 408
Chrysanthemum
  coronarium, 330
Chrysanthemum
  Tricolor, 287
Cicerbita
  bourgaei, 305
  macrophylla ssp. uralensis, 304
Cinquefoil
  Grey, 153
  Russian, 154
Cirsium
  oleraceum, 292
Clubmoss
  Krauss's, 1
CLUSIACEAE, 93
Cock's-eggs, 211
Cock's-foot
  Slender, 390
Comfrey
  bulbous, 231

Coneflower, 353
Consolida
  ajacis, 11
CONVOLVULACEAE, 222
Conyza
  bonariensis, 324
  sumatrensis, 323
Cord-grass
  Smooth, 410
Corispermum
  leptospermum, 56
CORNACEAE, 180
Cornflower
  Perennial, 294
Cornus
  sericea, 180
Coronilla
  valentina ssp. glauca, 167
Corydalis
  cava, 24
  solida, 23
Corydalis
  Pale, 25
Cosmos
  bipinnatus, 361
Costmary, 326
Cotula
  australis, 334
  dioica, 335
  squalida, 336
Cowherb, 77
Cowslip
  Sikkim, 130
Cranberry
  American, 129
Crane's-bill
  Alderney, 190
  Purple, 191
  Rock, 192
Crassula
  decumbens, 137
  helmsii, 136
CRASSULACEAE, 136
Cress
  Rosy, 107
  Tower, 106
  Trefoil, 104
Crocosmia
  × crocosmiiflora, 441
Crocus
  tommasinianus, 438
Crocus
  Early, 438

Crosswort
  Caucasian, 270
Cuckooflower
  Greater, 105
Cucumber
  Squirting, 98
CUCURBITACEAE, 98
Cudweed
  Cape, 309
Cuscuta
  campestris, 227
CUSCUTACEAE, 227
Cydonia
  oblonga, 161
Cymbalaria
  hepaticifolia, 247
  pallida, 246
Cynodon
  incompletus, 409
Cynorkis
  fastigiata, 444
CYPERACEAE, 374
Cyperus
  eragrostis, 374
Cyrtomium
  falcatum, 5
Dactylis
  polygama, 390
Daffodil
  Winter, 430
Daisy
  Bur, 312
  Crown, 330
  Seaside, 320
  Shasta, 332
Darmera
  peltata, 142
Daucus
  glochidiatus, 207
Delairia
  odorata, 341
Dewplant
  Deltoid-leaved, 41
  Pale, 43
  Purple, 42
  Shrubby, 40
Dianthus
  barbatus, 79
  gallicus, 78
Dicentra
  formosa, 22
Dichondra
  micrantha, 222

DIPSACACEAE, 283
Dipsacus
  sativus, 283
Disphyma
  crassifolium, 42
Dittrichia
  graveolens, 310
Dock
  Aegeaen, 92
  Greek, 90
  Hooked, 91
  Willow-leaved, 89
Dodder
  Yellow, 227
Dog's-tail
  Golden, 388
Dogwood
  Red-osier, 180
Doronicum
  pardalianches, 343
Drosanthemum
  floribundum, 43
DRYOPTERIDACEAE, 5
Duchesnea
  indica, 155
Ecballium
  elaterium, 98
Echinochloa
  colona, 413
Echinops
  bannaticus, 290
  exaltus, 289
  sphaerocephalus, 288
Echium
  pininana, 229
  rosulatum, 228
Elder
  American, 276
  Red-berried, 273 – 275
Eleusine
  indica ssp. **africana**, 407
Elodea
  nuttalii, 365
Epilobium
  komarovianum, 175
  pedunculare, 174
Eragrostis
  cilianensis, 405
  parviflora, 406
Erepsia
  heteropetala, 44

INDEX

Erica
 lusitanica, 128
 terminalis, 127
ERICACEAE, 120
Erigeron
 glaucus, 320
 karvinskianus, 322
 philadelphicus, 321
Erodium
 botrys, 193
 brachycarpum, 194
 crinitum, 195
 cygnorum, 196
Eruca
 vesicaria ssp. **sativa**, 117
Eryngium
 planum, 200
Eryngo
 Blue, 200
Escallonia
 macrantha, 135
Escallonia, 135
Eschscholzia
 californica, 20
Euphorbia
 corallioides, 181
EUPHORBIACEAE, 181
Evening-primrose
 Renner's, 177
 Small-flowered, 178
Everlasting-pea
 Two-flowered, 169
FABACEAE, 165
FAGACEAE, 32
Fagopyrum
 dibotrys, 83
Fallopia
 baldschuanica, 87
 japonica var. **compacta**, 85
 sachalinensis, 86
Fargesia
 spathacea, 378
Fennel-flower, 9
Fern
 Kangaroo, 3
 Ostrich, 4
 Ribbon, 2
Fescue
 Bearded, 387
Ficus
 carica, 28
Field-speedwell
 Crested, 252

Fig, 28
Firethorn, 164
Fleabane
 Argentine, 324
 Guernsey, 323
 Mexican, 322
 Stinking, 310
Forget-me-not
 Bur, 236
Fountain-bamboo
 Chinese, 378
 Indian, 377
Fox-sedge
 American, 375
Freesia
 x **hybrida**, 440
Freesia, 440
Fringecups, 147
Fuchsia
 magellanica, 179
 'Riccartonii', 179
FUMARIACEAE, 22
Gaillardia
 x **grandiflora**, 364
Galanthus
 elwesii, 432
 plicatus ssp. **plicatus**, 431
Galega
 officinalis, 165
Galingale
 Pale, 374
Garlic
 Few-flowered, 425
 Hairy, 423
 Honey, 426
 Neapolitan, 422
 Rosy, 421
 Yellow, 424
Gaultheria
 mucronata, 126
 procumbens, 125
 shallon, 124
GERANIACEAE, 190
Geranium
 macrorrhizum, 192
 x **magnificum**, 191
 submolle, 190
German-ivy, 341
Geum
 macrophyllum, 156
Giant-rhubarb, 171
Gladiolus
 communis ssp. **byzantinus**, 439
Gladiolus
 Eastern, 439

454 INDEX

Globe-thistle, 289
　Blue, 290
　Glandular, 288
Gnaphalium
　undulatum, 309
Goat's-rue, 165
Goldenrod
　Canadian, 313
　Early, 314
　Grass-leaved, 315
Goosefoot
　Clammy, 48
　Crested, 51
　Keeled, 49
　Probst's, 53
　Swedish, 52
Grape-hyacinth
　Garden, 419
Grass-poly
　False, 172
GROSSULARIACEAE, 135
Guizotia
　abyssinica, 350
Gunnera
　tinctoria, 171
GUNNERACEAE, 171
Hair-grass
　Mediterranean, 391
Hard-fern
　Chilean, 6
Hare's-ear
　Shrubby, 202
Hawk's-beard
　Tuberous, 302
Heath
　Corsican, 127
　Portuguese, 128
　Prickly, 126
Hebe
　x **franciscana**, 254
　salicifolia, 253
Hedera
　colchica, 197
Helianthus
　annuus, 354
　x **laetiflorus**, 355
　tuberosus, 356
Helleborus
　orientalis, 8
Hemp, 27
Herniaria
　hirsuta, 72

Hibiscus
　trionum, 97
HIPPOCASTANACEAE, 183
Hollowroot, 24
Holly-fern
　House, 5
Hollyhock, 95
Holodiscus
　discolor, 150
Honesty, 109
Honeysuckle
　Californian, 279
　Henry's, 280
　Himalayan, 278
　Japanese, 281
　Perfoliate, 282
Hordeum
　geniculatum, 403
　jubatum, 401
　murinum ssp. **glaucum**, 400
　pubiflorum, 402
Horse-chestnut, 183
Hottentot-fig, 46
Hyacinth
　Tassel, 420
HYDROCHARITACEAE, 365
Hydrocotyle
　moschata, 198
　novae-zeelandiae, 199
Hypericum
　hircinum, 94
　x **inodorum**, 93
Iberis
　umbellata, 112
Ice-plant
　Heart-leaf, 39
Indian-rhubarb, 142
Iochroma
　australe, 210
Ipomoea
　hederacea var. **hederacea**, 225
　hederacea var. **integriuscula**, 225
　lacunosa, 226
　purpurea, 224
IRIDACEAE, 434
Iris
　versicolor, 436
Iris
　Purple, 436
Ismelia
　carinata, 287
Ivy
　Persian, 197

Japanese-lantern, 212
Johnson-grass, 416
JUGLANDACEAE, 30
Juglans
    regia, 30
JUNCACEAE, 371
Juncus
    pallidus, 373
    planifolius, 371
    subulatus, 372
Juneberry, 163
Kalmia
    angustifolia, 123
    polifolia, 122
Kangaroo-apple, 219
Karo, 134
Ketmia
    Bladder, 97
Kidneyweed, 222
Knapweed
    Jersey, 297
    Russian, 293
Knotweed
    Giant, 86
    Japanese, 85
    Lesser, 80
Koromiko, 253
Lactuca
    tatarica, 303
Lady's-mantle, 160
Lagarosiphon
    major, 366
Lamarckia
    aurea, 388
LAMIACEAE, 237
Lappula
    squarrosa, 236
Lapsana
    communis ssp. **intermedia**, 300
Larkspur, 11
Lathraea
    clandestina, 256
Lathyrus
    grandiflorus, 169
Laurustinus, 277
Lenten-rose, 8
Leopard's-bane, 343
Lepidium
    hyssopifolium, 114
    virginicum, 113
Leptinella, 336
    Hairless, 335

Leptochloa
    fusca, 404
Lettuce
    Blue, 303
Leucanthemella
    serotina, 331
Leucanthemum
    x **superbum**, 332
Levisticum
    officinale, 206
Leycesteria
    formosa, 278
Libertia
    formosa, 434
Ligustrum
    ovalifolium, 241
Lilac, 240
LILIACEAE, 419
Lily
    African, 427
    Jersey, 429
    Peruvian, 433
Linaria
    dalmatica, 248
    maroccana, 249
Lobelia
    Lawn, 268
Lolium
    rigidum, 386
Londonpride, 145
London-rocket
    False, 101
Lonicera
    caprifolium, 282
    henryi, 280
    involucrata, 279
    japonica, 281
Loosestrife
    Dotted, 132
    Fringed, 131
    Lake, 133
Lovage, 206
Love-grass
    Weeping, 406
Luma
    apiculata, 173
Lunaria
    annua, 109
Lungwort
    Red, 230
Lycopersicon
    esculentum, 214

Lysichiton
  americanus, 367
Lysimachia
  ciliata, 131
  punctata, 132
  terrestris, 133
LYTHRACEAE, 172
Lythrum
  junceum, 172
Macleaya
  x **kewensis**, 21
Madia
  sativa, 286
Mahonia
  aquifolium, 17
MALVACEAE, 95
Maple
  Norway, 184
Marigold
  Dwarf, 363
  Field, 346
  Pot, 345
  Southern, 362
Marvel-of-Peru, 38
Matteuccia
  struthiopteris, 4
Matthiola
  longipetala ssp. **bicornis**, 103
Medicago
  praecox, 170
Medick
  Early, 170
Mexican-stonecrop
  Greater, 138
Michaelmas-daisy
  Hairy, 317
  Late, 318
  Narrow-leaved, 319
  Nettle-leaved, 316
Mignonette
  Corn, 119
Millet
  Common, 412
  Great, 417 – 418
  Shama, 413
Mind-your-own-business, 29
Mirabilis
  jalapa, 38
Monk's-hood
  Hybrid, 10
Montbretia, 441
MORACEAE, 28

Morning-glory
  Common, 224
  Ivy-leaved, 225
  White, 226
Mouse-ear-hawkweed
  Tall, 306 – 307
Mousetailplant, 370
Muehlenbeckia
  complexa, 88
Mugwort
  Slender, 327
Mullein
  Nettle-leaved, 243
  Purple, 242
Muscari
  armeniacum, 419
  comosum, 420
Mustard
  Ball, 110
  Chinese, 116
Myrica
  pensylvanica, 31
MYRICACEAE, 31
MYRTACEAE, 173
Myrtle
  Chilean, 173
Nectaroscordum
  siculum ssp. **siculum**, 426
Nertera
  granadensis, 269
Neslia
  paniculata, 110
Nicandra
  physaloides, 209
Nigella
  hispanica, 9
Niger, 350
Nightshade
  Green, 216
  Leafy-fruited, 217
  Small, 218
  Tall, 215
Ninebark, 148
Nipplewort
  Large, 300
NYCTAGINACEAE, 38
Oak
  Evergreen, 33
  Red, 34
  Turkey, 32
Oceanspray, 150

Odontites
    jaubertianus ssp. **chrysanthus**, 255
Oenothera
    cambrica, 178
    renneri, 177
    rubricaulis, 176
OLEACEAE, 240
ONAGRACEAE, 174
Oniongrass, 437
Orache
    Shrubby, 57
ORCHIDACEAE, 443
Oregon-grape, 17
Ornithopus
    compressus, 166
OROBANCHACEAE, 256
Oryzopsis
    miliacea, 385
Oscularia
    deltoides, 41
OXALIDACEAE, 186
Oxalis
    articulata, 186
    debilis var. **corymbosa**, 187
    incarnata, 189
    pes-caprae, 188
Oxeye
    Autumn, 331
    Yellow, 311
Panicum
    miliaceum, 412
Papaver
    atlanticum, 18
PAPAVERACEAE, 18
Parthenocissus
    inserta, 182
Pearlwort
    Heath, 71
Penny-cress
    Caucasian, 111
Pennywort
    Hairy, 198
    New Zealand, 199
Pepperwort,
    African, 114
    Least, 113
Periwinkle
    Greater, 208
Persicaria
    amplexicaulis, 81
    campanulata, 80
    sagittata, 82

Petasites
    japonicus, 344
Phalaris
    aquatica, 392
    paradoxa, 393
Phuopsis
    stylosa, 270
Phymatosorus
    diversifolius, 3
Physalis
    alkengi, 212
    ixocarpa, 213
Physocarpus
    opulifolius, 148
Phyteuma
    scheuchzeri, 267
Phytolacca
    acinosa, 37
PHYTOLACCACEAE, 37
Pick-a-back-plant, 146
Pigmyweed
    New Zealand, 136
    Scilly, 137
Pigweed
    Cape, 66
    Guernsey, 63
    Perennial, 62
    Thunberg's, 65
    White, 64
Pilosella
    praealta ssp. **praealta**, 306
    praealta ssp. **thaumasia**, 307
Pink
    Jersey, 78
Pink-sorrel, 186
    Large-flowered, 187
    Pale, 189
Pirri-pirri-bur
    Bronze, 157
PITTOSPORACEAE, 134
Pittosporum
    crassifolium, 134
Plane
    London, 26
PLANTAGINACEAE, 238
Plantago
    arenaria, 238
Plantain
    Branched, 238
PLATANACEAE, 26
Platanus
    x **hispanica**, 26

Pleioblastus
  humilis, 380
  simonii, 379
Plume-poppy
  Hybrid, 21
POACEAE, 376
Pokeweed
  Indian, 37
POLYGONACEAE, 80
Polygonum
  arenarium ssp. **pulchellum**, 84
POLYPODIACEAE, 3
Polypogon
  viridis, 397
Poppy
  Atlas, 18
  Californian, 20
  Mexican, 19
Portulaca
  oleraceae, 68
PORTULACACEAE, 68
Potentilla
  inclinata, 153
  intermedia, 154
Pratia
  angulata, 268
Primula
  sikkimensis, 130
PRIMULACEAE, 130
Privet
  Garden, 241
Pseudofumaria
  alba, 25
Pseudosasa
  japonica, 384
PTERIDACEAE, 2
Pteris
  cretica, 2
Pulmonaria
  rubra, 230
Purslane
  Common, 68
Pyracantha
  coccinea, 164
Quaking-grass
  Greater, 389
Quercus
  cerris, 32
  ilex, 33
  rubra, 34
Quince, 161
  Hybrid, 162

Ragweed, 347
  Giant, 349
  Perennial, 348
Ragwort
  Chinese, 342
  Golden, 339
  Magellan, 340
  Narrow-leaved, 337
  Wood, 338
Rampion
  Oxford, 267
RANUNCULACEAE, 8
Ranunculus
  marginatus var. **trachycarpus**, 14
  muricatus, 15
Rapistrum
  perenne, 118
Red-knotgrass
  Lesser, 84
Reseda
  phyteuma, 119
RESEDACEAE, 119
Rhododendron
  luteum, 121
  ponticum, 120
Rhododendron, 120
Robin's-plantain, 321
Rocket
  French, 102
  Garden, 117
Romulea
  rosea var. **australis**, 437
ROSACEAE, 148
Rostraria
  cristata, 391
RUBIACEAE, 269
Rubus
  laciniatus, 152
  spectabilis, 151
Rudbeckia
  laciniata, 353
Rumex
  brownii, 91
  cristatus, 90
  dentatus, 92
  salicifolius ssp. **triangulivalvis**, 89
Rupturewort
  Hairy, 72
Ruschia
  caroli, 40

Rush
　Broad-leaved, 371
　Somerset, 372
Russian-vine, 87
Rye-grass
　Mediterranean, 386
Safflower, 298
　Downy, 299
Sagina
　subulata 'Aurea', 71
SALICACEAE, 99
Salix
　× calodendron, 100
　daphnoides, 99
Sally-my-handsome, 45
Salmonberry, 151
Salpichroa
　origanifolia, 211
Salsify
　Slender, 301
Salsola
　kali ssp. **ruthenica**, 59
Saltwort
　Spineless, 59
Sambucus
　canadensis, 276
　racemosa, 273
　racemosa var. **pubens** 'Dissecta', 274
　racemosa var. **sieboldiana**, 275
Sandwort
　Mossy, 69
Sasa
　palmata, 381
　veitchii, 382
Sasaella
　ramosa, 383
Saxifraga
　cymbalaria var. **huetiana**, 143
　rotundifolia, 144
　× urbium, 145
SAXIFRAGACEAE, 142
Saxifrage
　Celandine, 143
　Round-leaved, 144
Scabiosa
　atropurpurea, 285
Scabious
　Giant, 284
　Sweet, 285
Schkuhria
　pinnata, 363
Scorpion-vetch
　Shrubby, 167

SCROPHULARIACEAE, 242
Sea-fig
　Lesser, 44
Sedum
　hispanicum, 141
　lydium, 140
　prealtum, 138
　spurium, 139
Selaginella
　kraussiana, 1
SELAGINELLACEAE, 1
Semiarundinaria
　fastuosa, 376
Senecio
　doria, 339
　inaequidens, 337
　ovatus ssp. **alpestris**, 338
　smithii, 340
Serradella
　Yellow, 166
Setaria
　italica, 415
　parviflora, 414
Shallon, 124
Sheep-laurel, 123
Sigesbeckia
　orientalis, 351
　serrata, 352
Silene
　armeria, 76
　dichotoma, 74
　fimbriata, 75
SIMAROUBACEAE, 185
Sinacalia
　tangutica, 342
Sisymbrium
　erysimoides, 102
　loeselii, 101
Sisyrinchium
　striatum, 435
Skunk-cabbage
　American, 367
Slipperwort, 244
Smearwort, 7
Smilo-grass, 385
Smyrnium
　perfoliatum, 201
Snow-in-summer, 70
Snowdrop
　Greater, 432
　Pleated, 431

Soft-rush
  Great, 373
SOLANACEAE, 209
Solanum
  chenopodioides, 215
  laciniatum, 219
  physalifolium var. **nitibaccatum**, 216
  rostratum, 221
  sarachoides, 217
  sisymbriifolium, 220
  triflorum, 218
Soleirolia
  soleirolii, 29
Solidago
  canadensis, 313
  gigantea ssp. **serotina**, 314
  graminifolia, 315
Sorghum
  bicolor, 417 – 418
  halepense, 416
Spanish-needles, 359
Spartina
  alternifolia var. **glabra**, 410
Speedwell
  American, 251
  French, 250
Spergula
  morisonii, 73
Spinach
  New Zealand, 47
  Tree, 54
Spiraea
  alba, 149
Spiranthes
  lancea, 443
Spurge
  Coral, 181
Spurrey
  Pearlwort, 73
St Paul's-wort
  Eastern, 351
  Western, 352
Stachys
  recta ssp. **labiosa**, 237
Star-thistle
  Lesser, 296
  Maltese, 295
Starflower
  Spring, 428
Sternbergia
  lutea, 430
Stink-grass, 405

Stock
  Night-scented, 103
Stonecrop
  Least, 140
  Spanish, 141
Stork's-bill
  Eastern, 195
  Hairy-pitted, 194
  Mediterranean, 193
  Western, 196
Strawberry
  Yellow-flowered, 155
Sunflower, 354
  Perennial, 355
Sweet-William, 79
Symphytum
  bulbosum, 231
Syringa
  vulgaris, 240
Tagetes
  minuta, 362
Tanacetum
  balsamita, 326
  macrophyllum, 325
Tansy
  Rayed, 325
Tarweed
  Coast, 286
Tear-thumb
  American, 82
Teasel
  Fuller's, 283
Telekia
  speciosa, 311
Tellima
  grandiflora, 147
Tetragonia
  tetragonioides, 47
Thistle
  Cabbage, 292
  Plymouth, 291
Thlaspi
  macrophyllum, 111
Thorow-wax
  False, 203
Throatwort, 266
Toadflax
  Annual, 249
  Balkan, 248
  Corsican, 247
  Italian, 246
  Malling, 245

Tolmeia
  menziesii, 146
Tomatillo, 213
Tomato, 214
Toothpick-plant, 205
Toothwort
  Purple, 256
Trachelium
  caeruleum, 266
Trachystemon
  orientalis, 235
Tragopogon
  hybridus, 301
Tragus
  racemosus, 411
Treasureflower
  Plain, 308
Tree-of-heaven, 185
Tripleurospermum
  decipiens, 333
Tristagma
  uniflorum, 428
Tutsan
  Stinking, 94
  Tall, 93
URTICACEAE, 29
Vaccaria
  hispanica, 77
Vaccinium
  macrocarpon, 129
Velvetleaf, 96
Verbascum
  chaixii, 243
  phoeniceum, 242
Veronica
  acinifolia, 250
  crista-galli, 252
  peregrina, 251
Veronica
  Hedge, 254
Vetch
  Fine-leaved, 168
Viburnum
  tinus, 277
Vicia
  tenuifolia, 168
Vinca
  major var. **oxyloba**, 208
Violet-willow
  European, 99
Viper's-bugloss
  Giant, 229
  Lax, 228

Virginia-creeper
  False, 182
VITACEAE, 182
Vulpia
  ciliata ssp. **ciliata**, 387
Walnut, 30
Waterweed
  Curly, 366
  Nuttall's, 365
Willow
  Holme, 100
Willowherb
  Bronzy, 175
  Rockery, 174
Windmill-grass, 408
Wireplant, 88
Woodruff
  Blue, 272
  Pink, 271
WOODSIACEAE, 4
Yard-grass, 407
Yarrow
  Southern, 328
Yellow-eyed-grass
  Pale, 435
Yellow-woundwort
  Perennial, 237
Yushania
  anceps, 377
Zantedeschia
  aethiopica, 369

BOTANICAL SOCIETY OF THE BRITISH ISLES

The BSBI traces its origin to the Botanical Society of London founded in 1836 and has a membership of some 2,850. It is the major source of information on the status and distribution of British and Irish flowering plants and ferns. This information, which is gathered through a network of vice-county recorders, is the basis for plant atlases and for publications on rare and scarce species and is vital to botanical conservation. The Society published Atlas of the British flora in 1962 and was a major partner in the production of New atlas of the British & Irish flora and a related CD-ROM, published by Oxford University Press in September 2002 (available from Summerfield Books: see below). It organises plant distribution surveys, publishes handbooks on difficult groups of plants and has a panel of referees available to members to name problematic specimens. The BSBI arranges conferences and field meetings throughout the British Isles and, occasionally, abroad. It welcomes as members all botanists, professional and amateur alike.

Details of membership and other information about the Society may be obtained from:
The Hon. General Secretary,
Botanical Society of the British Isles,
c/o Department of Botany,
The Natural History Museum,
Cromwell Road,
London SW7 5BD.

BSBI handbooks

Each handbook deals in depth with one or more difficult groups of British and Irish plants.

No. 1   *Sedges of the British Isles*
A. C. Jermy, A. O. Chater & R. W. David. Revised edition, 1982. 272 pp., with descriptions, drawings and distribution maps for all 73 species of Carex. Paperback.

No. 2   *Umbellifers of the British Isles*
T. G. Tutin. 1980. 200 pp., with descriptions and drawings of 73 species of Apiaceae (Umbelliferae). Paperback.

No. 3   *Docks and knotweeds of the British Isles*
J. E. Lousley & D. H. Kent. 1981. 208 pp., with descriptions and drawings of about 80 native and alien taxa of Polygonaceae. Paperback. Out of print. New edition with distribution maps in preparation; orders recorded.

No. 4   *Willows and poplars of Great Britain and Ireland*
R. D. Meikle. 1984. 200 pp., with descriptions and drawings of 65 species, subspecies, varieties and hybrids of Salix and Populus. Paperback.

No. 5   *Charophytes of Great Britain and Ireland*
J. A. Moore. 1986. 144 pp., with descriptions and drawings of 39 species and varieties of Characeae and 17 distribution maps. Paperback.

No. 6   *Crucifers of Great Britain and Ireland*
T. C. G. Rich. 1991. 344 pp., with descriptions of 148 taxa of Brassicaceae (Cruciferae), 129 of them with drawings, and 60 distribution maps. Paperback.

No. 7   *Roses of Great Britain and Ireland*
G. G. Graham & A. L. Primavesi. 1993. 208 pp., with descriptions and drawings of 13 native and nine introduced taxa of Rosa, descriptions of 76 hybrids, and 33 maps. Paperback.

No. 8   *Pondweeds of Great Britain and Ireland*
        C. D. Preston. 1995. 352 pp., with descriptions and drawings of all 50 species and hybrids of Potamogeton, Groenlandia and Ruppia, most of them with distribution maps; detailed introductory material and bibliography. Paperback.

No. 9   *Dandelions of Great Britain and Ireland*
        A. A. Dudman & A. J. Richards. 1997. 344 pp., with descriptions of 235 species of Taraxacum, most of them illustrated by silhouettes of herbarium specimens; drawings of bud involucres of 139 species and 178 distribution maps. Paperback.

No. 10 *Sea beans and nickar nuts*
        E. Charles Nelson. 2000. 156 pp., with descriptions of nearly 60 exotic seeds and fruits found stranded on beaches in north-western Europe (many illustrated by Wendy Walsh) and of the mature plants (some with drawings by Alma Hathway), accounts of their history and folklore, growing instructions, etc. Paperback.

Other publications

List of vascular plants of the British Isles
   D. H. Kent. 1992. 400 pp. Nomenclature and sequence as in Clive Stace's New Flora of the British Isles (1991, 1997), with selected synonyms. Paperback. Supplied with errata lists and two supplements.

Alien plants of the British Isles
   E. J. Clement & M. C. Foster. 1994. 616 pp. Lists 3,586 recorded non-native species (of which 885 are established), with English names, frequency, status, origin, references to descriptions and illustrations, and selected synonyms. Paperback.

Alien grasses of the British Isles
   T. B. Ryves, E. J. Clement & M. C. Foster. 1996. 234 pp. A companion volume to the last, listing over 700 non-native grasses; includes keys to bamboos and eight of the larger and more difficult genera and 29 pp. of drawings. Paperback.

Plant crib 1998
   T. C. G. Rich & A. C. Jermy. 1998. 400 pp. An identification guide for some 325 difficult taxonomic groups, with explanations, keys and illustrations of plant details. A4 paperback.

Aquatic plants in Britain and Ireland
   C. D. Preston & J. M. Croft. 1997. 365 pp. Accounts and distribution maps of 200 aquatic plants in 72 genera, with 72 line drawings. Paperback reprint, published by Harley Books.

New atlas of the British & Irish flora
   C. D. Preston, D. A. Pearman & T. D. Dines (eds). 2002. xi + 910 pp. Distribution maps and accompanying text for 2,412 plants, with introductory chapters. Large hardback with CD-ROM, published by Oxford University Press.

Vice-County Census Catalogue of the Vascular Plants of Great Britain the Isle of Man and the Channel Islands
   C.A. Stace, et al. 2003. 405pp. Listing of plant species found in Great Britain, giving v-c. numbers in which they have been recorded; coded to show if native, archaeophyte, neophyte or casual; also if recorded since 1969 or known to be extinct. Paperback.

Cumulative Index to Watsonia Journal of the BSBI Volumes 1-20, 1949-1995
   C.R. Boon. 2004. 246pp. The Index includes several major sections, incorporated within the A-Z listing, classified as follows: Bibliography, Biography, Book reviews, BSBI (reports of meetings, etc.), Chromosome numbers, Computing, Herbaria, Keys. Obituaries, Photography, v-c distribution (for taxa where lists for more than 12 v-cs are given), World botany. Paperback

Atlas of British and Irish Brambles

A. Newton & R.D. Randall. 2004. A phytogeographical analysis of microspecies of *Rubus* sect. *Rubus* & sect. *Corylifolii*. 98pp, 330 distribution maps with summaries of distribution and notes on changes. Paperback.

Altitudinal Limits of British and Irish Vascular Plants

D.A. Pearman & R Corner. Updated and revised 2004 (2003). 40pp. A provisional listing, alphabetically, giving maximum or minimum altitude, v.c., locality, grid reference, status, source, recorder, year found, and notes, all where known and appropriate. Booklet.

First records of Alien Plants in the wild in Britain and Ireland

D.A. Pearman & C.D. Preston. 2003. 44pp. A provisional listing, alphabetically, of species, subspecies and hybrids treated as neophytes in the *New Atlas* (2002) with year of introduction, source of record, v-c, and notes. Booklet.

Hybridization and the flora of the British Isles: collecting records for a new edition of Clive Stace's 1975 work

T.D. Dines, D.A. Pearman & C.D. Preston. 2005. 24pp. A provisional listing, alphabetically, of hybrids included in the project, with status and BRC number. Booklet.

Available from the official agents for BSBI Publications,
Summerfield Books (Jon & Sue Atkins), Summerfield House, High Street, Brough, Kirkby Stephen, Cumbria CA17 4BX (Telephone: 017683 41577. Fax: 017683 41687. e-mail: bsbipubs@beeb.net).

VICE-COUNTIES OF THE BRITISH ISLES

	CHANNEL ISLES					
S (J)	Jersey	49	Caerns	102	S. Ebudes
S (G)	Guernsey	50	Denbs	103	M. Ebudes
S (A)	Alderney	51	Flints	104	N. Ebudes
S (S)	Sark	52	Anglesey	105	W. Ross
	ENGLAND I		**ENGLAND II**	106	E. Ross
1	W. Cornwall	53	S. Lincs	107	E. Sutherland
2	E. Cornwall	54	N. Lincs	108	W. Sutherland
3	S. Devon	55	Leics	109	Caithness
4	N. Devon	56	Notts	110	Outer Hebrides
5	S. Somerset	57	Derbys	111	Orkney
6	N. Somerset	58	Cheshire	112	Shetland
7	N. Wilts	59	S. Lancs		**IRELAND**
8	S. Wilts	60	W. Lancs	H1	S. Kerry
9	Dorset	61	S.E. Yorks	H2	N. Kerry
10	Wight	62	N.E. Yorks	H3	W. Cork
11	S. Hants	63	S.W. Yorks	H4	M. Cork
12	N. Hants	64	M.W. Yorks	H5	E. Cork
13	W. Sussex	65	N.W. Yorks	H6	Co. Waterford
14	E. Sussex	66	Co. Durham	H7	S. Tipperary
15	E. Kent	67	S. Northumb	H8	Co. Limerick
16	W. Kent	68	Cheviot	H9	Co. Clare
17	Surrey	69	Westmorland	H10	N. Tipperary
18	S. Essex	70	Cumberland	H11	Co. Kilkenny
19	N. Essex		**ISLE OF MAN**	H12	Co. Wexford
20	Herts	71	Man	H13	Co. Carlow
21	Middlesex		**SCOTLAND**	H14	Laois
22	Berks	72	Dumfriess	H15	S.E. Galway
23	Oxon	73	Kirkcudbrights	H16	W. Galway
24	Bucks	74	Wigtowns	H17	N.E. Galway
25	E. Suffolk	75	Ayrs	H18	Offaly
26	W. Suffolk	76	Renfrews	H19	Co. Kildare
27	E. Norfolk	77	Lanarks	H20	Co. Wicklow
28	W. Norfolk	78	Peebless	H21	Co. Dublin
29	Cambs	79	Selkirks	H22	Meath
30	Beds	80	Roxburghs	H23	Westmeath
31	Hunts	81	Berwicks	H24	Co. Longford
32	Northants	82	E. Lothian	H25	Co. Roscommon
33	E. Gloucs	83	Midlothian	H26	E. Mayo
34	W. Gloucs	84	W. Lothian	H27	W. Mayo
35	**see Wales**	85	Fife	H28	Co. Sligo
36	Herefs	86	Stirlings	H29	Co. Leitrim
37	Worcs	87	W. Perth	H30	Co. Cavan
38	Warks	88	M. Perth	H31	Co. Louth
39	Staffs	89	E. Perth	H32	Co. Monaghan
40	Salop	90	Angus	H33	Fermanagh
	WALES	91	Kincardines	H34	E. Donegal
35	Mons	92	S. Aberdeen	H35	W. Donegal
41	Glam	93	N. Aberdeen	H36	Tyrone
42	Brecs	94	Banffs	H37	Co. Armagh
43	Rads	95	Moray	H38	Co. Down
44	Carms	96	Easterness	H39	Co. Antrim
45	Pembs	97	Westerness	H40	Co. Londonderry
46	Cards	98	Argyll		
47	Monts	99	Dunbarton		
48	Merioneth	100	Clyde Is		
		101	Kintyre		

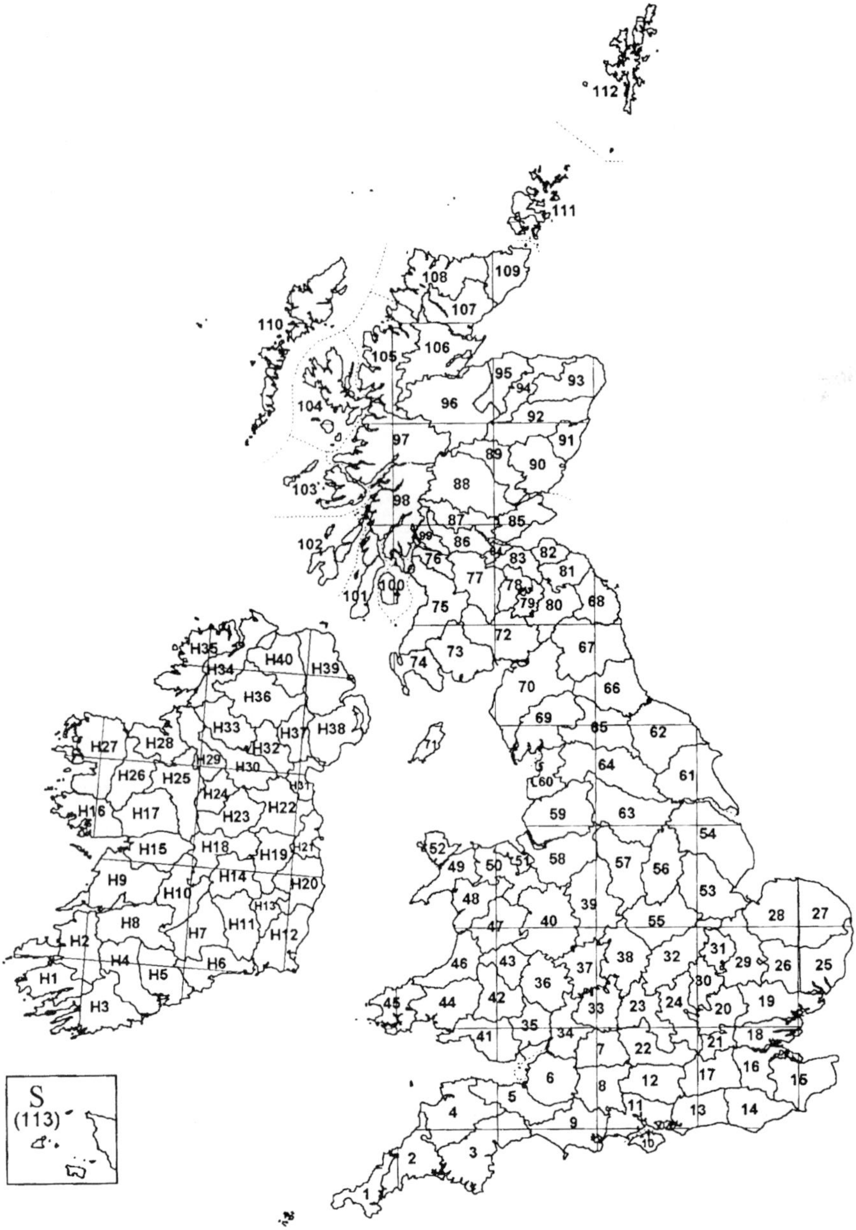
112
111
109
108
107
110
106
105
95
93
94
104
96
92
97
91
89
90
88
103
98
87
85
102
99
86
84
83
82
76
81
77
78
100
101
79
80
68
75
72
H35
H40
H39
H34
74
73
67
70
66
H36
69
65
62
H33
H37
H38
H27
H28
H32
71
H29
H30
64
61
H26
H25
60
H31
H24
H22
H16
H17
H23
59
63
54
H15
H18
H21
52
H19
58
51
50
49
57
56
H9
H14
H20
53
H10
48
39
H13
47
40
55
28
27
H8
H11
H2
H7
H12
31
H4
H6
46
43
37
38
32
29
26
25
H1
H5
36
30
H3
45
44
42
33
23
24
20
19
35
34
18
41
7
22
21
6
16
17
12
15
5
8
S
(113)
4
11
13
14
9
10
3
2
1